Modern Pupillometry

Megan H. Papesh • Stephen D. Goldinger
Editors

Modern Pupillometry

Cognition, Neuroscience, and Practical Applications

Springer

Editors
Megan H. Papesh
Department of Psychology
University of Massachusetts Lowell
Lowell, MA, USA

Stephen D. Goldinger
Department of Psychology
Arizona State University
Tempe, AZ, USA

ISBN 978-3-031-54895-6 ISBN 978-3-031-54896-3 (eBook)
https://doi.org/10.1007/978-3-031-54896-3

This Springer imprint is published by the registered company Springer Nature Switzerland AG
The registered company address is: Gewerbestrasse 11, 6330 Cham, Switzerland

Paper in this product is recyclable.

For Eggo.

Preface

Introduction

In recent years, the use of pupil size in psychological research has expanded tremendously. A Google Scholar keyword search for "pupillometry" restricted to the year 2010 yields 324 papers. Changing the year to 2020, however, yields 1890 results. Despite this massive increase in use, the most recent compendium about pupillometry was published by Michel Pierre Janisse in 1977. Since that time, several excellent tutorial and review articles have been published (e.g., Laeng et al., 2012; Mathôt et al., 2018; Reilly et al., 2019; Sirois & Brisson, 2014, among others), but no source provides a thorough accounting of the diverse topics investigated via pupillometry and, importantly, the diverse approaches to analyzing pupil data. With this volume, we bring together experts from many psychological science domains with the goal of organizing the vast landscape of pupillometry research in a single place. As experts in their domains, the authors have been given tremendous latitude in terms of what to cover and how. The result, we believe, is a collection of chapters most representative of the current state of the art in each discipline. Given such a vast landscape, of course, we are certain that this volume does not represent a fully exhaustive accounting of how pupillometry has been applied across psychology. If your specific area of interest is not represented, please contact us so that we might include it in a future edition.

Our goal in structuring this volume was to make it interesting and useful to experts, while remaining accessible to researchers unfamiliar with pupillometry. Whether you are one of the contributing experts or a new student in the field, we hope that you can take away meaningful, actionable (practical?) information. Specifically, our aim is to provide readers enough neurophysiological background to understand *why* pupil size is studied, enough discipline-specific information to understand *what* pupillometry can bring to their own research, and enough methodological and analytical detail to know *how* to conduct rigorous pupillometry studies of their own. To that end, this volume is organized into three parts: Neurophysiology, Psychological Science, and Practical Guides.

Neurophysiology

Although only two chapters appear in the Neurophysiology part of this book, each chapter is a phenomenal source containing current perspectives focused on different brain structures. Huang and Clewett focus their attention on the locus coeruleus (LC), a deep brainstem nucleus long appreciated as the source of norepinephrine (NE) for virtually the entire brain. As readers will see, the LC-NE system is mentioned in most chapters in this volume, and indeed in most empirical papers in which pupil size is recorded. Huang and Clewett clearly lay out why this is the case, providing a tour de force review of the connections between the LC-NE system and pupil size. The chapter includes a novel framework unifying information on the LC-NE system and arousal-attention interactions.

Wang and Munoz highlight the importance of the superior colliculus (SC), a midbrain structure that receives multisensory information from many cortical and subcortical regions. Readers familiar with pupillometry may have noticed that single-unit studies linking pupil size to neural activity will occasionally record from the SC, but most often, papers emphasize LC activity. Wang and Munoz provide a compelling review of the evidence for the involvement of the SC, and why students of pupillometry should be more familiar with it.

Psychological Science

The section on Psychological Science is the lengthiest of the volume, but still not an exhaustive collection of the various domains to which pupil size has been applied to better understand psychological behaviors. This part opens with several chapters covering core concepts in cognitive science, starting with the application of pupillometry to developmental psychology. Hepach provides a "deep dive" into pupillometric studies of cognitive development, including studies of expectation violations, neural activity, and arousal states. A particular strength of Hepach's analysis can be found in his overview of the diversity of approaches to studying pupil size (interested readers should also check out Part III, *Practical Guides*). Robison's coverage of how pupillometry can be harnessed to provide insights into attention control and working memory provides both rich theoretical background and practically useful information. For example, his summary of the various paradigms and findings can give readers an easy appreciation for how to incorporate pupillometry and individual differences methods into their own work. Guevara Pinto and Hout take attention a step further, exploring how pupillometry has already and can continue to inform research on visual search. Cairney et al. provide an overview of how pupillometry has been used to study human memory, and Schmidtke and Tobin cover language research. Both memory and language are incredibly dense topics, each deserving more than a single chapter to be sure, but these authors nonetheless provide readers with clear, thorough overviews. We anticipate that all readers will find something useful in their chapters.

After covering core topics in cognitive science, the Psychological Science part branches into other realms with clinical and applied relevance. Kuchinsky and Milvae discuss the importance of studying listening effort, and how pupillometry can be used to reveal individual and group-level differences that may otherwise be obscured by standard intelligibility measures. Real-world applications are further explored by Cohen Hoffing and Thurman, whose insights into the temporal dynamics of the pupil measure and the challenges of studying it "in the wild" promise to give readers much to think about. Clinical and social psychology are covered by McGovern and Cash et al., respectively. McGovern discusses the utility of pupillometry for schizophrenia research, and she maps out both the past and the future, clearly conveying the practical and theoretical relevance of the research. Cash et al. explore how pupil size can be applied to interpersonal communication, with several fun insights for future students of pupillometry to consider. The Psychological Science part closes with a fascinating chapter by Cook et al., who discuss the use of pupillometry for deception detection. Their ocular-motor deception test is sure to interest readers from wide-ranging backgrounds.

Practical Guides

As discussed by Hepach, and several other chapter authors, pupillometry is characterized by highly variable methods, in experimental designs, data collection, and analyses. On the one hand, this variety speaks to a vibrant and vast field of research topics and techniques. On the other hand, it can also make the discipline challenging to enter or to develop expertise. Our goal with the third and final part of the volume is to provide readers with a "one stop shop" to locate information about data collection, data processing, and data analysis. Laeng and Mathôt open Part III with methodological considerations for pupillometry research, using their own data to walk readers through the variable and methods for processing it. Reilly et al. then provide an accessible overview of the major considerations that will impact pupillometry studies, including artifact/outlier detection, response scaling, and statistical approaches. Lastly, Hershman et al. cover analysis methods with step-by-step instructions supplemented by online resources so readers can follow along. By the end of Part III, readers should have a solid appreciation for the diversity of approaches within pupillometry, which approach is best suited to their research question, and, importantly, how to get started with their own rigorous pupillometry research.

Concluding Remarks

The collection of chapters in this volume represents a passion project for everyone involved. We, the volume editors, could not be more grateful to the chapter contributors for dedicating their time and expertise to this project. We hope that the chapters contained in this volume inspire new research and new researchers, and we look forward to reading your work.

Lowell, MA, USA Megan H. Papesh
Tempe, AZ, USA Stephen D. Goldinger

References

Janisse, M. P. (1977). *Pupillometry: The psychology of the pupillary response.* Wiley.
Laeng, B., Sirois, S., & Gredebäck, G. (2012). Pupillometry: A window to the preconscious? *Perspectives on Psychological Science*, *7*(1), 18–27.
Mathôt, S., Fabius, J., Van Heusden, E., & Van der Stigchel, S. (2018). Safe and sensible preprocessing and baseline correction of pupil-size data. *Behavior Research Methods*, *50*, 94–106.
Reilly, J., Kelly, A., Kim, S. H., Jett, S., & Zuckerman, B. (2019). The human task-evoked pupillary response function is linear: Implications for baseline response scaling in pupillometry. *Behavior Research Methods*, *51*, 865–878.
Sirois, S., & Brisson, J. (2014). Pupillometry. *Wiley Interdisciplinary Reviews: Cognitive Science*, *5*(6), 679–692.

Acknowledgement

We are grateful for the hard work, dedication, and patience of all contributing authors, many of whom were kind enough to agree to this project during the height of the COVID-19 pandemic. This volume would not exist without them. We are also grateful to the team at Springer Nature for all the same reasons, especially their patience.

Contents

Part I
Neurophysiology

The Locus Coeruleus: Where Cognitive and Emotional Processing Meet the Eye

Ringo Huang and David Clewett

1 Overview

In recent years, pupillometry has experienced a resurgence in popularity as a technique for studying arousal and its complex effects on cognition. Arousal is often characterized as a combination of psychological and physiological responses to external stimulation (e.g., a danger signal) or internal mental processes (e.g., solving a math problem). It also relates to patterns of autonomic activity and wakefulness, both local and large-scale brain network dynamics, and a wide array of behavioral outcomes. Thanks in large part to pupillometry, it is now understood that fluctuations in arousal states help supply the energy required to meet processing demands (capacity), tune attention to task-relevant information (selectivity), and determine which sources of information are preferentially processed and remembered (quality). In this chapter, we dissect the meaning of "pupil as an index of arousal" with respect to different aspects of information processing in the mind and brain. We also focus on the critical role of one of the brain's core arousal systems, the *locus coeruleus–norepinephrine (LC–NE) system*, in regulating these complex processes.

To begin, we review findings converging on the idea that non-luminance-mediated changes in pupil size are regulated by arousal and mental processes. We introduce *classic models of arousal–cognition interactions* and present evidence showing that pupil-linked arousal signals index what we perceive, attend to, and remember. We then synthesize cross-species research establishing a strong link between the LC–NE system and attentional control of the pupil.

Ringo Huang and David Clewett contributed equally with all other contributors.

R. Huang · D. Clewett (✉)
Department of Psychology, University of California, Los Angeles, Los Angeles, CA, USA
e-mail: david.clewett@psych.ucla.edu

M. H. Papesh, S. D. Goldinger (eds.), *Modern Pupillometry*,
https://doi.org/10.1007/978-3-031-54896-3_1

In the second part of the chapter, we shift our discussion to how the LC–NE system influences *mental and neural selectivity* by implementing *neural gain*, a computational process by which strong patterns of brain activity are amplified and weaker patterns of brain activity are suppressed. We begin by describing influential computational and neurophysiological models of LC–arousal interactions, including Adaptive Gain Theory (AGT). In this discussion, we explain how these frameworks capture the important contributions of distinct LC processing modes to task performance.

Extending this work, we describe a neurobiological account of how LC–NE system activation implements neural gain through its influence on local patterns of brain activity. We then interpret this new model through the lens of classic theories of arousal–performance interactions to explain how the priority of different inputs—and patterns of activity in the regions and functional networks that support those representations—may shift according to the overall level of arousal.

Moving from there, we discuss how the LC–NE system serves to energize adaptive behaviors, including regulating action selection, action execution, and effort. We explore how fluctuations in arousal and LC activity enable organisms to meet task demands by influencing *attentional capacity*, or the availability of mental resources. We trace a common theme that the LC helps mobilize resources at the mental, physical, and physiological levels. In addition, we examine how adaptive adjustments in global arousal levels and sustained LC activity may also determine *attentional intensity*, or the amount of mental resources that are specifically devoted to processing task-relevant information.

At the end of this chapter, we propose a unifying framework of the LC–NE system function to reconcile predictions from multiple models of arousal and attention. We conclude by summarizing different pupil measures and what they reveal about the LC's strong influence over the selectivity, capacity, and quality of information processing in the brain.

2 Foundational Links Between Arousal, Cognition, and the Pupillary Response

Arousal is at the core of our mental and physical lives. In addition to promoting general wakefulness, the ebb and flow of arousal states enable organisms to adapt to their constantly changing environments. Early theoretical work endeavored to capture these effects by formulating models of arousal and attention. These theories offered elegant solutions to some challenging problems in cognitive psychology, including the issue of how arousal tunes the allocation policy, or selectivity, of attention. They also laid a strong foundation for future neurocomputational models that formalized the effects of arousal on our ability to sustain and direct limited mental resources to process important information. To begin, we briefly summarize some of these key theories and how they have promoted a deeper understanding of arousal–cognition interactions.

2.1 Relationships Between Arousal and Task Performance

As it pertains to cognition, arousal is thought to be a domain-general construct that regulates the recruitment and allocation of attentional resources. One influential framework, the Capacity Model of attention, characterizes these effects of arousal in terms of its dual influences on attentional capacity and the allocation policy of attention (Kahneman, 1973). Attentional capacity can be conceptualized as the energy that is available for *any* information processing. A low arousal level, characterized by drowsiness or listlessness, is associated with a diminished attentional capacity. As a result, less energy is available for information processing. Increasing the level of arousal promotes wakefulness and increases the available attentional capacity for cognitive processing. Thus, the first part of the Capacity Model of attention captures how arousal might regulate the amount of mental resources that are available at any given moment. The second part of the Capacity Model proposes that arousal exerts control over the allocation of this pool of resources. An increase in arousal narrows the breadth of attention, restricting the amount of cognitive processes that receive these valuable resources.

One of the litmus tests for the Capacity Model and other cognitive theories is how well they can explain the relationship between arousal and behavioral performance. In a classic experiment, researchers examined how mice performed on a luminance discrimination task as a function of the intensity of an electric shock administered for incorrect responses (Yerkes & Dodson, 1908). The results showed that the relationship between performance and arousal (i.e., shock intensity) depended on the difficulty of the task condition. When the task was simple (i.e., discriminating between dark versus bright luminance), the arousal-performance curve increased monotonically. By contrast, when the task was difficult (i.e., discriminating between dark versus gray), the arousal's relationship with performance followed an "inverted-U" function, whereby performance peaked for intermediate levels of shock. This famous characterization of the arousal–performance relationship has since been replicated in other animal and human research studies across many different paradigms (Broadhurst, 1957).

Despite its massive influence on the fields of cognitive psychology and neuroscience, the classic arousal–performance relationship is often misrepresented (Diamond et al., 2007). Many researchers mistakenly ascribe the "inverted-U" function to all behavioral tasks despite the presence of an additional monotonic pattern. In actuality, different tasks appear to follow different arousal–performance curves, with the shape of these functions being largely determined by the difficulty and nature of the task (Yerkes & Dodson, 1908). The arousal–performance curve for "difficult" tasks can be appropriately described by an "inverted-U" function, which has received the most attention in cognitive psychology research. However, this preoccupation with the inverted-U function has also led researchers to neglect the presence of the "easy" curve, which is characterized as a linear relationship between arousal and task performance. That is, for tasks that engage habitual or emotional responses that are more reflexive and require little cognitive effort, arousal will continue to benefit rather than impair performance as it increases. This distinction between easy and hard tasks has critical implications for how we interpret the role of arousal in influencing different brain and behavioral outcomes.

2.2 Cue Utilization Hypothesis and the Rigidity–Lability Paradox

Why do some arousal–performance curves follow the "inverted-U" function? And why do different tasks produce different arousal–performance curves? One of the early attempts to address these questions is the cue utilization hypothesis (Easterbrook, 1959), which elegantly explains how the relationship between the level of arousal and performance is modulated by task or response demands. According to this seminal hypothesis, at a low level of arousal, attention is broadly allocated among a large pool of cues, or cognitive processes, which may be task-relevant or task-irrelevant. This low arousal state is frequently described as a state of inattentiveness in which performance on any task is generally poor.

As arousal increases to moderate levels, however, the focus of attention also narrows and filters out task-irrelevant cues while preserving task-relevant cues. By promoting attentional focus on task-relevant information processing, moderate levels of arousal facilitate a state of task engagement in which task performance is generally optimized. As arousal levels increase past this task-optimal zone, the focus of attention may become too narrow to support some task-relevant cues, thus impairing task performance. In other words, the task-optimal level of arousal (i.e., the "peak" of the "inverted-U" function) is achieved when the focus of attention is large enough to encompass all task-relevant cues but also small enough to exclude all task-irrelevant cues (Easterbrook, 1959).

A corollary to this explanation is that the task-optimal level of arousal depends on the number of cues demanded by the task. Because difficult tasks demand more cue utilization, the narrowing of attention filters out task-relevant cues at lower levels of arousal, resulting in an earlier "inverted-U" peak. In contrast, tasks that demand less cue utilization are impaired by the narrowing of attention at higher levels of arousal. This difference in cue utilization demands between difficult and easy tasks helps account for the existence of dual arousal–performance curves: the task-optimal peak occurs at moderate levels of arousal for more difficult tasks, and performance peaks at higher levels of arousal until it eventually plateaus for easier tasks (Diamond et al., 2007).

Based on the findings of this time, the cue utilization hypothesis provided a compelling, high-level account of the arousal–performance relationship by describing how arousal influences cue utilization. However, while the key assumption that increasing arousal narrows the focus of attention was well supported by many studies (Bahrick et al., 1952; Broadbent, 1971; Bursill, 1958; Callaway & Dembo, 1958; Hockey, 1970), other studies had linked high levels of arousal to increased distractibility (Korchin, 1964; Woodhead, 1964). Thus, arousal appears to have the paradoxical effect of simultaneously narrowing attentional focus while also increasing attentional distractibility. Simply put, arousal makes attention more rigid yet more labile (Callaway & Stone, 1960; Kahneman, 1973; Wachtel, 1967).

Wachtel (1967) addressed this rigidity–lability paradox by dissociating two independent properties of attentional selectivity: its breadth and its stability. Using the metaphor of attention as a beam of light (Hernández-Peón, 1964), the breadth refers to the width of the flashlight beam while the stability refers to the amount of scanning across a field. Thus, in line with the cue utilization hypothesis, increasing arousal narrows the width of the flashlight beam such that fewer cues are illuminated by the beam of light (Fig. 1). However, high arousal also reduces the stability of the flashlight beam, leading to more movement and scanning across the array of cues (Fig. 1c). By narrowing the attentional beam and inducing more scanning, high levels of arousal simultaneously increase the likelihood of task cues being dropped and non-task cues being captured by the attention. This conceptual description of arousal's dual effects on the rigidity and lability of attentional selectivity is an important foundation through which later researchers tested and developed neurocomputational models.

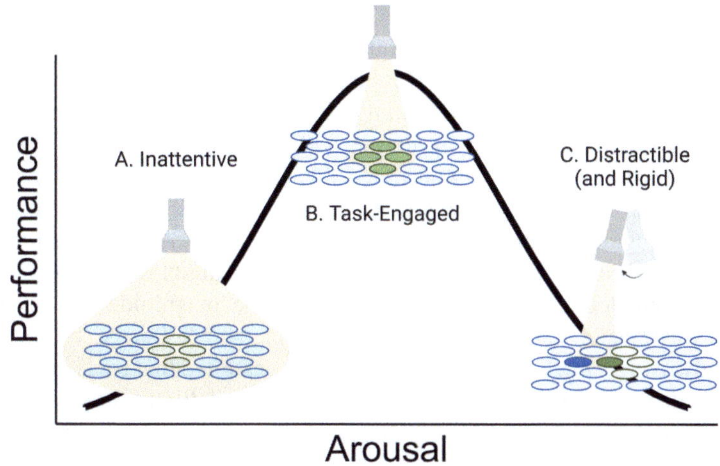

Fig. 1 The attentional flashlight and the "inverted-U" arousal-performance curve. Task performance varies as a function of overall levels of global arousal, with the attentional beam narrowing with increasing arousal. (A) At low levels of arousal, the beam is diffuse and attentional capacity is low, capturing a wide range of sensory inputs, or cues (circles). (B) At moderate levels of arousal, the attentional beam focuses on the set of task cues and enhances their processing (dark green circles), corresponding to an optimal, task-engaged state and optimal performance. (C) At high levels of arousal, the attentional beam remains narrow while scanning more often across the field of task cues (circles). This results in simultaneously more rigid and labile attentional processes that encourage the selection of task-irrelevant cues, or distractors (blue circles). One consequence of this relationship between arousal and the "attentional flashlight" is its effect on task performance that follows an "inverted-U" function for cognitively demanding tasks. (Image created with BioRender.com)

2.3 The Pupil as a Window into Arousal–Cognition Interactions

The influential models we have discussed so far drew important links between background levels of arousal and fluctuations in attentional states. They also provided a framework for understanding how transient, or phasic, arousal responses signal mental workloads and the amount of information devoted to processing a behaviorally relevant event. While these ideas provided testable hypotheses about arousal–attention interactions, they still required a tool or biomarker that could help quantify cognitive and arousal effects in the laboratory. For this endeavor, researchers turned to pupillometry, a noninvasive technique that measures physiological and mental activity through changes in pupil diameter. Pupil size is an easily measurable physiological variable that has frequently been used to quantify both tonic and phasic arousal. Beginning in the 1960s, this inspired researchers to use pupillometry to study how arousal modulates attention and mental operations during a wide range of cognitive and affective processes, including effort, affect, salience, novelty, and surprise. Here, we showcase just a few of the key findings regarding the functional significance of pupil responses to behavior.

One influential program of research examined how pupillary response tracks information processing load, or mental effort, during a demanding cognitive task (Beatty & Lucero-Wagoner, 2000; Simpson & Hale, 1969; van der Wel & van Steenbergen, 2018). For example, Hess & Polt (1964) observed greater pupillary responses when subjects were performing difficult arithmetic problems (e.g., complex multiplication) compared to simple addition problems. A few years later, Kahneman & Beatty (1966) demonstrated that pupil size tracks working memory load in a digit span task, suggesting that a larger pupil reflects the amount of information being juggled in mind and/or the complexity of cognitive operations needed to maintain task performance. Pupil size increased as digits were sequentially presented and encoded into working memory, and the magnitude of this pupillary response was greater for longer digit spans (Kahneman & Beatty, 1966). A subsequent study using a pitch discrimination task demonstrated that trials in which tones were more difficult to discriminate elicited greater pupil dilations (Kahneman et al., 1967). Taken together, these findings indicated that demanded greater mental effort elicits larger task-evoked pupillary responses.

According to some theories of arousal, an increased demand for task-related information processing, or mental effort, is thought to be supplied by an increase in arousal (Kahneman, 1973). Models of arousal and attention suggest that the continuous monitoring and evaluation of task demands adjust mental effort, leading to an appropriate adjustment in arousal and the capacity and selectivity of attention (Aston-Jones & Cohen, 2005; Kahneman, 1973). Thus, the pupillary response may reflect fluctuations in arousal that support the information processing load demanded by the current task (Beatty & Lucero-Wagoner, 2000; Bradshaw, 1967; Pribram & McGuinness, 1975; Zénon, 2019). Indeed, a large body of work demonstrates that pupil dilations are elicited by an increase in cognitive control, a topic we discuss in more detail in Sect. 5.3.

Another extensive program of research focuses on pupillary responses that are evoked in a bottom-up, rather than goal-directed, manner. The idea that pupillary responses reflect affect in emotional stimuli was first explored by Hess & Polt (1960). Although they reported differential pupillary responses to emotional stimuli of different valences (i.e., pupil constriction to negative stimuli and dilation to positive stimuli), numerous subsequent studies have disputed this result (Bradley et al., 2008; Partala & Surakka, 2003). Using images with validated arousal and valence ratings, Bradley et al. (2008) demonstrated that the pupil dilates to arousing stimuli, irrespective of their valence. Such studies have converged on the notion that the magnitude of the pupil dilation to an emotional stimulus varies as a function of arousal, irrespective of valence (see also Zekveld et al., 2018). In light of its close link to autonomic arousal responses, pupil dilation has also become a popular measure of fear conditioning (de Voogd et al., 2016; Reinhard & Lachnit, 2002).

Exogenous elicitors of arousal do not need to be inherently emotional. The pupillary response has also been shown to signal both contextual and absolute novelty. Importantly, these two forms of novelty differ in that contextual novelty refers to stimuli that are unexpectedly encountered within a certain context, while absolute novelty refers to stimuli that were never encountered before (Kafkas & Montaldi, 2018). One of the most robust demonstrations of a link between the pupil and contextual novelty is the oddball task. This commonly used paradigm manipulates novelty through the infrequent presentation of an oddball stimulus, which stands out from its surrounding stimuli by virtue of its semantic, perceptual, or emotional features. Many studies using the oddball detection paradigm across different sensory modalities have demonstrated that the pupil dilates more to oddball stimuli than to standard stimuli, or information that is presented more frequently (Gilzenrat et al., 2010; Murphy et al., 2011, 2014). This novel versus standard effect is also observed when the contextually novel stimuli does not require an overt button response, although the effect is slightly diminished (Krebs et al., 2018; Mather et al., 2020). Moreover, the dynamics of the pupillary fluctuations during the task covaried with brainstem arousal regions, such as the locus coeruleus, thought to mediate the arousal response (Murphy et al., 2014). Together, these findings suggest that the contextual novelty is associated with an arousal response that can be tracked with pupillometry.

It is worth pointing out that the other form of novelty, absolute novelty, yields different pupillary effects. When new stimuli versus familiar stimuli is shown, participants exhibited either more pupil constriction or diminished pupil dilation (Kafkas & Montaldi, 2014; Naber et al., 2013; Võ et al., 2007). The contrasting effects of absolute versus contextual novelty may suggest that contextual novelty is more closely tied to the activation of noradrenergic pathways (Mather et al., 2020; Murphy et al., 2014), while absolute novelty is more closely associated with the activation of cholinergic pathways in the brain (Duszkiewicz et al., 2019; Kafkas & Montaldi, 2018). It is also noteworthy that pupil constriction effects observed in old/new recognition paradigms are in response to "common" absolute novelty (e.g., first time seeing *this* apple) rather than "distinct" absolute novelty (e.g., first time seeing *any* apple; Duszkiewicz et al., 2019). Whether pupil dilations to these different types of novelty reflect the same underlying process remains an open question, given that distinct absolute novelty is very challenging to study in humans.

Pupil dilations also track surprise, which is a function of a mismatch between one's expectations about an outcome of an event and what actually occurs (Ekman & Davidson, 1994). When individuals have higher confidence in their incorrect expectations, they encounter a greater degree of surprise. In other words, surprise is a prediction error of certainty that signals the presence of unexpected information and elicits an increase in arousal. Several studies have demonstrated the link between surprise and arousal by showing that the pupillary response to feedback is modulated by surprise (Braem et al., 2015; de Gee et al., 2014; Nassar et al., 2012). Moreover, in an experiment that dissociated reward and uncertainty prediction errors, Preuschoff (2011) demonstrated that the pupil dilation response varied as a function of uncertainty prediction error but not reward prediction error. While it is well known that the dopaminergic system signals reward prediction errors, the pupillary findings suggest that the noradrenergic arousal system may signal the uncertainty-related prediction error (Preuschoff, 2011).

Updating expectations based on feedback is an important component of learning and decision-making. Surprise serves as a strong signal that increases arousal in response to unexpected information. This modulates the capacity and selectivity for the processing of surprising information, which can rapidly facilitate learning and benefit future decision-making (de Gee et al., 2020). It can also facilitate evidence accumulation in dynamic environments and may bias perceptual decision-making based on priors (Krishnamurthy et al., 2017) or motivational effects (Leong et al., 2021).

Similarly, context shifts, such as changes in one's thoughts, feelings, or surroundings, signal the presence of new information and are important for updating one's internal model of the world. Pupil dilations are thought to track context shifts that are both internal (e.g., when switching between cognitive tasks) and external (e.g., when encountering a change in the environment). For example, research in both monkeys and rodents shows that the pupil dilates when environmental contingencies change (Rajkowski et al., 1994; Sara, 2009; Sara & Segal, 1991). Similar results have been obtained in human pupillometry research, namely, transitions from highly regular and structured auditory sequences to new variable, irregular, or regular sequences elicit robust pupil dilations, suggesting the pupil tracks the statistics of the environment (Zhao et al., 2019).

Shifts from contextual stability to instability, or change, also have consequences for how we encode temporally extended sequences of information. In one pupillometry study, Clewett et al. (2020) manipulated the context during item sequence learning by presenting auditory tones in either the left or the right ear during a block of images. The event boundary, or the shift in context from the auditory tone presented in the left ear to the right ear, reliably elicited pupil dilation. Distinct temporal characteristics of this boundary-evoked pupil dilation were also related to temporal memory measures of episodic memory formation. This finding suggests that fluctuations in pupil size signal arousal processes that construct event representations in long-term memory.

In a similar vein, shifts in internal belief states, such as a violation of expected statistical regularity, may also produce an increase in arousal and pupil dilation. In a predictive inference task, subjects were asked to guess numbers that were randomly sampled from a Gaussian distribution of either high or low variance about an unknown mean (Nassar et al., 2012). After a series of trials, participants develop an internal belief state about this unknown mean. When the underlying mean switches, participants experience a violation of their internal belief states, which is followed by a period of elevated pupil size. The increased pupil size may reflect an increase in arousal after the change point, which may be adaptive in situations where a rapid updating of internal belief states is required. Critically, because no perceptual change occurs when the underlying mean switches, this study supports the idea that an increase in arousal occurs for general contextual changes, whether they are internal (e.g., internal expectations) or external (e.g., a change in environments).

The highly selective effects of arousal on memory are indexed by pupil dilations. For example, when viewing a series of overlapping image pairs, threat-induced arousal amplifies the memory benefit for the prioritized image at the cost of the non-prioritized image, revealing the influence of arousal on the selectivity of information processing (Clewett et al., 2018). Pupil dilations were associated with greater memory selectivity and greater activity in arousal regions of the brain, connecting the behavioral effects with physiological markers of arousal. Interestingly, pupil dilation can also serve as a predictor of successful retrieval of emotionally arousing memories. Larger pupil dilations during encoding relate to greater activity in arousal-related brain regions, such as the LC, during retrieval (Sterpenich et al., 2006). Thus, pupil dilation appears to be an indicator of the selective benefit of emotional arousal during both encoding and retrieval. As emotional stimuli tend to be better remembered than neutral stimuli, these findings also align with the broader literature linking larger stimulus-evoked pupil dilations to stronger encoding processes (Goldinger & Papesh, 2012; Papesh et al., 2012).

Finally, studies have used pupillometry to study how arousal influences the waxing and waning of attention across time. Attentional states shape cognitive processing and are important variables for explaining variability in behavioral performance and task engagement over time. In using pupil diameter to infer tonic levels of arousal, researchers identified ranges of baseline pupil diameter corresponding to an "exploitative" state (more task engagement) and an "explorative" state (less task engagement; Gilzenrat et al., 2010; Jepma & Nieuwenhuis, 2011; Van Den Brink et al., 2016). The "exploitative" state is not only characterized by improved task performance but also facilitates larger phasic task-evoked pupil dilations. The neural mechanisms connecting baseline, or non-stimulus-related, changes in pupil diameter to these attentional states will be a primary focus in later sections.

In summary, a large body of work has used pupillometry to study how arousal modulates information processing across multiple cognitive domains. While the pupil can reflect state changes in attention and arousal due to top-down factors like mental effort or fatigue, it can also reflect phasic increases in arousal elicited by salient events in the environment, including shifts in environmental contingencies or biological imperatives. Different pupil dynamics can thereby characterize and quantify the varied influences of arousal processes on mental selectivity and capacity, as predicted by earlier theoretical frameworks.

2.4 Challenges of Understanding Arousal: Multiple Pathways, Multiple Effects

Pupil changes are highly sensitive to changes in arousal and relate to key parameters of attention. But what neurophysiological mechanisms support these complex relationships? Is arousal merely a readout of different cognitive processes? Or does it reveal causal relationships between specific brain systems and information processing?

While researchers have uncovered much about the neurocircuitry that regulates pupil size, the functional significance of these pathways to cognitive processes is less clear. Part of the reason for this under-specification is that central arousal systems are very complex and comprised of multiple brainstem neuromodulatory pathways, including the noradrenergic, cholinergic, serotonergic, dopaminergic, and reticular activating systems (Jones, 2003; Larsen & Waters, 2018). Although some of these neuromodulatory pathways may have redundant functions (e.g., promoting wakefulness), they likely exert different effects on their ascending brain targets. Arousal is also very multifaceted and contains affective, autonomic, and wakeful components (Satpute et al., 2019). Furthermore, the literature is riddled with perplexing findings like directional fractionation, a term referring to the low correlations between pupil size and other autonomic markers of arousal, such as heart rate (Lacey & Lacey, 2007).

Although it is worth noting the complex, multidimensional nature of the arousal systems (Thayer, 1978), a clear consensus is beginning to emerge that one neuromodulatory system is a key mediator of cognitive control over the pupil: the locus coeruleus–norepinephrine (LC–NE) system. Indirect evidence of this relationship comes from work showing that pupil dilations are sensitive to the same mental processes and salient events that engage the LC–NE system. In fact, many of the studies outlined in the previous section report corroborating evidence of LC activity occurring alongside task-related pupil changes (Larsen & Waters, 2018). Pharmacological data indicate that drugs that modulate activity in the LC–NE system elicit changes in pupil size. Perhaps most convincingly, there is now direct evidence from neuroanatomical tracings as well as optogenetic stimulation and electrode recording work in animals demonstrating a causal link between LC activity and the pupil. Throughout the rest of this chapter, we focus on what this relationship reveals about the essential and multifaceted roles of LC activity in mediating arousal's effects on attention, neural processing, and behavior (Aston-Jones & Cohen, 2005; Berridge & Waterhouse, 2003).

3 LC–NE Arousal System and Pupillometry

3.1 Functional Neuroanatomy of the LC–NE System

The locus coeruleus (LC) is a small brainstem nucleus bilaterally positioned on the rostral aspect of the pons and serves as the primary supplier of norepinephrine (NE) to most of the brain (Berridge & Waterhouse, 2003). As a key hub region of ascending arousal systems, the LC plays an important role in regulating global arousal states, including overall levels of wakefulness and alertness (Carter et al., 2010). Remarkably, despite its small size—approximately 30,000–50,000 neurons in the human brain—the LC also has far-reaching and robust effects on attention and memory processes (Fig. 2). Much cross-species work demonstrates that dynamic patterns of LC–NE system activation regulate executive function, working memory, and behavioral flexibility (Poe et al., 2020; Sara, 2009). Activation of the LC also helps direct mental resources processes where they are most needed, either by facilitating focused attention or reorienting attention toward unexpected or salient stimuli (Corbetta et al., 2008; Sara, 2009). LC activity furthermore facilitates the encoding and consolidation of emotionally salient and novel information, ensuring that organisms can store and recall memories of motivationally relevant experiences (Mather et al., 2015; McGaugh, 2000; Poe et al., 2020; Sara, 2009). Such enhancements may be accomplished through the regulation of synaptic plasticity, including modulatory effects on amygdala and hippocampal circuitry (Bergado et al., 2011; Berridge & Waterhouse, 2003; Hansen, 2017; Harley, 1987; Mather et al., 2015; Tully & Bolshakov, 2010).

The LC sends dense efferent projections to most regions of the brain, enabling it to regulate both local patterns of brain activity as well as brain-wide communication between different functional networks (Berridge & Waterhouse, 2003; Poe et al., 2020). Under arousal, NE is diffusely released at its cortical and subcortical targets, where it engages various adrenoreceptors that have distinct functional properties, spatial distributions, and binding affinities for NE (Berridge & Waterhouse, 2003). Broadly, this includes alpha1-adrenoreceptors, alpha2-adrenoreceptors, and beta-adrenoreceptors. While a full description of these receptors is beyond the scope of this chapter, it is important to note that different receptors exert different effects on neural excitation and synaptic plasticity. Thus, different spatiotemporal patterns of adrenoreceptor activation elicit distinct effects on brain activity under arousal, including supporting increases in neural gain.

In addition to functional differences in adrenoreceptors, the LC can influence arousal and cognitive processing through changes in its own firing rate. LC neurons fire in two distinct modes: (1) phasic, or transient, activity that typically involves firing rates of ~8–10 Hz (Aston-Jones & Bloom, 1981) and (2) tonic, or background, activity that typically involves firing rates of ~1–6 Hz. Phasic LC responses are characterized by strong bursts of activities that occur in response to salient inputs, such as novel, unexpected, goal-relevant, or rewarding stimuli (Aston-Jones & Bloom, 1981; Bouret & Sara, 2005; Foote et al., 1980; Sara & Bouret, 2012). Research also shows that these transient LC responses are closely aligned with the appearance of target stimuli (Aston-Jones et al., 1994), consistent with the idea that LC activation acts as a temporal filter for processing task-relevant information (see also Nieuwenhuis et al., 2005). Phasic LC responses facilitate the encoding of salient information and promote optimal task performance, particularly on tasks that require focused attention and the detection of target information (Aston-Jones et al., 1997; Clayton et al., 2004; Sara, 2009). Even in the absence of an external stimulus, pairing optogenetic stimulation of LC phasic activity with low-intensity

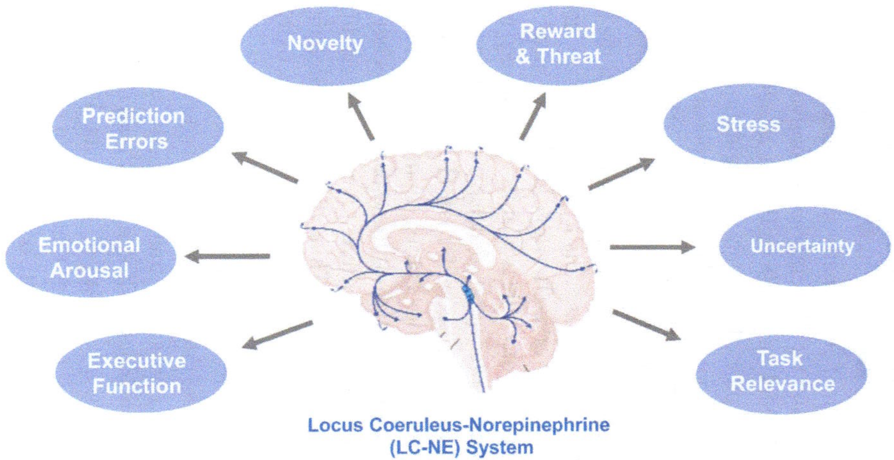

Fig. 2 The locus coeruleus–norepinephrine (LC–NE) system supports a wide variety of cognitive and emotional processes. The LC–NE system sends dense efferent projections across most of the brain, enabling it to modulate local brain activity patterns. Its wide reach also enables it to shape the functional dynamics of large-scale brain networks under arousal. (Image created with BioRender.com)

sensory stimulation can enhance the salience and encoding of those sensory representations (Vazey et al., 2018). This modulatory pattern mimics LC effects that are typically observed in response to intense or salient environmental stimuli.

By contrast to stimulus-linked bursts in LC activity, tonic LC activity is typically characterized by sustained—often described as "background"—levels of LC activation (Berridge & Waterhouse, 2003). Neurophysiological findings in monkeys indicate that spontaneous LC activity fluctuates over the course of a behavioral test session (Aston-Jones et al., 1996), which also seems to have important consequences for how attention and decision processes vary from trial to trial. High tonic levels of LC output are also closely tied to states of wakefulness and global arousal, such as during acute stress. Together, these different modes of LC activity are critical for regulating dynamic switching between different forms of learning and decision-making strategies (Aston-Jones & Cohen, 2005).

In many ways, the LC is blind to what is happening in the world. So, what determines how fast LC neurons will fire? Adaptive adjustments in LC firing are largely accomplished via top-down influences from higher cortical regions, including the prefrontal cortex (PFC) (Aston-Jones & Cohen, 2005), that signal and promote the prioritization of different stimuli. Once activated, the LC can reinforce those task-relevant or salience-related signals by modulating neural activity across the rest of the brain, including sensory cortical regions representing those inputs. Along with the lateral prefrontal cortex (LPFC), the orbitofrontal cortex (OFC) and anterior cingulate cortex (ACC) send dense projections to the LC (Arnsten & Goldman-Rakic, 1984; Jodo et al., 1998). At the subcortical level, the LC also receives dense afferents from the amygdala (Price & Amaral, 1981), which is known to activate the LC during emotionally arousing or stressful experiences (Valentino & Van Bockstaele, 2008). We suggest that the dominant locus of top-down inputs to the LC is largely determined by the current level of arousal, biasing which forms of priority will guide behavior and neural processing at a given moment. We will explore these arousal-dependent effects of LC modulation in more detail in Sect. 4.3.

Traditionally, the LC is characterized as a uniform cluster of neurons, each with widely diffuse projections to the neocortex. This anatomy implies that the LC controls a homogenous and global release of NE, which is consistent with the state-level modulatory effects of arousal (Sara & Bouret, 2012; Usher et al., 1999). However, a closer look at the subpopulations of LC neurons reveals a more modular organization, with evidence of specialization and specificity in its inputs (Uematsu et al., 2017), outputs (Chandler et al., 2013; Chandler & Waterhouse, 2012), molecular composition (Chandler et al., 2014), and neuronal firing patterns (Schwarz & Luo, 2015; Totah et al., 2018). Recent work suggests that this structural modularity also supports the functional modularity of the LC–NE system. For example, specific ensembles of LC neurons projecting to the amygdala have been linked to aversive learning, whereas a separate ensemble of LC neurons projecting to the medial prefrontal cortex has been linked to extinction learning (Uematsu et al., 2017). In mice, distinct spatiotemporal signatures of LC phasic responses are also associated with distinct aspects of learned behaviors during a go/no-go task (Breton-Provencher et al., 2022).

Considering these findings, many researchers now view the LC as a flexible system that can support both global states of arousal as well as more fine-tuned responses to salient information (Bouret & Sara, 2005; Foote & Berridge, 2019; Munn et al., 2021; Poe et al., 2020; Shine et al., 2018b; Wainstein et al., 2022). Thus, the LC–NE system may promote the selective processing of important information via a spatial specificity that enables a targeted release of NE *where* those representations are activated as well as a temporal specificity that enables the release of NE *when* those salient events occur.

3.2 Neuroanatomical Evidence Linking Pupil Dynamics to the LC–NE System

The autonomic circuitry underlying the pupillary light and dark reflexes has been relatively well established (Loewenfeld & Lowenstein, 1993). However, only recently has research begun to uncover the mechanisms that drive cognitive activity–related changes in the pupillary response. Researchers now recognize that both the cholinergic and noradrenergic pathways, key neuromodulatory systems underlying arousal, play a role in driving pupillary fluctuations that track arousal (Reimer et al., 2016). Yet, from a bird's-eye view, the bulk of the literature points to an especially strong link between the LC–NE system and the pupil. We review these key findings below.

Early direct evidence of the LC's involvement in pupillary fluctuations came from resting state recordings of the pupil and LC neurons in monkeys (Rajkowski, 1993). At rest, the pupillary fluctuations covaried with tonic LC fluctuations such that periods of elevated LC activity corresponded to larger pupil diameters. Later studies using microelectrode stimulation of LC neurons demonstrated a direct link between LC activation and a phasic pupil response (Joshi et al., 2016; Liu et al., 2017; Reimer et al., 2016; Fig. 3). These microstimulation studies further solidified

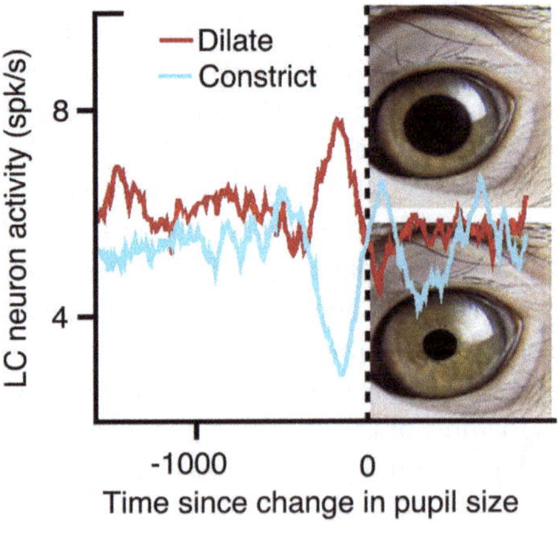

Fig. 3 Temporal fluctuations in pupil size track LC neuron activity. In simultaneous electrode recordings of monkey LC neurons and pupil diameter, a spike in LC activity precedes pupil dilation, whereas a dip in LC activity precedes pupil constriction. (Figure obtained from Costa and Rudebeck (2016))

the involvement of the LC in the circuitry underlying arousal-mediated pupillary response. Notably, these studies also implicated other structures, such as the intermediate layer of the superior colliculus (Joshi et al., 2016) and the basal forebrain (Reimer et al., 2016), suggesting that the noradrenergic system may influence the pupil via multiple pathways and that the cholinergic pathway also has a role in the control of the pupillary circuitry. However, compared to other structures, microstimulation of LC neurons produces especially robust and consistent results in terms of the magnitude and latency of the resulting pupil dilation (Joshi et al., 2016).

While the LC acts on both the parasympathetic and sympathetic limbs of the pupil circuitry, there may be differences in the laterality of these effects (Liu et al., 2017). In rats, unilateral LC stimulation resulted in a lateralized dilation of both pupils, meaning that the ipsilateral pupil dilated more than the contralateral pupil. When surgically removing the cells from the cervical spinal cord that are part of the sympathetic pupillary circuitry, the lateralization disappeared, suggesting that the LC's influence on the sympathetic limb is solely ipsilateral while its influence on the parasympathetic limb is bilateral.

Another technique that has produced evidence of LC's involvement in the regulation of pupil size is vagus nerve stimulation (VNS). The vagus nerve projects to the nucleus of the tractus solitarius, which in turn has projections to the LC among other target regions. Although not as direct as LC microstimulation experiments, VNS is an intriguing technique for indirectly probing the LC via projections from the nucleus tractus solitarius (Jodoin et al., 2015). Moreover, it has some potential for use in human studies, as VNS is an FDA-approved treatment for epilepsy and depression in humans and it has a related noninvasive procedure called transcutaneous vagal nerve stimulation (tVNS).

Emerging findings demonstrate that VNS treatment influences tonic LC–NE system activity. In rats, long-term VNS treatment resulted in elevated tonic LC firing rates (Dorr & Debonnel, 2006) and an increase in extracellular NE concentrations in the prefrontal cortex and hippocampus (Manta et al., 2012). In humans undergoing VNS therapy for major depression and epilepsy, average pupil diameter was larger during periods when chronic VNS was turned on compared to when it was turned off (Jodoin et al., 2015). In addition, some rodent studies have demonstrated that a phasic delivery of VNS evokes a pupil dilation that co-occurs with increases in other noradrenergic biomarkers (Collins et al., 2021; Mridha et al., 2021). Thus, VNS appears to exert both tonic and phasic influences on pupil size via the LC–NE system.

However, evidence linking the more indirect tVNS with LC–NE activation has thus far been mixed. In two studies comparing the effects of tVNS to sham stimulation, there were no effects of tVNS on baseline pupil diameter or task-evoked pupil change during an auditory oddball task (Keute et al., 2019) and an attentional blink task (Burger et al., 2020). The discrepancy between tVNS and VNS research may be due to differences in the stimulation strength of the vagus nerve (Mridha et al., 2021). Indeed, in systematically manipulating the intensity of the tVNS and the luminance of the environment, Capone et al. (2021) demonstrated that tonic tVNS induces an increase in the ipsilateral pupil diameter only under certain stimulation intensity (2 mA) and under low luminance conditions. Using a similar stimulation intensity (2.2 mA), Sharon et al. (2021) demonstrated that phasic tVNS pulses also induced pupil dilations. More work should be done to identify the optimal parameters for tVNS stimulation to induce changes in various LC–NE biomarkers.

Pharmacological studies have also provided some evidence that the LC–NE system is linked to pupillary fluctuations. Studies using clonidine (alpha-2 agonist) and yohimbine (alpha-2 antagonist) directly inhibit or stimulate LC neurons by targeting alpha-2 adrenoceptors on LC neurons (for review, see Szabadi, 2018). Clonidine is thought to preferentially bind to the alpha-2 autoreceptors on LC neurons (Berridge & Waterhouse, 2003). For example, clonidine has been shown to induce pupil constriction (Hou et al., 2005; Phillips et al., 2000), while yohimbine has been shown to induce pupil dilation (Phillips et al., 2000). Moreover, the administration of modafinil, a wakefulness-promoting drug that indirectly increases LC activity, has been found to increase pupil size along with other autonomic and behavioral markers of LC–NE activity (Hou et al., 2005).

Although it certainly appears that the LC plays a role in the arousal-mediated pupillary response, a complete view of the connections between the LC and the pupil circuitry is still being formulated. The circuitry responsible for the pupillary light and dark reflexes consists of a sympathetic limb, which dilates the pupil via its excitatory influence on the iris dilator muscles, and a parasympathetic limb, which constricts the pupil via its excitatory influence on the iris constrictor muscles (Loewenfeld & Lowenstein, 1993). Given the relatively delayed latency between LC microstimulation and the onset of pupil dilation (~500 ms; Joshi et al., 2016), researchers speculate that the LC has indirect connections to the pupil circuitry.

It is now thought that the LC modulates pupil size via multiple indirect pathways that either inhibit the parasympathetic limb or excite the sympathetic limb. One potential parasympathetic pathway is through the LC's inhibitory influences on the Edinger–Westphal nucleus, which is a key node in the parasympathetic limb of the pupillary reflex circuitry (Joshi & Gold, 2020; Szabadi, 2018). A second possible mechanism involves the LC sending excitatory signals to the sympathetic (dilation) limb via the intermediolateral cell column of the spinal cord (Joshi & Gold, 2020; Liu et al., 2017). Based on the finding that lesions to the parasympathetic or sympathetic pathways reduced normal patterns of LC-mediated pupil dilations (Liu et al., 2017), it is likely that both limbs of the pupil circuitry are involved in a coordinated effort to modulate pupil diameter.

In summary, accumulated evidence implicates the LC–NE system in supporting the arousal-mediated pupillary response. In addition to more causal manipulations of LC activity, functional magnetic resonance imaging (fMRI) methods have been used to examine the LC–pupil link. In a seminal fMRI study in humans, it was revealed that pupil dynamics covary with the blood oxygen level dependent (BOLD) signal in a brainstem region consistent with the LC region when subjects rested or performed an oddball task (Fig. 4; Murphy et al., 2014). This finding has since sparked a flurry of interest in using combinations of neuroimaging, pharmacology, and pupillometry to study the relationship between the pupil, cognition, and LC

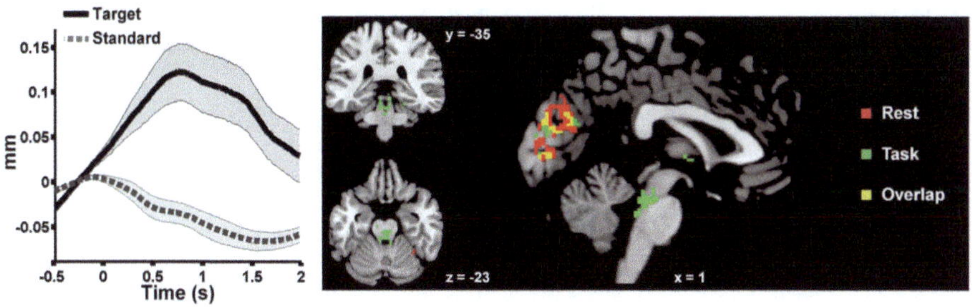

Fig. 4 Pupil dynamics covary with BOLD signal in a brainstem region that overlaps with the known location of the locus coeruleus. (Left) Contextually novel oddball targets induce a strong pupil dilation compared to standard, or more frequent, stimuli. (Right) In nonspatially smoothed functional images, pupil diameter covaries with the BOLD signal in a region of the brainstem that overlaps with the LC during an oddball detection task (green mask). This functional activation in the LC is not observed while participants rest (red mask). (Figure adapted from Murphy et al. (2014))

neuromodulation in humans. Wherever relevant, we highlight this important work throughout the rest of this chapter.

4 LC–NE System Modulates Neural Gain and the Breadth of Attention

4.1 Arousal Adaptively Amplifies Attentional Selectivity

The rigidity of attention, which describes how arousal narrows attentional focus (Easterbrook, 1959), is a useful characterization of one aspect of arousal's influence on mental selectivity. Specifically, arousal is often described as an *amplifier* of selectivity in which behaviorally relevant processes receive more mental resources at the cost of processing irrelevant information (Mather & Sutherland, 2011). In many contexts, this arousal mechanism adaptively tunes processing in favor of high-priority information, such as goal-relevant, novel, emotional, or perceptually salient stimuli to facilitate a winner-takes-more and loser-takes-less effect in perception and memory (Lee et al., 2014a,b; Mather & Sutherland, 2011; Sakaki et al., 2014; Sutherland & Mather, 2012).

Researchers have proposed a neurocomputational model based on neural gain to explain how arousal facilitates some (high-priority) processes while simultaneously suppressing other (low-priority) processes. Neural gain is a parameter of a neuron's responsivity function, which describes the relationship between the strength of input signals received by the neuron and the strength of the output signals sent by the neuron (Fig. 5, *the "S-shaped" functions*). An increase in neural gain increases the steepness of the responsivity function, meaning that the activity of a neuron receiving high excitatory inputs is further enhanced while the activity of a neuron

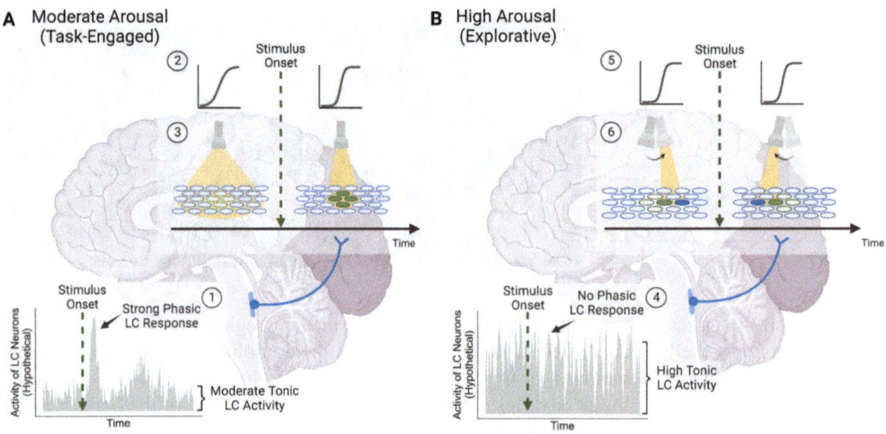

Fig. 5 LC–NE system activity, neural gain, and attentional selectivity in response to the onset of a task-relevant stimulus. (**a**) Under moderate arousal and tonic LC activity, (1) the onset of a task-relevant stimulus triggers a strong phasic LC response. (2) This LC response results in an increase in neural gain that (3) facilitates attentional selectivity for the set of task-relevant cues (green circles). (**b**) Under high arousal and high tonic LC activity, (4) the onset of a task-relevant stimulus fails to trigger a strong phasic LC response. (5) However, neural gain remains persistently high, even during periods when there is no task-relevant information processing. (6) This results in a rigid and labile attentional beam such that attentional selectivity for the set of task-relevant cues is impaired. Instead, attention becomes "distracted" and becomes grabbed by task-irrelevant cues. Note: the trajectory of the blue NE neuron is schematic; it would not pass over the cerebellum. (Image created with BioRender.com)

receiving low excitatory inputs is suppressed. Arousal may, therefore, modulate neural gain, with current levels of brain activity determining whether neuronal processing and accompanying mental representations are further enhanced or further suppressed (Aston-Jones & Cohen, 2005; Eldar et al., 2016).

It is now widely accepted that the LC–NE system helps regulate neural gain. The important role of LC activity in shaping neural selectivity first stemmed from evidence that NE enhances signal-to-noise processing in the brain (Hasselmo et al., 1997; Hirata et al., 2006; Waterhouse et al., 1980; Waterhouse & Navarra, 2019). The release of NE modifies the responsiveness of neurons, effectively sharpening the sensory receptive field to facilitate neuronal activity involved in signal processing while suppressing neuronal activity related to background noise (Manunta & Edeline, 2004). These physiological effects have also been formalized in several computational neural network models of LC function (Servan-Schreiber et al., 1990; Usher et al., 1999). Together, these physiological findings inspired the idea that NE promotes neural gain.

While an indiscriminate increase in neural gain may address the rigidity of arousal–attention interactions, it cannot fully explain why attention is also labile under high arousal. To address this issue, one influential theory highlighted the distinct computation and physiological functions of different modes of LC output. We review this model and its important predictions in the next section.

4.2 Adaptive Gain Theory: A Model of How Distinct LC Processing Modes Influence Behavior

How might the LC regulation of signal-to-noise processing in the brain influence mental selectivity or decision-making processes? According to Adaptive Gain Theory, the selective enhancement of task-relevant processes depends on the temporal specificity of stimulus-linked phasic LC activity (Aston-Jones & Cohen, 2005). Task cues, such as the appearance of a sensory target- or goal-relevant stimulus, elicit phasic bursts of LC firing, resulting in a momentary increase in neural gain for those active task representations. By contrast, task-irrelevant processes are relatively less active during this period and are suppressed with this phasic increase in arousal and phasic LC activation.

The magnitude of the task-evoked, phasic LC response is constrained by tonic, or background, levels of LC activity. Tonic LC activity fluctuates alongside changes in behavioral states, ranging from a low arousal state of drowsiness to a high arousal state of stress. Under low levels of tonic LC activity, task cues elicit low phasic LC responses, corresponding with the impaired task performance seen during low arousal, inattentive state. With intermediate levels of tonic LC activity, task-related phasic LC responses reach a peak magnitude, corresponding with optimal task performance seen during peak task engagement. Under high levels of tonic LC activity, phasic LC responses again diminish, corresponding with worse task performance and increased distractibility.

In addition to environmental influences like external stressors, arousal states and the related tonic LC activity may also be influenced by changes in utility over the course of performing the task. Dynamic transitions between the two modes of LC activity appear to be driven by cost–benefit evaluations carried out in higher cortical regions, such as the orbitofrontal cortex (OFC) and anterior cingulate cortex (ACC; Aston-Jones & Cohen, 2005). The dACC is a key region for supporting the metacognitive function of evaluating task utility. It also helps upregulate tonic LC activity as task utility diminishes. Such arousal-related adjustments in behavioral modes correspond with shifts in neural gain processes throughout the brain.

Notably, this relationship between the tonic LC activity and phasic LC activity resembles the inverted-U "difficult" curve that describes the arousal–performance relationship. As tonic LC activity is thought to reflect an individual's arousal state, this tonic–phasic LC relationship can be used to explain the arousal–task performance relationship. Generally, performance is poor during periods of low arousal and low tonic LC activity, which is characterized by diminished task-related phasic LC responses. In this low arousal state, individuals may be inattentive and prone to off-task or exploratory mind wandering because a lack of phasic neural gain prevents the selective facilitation of task-relevant processes (Mittner et al., 2016).

Under an intermediate arousal state with moderate tonic LC activity, relatively strong phasic LC responses boost task-relevant processing. By enabling the phasic increase in neural gain in response to task cues, moderate tonic LC activity provides the optimal conditions for goal-relevant behavior, target detection, and the exploitation of known rewards (Aston-Jones et al., 1997; Aston-Jones & Cohen, 2005; Rajkowski et al., 2004). While not directly linked to the LC, this is also in accordance with other work showing an inverted-U relationship between pupil-linked arousal and sensory cortical activity. In this study in mice, intermediate levels of arousal were associated with reduced cortical noise and optimal performance in a tone-in-noise detection task. These effects were also accompanied by enhanced activity in sensory-evoked thalamic and cortical responses, consistent with an increase in neural gain (McGinley et al., 2015). Given the strong LC–pupil link, it is plausible that the LC helps drive these gain effects under arousal.

At high levels of tonic LC activity, task performance is impaired due to sustained and high levels of neural gain reducing the ability of task cues to elicit a phasic LC response. As a result, these high-arousal states are more likely to induce noisier decision-making and attentional disengagement during goal-directed tasks (e.g., Aston-Jones & Cohen, 2005). Indeed, empirical studies show that behavioral variability and exploratory behaviors increase under arousal. For example, Designer Receptors Exclusively Activated by Designer Drugs (DREADD)-induced LC tonic activity in rodents leads to greater decision noise in a path foraging task, leading to disengagement from patch exploitation (Kane et al., 2017). Activating LC–NE system inputs to the ACC has also been shown to increase decision noise (i.e., stochastic choices) and consequently greater behavioral variability in rodents (Tervo et al., 2014). In monkeys performing a pitch discrimination task, increased distractibility coincided with periods of elevated baseline pupil diameter, an indirect index of tonic LC activity (Rajkowski, 1993). Similar effects have been shown in human studies, whereby larger baseline pupil diameters track task utility in a foraging-like task (Gilzenrat et al., 2010) and decisions to explore unknown rewards (Jepma & Nieuwenhuis, 2011). In addition, larger pupil diameters correlate with periods of high belief (Nassar et al., 2012) and periods of decision uncertainty that predict subsequent shifts in serial choices (Urai et al., 2017). Together, these studies are consistent with prior computational models positing a relationship between tonic LC activity and inverted-U arousal–performance function (Aston-Jones & Cohen, 2005). They also highlight the tight regulation of the dACC function by LC activation (e.g., Aston-Jones & Cohen, 2005; Joshi et al., 2016; Joshi & Gold, 2022), with high levels of activation predicting a more exploratory behavior, larger pupil diameters, and persistently high neural gain.

As previously mentioned, arousal's general effect of increasing neural gain manifests as increased "attentional rigidity," or a narrower focus of attention on strong, high-priority inputs (Mather et al., 2015). However, if increasing neural gain narrows the focus of our attention, why do we become more distractible under very high levels of arousal? One solution to this puzzle is provided by Adaptive Gain Theory, which accounts for the "attentional lability" side of the rigidity–lability paradox by characterizing high levels of arousal as a state of persistently high neural gain (Aston-Jones & Cohen, 2005). This persistently high state of neural gain indiscriminately tunes the responsiveness of neurons, even when task-relevant cues are absent (e.g., during an interstimulus interval). Persistently high neural gain thereby facilitates information processing for *any* cue, resulting in more exploratory decisions and behavioral variability. These effects are akin to the metaphor of a "labile" beam of attention that frequently switches its focus among competing task-irrelevant processes (*see* Fig. 5b; Gilzenrat et al., 2010).

Importantly, an elevated state of tonic LC activity diminishes the phasic LC response to task-relevant cues, preventing a discrete and temporally specific increase in neural gain. Consequently, attention becomes not only more sensitive to bottom-up and distracting cues but also less sensitive to the onset of goal-relevant information. The seemingly paradoxical nature of attention as being both rigid and labile under high arousal can thereby be attributed to the constraints placed on phasic LC firing by elevated levels of tonic LC activity.

4.3 *Arousal and Informational Gain in the Brain: The Glutamate Amplifies Noradrenergic Effects (GANE) Model*

In the previous section, we reviewed computational models of how fluctuations in LC activity adaptively influence the gain of information processing. Here, we will discuss a recent neurobiological model of how the LC–NE system implements this process of gain in the brain. The highly influential AGT model makes strong predictions about the role of different modes of LC activity in mediating the relationship between arousal and behavioral flexibility (Aston-Jones & Cohen, 2005). For many years, however, it was unclear how the all-or-none firing pattern of the LC could elicit opposite neural and behavioral outcomes for competing sensory inputs. That is, if task-relevant inputs are constantly competing with distracting stimuli for a foothold in awareness, how would the LC "know" which inputs to enhance and which inputs to suppress?

To address this longstanding question, the Glutamate Amplifies Noradrenergic Effects (GANE) model posits that arousal-induced phasic activation of the LC biases sensory processing and memory formation to favor high-priority representations at the expense of processing low-priority representations (Mather et al., 2015). According to GANE, the effects of NE release on neural excitation differ as a function of local levels of brain activity that are regulated by glutamate, the brain's primary excitatory neurotransmitter. At the core of this model is the idea that glutamate provides the neural substrate of priority signals in the brain, with priority being determined by the goal relevance, perceptual salience, novelty, or emotional significance of a stimulus (Fig. 6). By modulating NE levels locally, high glutamate levels are self-regulating: they recruit additional NE to strengthen their corresponding representations even further. As a result, winning mental representations "win more" under arousal, while losing representations take even less.

Importantly, neural gain is predominantly thought to occur during strong phasic LC responses to salient inputs because phasic LC responses yield greater volume release of NE than tonic activation. In this way, transient LC responses are time-locked to external demands, ensuring that behaviorally relevant stimuli receive the boost they may need to garner valuable mental resources.

Arousal-related increases in neural gain rely on two concentration-dependent functions of NE: the ability to further weaken low-priority inputs and the ability to further strengthen high-priority inputs. While LC activity is typically associated with states of intense mental activation or stress (Berridge & Waterhouse, 2003; Sara, 2009), its predominant function is to suppress most patterns of brain activity. In brain regions processing task-irrelevant or noisy information, NE's widespread release inhibits weak patterns of neural activity by engaging inhibitory alpha2-adrenoreceptors. As alpha2-adrenoreceptors have a higher affinity for NE than other adrenoreceptor subtypes, they will be engaged first and inhibit most patterns of brain activity under low-to-moderate levels of arousal. This widespread inhibition, in turn, creates ideal neurochemical conditions for enhancing signal-to-noise processing.

In contrast to regions processing irrelevant information, NE has the opposite modulatory effect on areas with strong activity. In these brain regions with high-priority representations, levels of local glutamate are high enough to spillover from synapses and bind to nearby NE varicosities, large swellings along the length of LC fibers that release NE into extracellular space. This additional local release of NE is thought to generate NE levels that are high enough to engage low-affinity beta-adrenergic receptors in neural pathways transmitting salient information. Beta-adrenoreceptors have potentiating rather than inhibitory effects on neural activity, thereby ensuring that high-priority representations gain privileged access to limited mental and metabolic resources (Mather et al., 2015).

The processing benefits of beta-adrenoreceptor engagement are manifold. First, engaging beta-adrenoreceptors enhances neuronal excitability even further by stimulating even greater local release of glutamate (Fig. 6). This, in turn, creates a positive glutamate–NE feedback loop, or "NE hotspot," that upregulates strong mental representations even further (Mather et al., 2015). Highly salient mental representations are, therefore, self-selecting in that they can recruit phasic LC responses to upregulate their own level of activity.

Second, the engagement of beta-adrenoreceptors has been shown to enhance long-term potentiation and synaptic plasticity. Thus, beyond initial perceptual and attentional processing, beta-adrenoreceptors may selectively enhance encoding and consolidation of high-priority inputs (Salgado et al., 2012). Meanwhile, the inhibitory effects of alpha2-adrenoreceptor activation prevail in regions with lower activity, and this may reduce synaptic plasticity and prevent long-term memory consolidation (Salgado et al., 2012).

Third, engaging beta-adrenoreceptors recruits metabolic resources, including glucose and oxygen, to active brain regions. This helps ensure that active neurons receive the resources needed to continue to process high-priority information (Bekar et al., 2012). Interestingly, this ability of NE to direct energetic resources to meet sensory or task demands may help account for the effects of arousal on regulating levels of mental energy and attentional capacity (as reviewed in detail in Sect. 5.4). The GANE model expands upon this link by suggesting that mental and physical

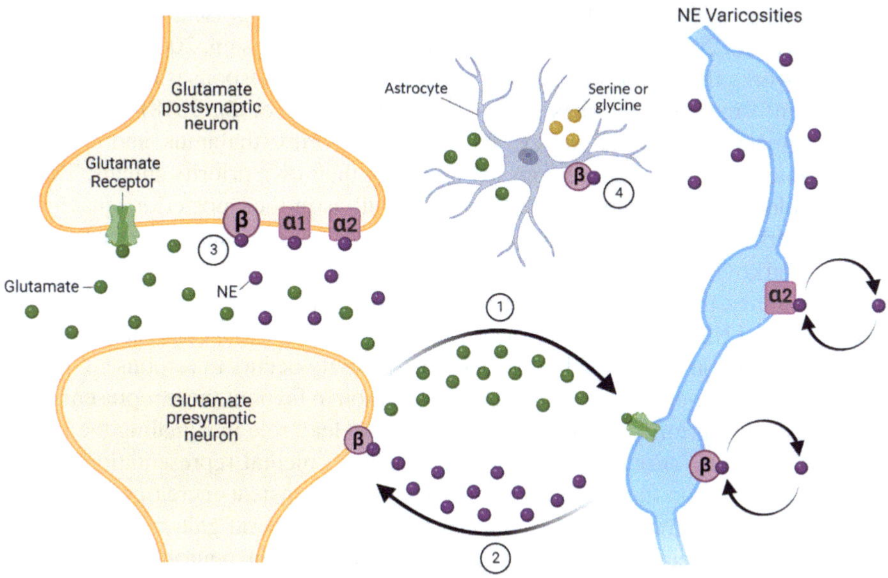

Fig. 6 Glutamate Amplifies Noradrenergic Effects (GANE) model of how LC activation implements neural gain. According to the GANE model, a surge in LC activity under arousal will lead to the diffuse release of NE from varicosities at target brain sites (large blue shape on right). High levels of glutamate signal the priority, or salience, of a stimulus, such that higher concentrations of glutamate correspond with strong sensory inputs or active mental representations. (1) If local glutamate levels are high enough, they will spill over into extracellular space and bind to their corresponding receptors along NE axons. This binding stimulates even greater release of NE from nearby varicosities, thereby upregulating local NE concentrations in these highly active areas. (2) In turn, elevated NE levels will bind to presynaptic glutamate axon terminals and trigger even greater release of glutamate in those synaptic pathways. (3) In these synaptic pathways transmitting strong, high-priority information, NE levels will also be sufficiently elevated to gain access to beta-adrenergic receptors on the postsynaptic glutamate neuron. Although NE engages all receptor subtypes, the beneficial effects of beta-adrenoreceptors will prevail. Specifically, binding to these specific adrenoreceptors enhances the throughput of strong signals and enhances synaptic plasticity, which supports stronger memory consolidation. (4) Elevated NE will also engage beta-adrenoreceptors on nearby astrocytes, which support the excitability of nearby neurons. Together, these self-regulating positive feedback loops, or "NE hotspots," upregulate the excitation of already highly active synaptic pathways. As a result, phasic LC activation increases neural gain and amplifies perceptual and memory selectivity. For additional details, see Mather et al. (2015). (Figure created by Ziyuan Chen with BioRender.com)

energy will be focused on select regions processing high-priority information. Thus, the gain of prioritized representations and/or task-relevant motor responses in the brain is supported by synergistic interactions between LC activity and both attentional and physiological resource allocation.

Finally, the local upregulation of neuronal activity at NE hotspots has larger ramifications for global communication patterns across the brain. Activation of beta-adrenoreceptors modulates alpha, gamma, and theta oscillations both locally and across brain regions (Mather et al., 2015). Engaging these receptors may also preferentially route prioritized information through large-scale frontoparietal (Corbetta et al., 2008; T.-H. Lee et al., 2018; Robbins & Arnsten, 2009) and salience brain networks that coordinate attentional selectivity and responses to behaviorally relevant information (Hermans et al., 2014b; T.-H. Lee et al., 2018). In addition, direct connections between the LC and prefrontal cortex, thalamus, and amygdala may enable activity in these regions to upregulate their own priority signals. This, in turn, would fuel NE hotspots both locally as well as in their posterior sensory and parietal cortical targets (e.g., amygdala projections to early visual cortex or parietal cortex; Markovic et al., 2014).

In summary, the GANE model proposes that local interactions between neuronal excitation and NE release regulate a delicate balance between local neuronal excitation and inhibition across the brain. Phasic LC activity occurs in response to behaviorally relevant inputs and tips the scale of excitation in favor of these representations to increase their priority. Glutamate levels play the lead role in signaling the importance of incoming stimuli and determine whether a mental representation will be further enhanced or further suppressed under such transient increases in arousal. Throughout the brain, phasic LC activity implements neural gain by promoting a few hotspots of neuronal excitation amidst a backdrop of neuronal suppression. Ultimately, this stratification of brain activity under arousal ensures that strong, high-priority stimuli win the competition for limited mental resources when it matters the most.

4.4 Empirical Support for the GANE Model

In humans, combining neuroimaging methods with pupillometry has provided initial support for the GANE model. In one fMRI study, participants were motivated to encode a background scene while ignoring a transparent overlaid object (Fig. 7). Scenes thereby acquired high priority through their goal relevance. Physiological arousal, as indexed by cue-evoked pupil dilation, was induced on half the trials by threatening to deduct money if participants forgot the target scene during a subsequent recognition memory test. The threat of punishment enhanced memory for high-priority scenes and impaired memory for low-priority objects. Enhanced scene memory was also correlated with greater threat-evoked pupil dilations as well as increased activity in the parahippocampal place area (PPA), a region specialized to respond to the high-priority scene information (Clewett et al., 2018). Thus, a surge in arousal upregulated local activation of the high-priority mental representation. Critically, weighting brain activity by the magnitude of these pupil dilations revealed a link between LC activation and threat-enhanced memory for prioritized scenes.

Building on these findings, a separate fMRI study showed that phasic pupil dilations to threatening cues were correlated with the activation of a frontoparietal network that coordinates attentional selectivity and is highly sensitive to NE release (Lee et al., 2018; Robertson, 2014). This discrete increase in arousal also

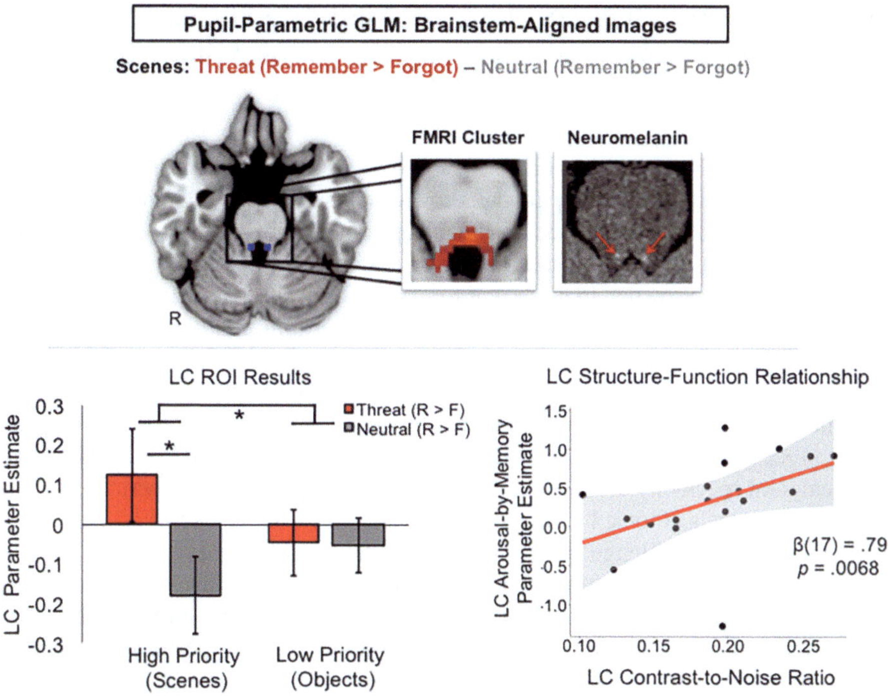

Fig. 7 Neuroimaging evidence that threat-evoked pupil dilations modulate LC-enhanced encoding of high-priority information. In this pupil-parametric fMRI analysis, participant's whole-brain statistical parametric maps were weighted according to the magnitude of pupil dilation when people were motivated to encode a goal-relevant scene and ignore a distracting object image. Qualitative comparisons revealed a substantial overlap between a cluster of activation that overlapped with the known location of the LC (red) and a standard LC mask from the literature (blue dots). The functional cluster from fMRI also overlapped with an area consistent with one participant's neuromelanin-sensitive-weighted image, a structural scan that reveals LC neurons (bright dots; top panel). A region-of-interest analysis using each participant's own neuromelanin-defined LC showed the direction of this memory effect, with pupil-weighted arousal-enhancing LC activation during threat trials but not neutral cue trials (bottom left). The degree to which participants engaged the LC during threat-related scene encoding was also correlated with his/her LC neuromelanin contrast-to-noise ratio (CNR), a putative biomarker of LC structural integrity (bottom right). *$p < 0.05$; R = scene remember; F = scene forgot. (Figure taken from Clewett et al. (2018))

corresponded with greater functional connectivity between the LC and the PPA in young adults. In addition, Lee et al. (2018) performed computational modeling to instantiate GANE using biologically plausible parameters of NE, glutamate, and GABA concentrations in the brain. As expected, the results of these simulations mirrored priority-dependent activation patterns observed in PPA. Together, these neuroimaging studies demonstrate LC-related amplification of local priority signals in the brain, an effect that can be indexed by task-relevant increases in pupil dilation.

5 Relationship Between Tonic LC Activity and Classic Arousal–Performance Curves: How Arousal States Shift the Locus of Priority Signals in the Brain

A key strength of the GANE model is that it provides considerable flexibility about the types of priority that can be amplified under arousal (see Mather & Sutherland, 2011). In so doing, this model encompasses a wide range of behavioral findings that emotional arousal can enhance cognitive selectivity in a wide range of contexts and for diverse inputs. It is noteworthy that in its current form, GANE posits that *phasic* LC activation amplifies neural gain and promotes the processing of prioritized information, irrespective of whether priority is determined by an input's emotional significance, perceptual salience, novelty, or goal relevance. All these factors are known to activate the LC (Mather et al., 2015; Sara, 2009), making this region well equipped to momentarily boost processing of important mental representations *when* they are active. However, this temporal specificity depends on strong phasic LC signals that are permitted under intermediate levels of tonic LC activity (Aston-Jones & Cohen, 2005).

An important challenge for the GANE model is addressing how norepinephrine can enhance neural gain at both intermediate and high levels of arousal. As we reviewed earlier, not all tasks or types of learning are impaired by this high tonic LC state, as suggested by the monotonic arousal–performance curve (Diamond et al., 2007; Yerkes & Dodson, 1908). This suggests that rather than impairing all stimulus-evoked responses, high arousal may instead lead to a qualitative shift in the type of information processing that benefits from increased neural gain.

It is important to stress that while GANE predicts that hotspots will emerge locally in sensory cortices processing high-priority information (e.g., a visual input), the priority signal itself often arises from top-down inputs from other regions, such as the LPFC (e.g., Lee et al., 2018). Thus, NE hotspots arise not only in regions representing high-priority sensory information but also in the brain regions that evaluate and modulate stimulus priority in the first place. We propose that viewing the GANE model through the "easy" (monotonic) and "difficult" (inverted-U) curves of the arousal–performance function may explain shifts in the neural locus of priority signals under arousal (Fig. 8). Our proposal builds on earlier work proposing that these two curves map onto the activation states of separate but interacting brain systems that are differentially sensitive to NE concentration and, in most cases, support different types of priority: the LPFC and the amygdala (Diamond et al., 2007).

5.1 *Curvilinear Component of the Arousal–Performance Curves: Modulation by Prefrontal Cortex and Frontoparietal Networks*

For tasks involving goal-directed behavior, arousal and LC activity modulate cognitive processing according to an inverted-U function (Diamond et al., 2007). Specifically, intermediate levels of arousal seem to represent a "sweet spot" wherein neural gain benefits processing of many types of prioritized information, especially those that demand focused attention (Fig. 8). One of the key targets of LC neuromodulation is the prefrontal cortex (PFC), which plays an essential role in supporting a wide range of executive functions, including working memory, inhibitory control, conflict monitoring, and decision-making (Funahashi & Andreau, 2013; Fuster, 1989; Petrides, 2005; Robbins, 2000; Robbins & Arnsten, 2009). In addition, the PFC coordinates priority via top-down inputs to posterior sensory cortical regions by tagging and amplifying goal-relevant processing while suppressing task-irrelevant processing (Gazzaley et al., 2005; Lee et al., 2018).

Converging evidence from humans and animals indicates that the LPFC is highly sensitive to the release of different catecholamines, including NE and DA (Arnsten, 2007; Robbins & Arnsten, 2009). Like the curvilinear component of the

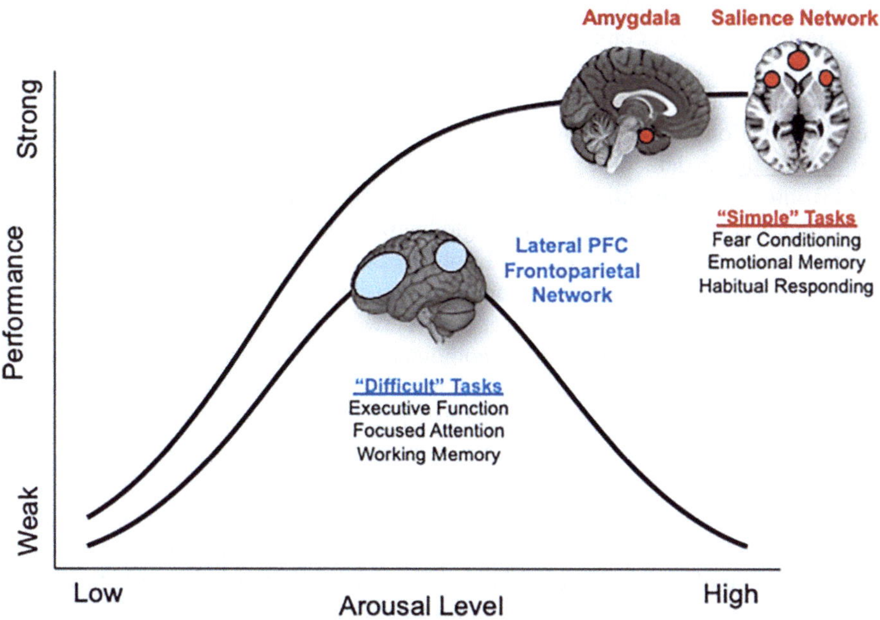

Fig. 8 Mapping different brain regions and functional networks to the classic arousal–performance curves. Tonic, or background, arousal levels may dictate whether performance and learning processes will be biased toward networks that support difficult mental operations, such as the frontoparietal network (blue dots), or toward regions and networks that support more simple tasks, such as the amygdala and salience network (red dots). Cognitive processes required for difficult tasks, such as focused attention and working memory, are optimal at intermediate levels of arousal when the prefrontal cortical function is intact. Different forms of priority supported by these regions, including top-down attention, will benefit from a surge in phasic LC activity and, as a result, an increase in phasic neural gain. However, these executive processes are impaired when arousal is too high due to impairments in prefrontal cortex function. On the other hand, amygdala and salience network activation gradually increase with arousal and peak under stress (far right of arousal axis). Under these high-arousal conditions, amygdala and salience network activation enhance attention and memory for bottom-up, emotional, or habit-related information via LC-mediated increases in persistently high states of neural gain. (Figure adapted from Diamond et al. (2007))

arousal–performance relationship, these neuromodulatory systems also regulate LPFC-mediated cognitive processes according to an inverted-U function (Arnsten, 2007; Arnsten & Li, 2005; Robbins & Arnsten, 2009). Too little NE leads to drowsiness, whereas too much NE induced by stress leads to impaired PFC function. Noradrenergic modulation of broader frontoparietal network function also plays an important role in optimizing selective attention under moderate arousal, including reorienting attention to salient inputs (Bouret & Sara, 2005; Corbetta et al., 2008), enhancing goal-relevant attention (Lee et al., 2018; Mather et al., 2020; Murphy et al., 2014), regulating cognitive effort and focus (Grueschow et al., 2020; Minzenberg et al., 2008; Raizada & Poldrack, 2008), and exploiting known sources of reward (Aston-Jones & Cohen, 2005).

Neuroimaging research in humans also supports the idea that task-related pupil dilations provide a window into LC–PFC interactions under moderate levels of arousal. Such modulation is not limited to local LPFC effects but rather extends to patterns of functional connectivity both within and between large-scale functional networks. For example, in one human fMRI study, isometric handgrip was shown to lower tonic states of arousal several minutes later, creating ideal conditions for phasic LC activation to emerge. Indeed, oddball-evoked pupil dilations were correlated with faster target detection processes, which were facilitated by the activation of right frontoparietal attention (Mather et al., 2020). In a similar finding, phasic LC activation was shown to be tightly coupled with oddball-evoked pupil dilations and frontoparietal network activation (Murphy et al., 2014). Pupil responses to goal-relevant targets might thereby provide a useful window into phasic LC activation and its modulation of LPFC- and executive-network-mediated cognitive processes.

Recent fMRI work in humans has combined pupillometry and graph measures to examine how phasic LC activation influences the dynamic configuration of frontoparietal networks. These studies reveal that pupil dilation is correlated with graph measures of network integration and neural gain, with these patterns being most pronounced in frontoparietal networks that are engaged during target processing (Shine et al., 2018a,b). Pharmacological manipulations lend additional support to the idea that the frontoparietal network is highly sensitive to phasic LC activation. Administration of atomoxetine, an NE transporter blocker that increases free levels of NE in the brain, promotes a neurochemical brain state that enables phasic LC responses to prevail (Shine et al., 2018b). Under atomoxetine administration, measures of neural gain and frontoparietal network integration are increased (Shine et al., 2018b). Likewise, the administration of modafinil, another NE transporter blocker drug that inhibits NE transporters and the reuptake of NE from synapses, induces a low-tonic/high-phasic mode of LC activity that leads to enhanced task-evoked LC activity during a mentally demanding task (Minzenberg et al., 2008). Such increases in the LC phasic mode also corresponded with increased LC–PFC functional connectivity and better task performance (Minzenberg et al., 2008).

Together, these imaging studies support the idea that moderate levels of global arousal—whether they are induced by drugs, acute exercise, or general wakefulness—promote brain states that are highly conducive to phasic LC activity. With LPFC function intact at intermediate levels of arousal and alertness, attentional resources are appropriately deployed to enhance task engagement and performance.

5.2 *Monotonic Component of the Arousal–Performance Curves: Modulation by the Amygdala and Salience Network*

In contrast to the inverted-U relationship between the LC–NE system, LPFC, and frontoparietal network connectivity, the linear arousal–performance function that is theorized to reflect simple and sometimes reflexive behaviors as well as enhanced encoding of highly emotional memories may be more closely tied to amygdala activation (Diamond et al., 2007; Fig. 8). The amygdala plays a central role in facilitating attention and memory for emotionally arousing stimuli (Cahill et al., 1994, 1995, 1996; LaBar & Cabeza, 2006). Much of these amygdala-mediated enhancements are driven by inputs from the noradrenergic system (Cahill et al., 1995; Strange & Dolan, 2004). For instance, the pharmacological administration of reboxetine enhances amygdala-mediated emotional memory enhancements and biases attention selectively toward emotional information (Markovic et al., 2014).

Importantly, much of these amygdala-mediated memory enhancements are mediated by beta-adrenoreceptors, suggesting the presence of local NE hotspots under emotional arousal. For example, beta-adrenoreceptor blockers reduce amygdala activity and reduce emotional memory enhancements (Strange & Dolan, 2004). Drugs that suppress amygdala activity can also disrupt the consolidation of emotionally arousing experiences (Ferry et al., 1999; Roozendaal et al., 2008).

The preferential enhancement of emotional representation is thought to occur in part through NE biasing the amygdala to modulate plasticity or processing in targets elsewhere in the brain, including the hippocampus and sensory cortex (McGaugh et al., 2002). This finding is further corroborated by fMRI evidence in humans showing that the amygdala strengthens memories of highly emotional scene images by modulating activity in the regions that process scene information and the hippocampus (Fastenrath et al., 2014).

From a neurophysiological perspective, we anticipate that intermediate arousal states should benefit the processing of emotional stimuli. For example, manipulating sustained arousal levels via simple exercises can modulate emotional memory enhancements. In one pupillometry study, isometric handgrip led to overall increases in pupil diameter, indexing an elevated tonic state of arousal. Importantly, this heightened arousal state was also related to a gain in memory selectivity for negative over positive and neutral information, whereby negative emotional memories were further enhanced following handgrip compared to images encoded following a no-handgrip control condition (Nielsen et al., 2015).

Genetic studies of the alpha-2B-adrenoreceptor (ADRA2B) allele deletion variant also offer indirect evidence supporting a link between enhanced emotional memory selectivity under tonically (or trait) elevated levels of NE. While the exact mechanism is unclear, it is thought that NE levels are high in these individuals. ADRA2B deletion carriers show increased availability of NE that may amplify negative emotion-related memory enhancements (Rasch et al., 2009). Individuals with this genetic deletion variant also show greater attention toward (Todd et al., 2013) and memory for emotionally arousing information (de Quervain et al., 2007).

Genotyping evidence lends additional support to the idea that noradrenergic activation of the amygdala enhances the preferential processing of emotionally arousing information against a backdrop of tonically elevated NE. But whether this ostensible gain relates to the midpoint or far-right of the monotonic arousal-performance function is somewhat unclear. Fortunately, the linearity of the amygdala-centric arousal–performance function helps circumvent this ambiguity about the tonic level of LC activation and NE in emotional arousal manipulations. Regardless of an individual's overall level of arousal, the presentation of emotional stimuli will typically enhance neural gain and selectivity for those stimuli in perception and memory.

5.3 Far Right of the Arousal–Performance Curves: Effects of Stress and Very High Tonic Arousal on LC and Brain Activity

Existing empirical findings dovetail with multiple models of LC–NE function and the curvilinear model of the arousal–performance relationship: at intermediate levels of tonic, or background, arousal, phasic LC activation enhances cognitive selectivity for bottom-up, top-down, or emotional priority. In this way, multiple brain regions that regulate the selective processing of important information, including the amygdala and LPFC, are poised to amplify neural gain and modulate the strength of NE hotspots both locally and at their target brain sites.

But what happens at extreme levels of arousal, such as stress? Are all sources of priority equally competitive or do some informational inputs gain an advantage in the competition for limited mental resources? Emerging findings suggest that not all sources of priority have equal stakes in the competition for mental resources and behavior control when arousal is very high. To be able to explain the full range of arousal and NE's effects on neural gain, the GANE model must also be able to address how elevated arousal states amplify attentional and memory selectivity in the absence of strong phasic LC responses. Of particular importance to applications of pupillometry to psychology research, GANE must also be able to reconcile how different pupil measures, including stimulus-evoked dilation and overall pupil diameter, can simultaneously account for context-dependent biases in learning and memory.

Why might this shift in stimulus priority occur under stress? One key answer might lie in how stress influences the function of different brain regions. Stress elicits robust impairments in LPFC executive functions, including working memory and goal-directed attention (Arnsten, 2009). By contrast, amygdala function generally remains intact at intermediate and high levels of arousal, with stress potentiating many amygdala processes (Roozendaal et al., 2009). Functional connectivity between the LC and amygdala also increases following acute psychological stress (van Marle et al., 2010). The degree to which the LPFC is impaired by a highly emotional or stressful event might thereby enable researchers to predict whether the arousal–performance curve will be linear or curvilinear (Diamond et al., 2007). Simply put, the relationship between arousal and prioritization can be distinguished by whether the task recruits LPFC activity (Diamond et al., 2007). Flexible regulation of higher-order executive functions peak at intermediate levels of arousal/tonic LC activity but are significantly impaired under high states of arousal or stress (i.e., the far right of the inverted-U). As the amygdala is potentiated by stress, its priority signals will likely prevail under these high-arousal conditions, leading to different behavioral and learning outcomes (Fig. 8).

At present, little work has examined relationships between LC activity, pupil diameter (as opposed to dilation), and behavior under stress. Across many stress studies, the pupil is often used to verify the successful induction of stress rather than used as a trial-level measure to predict task performance or memory. In some studies, acute stress has been shown to blunt pupil dilation responses during the encoding of different neutral and negative pictures (Henckens et al., 2009; Qin et al., 2012), consistent with the idea that phasic LC responses diminish under conditions of high tonic activity. While attenuation of pupil dilation was not associated with encoding performance under stress, it is possible that pre-stimulus baseline measures of pupil diameter would provide a more accurate measure of stress-related memory enhancements. It is also noteworthy that other work demonstrates that fMRI indices of neural gain, such as representational precision in occipitotemporal cortex, are reduced when cortisol levels are elevated, as would occur under stress (Warren et al., 2016). This gain reduction, however, was not linked to differences in a simple image categorization task, so it is unclear how these changes in neural gain relate to behavior.

At the core of the GANE model is the idea that selective and local activation of low-affinity beta-adrenoreceptors facilitate glutamatergic activity and synaptic plasticity in regions representing a high-priority stimulus. While such task- or stimulus-evoked neural gain is proposed to occur under conditions of phasic LC activity, it is important to acknowledge that stress elicits increases in tonic levels of LC activity (Mana & Grace, 1997). Under stress, the release of corticotropin-releasing factor (CRF) also increases the tonic firing rate of LC neurons, leading to elevated levels of NE being released across both cortical and subcortical brain regions (Valentino et al., 1998). Such increases in LC output can generate high enough NE concentrations to engage beta-adrenoreceptors and, by extension, trigger NE hotspots (Mather et al., 2015).

As the NE hotspots triggered by persistently high neural gain are not time-locked to the processing of task-relevant stimuli, *any* active mental process may be amplified by the increased neural gain, possibly manifesting as distractibility observed under high arousal states. Although this persistently high neural gain leads to task-irrelevant processing, evidence suggests that performance on "easy" tasks benefits, or at least do not suffer, from high arousal states (Yerkes & Dodson, 1908). Here, we argue that stress, or a state of extremely high arousal, induces a qualitative shift in attention and mental resource allocation toward specific sources of priority, namely, motivational significance driven by emotional salience or more automatic and habitual behaviors.

Supporting this idea, attention tends to be biased toward processing emotional stimuli under stress, which occurs in part due to activation of the amygdala and the salience network (Hermans et al., 2011, 2014a). Intriguingly, one study showed that inducing acute stress prior to studying emotional word lists leads to a subsequent emotional memory enhancement in ADRA2B deletion variant carriers, a group thought to have tonically elevated levels of NE (Zoladz et al., 2014). Thus, high tonic levels of arousal and NE concentration may lead to selective memory enhancements for emotional material. From the perspective that the amygdala drives emotional memory enhancements, this may provide indirect evidence of greater emotional selectivity under stress and very high NE levels.

In addition, human neuroimaging evidence shows that acute stress inductions lead to enhanced sensory cortical processing in response to emotional faces compared to neutral faces (van Marle et al., 2009). At the behavioral level, stress is associated with heightened attentional vigilance (van Marle et al., 2010), especially toward the source of threat (Rued et al., 2019). Furthermore, viewing negative pictures under stress relates to an increase in an EEG component that relates to increased elaborative processing (Weymar et al., 2012). Administration of propranolol, a beta-adrenoreceptor blocker, has also been shown to prevent stress-induced activation of large-scale salience brain network that supports hypervigilant attentional states and coordinates behavioral responses to threat (Hermans et al., 2011).

Interestingly, the behavioral and neural effects of stress are also mimicked by chemogenetic activation of the LC in mice (Zerbi et al., 2019). This elegant study combined chemogenetics with resting-state fMRI to show that strong global activation of the LC led to a drastic shift in brain-wide functional connectivity, with the strongest modulation occurring in the amygdala and salience networks. Strong, global activation of the LC was accompanied by a large increase in the overall pupil diameter, consistent with the idea that these effects were driven by a large increase in tonic LC activation that would typically be observed under stress (Zerbi et al., 2019).

5.4 *Stress-Induced Release of Cortisol May Fuel NE Hotspots and Persistently High Neural Gain*

One important open question is whether NE hotspots still emerge under conditions of high tonic LC activation, especially in the presence of other stress hormones that modulate activity in attentional and memory regions across the brain. Two defining features of the stress response are the release of cortisol, or glucocorticoids, across the brain as well as the activation of the sympathetic nervous system and LC (Joëls & Baram, 2009; McEwen, 2007). While glucocorticoids impair PFC function

(Arnsten, 2009), they may actually fuel the emergence of NE hotspots in other select brain regions. Both beta-adrenoreceptor activation (Ferry et al., 1997) and the administration of glucocorticoids (Reznikov et al., 2007) have also been shown to enhance glutamate release in the amygdala, providing the necessary fuel to generate an NE hotspot.

The co-release of NE and cortisol under emotional arousal or stress also plays an important role in promoting the formation of declarative memories. Pharmacological manipulations that elevate both NE and cortisol levels elicit a negative response bias toward fearful faces in the amygdala (Kukolja et al., 2008). This finding lends additional support to the idea that stress and LC activation will enhance neural gain processes mediated by the amygdala through a combination of cortisol and beta-adrenoreceptor activation. The upregulation of NE hotspots in the amygdala would then serve to further strengthen the throughput of emotionally salient information through both sensory and hippocampal networks (Markovic et al., 2014; Tully & Bolshakov, 2010).

Extending this idea, acute stress is known to activate the hippocampal–amygdala pathways that prioritize the encoding and storage of emotional information (Kukolja et al., 2011). Such enhancements occur alongside stress-induced impairments in PFC function (e.g., van Stegeren et al., 2010), with elevated levels of cortisol and NE in the PFC impairing goal-directed attentional processes (Schwabe et al., 2012). As a result, behavioral control tends to shift toward more habitual behaviors and prefrontal cortical regions become more insensitive to reward devaluation (Schwabe et al., 2012). This behavioral shift may be mediated, at least in part, by beta-adrenoreceptors (Schwabe et al., 2011). In sum, stress may shift the locus of priority signals in the brain from the prefrontal cortex to the amygdala. By activating beta-adrenoreceptors in a region that is responsive to threatening or emotional stimuli and is not compromised by cortisol release, NE hotspots can still emerge to promote some level of selectivity under stress.

5.5 *High Arousal Amplifies Hard-Wired or Habitual Learning Preferences via Brain-Wide LC Effects*

While not driven by stress, per se, recent neuroimaging work also aligns with the idea that elevated arousal states promote persistent neural gain. Unlike research focused specifically on one brain network or activity in one brain region, this influential study examined arousal-related changes in brain-wide network topography using a combination of graph metrics and pupillometry. Participants performed a reward learning task in which they had to determine whether rewards were associated with the categorical nature or perceptual features of a stream of images. Pupillometry was used to track fluctuations in arousal across the task and to examine how stimulus-driven changes in arousal were related to the functional integration of different brain networks. Importantly, this study also used average pre-stimulus pupil diameter, a measure of tonic LC activity, as opposed to task-evoked pupil dilations to index neural gain (Eldar et al., 2013). The results revealed that larger pre-stimulus pupil diameters were associated with greater neural gain across the brain but not overall task performance.

At first blush, these results seem to run counter to the idea that high tonic LC activation should impair perceptual and memory selectivity. A closer examination of the data, however, reveals that these global increases in pupil diameter and neural gain corresponded with better performance when trials were aligned with the participants' intrinsic learning preferences, namely, whether participants prefer to learn from the conceptual or perceptual features of target information. Like other high-arousal behaviors, such as habitual responding, these behavioral, neural, and pupil patterns align well with the linear "easy" component of the arousal–performance relationship. At high tonic levels of arousal, as indexed by higher pre-stimulus pupil diameters, learning and neural gain were preferentially biased toward more hard-wired, habit-related information and away from top-down attention/control.

Interestingly, these graph measures of neural gain were indexed by local patterns of functional integration amid more global increases in network segregation. On the surface, this brain network topology appears to be qualitatively different from the functional network connectivity patterns observed during phasic gain states (Shine et al., 2018a). In those lines of work, brain-wide networks become more functionally integrated with each other to support focused attention, which was also related to task-evoked pupil dilations and drug-induced brain states conducive to phasic LC responses (i.e., under atomoxetine administration; Shine et al., 2018b). It is possible that large pupil diameters may be a better metric of persistent gain states in the brain and bear their own signature of network integration and segregation patterns. On the other hand, phasic, stimulus-evoked pupil dilations during cognitive tasks may serve as a better metric of phasic LC activity and neural gain—that is, neural processes focused on representations of goal-relevant stimuli. In these instances, task-relevant gain would optimize task performance via LC modulation of more functionally integrated and cooperative brain states.

Network neuroscience studies have helped reveal the effects of arousal on large-scale network dynamics and their relation to cognitive processing (Shine, 2019; Shine et al., 2016, 2018a). They also support the critical role of the LC–NE system in regulating global arousal states and different types of neural gain, which may be indexed by pupil measures collected at different time points, namely, whether arousal is measured during pre-stimulus baseline periods or in response to task-relevant stimuli. Furthermore, converging evidence in this field points to diverse effects of the tonic and phasic modes of LC processing on information processing via the reorganization of dynamic brain networks (Wainstein et al., 2022). While additional work is needed to understand how these network dynamics relate to different behaviors, the application of graph theory measures to study the function of the LC–NE system is an exciting direction for future research.

5.6 Summary: LC Activation Patterns Shape the Quality of Information Processing in the Brain

Much research has focused on how task-evoked phasic LC activation can implement neural gain for a wide range of prioritized information. Yet other work suggests that high tonic levels of LC activity may facilitate a qualitatively different form of neural gain, leading to different behavioral outcomes. The key differences between these phasic and tonic neural gain states may depend on the temporal specificity of LC activation, background levels of arousal, and the differential sensitivity of brain regions to the local and concentration-dependent effects of NE on different adrenoreceptors.

Here, we suggest that the "easy" and "difficult" curves of the arousal–performance relationship may help explain some of these qualitative shifts in information processing. These two arousal–performance curves are thought to map onto two separate but interacting brain systems that show different sensitivities to elevated states of arousal and NE release. On the one hand, the lateral PFC and its broader cortical networks support complex executive functions, including goal-directed behavior and top-down attention. On the other hand, the amygdala supports the processing of "easy" cue-based or highly emotional information (Diamond et al., 2007). Consistent with the GANE model, both the amygdala and LPFC are activated by phasic LC activity and behaviorally relevant stimuli. As shown in Fig. 8, this may explain why stronger phasic LC activation at an intermediate level of tonic arousal can enhance neural gain irrespective of the type of priority. That is, top-down inputs from multiple brain regions can regulate processing of *task-relevant* and *temporally specific* inputs, reflecting the midpoint of both the monotonic and curvilinear features of arousal–performance relationship.

As tonic arousal continues to increase, however, neural gain may become decoupled from task demands as individuals begin to explore new behaviors or mind wander. Accordingly, increases in global arousal may promote the reconfiguration of large-scale functional networks to promote high persistent neural gain for distracting information and, in cases like stress, emotional or habit-related information. This pattern accords with the far-right portion of the inverted-U component of the arousal–performance curve, where performance on complex cognitive tasks becomes noisier and attention becomes less focused on the task at hand.

Under extremely high arousal and global NE release, the LPFC function becomes impaired and amygdala-centric processing will likely prevail. This should result in a qualitative shift in information processing toward emotional sources of priority. Beta-adrenoreceptors are more readily engaged under stress when global levels of NE are elevated, creating optimal conditions for NE hotspots to emerge in the amygdala and nodes of the salience network when emotional stimuli are encountered. The concomitant release of the stress hormone cortisol is also likely to fuel NE hotspots by amplifying glutamatergic signaling in the amygdala. Noradrenergic modulation of persistent neural gain and NE hotspots is thereby likely to obey the linear component of the arousal–performance function.

By interpreting the GANE model and large-scale network dynamics through the lens of the original account of the arousal–performance relationship (Yerkes & Dodson, 1908), we hope this section has illuminated how background or global arousal states could determine the nature of local and global patterns of brain activity. This would, in turn, lead to biases in the types of information that will be prioritized by top-down regions of the brain and whose processing will benefit from a corresponding increase in neural gain.

6 LC–NE System Regulation of Attentional Capacity and Effortful Behavior

Classic neurocognitive models of arousal dissociate separate aspects of attention. As we have reviewed throughout this chapter, phasic arousal and LC activity play lead roles in facilitating adaptive gain and amplifying selectivity in attention and memory. Background arousal levels may also bias information processing toward processing bottom-up or top-down information and privilege different brain systems to win the competition for behavioral control. In the following section, we will review evidence suggesting that in addition to influencing the selectivity and quality of information processing in the brain, tonic arousal and LC activity also regulate another key parameter of attention: capacity. We discuss findings showing that tonic LC–NE activity sets the overall attentional state and thereby determines if working memory performance and other executive functions, such as cognitive flexibility, will function efficiently (Chamberlain & Robbins, 2013; Coull et al., 2004; Minzenberg et al., 2008). In addition, we examine how LC activity helps supply organisms with the mental and physiological energy needed to respond to and encode behaviorally relevant events.

6.1 Evaluation of Task Utility and Task Performance Modulates Tonic LC–NE Activity

Attention is costly—we must "pay" attention to information we wish to process. This quantitative component of attention, the resources we invest into performing a specific task, is often described as "attentional intensity" (Robison & Unsworth, 2017). But limitations on the availability of mental resources challenge an organism's ability to regulate and sustain attention and energy levels sufficient to meet different task demands. As discussed in previous sections, cognitive demands can be met by modulating attentional selectivity or by the allocation of the limited cognitive resources toward task-relevant processes. In this section, we focus on a second means of meeting task demands—how arousal also modulates attentional capacity to increase the available pool of cognitive resources.

Situations that require intense focus or difficult choices are often accompanied by an increase in arousal, which is likened to the concept of recruiting mental effort (Kahneman, 1973; Shenhav et al., 2017). As reviewed throughout this chapter, the pupil serves as an especially powerful tool for measuring and evaluating the amount of effort needed to perform a difficult task (Just et al., 2003; Kahneman & Beatty, 1966; Laeng et al., 2012; van der Wel & van Steenbergen, 2018). Insofar as LC activity is directly linked to fluctuations in pupil size, the LC–NE system might, therefore, be essential for adjusting attentional intensity to match task demands via its modulation of both attentional capacity and selectivity (Unsworth & Miller, 2021; Unsworth & Robison, 2017b).

The LC–NE system is well suited to regulate attentional capacity, given its dense connections to brain networks that support both bottom-up and goal-driven attention, including the salience and frontoparietal networks (Corbetta et al., 2008; Hermans et al., 2011; T.-H. Lee et al., 2018; Mather et al., 2020; Unsworth & Robison, 2017a). Neuroenergetic models of attention and arousal propose that attentional capacity is constrained by background arousal states (Kahneman, 1973), which are set by levels of tonic LC activity (Aston-Jones & Cohen, 2005). This leads to the question of how tonic arousal and LC–NE activation levels are adjusted. Earlier models posit that an evaluation center assesses task demands and utility and drives adjustments in arousal levels to support task-related information processing (Kahneman, 1973). Extending this idea, Adaptive Gain Theory proposes that multiple evaluation centers in the brain, consisting of the dACC and OFC, integrate inputs about performance feedback and reward to adaptively adjust the tonic levels of LC–NE system activity (Aston-Jones & Cohen, 2005). As cost–benefit evaluations fluctuate during cognitive or decision-making tasks, the dACC and OFC may regulate tonic LC activity to bias arousal processes toward either task-engaged behavior or more exploratory behavior. Thus, Adaptive Gain Theory hypothesizes that the dACC and OFC coordinates changes in tonic LC–NE activity in response to fluctuating task utility and task demands.

An emerging body of evidence supports the existence of rich anatomical and functional connections between the LC–NE system and the anterior cingulate cortex. For instance, direct microstimulation of the ACC elicits pupil dilation, albeit weaker and more delayed compared to the stimulation of the LC (Joshi et al., 2016). This shared physiological function and bidirectional connections between the two structures (Koga et al., 2020) suggest that the ACC and the LC form one part of the arousal network. Researchers have also found that this LC–anterior cingulate cortex network is involved in resolving task conflict, which refers to the maintenance of goal-directed behavior in the presence of distracting stimuli (Ebitz & Platt, 2015). This study found that increased neuronal activity in the dorsal anterior cingulate cortex signaled the presentation of distracting stimuli, especially when performing a goal-directed task. Moreover, this activation of the dACC predicted a change in the baseline pupil diameter, indexing an increase in tonic LC–NE activity. Thus, the anterior cingulate cortex's role in maintaining goal-directed behavior in the face of distractors may be mediated by an increase in tonic LC activity.

In addition to signaling the presence of distractors that conflict with goal-directed behavior, the anterior cingulate cortex (ACC) has been implicated in signaling reward and plays a role in monitoring task utility (Bush et al., 2002; Kennerley et al., 2006). Recent work has linked this utility-monitoring function to changes in tonic LC levels (de Gee et al., 2022). In this study, periods of high task utility corresponded with periods of task-optimal tonic LC activity, as determined when prestimulus pupil diameters were stabilized at an intermediate level. This finding suggests that motivation is an important input into the evaluation centers that regulate tonic LC activity and that this process may mediate motivation-driven improvements in task performance. These findings align with the idea that the ACC monitors bottom-up (e.g., distractors) and top-down (e.g., reward) inputs and modulates tonic LC–NE arousal levels, thereby adaptively adjusting attentional intensity to match the situation.

6.2 Setting the Global Arousal State: LC Interactions with the Salience Network

Through its dense efferent projections, the LC is essential for regulating global arousal states and for promoting the functional integration or segregation of large-scale brain networks. There appears to be an especially strong relationship between pupil measures and fluctuations in activity within the salience network, a collection of brain regions primarily anchored in dACC and anterior insula (DiNuzzo et al., 2019; Joshi et al., 2016; Joshi & Gold, 2022; Seeley et al., 2007; Zerbi et al., 2019). For example, the change rate of pupil diameter (i.e., first-order temporal derivative of pupil size) is positively correlated with salience network activity both while individuals rest in an MRI scanner (Schneider et al., 2016), detect salient oddball stimuli (Murphy et al., 2014), must sustain attention (DiNuzzo et al., 2019), or perform a mentally demanding working memory task (Fietz et al., 2021). During fear learning, stimulus-evoked pupil dilations have also been shown to track salience network activation on a trial-by-trial basis (Leuchs et al., 2017). Salience network responses, including responses localized to both the anterior insula and dACC, also correspond with behavioral errors during continuous attention tasks (Kucyi & Parvizi, 2020) and Stroop interference tasks (Critchley et al., 2005).

Pupil-linked responses in the salience network appear to relate to multiple forms of priority. These relationships also seem to respect the monotonic arousal–performance curve, as pupil dilations are linearly correlated with the magnitude of salience network activation (Mäki-Marttunen, 2021). Interestingly, one neuroimaging experiment showed an inverted-U relationship between pupil dilation and accuracy on two cognitively demanding tasks (Mäki-Marttunen, 2021). In addition, Maki-Marttunen (2021) demonstrated that pupil dilations were positively correlated with functional integration within the salience network.

Because both resting (non-task-related) and task-related pupil dilations correspond with increases in salience network activity or connectivity, we believe that salience network activation likely represents the capacity of attention. Indeed, human neuroimaging work shows that salience network activation is associated with tonic alertness rather than selective attention (Sadaghiani & D'Esposito, 2014). This was theorized to index an effortful process of maintaining mental resources to meet current processing needs (Sadaghiani & D'Esposito, 2014). Within-network functional integration of the salience network is also linearly correlated with other measures of sympathetic nervous system activation, such as heart rate (Young et al., 2017), consistent with its role in representing tonic arousal levels. Together, these findings suggest that salience network engagement does not reflect the content or priority of information processing at a given moment. Instead, its activity reflects the number of domain-general resources that can be recruited through other regions or networks that signal priority, such as the frontoparietal network (Menon & Uddin, 2010).

As described earlier, monotonically increasing levels of arousal (and, therefore, salience network activity) may correspond with a qualitative shift in information processing and patterns of inter-network functional connectivity. That is, at very high levels of arousal, the salience network might prioritize emotional information via stronger functional integration of the amygdala (Hermans et al., 2011). Cohesion between executive control and the salience network has also been shown to be maximal at intermediate levels of arousal, suggesting that phasic LC activity enhances coordinated activity between brain networks that support alertness and task-focused attention (Bouret & Sara, 2005). However, when arousal levels were especially high, this pattern of executive and salience inter-network communication diminished.

Broadly, our view is also consistent with the role of the dACC in detecting errors and signaling the need for additional resources, effort, and behavioral adjustments. Following the detection of these salient events, descending inputs from the dACC adjust the LC tonic activity accordingly to meet these demands (Aston-Jones & Cohen, 2005; Sara, 2015). Increased functional coupling between the LC and other nodes of the salience network, namely, the anterior insula, has also been observed following negative or uncertain action outcomes (Clewett et al., 2013). Finally, the close link between the salience network and tonic arousal dovetails with evidence that insula and dACC activation drive exploration (e.g., Blanchard & Gershman, 2018) and hypervigilance (Hermans et al., 2011), two behaviors that are strongly associated with very high levels of tonic arousal and LC activation.

6.3 Pupil-Linked LC Activity Is Related to Task Difficulty and Effort

The principle that arousal supplies the energy, or capacity, for cognitive processes is a common thread that runs through many neuroenergetic models of arousal and attention (Kahneman, 1973). Recent research has now begun to explore the possibility that the LC implements this process in the brain by mobilizing attentional resources when an increase in mental effort is required (Killeen et al., 2013). Remarkably, this function appears to be domain-general and consistent across species, as the LC is reliably engaged by a wide variety of perceptual discrimination and decision-making tasks. For example, evidence in humans (Raizada & Poldrack, 2008) and rodents (Doucette et al., 2007; McBurney-Lin et al., 2019) shows that the LC–NE system activation increases during challenging sensory discriminations. In addition to facilitating perceptual decisions, the LC is recruited to support higher-order cognitive processes, including response conflict resolution (Grueschow et al., 2020) and cognitive control (Köhler et al., 2016). Pupil dilations during a mentally demanding multiple object tracking task correlate with activity in a brainstem region consistent with the location of the LC (Alnaes et al., 2014). This pupil-linked LC effect is also modulated by cognitive load, suggesting that arousal-related activation of the LC may track attentional resource allocation. Functional connectivity between the LC and prefrontal cortex has been shown to occur during successful conflict resolution on an emotional Stroop task (Grueschow et al., 2020). Conflict trials on this task were also associated with increased pupil dilation, an autonomic arousal effect that is reliably observed across many Stroop experiments (Brown et al., 1999; Hershman & Henik, 2019; Laeng et al., 2011).

The LC may also be involved in coordinating the supply of *physical* energy needed to fuel goal-driven actions. Indeed, spikes in LC activity have been reported across different species during motivated behavior. Electrode recordings in monkeys reveal that LC firing is related to the amount of physical effort that is necessary to receive a juice reward (Varazzani et al., 2015; Fig. 9). Researchers also found that LC activity not only spiked during presentation of the reward cue but also scaled with the amount of force that was necessary for lever pressing to acquire that reward. Importantly, phasic LC activity during the lever press was also correlated with the magnitude of pupil dilation, supporting the idea that transient pupil responses serve as an index of LC activity and effort.

Work in rodents also corroborates this link between phasic LC activity and successful task execution. Using a sophisticated combination of optogenetics, electrode recordings, and tracing techniques in rodents, Breton-Provencher et al. (2022) showed that phasic LC activity spikes in the period preceding a lever-press during a go/no-go task. Interestingly, they also observed specificity in the LC projections involved in this behavior. That is, pre-movement NE release primarily occurred in target motor regions and not in task-irrelevant cortical regions (Breton-Provencher et al., 2022). These results suggest that there may be specificity in LC modulation during cognitive tasks such that LC activity selectively enhances processing in task-relevant regions and drives task-relevant actions that emerge from decisions to act (see also Clayton et al., 2004; Pavlenko & Kulichenko, 2003). Likewise, phasic LC activity tracks effort production as well as difficulty in monkeys performing discounting tasks (Bornert & Bouret, 2021). Specifically, LC activation during action initiation scales not only with the level of force production but also with behavioral response times related to this decision period.

In addition to driving motivated behavior, LC-mediated task execution also appears to influence subsequent memories for those experiences. Neuroimaging evidence in humans shows that phasic LC activation facilitates task responses in ways that boost episodic encoding (Yebra et al., 2019). Using a combination of

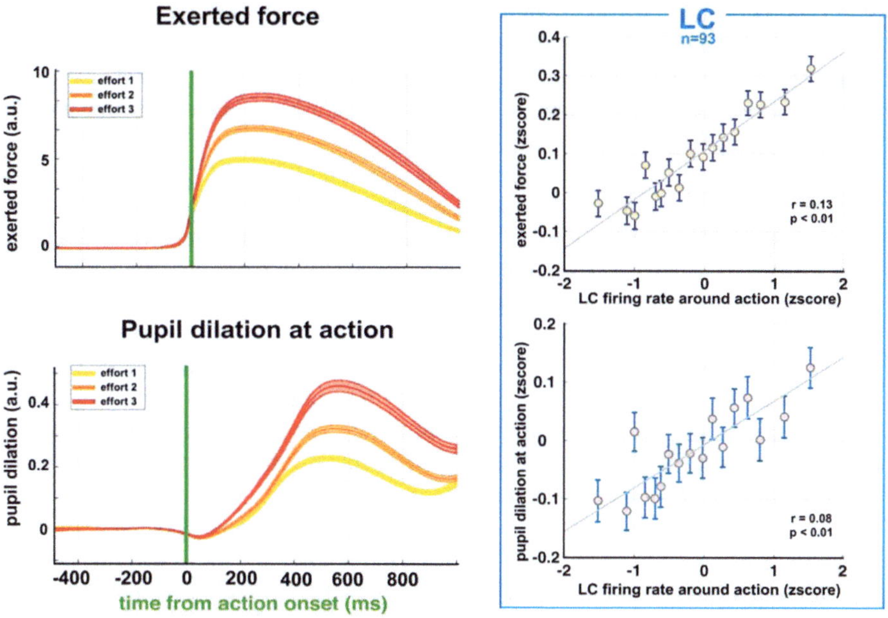

Fig. 9 Expected effort tracks with pupil dilation and LC firing rate. (Top left) In a lever press task, monkeys exerted the most force when the largest amount of effort was expected (effort 3). (Bottom left) Pupil dilation was greatest for trials in which monkeys were expected to spend the largest amount of effort. (Right) LC firing rate had a positive, linear relationship with both exerted force and pupil dilation. Together, these findings suggest a link between the pupil-linked LC–NE system and the recruitment of effort to meet expected task demands. (Figure adapted from Varazzani et al. (2015))

pupillometry and fMRI, researchers showed that LC activation selectively occurred during the successful encoding of images paired with "go" responses. This memory boost for action-related stimuli also corresponded with greater pupil dilation, reinforcing the link between phasic LC activation and action-evoked pupil responses.

6.4 LC Activation Coordinates the Supply of Metabolic Resources

Task-related responses in the LC may be essential for facilitating cognitive control, especially under high task demands and when stimuli or behaviors conflict. But does the LC also supply the physiological resources needed to meet these cognitive demands? Intriguingly, the answer appears to be yes. Empirical evidence supporting this view comes from a study examining the relationship between LC activity and neurovascular coupling, whereby LC–NE activation redistributed blood flow toward highly active brain regions (i.e., sensory regions being stimulated; Bekar et al., 2012). Interestingly, patterns of LC modulation also exhibited a gain-like effect in blood delivery, with vasoconstriction occurring more globally across less active brain regions and vasodilation occurring more locally in stimulated sensory regions. Thus, LC activity helps couple blood distribution changes with local demands for oxygen, which is a critical component of the oxidative metabolic pathways that generate energy in neurons.

In addition to modulating neurovascular dynamics, the LC also regulates the brain's utilization of glucose, the metabolic equivalent of fuel for neuronal activity (for review, see Mergenthaler et al., 2013). Studies have found that NE interactions with astrocytes trigger glycogenolysis, the process of breaking down glycogen stored into glucose (Dienel & Cruz, 2016; Magistretti et al., 1981; Sorg & Magistretti, 1991). NE–astrocyte interactions also induce glycolysis, the initial phase of glucose metabolism that yields the intermediate by-product pyruvate (Magistretti & Allaman, 2018). In astrocytes, this pyruvate is often converted to lactate, even under aerobic conditions (Magistretti & Allaman, 2018). Both pyruvate and lactate are then transported to neurons, where they serve as fuel for the final metabolic step that produces the bulk of the energy used in neurons.

Recent studies implicate the LC–NE arousal system in the engagement of these metabolic pathways. In mice, various arousal manipulations induced global cortical activation that corresponded to an increased production of lactate. Importantly, this increased lactate production corresponded with physiological markers of LC activity, including pupil dilation (Zuend et al., 2020). Intriguingly, NE-driven production of lactate not only fuels neuronal activity (Coggan et al., 2018) but may also serve as a signaling molecule that enhances the formation of spatial memory (Newman et al., 2011). Thus, NE is thought to play a critical role in modulating energy availability for neuronal use by triggering glucose metabolic pathways in astrocytes (Berridge & Waterhouse, 2003). As the local production of intermediate by-products lactate and pyruvate relies on beta-adrenoreceptor activation (Vaishnavi et al., 2010), it is tempting to speculate that local NE hotspots drive this selective allocation of energetic resources to enhance neural gain and mental selectivity (e.g., Mather et al., 2015).

6.5 Summary: LC Activation Supports Attentional Capacity and Goal-Directed Actions

In summary, noradrenergic system regulation of central blood flow, glucose utilization, and astrocytic activity provides a plausible biological substrate for linking LC activity to both physical and mental energy, or effort. Importantly, this mobilization of resources also exhibits spatiotemporal specificity. Through targeted projections across the brain, the LC may help amplify processing of task-relevant information under arousal by ensuring task-relevant brain regions receive the biological resources they need to sustain and mount appropriate responses to various challenges in the environment. Importantly, fluctuations in tonic LC activation also set attentional states, leading to both intra- and inter-subject variations in attentional control over time.

Tonic arousal and LC activation track with the functional dynamics of the salience network, a collection of brain regions that facilitate adaptive task switching, reorganize other functional networks, and promote appraisals of motivational significance. Increases in pupil size correlate with activity in central nodes of the salience network, with the dACC dynamically regulating LC output to modulate adaptive gain and enhance behavioral utility. We suspect that reciprocal LC interactions with the salience network reflect the amount of effort that is needed to implement goal-directed actions and meet homeostatic demands under stress. In this way, the salience functional network is well positioned to modulate tonic arousal and attentional capacity via its interactions with the LC and other large-scale functional networks.

7 Unifying Model of the LC–NE System and Arousal–Attention Interactions

For many years, researchers have sought to understand how arousal influences different parameters of attention and information processing. However, these psychological constructs are multifaceted and are not always easily reconciled. Back in the 1970s, the Capacity Model of attention laid important groundwork for studying arousal–attention interactions by describing cognitive constructs like "capacity," "arousal," and "effort" (Kahneman, 1973). Since then, neurobiological and computational models have uncovered mechanisms that ground these ideas in the brain. Owing in large part to pupillometry research, it appears that the LC–NE system may be a linchpin between different neuroenergetic and cognitive models of arousal and attention.

In this section, we recap many of the topics covered earlier in this chapter. We describe a unifying framework of the LC–NE system that aims to account for different cognitive constructs, as outlined in Fig. 10. Broadly, our framework expands upon the seminal Capacity Model (Kahneman, 1973) by integrating theories and empirical work on the functional roles of phasic and tonic modes of LC activity in shaping neurocognitive processes (Aston-Jones & Cohen, 2005). We also highlight where tonic and phasic pupil biomarkers fit within the context of this expanded model of attention–arousal interactions.

7.1 Links Between the LC–NE System and Cognitive Constructs of Attention

The original Capacity Model sparked discussions about arousal–attention interactions by distinguishing between state-specific (attentional capacity) and task-specific (later called attentional intensity) attentional resources (Kahneman, 1973). Attentional capacity, which has been likened to the idea of mental energy, represents the available pool of attentional resources that can be deployed for any cognitive process, whether task-relevant or not. Attentional capacity fluctuates with changes in arousal, which is in turn induced by environmental cues or from top-down cognitive and metacognitive factors. Of this attentional capacity, an allocation policy (here, referred to as "attentional selectivity") determines attentional intensity by controlling how much attentional resources are devoted to task-relevant information processing (Fig. 10a). Presumably, the remaining attentional capacity that is unused by task-related processing is automatically allocated toward any task-irrelevant information processing, a core tenet within Perceptual Load Theory's explanation of selectivity (Lavie & Tsal, 1994; Fig. 10b).

An enduring feature of the Capacity Model is its proposal that arousal regulates both attentional capacity and attentional selectivity. Building on this foundation, leading neurocomputational, cognitive, and neurobiological models aim to capture how modulation of attention by different extrinsic (e.g., salience, surprise, and novelty) and intrinsic (e.g., mental effort and motivation) factors are mediated by arousal (Aston-Jones & Cohen, 2005; Kahneman, 1973; Richter et al., 2016; Unsworth & Miller, 2021; Unsworth & Robison, 2017a). As displayed at the center of Fig. 10, the LC–NE system has been proposed as the mechanism that implements state arousal's dual effects on attentional capacity and selectivity (Aston-Jones & Cohen, 2005). LC activity, and its release of NE, coordinates the delivery of oxygen and glucose from the bloodstream and induces astrocytic metabolic activity, thus regulating the supply of energy for local neuronal activity (Bekar et al., 2012; Zuend et al., 2020; Fig. 10c). Therefore, an increase in tonic LC–NE activity sustains metabolic activity that fuels information processing, paralleling the hypothesized influence of state arousal on attentional capacity. What triggers this energizing function of the LC–NE system? The original Capacity Model posited that one important factor was mental effort, which reflected an individual's evaluation of task demands. When performing the task becomes more effortful, a metacognitive assessment of the situation triggers an increase in state arousal, leading to increased attentional capacity. From this idea, Adaptive Gain Theory proposed that dACC and OFC evaluate task demands and task utility and drive an appropriate adjustment in tonic LC activity (Aston-Jones & Cohen, 2005; Fig. 10d). These higher-order regions are important for monitoring task performance (Fig. 10e) and integrating inputs from regions that also evaluate reward, cognitive conflict, and salience (Fig. 10f). Thus, state factors that affect task utility, like fatigue and motivation, feed into these pathways to indirectly influence changes in attentional capacity and selectivity (Unsworth & Miller, 2021).

Through the neural gain framework, Adaptive Gain Theory explains how the phasic LC–NE component, the substrate for "induced arousal," acts as a temporal filter for boosting the processing and prioritization of important information. As described in Sect. 3.2, the LC receives dense inputs from regions that evaluate the task relevance and the biological significance of stimuli, including the OFC, dACC, and amygdala. When salient events occur, activation of these brain regions recruits a phasic LC response (Fig. 10g) to enhance neural gain in target regions throughout the brain (Fig. 10h). Such neural gain may be implemented through complementary

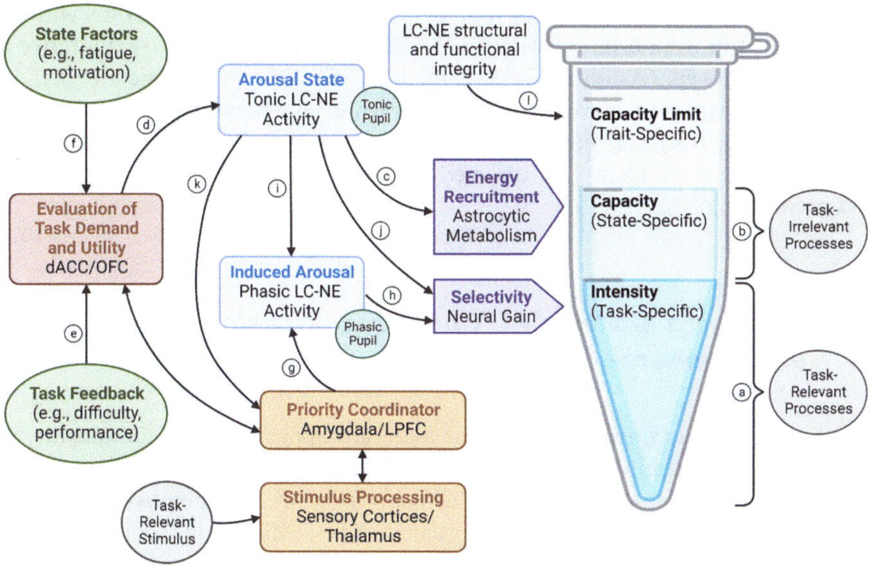

Fig. 10 A unifying LC–NE system framework for integrating different models of arousal-attention interactions. This expanded model illustrates the underlying mechanisms that determine the attentional intensity (bottom layer of the vial with liquid) or the amount of attentional resources allocated to task-relevant processes. The LC–NE system influences state-dependent attentional capacity (middle layer of the vial) and intensity by mobilizing the delivery of energetic resources across the brain. The structural and functional integrity of the LC–NE system also place upper limits of the maximum number of mental resources that are ever available to process information (top mark of the vial). To modulate the capacity, selectivity, and intensity of attention, distinct modes of LC activity interact with higher-order brain regions that evaluate stimulus salience and task utility. Further details corresponding to the numbered labels are described in the main text. (Figure created with BioRender.com)

local and global processes, including the emergence of local NE hotspots (i.e., GANE model) within highly active regions. The phasic LC responses can also issue a brain-wide "reset" signal that promotes the reorganization of functional networks to prioritize behaviorally relevant information (Bouret & Sara, 2005). In general, phasic LC effects seem to target activity and functional connectivity within right-lateralized PFC regions and functional networks, including the ventral attention network (Corbetta et al., 2008; Sara, 2015). Together, these noradrenergic effects work in concert to focus whatever limited mental resources are available to process what matters most when it matters most.

Earlier models of attention noted some perplexing findings about elevated state arousal's effects on selectivity, namely, the rigidity–lability paradox (Easterbrook, 1959; Wachtel, 1967). Adaptive Gain Theory offered a solution to this puzzle by describing how elevated tonic LC activity constrains phasic task-evoked LC activity (Fig. 10i) while sustaining a state of high neural gain (Fig. 10j; Aston-Jones & Cohen, 2005; Eldar et al., 2016). Together, these effects deny the temporal specificity of task-focused attention while enabling non-task cues to capture the enhancing effects of a persistently high neural gain. This account accords with the behavioral shift from task-exploitative at moderate arousal levels to explorative at high arousal levels (Aston-Jones & Cohen, 2005). Moreover, tonic LC activity boosts activity in amygdala-centric networks while also suppressing LPFC-centric functions (Fig. 10k), shifting the balance of priority signaling—and by extension, behavioral control—from goal-directed to emotional or bottom-up information processing.

7.2 *Aging as a Model of LC Changes and Their Relation to Parameters of Attention*

Another important consideration is that there are upper limits on the quantity of mental resources within a given individual. While many models focus on how attentional capacity is state-dependent and fluctuates over time, it is intuitive that there is a ceiling to how many mental resources can ever be available. In theory, these biological constraints would manifest individual differences in cognitive abilities. Indeed, pupillometry studies show that baseline pupil diameters relate to trait-level differences in fluid intelligence and working memory capacity (Tsukahara et al., 2016; Tsukahara & Engle, 2021; Unsworth & Robison, 2020). Moreover, within-subject measures of baseline pupil diameter are stable over very long periods of time, further supporting the idea that this may be an effective physiological marker of trait differences in attentional capacity and cognitive abilities (Robison et al., 2022; but see Tsukahara & Engle, 2021). Pupil diameter differences have also been reported across different disorders of attention, such as attention-deficit/hyperactivity disorder (ADHD) (Anderson & Colombo, 2009; Unsworth et al., 2019).

In recent years, there has been a surge of interest in examining how age-related shifts in LC–NE functional and structural integrity relate to cognitive decline and the neuropathology of Alzheimer's disease (AD; Betts et al., 2019; James et al., 2021; Mather & Harley, 2016). Seminal work on the etiology of AD demonstrated that tau pathology is observable in the LC even before the emergence of tau in the medial temporal lobe (MTL), a hallmark feature of the disease (Braak et al., 2011). Pupil dilation responses during digit span working memory tasks (Granholm et al., 2017) and salient auditory stimuli (Fletcher et al., 2015) are also blunted in individuals with AD, suggesting that LC phasic responses are compromised in age-related disorders. Diminished phasic LC responses may be exacerbated by age-related changes in the structure and/or function of higher-order brain regions, such as the LPFC, that signal priority and recruit transient LC activity (Moscovitch & Winocur, 1995). In addition, altered LC interactions with the frontoparietal network could account for decrements in task-focused attentional processes and distractor inhibition, which are both characteristic of normal cognitive aging (Kennedy & Mather, 2019; Lee et al., 2018; Mather et al., 2020; Robertson, 2013).

Deficits in the LC–NE structure and/or function may result in a reduced range in which the arousal–attention system can operate (Fig. 10l). When demands on attentional resources exceed the overall supply, or trait attentional capacity, behavioral performance may fail. One compelling example of this is normal aging. Specifically, age-related issues with this supply-and-demand process help account for lower working memory capacities in older adults (Schneider-Garces et al., 2010). While seemingly counterintuitive, age-related changes in the LC–NE system may relate to and/or be driven by a tonic increase rather than a decrease in baseline arousal levels (Mather et al., 2020). Elevated tonic arousal and LC activation appear to be a compensatory mechanism for the decline in LC neurons (Mather et al., 2020). Heightened arousal also diminishes phasic LC discharges that are important for regulating neuroinflammation and clearing pathological markers of AD (Janitzky, 2020; Robertson, 2013).

If tonic LC activation increases, would not this mean attentional capacity is also reduced from moment-to-moment? Turns out, this might not be the case: age-related increases in tonic LC activity might enhance *state*-dependent attentional capacity despite *trait* attentional capacity being lower. Models of arousal–attention interactions would suggest that beyond overall decrements in performance in older adults, elevated tonic LC activity would allow both task-relevant and task-irrelevant, or distracting, cues to consume whatever cognitive resources remain (Hasher & Zacks, 1988; Lavie & Tsal, 1994). Indeed, failures to inhibit distractions and maintain focused attention are some of the defining features of cognitive aging, and these failures appear to be exacerbated by stimulus-evoked increases in arousal (Lee et al., 2018).

To summarize, aging may be a good model of how trait levels of attentional capacity are set by neurobiological constraints. Trait levels of attentional capacity may decrease with age due to the degeneration of LC neurons (Clewett et al., 2016; Mather & Harley, 2016; Wilson et al., 2013). However, a disproportionate amount of these attentional resources may remain available due to a compensatory increase in tonic LC activity. The consequence of these resource increases is that older adults may become more distracted by task-relevant cues and exhibit more generalized increases in LC-mediated gain, especially under exogenous manipulations of arousal (e.g., Lee et al., 2018). Due to changes in higher-order brain regions like the LPFC with age, attentional intensity may also diminish along with a decrease in task-relevant phasic LC responses. There has been a rapid increase in the number of studies examining links between the LC and both pathological and healthy cognitive aging. We anticipate that this relationship will continue to be an area of intense interest in the coming years.

8 Methodological Considerations for Studying the LC–Pupil Link

Pupillometry is a highly effective technique for testing predictions about how the LC influences distinct aspects of neural processing and behavior because tonic and phasic modes of LC activity can be linked to different pupil metrics. Here, we review how pupil measures reveal the influence of tonic and phasic patterns of LC activation on key parameters of arousal and attention. We also discuss how a careful consideration of the behavioral/arousal context may deepen our understanding of the link between pupil fluctuations and LC function.

With respect to the arousal and attention models described in this chapter, how can researchers benefit from acquiring pupillary measures of tonic and phasic modes of LC activity? Both tonic and phasic pupillary measures may capture latent attentional factors that explain trial-by-trial variability in behavioral performance. Tonic pupil measures index fluctuations in the background levels of LC–NE system activity throughout an experiment session (Rajkowski et al., 1994). This can provide researchers with insight into changes in the state of task engagement, ranging from inattentive to engaged to distractible, and to relate task performance to fluctuations in attentional state. In contrast, phasic pupil measures are task-evoked, meaning that they are specifically tied to the onset of a stimulus, a decision, or an action. Phasic measures may, therefore, capture the temporally specific component of attention related to the processing of a task-relevant event.

To reemphasize, a key distinction is that the tonic pupil aims to measure the attentional state in which certain task events occur, whereas phasic changes in pupil diameter aim to measure the attentional resources allocated toward a motivationally relevant stimulus. Thus, the tonic pupil is usually measured before an overt task event (e.g., during inter-trial intervals), reflecting the conditions that may predict behavioral responses to the ensuing task event. In contrast, the phasic pupil is often measured after a task event, reflecting the attentional or decision outcome related to that event.

How do we define tonic and phasic pupil measures? In general, researchers distinguish the two measures based on their temporal dynamics, with tonic measures capturing the lower frequency pupillary fluctuations and phasic measures reflecting the transient pupillary responses to a task event (Beatty, 1982). Many studies employ the intuitive approach of using a stimulus-evoked pupil dilation to index phasic activity and a pre-trial baseline pupil size to index tonic activity (Gilzenrat et al., 2010; Hong et al., 2014; Murphy et al., 2011). However, a quick survey of the literature reveals many variations to this basic approach due to different theoretical and practical considerations (for a thorough overview of pupil methods, see Steinhauer et al., 2022). In the following sections, we review several approaches to measuring tonic and phasic LC–NE system activities from pupil data and discuss the validity and discrepancies surrounding these methods (Fig. 11).

8.1 Measuring Tonic LC Activity Using Pupillometry

One of the main objectives of measuring pupil size is to index tonic LC–NE activity, which is in turn related to attentional and arousal states. This motivation stems from evidence that temporally prolonged fluctuations in LC firing rates correspond to low-frequency fluctuations in pupil size (Rajkowski, 1993). Fluctuations in the size

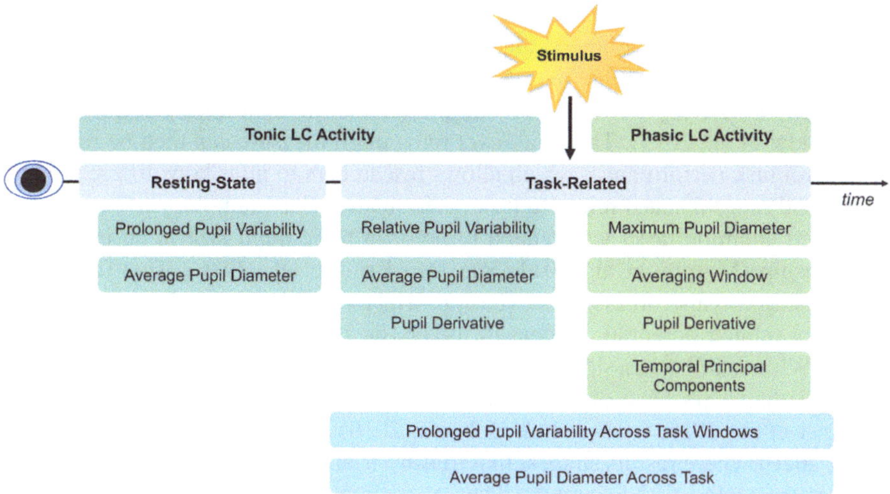

Fig. 11 Pupillometry methods can be used to index phasic and tonic modes of LC activity. Different pupil measures can be used to infer tonic (turquoise colors) or phasic LC (green colors) activity depending on when the pupil is measured, how it is computed, and whether it is time-locked to the onset of a stimulus. These pupil measures can be acquired during a resting-state block or during different task periods (pre-stimulus baseline, post-stimulus, or across multiple trials). Most tonic LC estimates involve measuring pupil diameter, or size. On the other hand, task-evoked phasic LC and pupil dilation responses are linked to the onset of a stimulus; in some cases, these estimates involve normalizing an average pupil response by subtracting a pre-stimulus baseline estimate. In other cases, pupil dilation estimates are measured agnostically with respect to baseline (e.g., pupil derivative or temporal principal components). Pupil size and dynamics can also be measured across more prolonged periods of time by taking advantage of the high temporal resolution of pupillometry data. Measures of pupil diameter variability or average size across a task or at rest may index a mix of tonic and phasic LC activities that relate to state- or trait-level differences in cognitive flexibility and learning, respectively

of the pupil also track varying states of arousal, where low arousal states correspond to smaller baseline pupil size (Hou et al., 2005; Morad et al., 2000) and high arousal states correspond to larger baseline pupil size (Phillips et al., 2000).

A common approach to acquiring a tonic pupil measure is to calculate the mean pupil size over a baseline, or non-stimulus-related, period. It is important to point out that the precise definition of this "tonic baseline" varies from study to study. For example, the baseline can refer to a "resting-state" period that is separate from the main task and usually lasts several minutes. This "resting-state" baseline can be acquired before the task, after the task, or interspersed between blocks of the task. This approach defines a baseline that is completely separate from the main task, ensuring that this tonic pupil measure is uncontaminated by task-evoked pupillary changes. Researchers may use this form of tonic pupil to measure changes in attentional and/or arousal states before and after an intervention, such as following isometric handgrip (Nielsen et al., 2015), or if they want to relate attentional states to a block of task performance (Mather et al., 2020).

Other studies define a "tonic baseline" at the trial level during a cognitive task, enabling researchers to query changes in attention and task engagement from trial to trial. This approach often involves calculating the mean pupil size during a pre-trial baseline, usually during a preceding inter-trial interval lasting a few seconds (e.g., Eldar et al., 2013). This trial-level tonic pupil measure can then be related to trial-level task performance, which allows researchers to infer how different attentional and arousal states relate to task performance. For example, performance on sustained attention tasks was related to pre-trial tonic pupil size in an inverted-U relationship (Murphy et al., 2011; Van Den Brink et al., 2016). This inverted-U relationship was also observed between pre-trial tonic pupil size and the task-evoked P3 event-related potential, a measure of phasic LC–NE activity (Murphy et al., 2011). While a full discussion is beyond the scope of this chapter, pre-trial measures of pupil diameters have also been linked to different forms of mind wandering (Mittner et al., 2016; Unsworth & Robison, 2016) and attentional lapses (Madore et al., 2020). These results suggest that trial-by-trial tonic pupil measures are useful proxies for tonic LC–NE activity. They also align with the proposed relationship between tonic LC–NE system activity and an individual's current level of task engagement (Aston-Jones & Cohen, 2005).

While arousal modulations during sustained attention tasks span the full scale of the arousal–performance relationship, variations in arousal for most experiments cover only a segment of the full inverted-U curve (Joshi & Gold, 2020). Thus, for most cognitive tasks, the conventional approach of using a baseline to acquire tonic pupil measures forces researchers to assume the shape of the task's arousal–performance curve and the range of arousal that occurs during the task. To work around these assumptions, some researchers adopt less conventional measures of the tonic pupil. For instance, rather than inferring tonic LC activity per se, researchers instead use tonic pupil as a measure of task engagement (Gilzenrat et al., 2010; Jepma & Nieuwenhuis, 2011). This subtle reframing of the tonic pupil enables researchers to make behavioral predictions based on the level of task engagement, which relates to task performance in a monotonic manner, rather than the level of tonic LC–NE activity, which relates to task performance as an "inverted-U" function.

Following this line of reasoning, researchers have proposed using a reverse inference approach to interpreting pupil effects on behavior—that is, using the magnitude of task-evoked pupil dilations to infer the state of task engagement (Eldar et al., 2013, 2016; Hopstaken et al., 2015). In terms of Adaptive Gain Theory, a period of strong LC–NE phasic responses indicates that tonic LC activity is in a "phasic mode" that facilitates a state of task engagement. Therefore, a period characterized by larger pupil dilations is thought to reflect conditions that permit robust phasic LC–NE responses—that is, a state of greater task engagement. While this reverse inference approach facilitates the interpretation of pupil-behavioral results, it also makes it challenging to draw interpretations regarding the tonic versus phasic dynamics of the LC–NE system (as both tonic and phasic measures would be derived from the same pupil dilation).

Rather than using *task-evoked* pupil dilation to infer the state of task engagement, some researchers index the magnitude of task engagement using pupil size during a pre-stimulus baseline. For instance, Van Den Brink et al. (2016) used the temporal derivative of the pre-stimulus pupil time series as an indicator of task engagement. Baselines with larger temporal derivatives are driven by larger non-stimulus-evoked fluctuations in pupil dilations, perhaps reflecting a tonic LC–NE mode that is more permissive to larger phasic responses. In this sustained attention task, they found that baseline pupil size was tracked with task performance according to an inverted-U relationship, in line with the expected relationship between tonic LC–NE activity (or arousal level) and task performance. However, they reported that the temporal derivative of baseline pupil tracked with task performance in a linear relationship, in line with the interpretation that a greater state of task engagement, irrespective of its place on the arousal–performance curve, promotes greater task performance.

Recent work suggests that a sudden increase in pre-stimulus baseline pupil size may index a state of persistently high and indiscriminate gain. A combination of pupillometry and computational modeling was used to examine how neural gain may influence cognitive control processes during a classic Stroop task (Tromp et al., 2022). In this study, researchers measured gain by taking the pre-trial pupil derivative (i.e., rate of change) as opposed to the stimulus conflict-evoked pupil dilation. Pupil analyses revealed that increased gain, or a larger increase in the pupil derivative, was related to faster responses but greater Stroop interference. One possible interpretation of these findings is that pre-trial gain amplifies any active input, that is, a higher pupil state may reflect a persistent and indiscriminate increase in neural gain. This result also underscores that tonic arousal levels and neural gain can be regulated by many factors during cognitive tasks, some of which occur prior to the onset of a new stimulus.

Another "tonic" measure that has garnered interest is variability in pupil size either across a task or while participants rested. Even here, however, there are some discrepancies between how and when this variability is measured. For example, some studies measure pupil variability as deviations in pre-trial pupil size across all the trials in a task (Madore et al., 2020; Unsworth & Robison, 2017a). In essence, this provides a "snapshot" of participants' relative state of engagement prior to processing a relevant task stimulus. In most cases, greater variability reflects lapses in attention or greater mind wandering.

Other pupillometry work leverages the full temporal resolution of pupil sampling to examine how dynamic pupil fluctuations relate to behavior across a meaningful analysis window. This approach differs from examining pre-trial baselines in that it is agonistic to what drives pupil fluctuations across time. For example, it has been shown that moment-to-moment variability in pupil size across a 20+ second window relates to temporal memory for pairs of items that spanned that exact window of time (Clewett et al., 2020; Fig. 12). Specifically, during sequence encoding, continuous variability during the window of temporal encoding was related to worse temporal order memory and more expanded retrospective

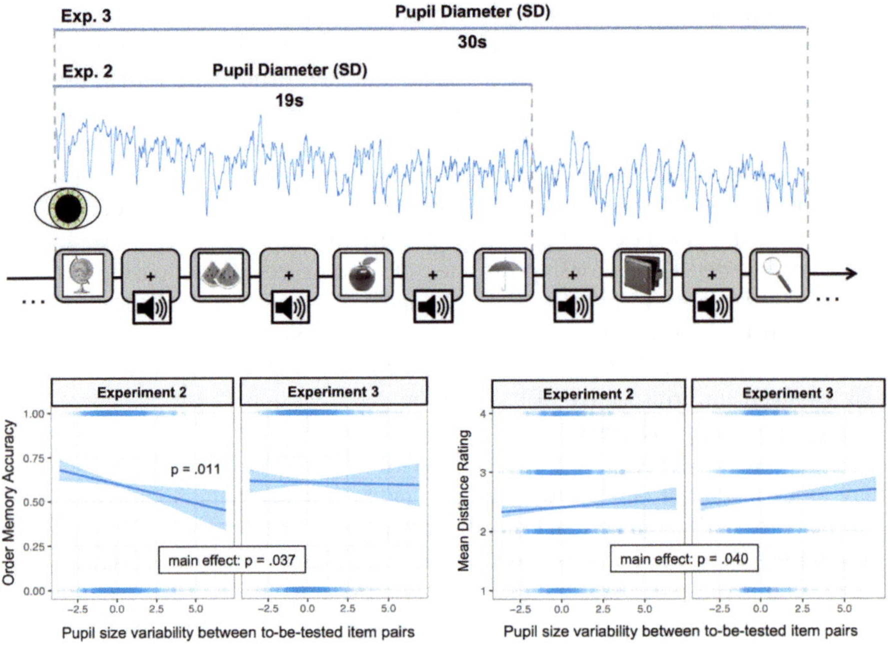

Fig. 12 The temporal stability of pupil-linked arousal states facilitates temporal memory integration during sequence encoding. (Top Panel) To assess variability in pupil-linked arousal fluctuations across each sequence, the standard deviation (SD) of pupil diameter values was measured across continuous pupil samples between a pair of items that would later be tested for temporal order memory (bottom left; y-axis) and subjective temporal distance (bottom right; y-axis). This period was measured from the onset of the first image from a pair to the offset of the second image from that pair to capture the window during which inter-item associations were encoded and linked together. In Experiment 2, the temporal distance between item pairs was 19 seconds, and in Experiment 3, the temporal distance between item pairs was 30 seconds. (Left Bottom Panel) Mixed-effect linear modeling revealed a significant relationship between temporal recency discrimination and pupil size variability, such that participants were better at remembering the order of item pairs if there had been less variable changes in pupil size between those item pairs at encoding. (Right Bottom Panel) Linear regression models revealed a significant relationship between temporal distance ratings and pupil size variability, such that participants were more likely to remember item pairs as having appeared closer together in time if there had been less variability in pupil diameter between those pairs at encoding. These results demonstrated that variability in pupil fluctuations can be meaningfully related to how information is processed and encoded over time. (Figure taken from Clewett et al. (2020))

estimates of temporal distance between pairs of items spanning that window (Clewett et al., 2020). This finding suggests that temporally prolonged fluctuations in pupil size may reflect behaviorally relevant changes in arousal and attentional states that bind sequential representations in long-term memory. While more work is needed in this area of research, we believe it is likely that these pupil fluctuations relate to cognitive flexibility and the ability to adaptively engage the phasic mode of LC activity when necessary (i.e., to process a goal-relevant stimulus). Using a similar approach, some studies measure resting-state (i.e., not explicitly related to a cognitive task) variability in pupil size as a trait measure of behavioral flexibility. For example, greater resting-state pupil variability has been linked to higher trait working memory capacity, as measured by the Letter-Number Sequencing (LNS) task (Aminihajibashi et al., 2019). The use of pupil variability to study tonic LC function is still a nascent area of research. However, we believe this approach holds great promise for understanding how we perceive and remember continuous experiences. It may also help reveal individual differences in people's ability to self-regulate their levels of arousal to meet shifting task demands, with implications for diagnosing different mental health disorders or assessing cognitive abilities.

8.2 *Measuring Phasic LC Activity Using Pupillometry*

One of the motivations for measuring the phasic LC response is to estimate the attentional intensity paid toward processing task-relevant stimuli. Researchers may use this estimate to explain differences in behavioral performance from trial-to-trial, between conditions, or between subjects. For example, a well-documented finding from auditory oddball experiments is that the pupil dilates more to tones that are presented infrequently (oddballs) compared to tones that are presented more frequently (the standard; Mather et al., 2020; Murphy et al., 2014). A basic interpretation of this finding is that the oddball tones draw more attentional intensity than standard tones. This dilation estimate can also be used to explain between-subject variability in behavioral performance. Participants who exhibit greater oddball-evoked pupil dilations also detect oddballs more quickly, suggesting that individuals who paid more attention to salient stimuli tended to perform better on this task (Mather et al., 2020).

Phasic pupil measures can also be used to account for trial-by-trial fluctuations in behavioral performance and brain activity within an experimental condition. For example, researchers model the neuromodulation due to phasic LC–NE responses by weighting EEG and BOLD activity by the magnitude of trial-by-trial pupil dilations (Clewett et al., 2018; Murphy et al., 2011; Sterpenich et al., 2006). Using this approach in an fMRI study, we found that threat-induced arousal selectively enhanced successful encoding-related activity in sensory regions specialized to process goal-relevant information (Clewett et al., 2018).

Increasing evidence supports the notion that phasic pupil dilations can be used to index phasic LC–NE system activity. In animal models, studies have shown that spontaneous phasic LC firing and direct microelectrode stimulation of the LC reliably elicit a phasic pupil dilation (Joshi et al., 2016; Liu et al., 2017; Rajkowski et al., 1994; Varazzani et al., 2015). Similar findings using analogous stimulation techniques (e.g., tVNS) have been reported in humans (Capone et al., 2021).

In the context of an experiment with a single stimulus per trial interval, pupil dilation is conceptualized as a change in pupil diameter triggered by the onset of the stimulus. However, discrepancies in calculating pupil dilation may arise from how the pre-stimulus and post-stimulus windows are chosen. Sometimes, these windows are defined a priori based on the temporal dynamics of the task-evoked pupillary response and its return to baseline (Aminihajibashi et al., 2020; Mathôt et al., 2015). In practice, as temporal dynamics may be influenced by environmental factors that vary between experiment settings, researchers often opt for the empirical approach of defining a pupil dilation window based on the observed elements of their pupil time series. For example, some researchers choose a window based on the peak pupillary response in a group-averaged time series (Clewett et al., 2018). A similar approach defines pupil dilation as the peak change in pupillary response from the pre-stimulus baseline. Although this approach gets around the issue that the temporal dynamics of the pupillary response are not the same across all trials, it still requires choosing a sensible duration for the moving window used to calculate the peak pupil diameter.

Oftentimes, the time windows used for pupil dilation analyses are limited by different experimental constraints. For example, the duration of the inter-trial interval that may serve as a baseline or the duration of inter-stimulus intervals may not be long enough to capture the full time course of a stimulus-evoked pupillary response (e.g., rapid serial visual presentation designs). This latter issue is especially notable because while some studies define pupil dilation as the pupillary response evoked by a single stimulus, other studies define pupil dilation as the cumulative pupillary response evoked by a rapid succession of stimuli.

In some instances, pupil dilation is not defined as a discrete stimulus-evoked response. For example, in the classic digit span tasks, although the presentation of each digit leads to an increase in pupil size, it is common to define pupil dilation as the difference in pupil size during the delay period (i.e., after all digits have been played) minus the baseline period (e.g., Kahneman & Beatty, 1966; Piquado et al., 2010). However, this definition of pupil dilation actually captures the influence of working memory maintenance processes during the delay period on pupil size and does not isolate the phasic pupil dilations elicited by each digit stimulus. Thus, the digit span pupil dilation arguably contains a mix of both phasic and tonic components of arousal.

Some researchers have adopted an alternative approach to inferring task-evoked phasic LC–NE system activity from pupillometry. These measures theoretically capture the critical component of the phasic pupillary response by looking at its temporal dynamics, such as its first derivative or slope (e.g., de Gee et al., 2020). There is empirical evidence that measures derived from the velocity of the pupil time series are more specific indices of the noradrenergic system than other neuromodulatory influences over the pupil, such as the basal cholinergic system (Reimer et al., 2016). Measuring pupil derivatives or slopes also holds several practical advantages over window-averaging techniques. Quantifying the slope of pupil change immediately after the onset of a stimulus avoids the issue of arbitrarily defining averaging windows to calculate a pupil dilation score. It is also agnostic to the pre-stimulus baseline and is proposed to be more robust to artifacts and more reliable for tasks with short inter-trial intervals (de Gee et al., 2020).

Another important challenge to studying the links between the LC, pupil, and behavior is that the pupil is regulated by multiple brain regions and neuromodulatory pathways (Joshi & Gold, 2020). Fortunately, recent evidence shows that temporal principal component analysis (PCA) may be a useful way of dissociating the unique contributions of different autonomic and neuromodulatory pathways to pupil size. For example, pharmacological work shows that distinct characteristics of pupil dilations, or components, may be regulated by different autonomic pathways (Steinhauer & Hakerem, 1992). The degree of loading on these pupil components has also been associated with distinct cognitive and emotional processes, including sustained cognitive processing, conflict resolution, emotional responses, stimulus anticipation, motor responses, and salience detection (Clewett et al., 2020; Johansson et al., 2018; Steinhauer et al., 2004; Steinhauer & Hakerem, 1992; Verney et al., 2004; Wetzel et al., 2016; Widmann et al., 2018). For instance, oddball stimuli modulate an early-peaking (~800 ms) component under moderate light but not under darker conditions when parasympathetic tone is low (Steinhauer et al., 2004; Widmann et al., 2018). This finding suggests that this early-peaking aspect of pupil dilation may specifically reflect parasympathetic inhibition of the pupil, which elicits dilation via relaxation of the sphincter muscle in the eye (Loewenfeld & Lowenstein, 1993).

Of relevance to this chapter, a separate pupil component that peaks around 1.5 s post-stimulus has instead been linked to sympathetic activation (Steinhauer & Hakerem, 1992). Loading on this specific component relates to the onset of highly emotional stimuli (Widmann et al., 2018), consistent with the range of behaviorally relevant events known to activate the LC–NE system. Novel sounds also elicit a concomitant increase in loading on this component and the magnitude of the P3 event-related potential, another ostensible marker of LC activity (Nieuwenhuis et al., 2011; Widmann et al., 2018). The application of PCA to pupil data may serve as an effective method for pinpointing and quantifying the effects of LC neuromodulation on cognitive and emotional processing.

9 Closing Remarks

As a meaningful proxy for LC activity, the pupil offers a unique window into how arousal systems regulate what we perceive, feel, and remember. Novel combinations of advanced imaging techniques and pupillometry continue to lead to new revelations about the LC–NE system's role in modulating the capacity, selectivity, and quality of information processing in the brain. Based on this compelling body of work, we propose that the LC–NE system is a through line for integrating many influential models of arousal and attention.

Pupillometry promises to be an important technique for studying the links between the LC–NE system and mental health. For example, the buildup of neurofibrillary tangles in LC neurons is an early feature of Alzheimer's disease and presumably leads to functional deficits in the LC–NE system as well as its cortical and subcortical targets. Researchers are already exploring if pupillometry can be used as a biomarker of LC–NE dysfunction in individuals with mild cognitive impairment and early Alzheimer's disease (Elman et al., 2017; Granholm et al., 2017; Kremen et al., 2019). Pupil-linked arousal signals have also been used to track different factors and conditions associated with altered noradrenergic signaling, including age (Morris et al., 1997; Piquado et al., 2010), stress (Kimble et al., 2010), depression (Siegle et al., 2001), post-traumatic stress disorder (Mckinnon et al., 2020), and attention-deficit/hyperactivity disorder (Nobukawa et al., 2021; Wainstein et al., 2017). In time, pupillometry may help facilitate earlier diagnoses of these conditions and be used to assess the efficacy of various therapeutic interventions.

Through its broad neuromodulatory effects, LC activity is essential for amplifying attention, prioritizing task-appropriate decisions, and ensuring important experiences are encoded and stored as enduring memories. Pupillometry provides a powerful tool for uncovering these core functions of the LC–NE system, with an eye toward identifying new strategies for promoting healthy cognitive function and well-being.

References

Alnaes, D., Sneve, M. H., Espeseth, T., Endestad, T., van de Pavert, S. H. P., & Laeng, B. (2014). Pupil size signals mental effort deployed during multiple object tracking and predicts brain activity in the dorsal attention network and the locus coeruleus. *Journal of Vision, 14*(4), 1–1. https://doi.org/10.1167/14.4.1

Aminihajibashi, S., Hagen, T., Foldal, M. D., Laeng, B., & Espeseth, T. (2019). Individual differences in resting-state pupil size: Evidence for association between working memory capacity and pupil size variability. *International Journal of Psychophysiology, 140*, 1–7. https://doi.org/10.1016/j.ijpsycho.2019.03.007

Aminihajibashi, S., Hagen, T., Laeng, B., & Espeseth, T. (2020). Pupillary and behavioral markers of alerting and orienting: An individual difference approach. *Brain and Cognition, 143*. https://doi.org/10.1016/j.bandc.2020.105597

Anderson, C. J., & Colombo, J. (2009). Larger tonic pupil size in Young children with autism Spectrum disorder. *Developmental Psychobiology, 51*(2), 207–211. https://doi.org/10.1002/dev.20352

Arnsten, A. F. T. (2007). Catecholamine and second messenger influences on prefrontal cortical networks of "representational knowledge": A rational bridge between genetics and the symptoms of mental illness. *Cerebral Cortex, 17*(suppl_1), i6–i15. https://doi.org/10.1093/cercor/bhm033

Arnsten, A. F. T. (2009). Stress signalling pathways that impair prefrontal cortex structure and function. *Nature Reviews Neuroscience, 10*(6), 410–422. https://doi.org/10.1038/nrn2648

Arnsten, A. F. T., & Goldman-Rakic, P. S. (1984). Selective prefrontal cortical projections to the region of the locus coeruleus and raphe nuclei in the rhesus monkey. *Brain Research, 306*(1), 9–18. https://doi.org/10.1016/0006-8993(84)90351-2

Arnsten, A. F. T., & Li, B. M. (2005). Neurobiology of executive functions: Catecholamine influences on prefrontal cortical functions. *Biological Psychiatry, 57*(11), 1377–1384.

Aston-Jones, G., & Bloom, F. (1981). Activity of norepinephrine-containing locus coeruleus neurons in behaving rats anticipates fluctuations in the sleep-waking cycle. *The Journal of Neuroscience, 1*(8), 876–886.

Aston-Jones, G., & Cohen, J. D. (2005). An integrative theory of locus coeruleus-norepinephrine function: Adaptive gain and optimal performance. *Annual Review of Neuroscience, 28*, 403–450.

Aston-Jones, G., Rajkowski, J., Kubiak, P., & Alexinsky, T. (1994). Locus coeruleus neurons in monkey are selectively activated by attended cues in a vigilance task. *The Journal of Neuroscience, 14*(7), 4467–4480. https://doi.org/10.1523/JNEUROSCI.14-07-04467.1994

Aston-Jones, G., Rajkowski, J., Kubiak, P., Valentino, R. J., & Shipley, M. T. (1996). Role of the locus coeruleus in emotional activation. In G. Holstege, R. Bandler, & C. B. Saper (Eds.), *Progress in brain research* (Vol. 107, pp. 379–402). Elsevier. https://doi.org/10.1016/S0079-6123(08)61877-4

Aston-Jones, G., Rajkowski, J., & Kubiak, P. (1997). Conditioned responses of monkey locus coeruleus neurons anticipate acquisition of discriminative behavior in a vigilance task. *Neuroscience, 80*(3), 697–715. https://doi.org/10.1016/S0306-4522(97)00060-2

Bahrick, H. P., Fitts, P. M., & Rankin, R. E. (1952). Effect of incentives upon reactions to peripheral stimuli. *Journal of Experimental Psychology, 44*(6), 400–406. https://doi.org/10.1037/h0053593

Beatty, J. (1982). Task-evoked pupillary responses, processing load, and the structure of processing resources. *Psychological Bulletin, 91*(2), 276.

Beatty, J., & Lucero-Wagoner, B. (2000). The pupillary system. *Handbook of Psychophysiology, 2*(142–162), 319.

Bekar, L. K., Wei, H. S., & Nedergaard, M. (2012). The locus coeruleus-norepinephrine network optimizes coupling of cerebral blood volume with oxygen demand. *Journal of Cerebral Blood Flow and Metabolism, 32*, 2135–2145.

Bergado, J. A., Lucas, M., & Richter-Levin, G. (2011). Emotional tagging: A simple hypothesis in a complex reality. *Progress in Neurobiology, 94*(1), 64–76. https://doi.org/10.1016/j.pneurobio.2011.03.004

Berridge, C. W., & Waterhouse, B. D. (2003). The locus coeruleus–noradrenergic system: Modulation of behavioral state and state-dependent cognitive processes. *Brain Research Reviews, 42*(1), 33–84. https://doi.org/10.1016/S0165-0173(03)00143-7

Betts, M. J., Kirilina, E., Otaduy, M. C. G., Ivanov, D., Acosta-Cabronero, J., Callaghan, M. F., Lambert, C., Cardenas-Blanco, A., Pine, K., Passamonti, L., Loane, C., Keuken, M. C., Trujillo, P., Lüsebrink, F., Mattern, H., Liu, K. Y., Priovoulos, N., Fliessbach, K., Dahl, M. J., et al. (2019). Locus coeruleus imaging as a biomarker for noradrenergic dysfunction in neurodegenerative diseases. *Brain, 142*(9), 2558–2571. https://doi.org/10.1093/brain/awz193

Blanchard, T. C., & Gershman, S. J. (2018). Pure correlates of exploration and exploitation in the human brain. *Cognitive, Affective, & Behavioral Neuroscience, 18*(1), 117–126. https://doi.org/10.3758/s13415-017-0556-2

Bornert, P., & Bouret, S. (2021). Locus coeruleus neurons encode the subjective difficulty of triggering and executing actions. *PLoS Biology, 19*(12), e3001487. https://doi.org/10.1371/journal.pbio.3001487

Bouret, S., & Sara, S. J. (2005). Network reset: A simplified overarching theory of locus coeruleus noradrenaline function. *Trends in Neurosciences, 28*(11), 574–582. https://doi.org/10.1016/j.tins.2005.09.002

Braak, H., Thal, D. R., Ghebremedhin, E., & Del Tredici, K. (2011). Stages of the pathologic process in alzheimer disease: Age categories from 1 to 100 years. *Journal of Neuropathology & Experimental Neurology, 70*(11), 960–969. https://doi.org/10.1097/NEN.0b013e318232a379

Bradley, M. M., Miccoli, L., Escrig, M. A., & Lang, P. J. (2008). The pupil as a measure of emotional arousal and autonomic activation. *Psychophysiology, 45*(4), 602–607. https://doi.org/10.1111/j.1469-8986.2008.00654.x

Bradshaw, J. (1967). Pupil size as a measure of arousal during information processing. *Nature, 216*(5114), 515–516. https://doi.org/10.1038/216515a0

Braem, S., Coenen, E., Bombeke, K., van Bochove, M. E., & Notebaert, W. (2015). Open your eyes for prediction errors. *Cognitive, Affective, & Behavioral Neuroscience, 15*(2), 374–380.

Breton-Provencher, V., Drummond, G. T., Feng, J., Li, Y., & Sur, M. (2022). Spatiotemporal dynamics of noradrenaline during learned behaviour. *Nature, 606*(7915), 732–738. https://doi.org/10.1038/s41586-022-04782-2

Broadbent, D. E. (1971). *Decision and stress* (p. 522). Academic Press.

Broadhurst, P. L. (1957). Emotionality and the Yerkes-Dodson law. *Journal of Experimental Psychology, 54*(5), 345–352. https://doi.org/10.1037/h0049114

Brown, G. G., Kindermann, S. S., Siegle, G. J., Granholm, E., Wong, E. C., & Buxton, R. B. (1999). Brain activation and pupil response during covert performance of the stroop color word task. *Journal of the International Neuropsychological Society, 5*(4), 308–319. https://doi.org/10.1017/S1355617799544020

Burger, A. M., Van der Does, W., Brosschot, J. F., & Verkuil, B. (2020). From ear to eye? No effect of transcutaneous vagus nerve stimulation on human pupil dilation: A report of three studies. *Biological Psychology, 152*, 107863. https://doi.org/10.1016/j.biopsycho.2020.107863

Bursill, A. E. (1958). The restriction of peripheral vision during exposure to hot and humid conditions. *Quarterly Journal of Experimental Psychology, 10*(3), 113–129. https://doi.org/10.1080/17470215808416265

Bush, G., Vogt, B. A., Holmes, J., Dale, A. M., Greve, D., Jenike, M. A., & Rosen, B. R. (2002). Dorsal anterior cingulate cortex: A role in reward-based decision making. *Proceedings of the National Academy of Sciences of the United States of America, 99*(1), 523–528.

Cahill, L., Prins, B., Weber, M., & McGaugh, J. L. (1994). Beta-adrenergic activation and memory for emotional events. *Nature, 371*(6499), 702–704.

Cahill, L., Babinsky, R., Markowitsch, H. J., & McGaugh, J. L. (1995). The amygdala and emotional memory. *Nature, 377*(6547), 295–296.

Cahill, L., Haier, R. J., Fallon, J., Alkire, M. T., Tang, C., Keator, D., Wu, J., & McGaugh, J. L. (1996). Amygdala activity at encoding correlated with long-term, free recall of emotional information. *Proceedings of the National Academy of Sciences of the United States of America, 93*(15), 8016–8021.

Callaway, E., & Dembo, D. (1958). Narrowed attention: A psychological phenomenon that accompanies a certain physiological change. *Archives of Neurology and Psychiatry, 79*(1), 74–90.

Callaway, E., & Stone, G. (1960). Re-evaluating focus of attention. In *Drugs and behavior* (pp. 393–398). NIH.

Capone, F., Motolese, F., Di Zazzo, A., Antonini, M., Magliozzi, A., Rossi, M., Marano, M., Pilato, F., Musumeci, G., Coassin, M., & Di Lazzaro, V. (2021). The effects of transcutaneous auricular vagal nerve stimulation on pupil size. *Clinical Neurophysiology, 132*(8), 1859–1865. https://doi.org/10.1016/j.clinph.2021.05.014

Carter, M. E., Yizhar, O., Chikahisa, S., Nguyen, H., Adamantidis, A., Nishino, S., Deisseroth, K., & de Lecea, L. (2010). Tuning arousal with optogenetic modulation of locus coeruleus neurons. *Nature Neuroscience, 13*(12), 1526–1533. https://doi.org/10.1038/nn.2682

Chamberlain, S. R., & Robbins, T. W. (2013). Noradrenergic modulation of cognition: Therapeutic implications. *Journal of Psychopharmacology, 27*(8), 694–718. https://doi.org/10.1177/0269881113480988

Chandler, D., & Waterhouse, B. (2012). Evidence for broad versus segregated projections from cholinergic and noradrenergic nuclei to functionally and anatomically discrete subregions of prefrontal cortex. *Frontiers in Behavioral Neuroscience, 6*, 20. https://doi.org/10.3389/fnbeh.2012.00020

Chandler, D. J., Lamperski, C. S., & Waterhouse, B. D. (2013). Identification and distribution of projections from monoaminergic and cholinergic nuclei to functionally differentiated subregions of prefrontal cortex. *Brain Research, 1522*, 38–58. https://doi.org/10.1016/j.brainres.2013.04.057

Chandler, D. J., Gao, W.-J., & Waterhouse, B. D. (2014). Heterogeneous organization of the locus coeruleus projections to prefrontal and motor cortices. *Proceedings of the National Academy of Sciences, 111*(18), 6816–6821.

Clayton, E. C., Rajkowski, J., Cohen, J. D., & Aston-Jones, G. (2004). Phasic activation of monkey locus ceruleus neurons by simple decisions in a forced-choice task. *Journal of Neuroscience, 24*(44), 9914–9920. https://doi.org/10.1523/JNEUROSCI.2446-04.2004

Clewett, D., Schoeke, A., & Mather, M. (2013). Amygdala functional connectivity is reduced after the cold pressor task. *Cognitive, Affective, & Behavioral Neuroscience, 13*(3), 501–518.

Clewett, D. V., Lee, T.-H., Greening, S., Ponzio, A., Margalit, E., & Mather, M. (2016). Neuromelanin marks the spot: Identifying a locus coeruleus biomarker of cognitive reserve in healthy aging. *Neurobiology of Aging, 37*, 117–126. https://doi.org/10.1016/j.neurobiolaging.2015.09.019

Clewett, D. V., Huang, R., Velasco, R., Lee, T.-H., & Mather, M. (2018). Locus coeruleus activity strengthens prioritized memories under arousal. *The Journal of Neuroscience, 38*(6), 1558–1574. https://doi.org/10.1523/JNEUROSCI.2097-17.2017

Clewett, D., Gasser, C., & Davachi, L. (2020). Pupil-linked arousal signals track the temporal organization of events in memory. *Nature Communications, 11*(1), 4007. https://doi.org/10.1038/s41467-020-17851-9

Coggan, J. S., Keller, D., Calì, C., Lehväslaiho, H., Markram, H., Schürmann, F., & Magistretti, P. J. (2018). Norepinephrine stimulates glycogenolysis in astrocytes to fuel neurons with lactate. *PLoS Computational Biology, 14*(8), e1006392. https://doi.org/10.1371/journal.pcbi.1006392

Collins, L., Boddington, L., Steffan, P. J., & McCormick, D. (2021). Vagus nerve stimulation induces widespread cortical and behavioral activation. *Current Biology, 31*(10), 2088–2098.e3. https://doi.org/10.1016/j.cub.2021.02.049

Corbetta, M., Patel, G., & Shulman, G. L. (2008). The reorienting system of the human brain: From environment to theory of mind. *Neuron, 58*(3), 306–324.

Costa, V. D., & Rudebeck, P. H. (2016). More than meets the eye: The relationship between pupil size and locus coeruleus activity. *Neuron, 89*(1), 8–10. https://doi.org/10.1016/j.neuron.2015.12.031

Coull, J. T., Jones, M. E., Egan, T. D., Frith, C. D., & Maze, M. (2004). Attentional effects of noradrenaline vary with arousal level: Selective activation of thalamic pulvinar in humans. *NeuroImage, 22*(1), 315–322.

Critchley, H. D., Tang, J., Glaser, D., Butterworth, B., & Dolan, R. J. (2005). Anterior cingulate activity during error and autonomic response. *NeuroImage, 27*(4), 885–895.

de Gee, J. W., Knapen, T., & Donner, T. H. (2014). Decision-related pupil dilation reflects upcoming choice and individual bias. *Proceedings of the National Academy of Sciences, 111*(5), E618–E625.

de Gee, J. W., Tsetsos, K., Schwabe, L., Urai, A. E., McCormick, D., McGinley, M. J., & Donner, T. H. (2020). Pupil-linked phasic arousal predicts a reduction of choice bias across species and decision domains. *eLife, 9*, e54014. https://doi.org/10.7554/eLife.54014

de Gee, J. W., Mridha, Z., Hudson, M., Shi, Y., Ramsaywak, H., Smith, S., Karediya, N., Thompson, M., Jaspe, K., Zhang, W., & McGinley, M. J. (2022). *Mice regulate their attentional intensity and arousal to exploit increases in task utility* (p. 482962). bioRxiv. https://doi.org/10.1101/2022.03.04.482962

de Quervain, D. J., Kolassa, I.-T., Ertl, V., Onyut, P. L., Neuner, F., Elbert, T., & Papassotiropoulos, A. (2007). A deletion variant of the α2b-adrenoceptor is related to emotional memory in Europeans and Africans. *Nature Neuroscience, 10*(9), 1137–1139.

de Voogd, L. D., Fernández, G., & Hermans, E. J. (2016). Disentangling the roles of arousal and amygdala activation in emotional declarative memory. *Social Cognitive and Affective Neuroscience, 11*(9), 1471–1480. https://doi.org/10.1093/scan/nsw055

Diamond, D. M., Campbell, A. M., Park, C. R., Halonen, J., & Zoladz, P. R. (2007). The temporal dynamics model of emotional memory processing: A synthesis on the neurobiological basis of stress-induced amnesia, flashbulb and traumatic memories, and the Yerkes-Dodson law. *Neural Plasticity, 2007*, e60803. https://doi.org/10.1155/2007/60803

Dienel, G. A., & Cruz, N. F. (2016). Aerobic glycolysis during brain activation: Adrenergic regulation and influence of norepinephrine on astrocytic metabolism. *Journal of Neurochemistry, 138*(1), 14–52. https://doi.org/10.1111/jnc.13630

DiNuzzo, M., Mascali, D., Moraschi, M., Bussu, G., Maugeri, L., Mangini, F., Fratini, M., & Giove, F. (2019). Brain networks underlying eye's pupil dynamics. *Frontiers in Neuroscience, 13*, 965. https://doi.org/10.3389/fnins.2019.00965

Dorr, A. E., & Debonnel, G. (2006). Effect of Vagus nerve stimulation on serotonergic and noradrenergic transmission. *Journal of Pharmacology and Experimental Therapeutics, 318*(2), 890–898. https://doi.org/10.1124/jpet.106.104166

Doucette, W., Milder, J., & Restrepo, D. (2007). Adrenergic modulation of olfactory bulb circuitry affects odor discrimination. *Learning & Memory, 14*(8), 539–547. https://doi.org/10.1101/lm.606407

Duszkiewicz, A. J., McNamara, C. G., Takeuchi, T., & Genzel, L. (2019). Novelty and dopaminergic modulation of memory persistence: A tale of two systems. *Trends in Neurosciences, 42*(2), 102–114. https://doi.org/10.1016/j.tins.2018.10.002

Easterbrook, J. A. (1959). The effect of emotion on cue utilization and the organization of behavior. *Psychological Review, 66*(3), 183–201. https://doi.org/10.1037/h0047707

Ebitz, R. B., & Platt, M. L. (2015). Neuronal activity in primate dorsal anterior cingulate cortex signals task conflict and predicts adjustments in pupil-linked arousal. *Neuron, 85*(3), 628–640. https://doi.org/10.1016/j.neuron.2014.12.053

Ekman, P., & Davidson, R. J. (1994). The nature of emotion. In *Fundamental questions* (p. 496). Oxford University Press.

Eldar, E., Cohen, J. D., & Niv, Y. (2013). The effects of neural gain on attention and learning. *Nature Neuroscience, 16*(8), 1146–1153. https://doi.org/10.1038/nn.3428

Eldar, E., Cohen, J. D., & Niv, Y. (2016). Amplified selectivity in cognitive processing implements the neural gain model of norepinephrine function. *Behavioral and Brain Sciences, 39*, e206. https://doi.org/10.1017/S0140525X15001776

Elman, J. A., Panizzon, M. S., Hagler, D. J., Eyler, L. T., Granholm, E. L., Fennema-Notestine, C., Lyons, M. J., McEvoy, L. K., Franz, C. E., Dale, A. M., & Kremen, W. S. (2017). Task-evoked pupil dilation and BOLD variance as indicators of locus coeruleus dysfunction. *Cortex, 97*, 60–69. https://doi.org/10.1016/j.cortex.2017.09.025

Fastenrath, M., Coynel, D., Spalek, K., Milnik, A., Gschwind, L., Roozendaal, B., Papassotiropoulos, A., & de Quervain, D. J. (2014). Dynamic modulation of amygdala–hippocampal connectivity by emotional arousal. *The Journal of Neuroscience, 34*(42), 13935–13947.

Ferry, B., Magistretti, P. J., & Pralong, E. (1997). Noradrenaline modulates Glutamate-mediated neurotransmission in the rat basolateral amygdala in vitro. *European Journal of Neuroscience, 9*(7), 1356–1364. https://doi.org/10.1111/j.1460-9568.1997.tb01490.x

Ferry, B., Roozendaal, B., & McGaugh, J. L. (1999). Basolateral amygdala noradrenergic influences on memory storage are mediated by an interaction between beta- and alpha(1)-adrenoceptors. *Journal of Neuroscience, 19*(12), 5119–5123.

Fietz, J., Pöhlchen, D., Binder, F. P., Czisch, M., Sämann, P. G., & Spoormaker, V. I. (2021). Pupillometry tracks cognitive load and salience network activity in a working memory functional magnetic resonance imaging task. *Human Brain Mapping, 43*(2), 665–680. https://doi.org/10.1002/hbm.25678

Fletcher, P. D., Nicholas, J. M., Shakespeare, T. J., Downey, L. E., Golden, H. L., Agustus, J. L., Clark, C. N., Mummery, C. J., Schott, J. M., Crutch, S. J., & Warren, J. D. (2015). Dementias show differential physiological responses to salient sounds. *Frontiers in Behavioral Neuroscience, 9*, 73. https://doi.org/10.3389/fnbeh.2015.00073

Foote, S. L., & Berridge, C. W. (2019). New developments and future directions in understanding locus coeruleus—Norepinephrine (LC-NE) function. *Brain Research, 1709*, 81–84. https://doi.org/10.1016/j.brainres.2018.09.033

Foote, S. L., Aston-Jones, G., & Bloom, F. E. (1980). Impulse activity of locus coeruleus neurons in awake rats and monkeys is a function of sensory stimulation and arousal. *Proceedings of the National Academy of Sciences, 77*(5), 3033–3037. https://doi.org/10.1073/pnas.77.5.3033

Funahashi, S., & Andreau, J. M. (2013). Prefrontal cortex and neural mechanisms of executive function. *Journal of Physiology, Paris, 107*(6), 471–482. https://doi.org/10.1016/j.jphysparis.2013.05.001

Fuster, J. M. (1989). *The prefrontal cortex*. Raven Press.

Gazzaley, A., Cooney, J. W., Rissman, J., & D'Esposito, M. (2005). Top-down suppression deficit underlies working memory impairment in normal aging. *Nature Neuroscience, 8*(10), 1298–1300.

Gilzenrat, M. S., Nieuwenhuis, S., Jepma, M., & Cohen, J. D. (2010). Pupil diameter tracks changes in control state predicted by the adaptive gain theory of locus coeruleus function. *Cognitive, Affective, & Behavioral Neuroscience, 10*(2), 252–269. https://doi.org/10.3758/CABN.10.2.252

Goldinger, S. D., & Papesh, M. H. (2012). Pupil dilation reflects the creation and retrieval of memories. *Current Directions in Psychological Science, 21*(2), 90–95. https://doi.org/10.1177/0963721412436811

Granholm, E. L., Panizzon, M. S., Elman, J. A., Jak, A. J., Hauger, R. L., Bondi, M. W., Lyons, M. J., Franz, C. E., & Kremen, W. S. (2017). Pupillary responses as a biomarker of early risk for Alzheimer's disease. *Journal of Alzheimer's Disease, 56*(4), 1419–1428. https://doi.org/10.3233/JAD-161078

Grueschow, M., Kleim, B., & Ruff, C. C. (2020). Role of the locus coeruleus arousal system in cognitive control. *Journal of Neuroendocrinology, 32*(12), e12890. https://doi.org/10.1111/jne.12890

Hansen, N. (2017). The longevity of hippocampus-dependent memory is orchestrated by the locus coeruleus-noradrenergic system. *Neural Plasticity, 2017*, 2727602. https://doi.org/10.1155/2017/2727602

Harley, C. W. (1987). A role for norepinephrine in arousal, emotion and learning?: Limbic modulation by norepinephrine and the kety hypothesis. *Progress in Neuro-Psychopharmacology and Biological Psychiatry, 11*(4), 419–458. https://doi.org/10.1016/0278-5846(87)90015-7

Hasher, L., & Zacks, R. T. (1988). Working memory, comprehension, and aging: A review and a new view. *The Psychology of Learning and Motivation: Advances in Research and Theory, 22*, 193–225.

Hasselmo, M. E., Linster, C., Patil, M., Ma, D., & Cekic, M. (1997). Noradrenergic suppression of synaptic transmission may influence cortical signal-to-noise ratio. *Journal of Neurophysiology, 77*(6), 3326–3339.

Henckens, M., Hermans, E. J., Pu, Z. W., Joels, M., & Fernandez, G. N. (2009). Stressed memories: How acute stress affects memory formation in humans. *Journal of Neuroscience, 29*(32), 10111–10119. https://doi.org/10.1523/jneurosci.1184-09.2009

Hermans, E. J., van Marle, H. J. F., Ossewaarde, L., Henckens, M., Qin, S. Z., van Kesteren, M. T. R., Schoots, V. C., Cousijn, H., Rijpkema, M., Oostenveld, R., & Fernandez, G. (2011). Stress-related noradrenergic activity prompts large-scale neural network reconfiguration. *Science, 334*(6059), 1151–1153. https://doi.org/10.1126/science.1209603

Hermans, E. J., Battaglia, F. P., Atsak, P., de Voogd, L. D., Fernández, G., & Roozendaal, B. (2014a). How the amygdala affects emotional memory by altering brain network properties. *Neurobiology of Learning and Memory, 112*, 2–16. http://www.sciencedirect.com.idpproxy.reading.ac.uk/science/article/pii/S1074742714000380

Hermans, E. J., Henckens, M. J. A. G., Joëls, M., & Fernández, G. (2014b). Dynamic adaptation of large-scale brain networks in response to acute stressors. *Trends in Neurosciences, 37*(6), 304–314. https://doi.org/10.1016/j.tins.2014.03.006

Hernández-Peón, R. (1964). Attention, sleep, motivation, and behavior. In R. G. Heath (Ed.), *The role of pleasure in behavior* (Vol. 1964, pp. 195–219). Harper & Row.

Hershman, R., & Henik, A. (2019). Dissociation between reaction time and pupil dilation in the Stroop task. *Journal of Experimental Psychology: Learning, Memory, and Cognition, 45*(10), 1899–1909. https://doi.org/10.1037/xlm0000690

Hess, E. H., & Polt, J. M. (1960). Pupil size as related to interest value of visual stimuli. *Science, 132*(3423), 349–350.

Hess, E. H., & Polt, J. M. (1964). Pupil size in relation to mental activity during simple problem-solving. *Science, 143*(3611), 1190–1192.

Hirata, A., Aguilar, J., & Castro-Alamancos, M. A. (2006). Noradrenergic activation amplifies bottom-up and top-down signal-to-noise ratios in sensory thalamus. *Journal of Neuroscience, 26*(16), 4426–4436. https://doi.org/10.1523/jneurosci.5298-05.2006

Hockey, G. R. J. (1970). Effect of loud noise on attentional selectivity. *Quarterly Journal of Experimental Psychology, 22*(1), 28–36. https://doi.org/10.1080/14640747008401898

Hong, L., Walz, J. M., & Sajda, P. (2014). Your eyes give you away: Prestimulus changes in pupil diameter correlate with poststimulus task-related EEG dynamics. *PLoS One, 9*(3), e91321. https://doi.org/10.1371/journal.pone.0091321

Hopstaken, J. F., van der Linden, D., Bakker, A. B., & Kompier, M. A. J. (2015). The window of my eyes: Task disengagement and mental fatigue covary with pupil dynamics. *Biological Psychology, 110*, 100–106. https://doi.org/10.1016/j.biopsycho.2015.06.013

Hou, R. H., Freeman, C., Langley, R. W., Szabadi, E., & Bradshaw, C. M. (2005). Does modafinil activate the locus coeruleus in man? Comparison of modafinil and clonidine on arousal and autonomic functions in human volunteers. *Psychopharmacology, 181*(3), 537–549. https://doi.org/10.1007/s00213-005-0013-8

James, T., Kula, B., Choi, S., Khan, S. S., Bekar, L. K., & Smith, N. A. (2021). Locus coeruleus in memory formation and Alzheimer's disease. *The European Journal of Neuroscience, 54*(8), 6948–6959. https://doi.org/10.1111/ejn.15045

Janitzky, K. (2020). Impaired phasic discharge of locus coeruleus neurons based on persistent high tonic discharge—A new hypothesis with potential implications for neurodegenerative diseases. *Frontiers in Neurology, 11*, 371. https://doi.org/10.3389/fneur.2020.00371

Jepma, M., & Nieuwenhuis, S. (2011). Pupil diameter predicts changes in the exploration–Exploitation trade-off: Evidence for the adaptive gain theory. *Journal of Cognitive Neuroscience, 23*(7), 1587–1596. https://doi.org/10.1162/jocn.2010.21548

Jodo, E., Chiang, C., & Aston-Jones, G. (1998). Potent excitatory influence of prefrontal cortex activity on noradrenergic locus coeruleus neurons. *Neuroscience, 83*(1), 63–79. https://doi.org/10.1016/s0306-4522(97)00372-2

Jodoin, V. D., Lespérance, P., Nguyen, D. K., Fournier-Gosselin, M.-P., & Richer, F. (2015). Effects of vagus nerve stimulation on pupillary function. *International Journal of Psychophysiology, 98*(3, Part 1), 455–459. https://doi.org/10.1016/j.ijpsycho.2015.10.001

Joëls, M., & Baram, T. Z. (2009). The neuro-symphony of stress. *Nature Reviews. Neuroscience, 10*(6), 459–466. https://doi.org/10.1038/nrn2632

Johansson, R., Pärnamets, P., Bjernestedt, A., & Johansson, M. (2018). Pupil dilation tracks the dynamics of mnemonic interference resolution. *Scientific Reports, 8*(1), 4826.

Jones, B. E. (2003). Arousal systems. *Frontiers in Bioscience-Landmark, 8*(6), 438–451.

Joshi, S., & Gold, J. I. (2020). Pupil size as a window on neural substrates of cognition. *Trends in Cognitive Sciences, 24*(6), 466–480. https://doi.org/10.1016/j.tics.2020.03.005

Joshi, S., & Gold, J. I. (2022). Context-dependent relationships between locus coeruleus firing patterns and coordinated neural activity in the anterior cingulate cortex. *eLife, 11*, e63490. https://doi.org/10.7554/eLife.63490

Joshi, S., Li, Y., Kalwani, R. M., & Gold, J. I. (2016). Relationships between pupil diameter and neuronal activity in the locus coeruleus, colliculi, and cingulate cortex. *Neuron, 89*(1), 221–234. https://doi.org/10.1016/j.neuron.2015.11.028

Just, M. A., Carpenter, P. A., & Miyake, A. (2003). Neuroindices of cognitive workload: Neuroimaging, pupillometric and event-related potential studies of brain work. *Theoretical Issues in Ergonomics Science, 4*(1–2), 56–88. https://doi.org/10.1080/14639220210159735

Kafkas, A., & Montaldi, D. (2014). Two separate, but interacting, neural systems for familiarity and novelty detection: A dual-route mechanism: Familiarity and novelty detection processes. *Hippocampus, 24*(5), 516–527. https://doi.org/10.1002/hipo.22241

Kafkas, A., & Montaldi, D. (2018). How do memory systems detect and respond to novelty? *Neuroscience Letters, 680*, 60–68. https://doi.org/10.1016/j.neulet.2018.01.053

Kahneman, D. (1973). *Attention and effort*. Prentice-Hall.

Kahneman, D., & Beatty, J. (1966). Pupil diameter and load on memory. *Science, 154*(3756), 1583–1585.

Kahneman, D., Beatty, J., & Pollack, I. (1967). Perceptual deficit during a mental task. *Science, 157*(3785), 218–219. https://doi.org/10.1126/science.157.3785.218

Kane, G. A., Vazey, E. M., Wilson, R. C., Shenhav, A., Daw, N. D., Aston-Jones, G., & Cohen, J. D. (2017). Increased locus coeruleus tonic activity causes disengagement from a patch-foraging task. *Cognitive, Affective, & Behavioral Neuroscience, 17*(6), 1073–1083. https://doi.org/10.3758/s13415-017-0531-y

Kennedy, B. L., & Mather, M. (2019). Neural mechanisms underlying age-related changes in attentional selectivity. In *The aging brain: Functional adaptation across adulthood* (pp. 45–72). American Psychological Association. https://doi.org/10.1037/0000143-003

Kennerley, S. W., Walton, M. E., Behrens, T. E. J., Buckley, M. J., & Rushworth, M. F. S. (2006). Optimal decision making and the anterior cingulate cortex. *Nature Neuroscience, 9*(7), 940–947. https://doi.org/10.1038/nn1724

Keute, M., Boehrer, L., Ruhnau, P., Heinze, H.-J., & Zaehle, T. (2019). Transcutaneous Vagus nerve stimulation (tVNS) and the dynamics of visual Bistable perception. *Frontiers in Neuroscience, 13*, 227. https://doi.org/10.3389/fnins.2019.00227

Killeen, P. R., Russell, V. A., & Sergeant, J. A. (2013). A behavioral neuroenergetics theory of ADHD. *Neuroscience and Biobehavioral Reviews, 37*(4), 625–657. https://doi.org/10.1016/j.neubiorev.2013.02.011

Kimble, M. O., Fleming, K., Bandy, C., Kim, J., & Zambetti, A. (2010). Eye tracking and visual attention to threating stimuli in veterans of the Iraq war. *Journal of Anxiety Disorders, 24*(3), 293–299. https://doi.org/10.1016/j.janxdis.2009.12.006

Koga, K., Yamada, A., Song, Q., Li, X.-H., Chen, Q.-Y., Liu, R.-H., Ge, J., Zhan, C., Furue, H., Zhuo, M., & Chen, T. (2020). Ascending noradrenergic excitation from the locus coeruleus to the anterior cingulate cortex. *Molecular Brain, 13*(1), 49. https://doi.org/10.1186/s13041-020-00586-5

Köhler, S., Bär, K., & Wagner, G. (2016). Differential involvement of brainstem noradrenergic and midbrain dopaminergic nuclei in cognitive control. *Human Brain Mapping, 37*(6), 2305–2318.

Korchin, S. J. (1964). *Anxiety and cognition* (pp. 58–78). Theory, Research, Promise.

Krebs, R. M., Park, H. R. P., Bombeke, K., & Boehler, C. N. (2018). Modulation of locus coeruleus activity by novel oddball stimuli. *Brain Imaging and Behavior, 12*(2), 577–584. https://doi.org/10.1007/s11682-017-9700-4

Kremen, W. S., Panizzon, M. S., Elman, J. A., Granholm, E. L., Andreassen, O. A., Dale, A. M., Gillespie, N. A., Gustavson, D. E., Logue, M. W., Lyons, M. J., Neale, M. C., Reynolds, C. A., Whitsel, N., & Franz, C. E. (2019). Pupillary dilation responses as a midlife indicator of risk for Alzheimer's disease: Association with Alzheimer's disease polygenic risk. *Neurobiology of Aging, 83*, 114–121. https://doi.org/10.1016/j.neurobiolaging.2019.09.001

Krishnamurthy, K., Nassar, M. R., Sarode, S., & Gold, J. I. (2017). Arousal-related adjustments of perceptual biases optimize perception in dynamic environments. *Nature Human Behaviour, 1*, 0107.

Kucyi, A., & Parvizi, J. (2020). Pupillary dynamics link spontaneous and task-evoked activations recorded directly from human insula. *Journal of Neuroscience, 40*(32), 6207–6218. https://doi.org/10.1523/JNEUROSCI.0435-20.2020

Kukolja, J., Schlapfer, T. E., Keysers, C., Klingmuller, D., Maier, W., Fink, G. R., & Hurlemann, R. (2008). Modeling a negative response bias in the human amygdala by noradrenergic-glucocorticoid interactions. *Journal of Neuroscience, 28*(48), 12868–12876. https://doi.org/10.1523/jneurosci.3592-08.2008

Kukolja, J., Klingmüller, D., Maier, W., Fink, G., & Hurlemann, R. (2011). Noradrenergic-glucocorticoid modulation of emotional memory encoding in the human hippocampus. *Psychological Medicine, 41*(10), 2167–2176.

LaBar, K. S., & Cabeza, R. (2006). Cognitive neuroscience of emotional memory. *Nature Reviews Neuroscience, 7*(1), 54–64.

Lacey, B. C., & Lacey, J. I. (2007). *Studies of heart rate and other bodily processes in sensorimotor behavior*. Routledge.

Laeng, B., Ørbo, M., Holmlund, T., & Miozzo, M. (2011). Pupillary stroop effects. *Cognitive Processing, 12*(1), 13–21. https://doi.org/10.1007/s10339-010-0370-z

Laeng, B., Sirois, S., & Gredebäck, G. (2012). Pupillometry: A window to the preconscious? *Perspectives on Psychological Science, 7*(1), 18–27. https://doi.org/10.1177/1745691611427305

Larsen, R. S., & Waters, J. (2018). Neuromodulatory correlates of pupil dilation. *Frontiers in Neural Circuits, 12*, 21.

Lavie, N., & Tsal, Y. (1994). Perceptual load as a major determinant of the locus of selection in visual attention. *Perception & Psychophysics, 56*(2), 183–197. https://doi.org/10.3758/BF03213897

Lee, T.-H., Baek, J., Lu, Z.-L., & Mather, M. (2014a). How arousal modulates the visual contrast sensitivity function. *Emotion, 14*(5), 978.

Lee, T. H., Sakaki, M., Cheng, R., Velasco, R., & Mather, M. (2014b). Emotional arousal amplifies the effects of biased competition in the brain. *Social Cognitive and Affective Neuroscience, 9*(12), 2067–2077.

Lee, T. H., Greening, S. G., Ueno, T., Clewett, D., Ponzio, A., Sakaki, M., & Mather, M. (2018). Arousal increases neural gain via the locus coeruleus–noradrenaline system in younger adults but not in older adults. *Nature Human Behaviour, 2*(5), 356–366. https://doi.org/10.1038/s41562-018-0344-1

Leong, Y. C., Dziembaj, R., & D'Esposito, M. (2021). Pupil-linked arousal biases evidence accumulation toward desirable percepts during perceptual decision-making. *Psychological Science, 32*(9), 1494–1509. https://doi.org/10.1177/09567976211004547

Leuchs, L., Schneider, M., Czisch, M., & Spoormaker, V. I. (2017). Neural correlates of pupil dilation during human fear learning. *NeuroImage, 147*, 186–197.

Liu, Y., Rodenkirch, C., Moskowitz, N., Schriver, B., & Wang, Q. (2017). Dynamic lateralization of pupil dilation evoked by locus coeruleus activation results from sympathetic, not parasympathetic, contributions. *Cell Reports, 20*(13), 3099–3112. https://doi.org/10.1016/j.celrep.2017.08.094

Loewenfeld, I. E., & Lowenstein, O. (1993). *The pupil: Anatomy, physiology, and clinical applications* (Vol. 2). Wiley-Blackwell.

Madore, K. P., Khazenzon, A. M., Backes, C. W., Jiang, J., Uncapher, M. R., Norcia, A. M., & Wagner, A. D. (2020). Memory failure predicted by attention lapsing and media multitasking. *Nature, 587*(7832), 87–91. https://doi.org/10.1038/s41586-020-2870-z

Magistretti, P. J., & Allaman, I. (2018). Lactate in the brain: From metabolic end-product to signalling molecule. *Nature Reviews Neuroscience, 19*(4), 235–249. https://doi.org/10.1038/nrn.2018.19

Magistretti, P. J., Morrison, J. H., Shoemaker, W. J., Sapin, V., & Bloom, F. E. (1981). Vasoactive intestinal polypeptide induces glycogenolysis in mouse cortical slices: A possible regulatory mechanism for the local control of energy metabolism. *Proceedings of the National Academy of Sciences, 78*(10), 6535–6539.

Mäki-Marttunen, V. (2021). Pupil-based states of brain integration across cognitive states. *Neuroscience, 471*, 61–71. https://doi.org/10.1016/j.neuroscience.2021.07.016

Mana, M. J., & Grace, A. A. (1997). Chronic cold stress alters the basal and evoked electrophysiological activity of rat locus coeruleus neurons. *Neuroscience, 81*(4), 1055–1064. https://doi.org/10.1016/S0306-4522(97)00225-X

Manta, S., El Mansari, M., & Blier, P. (2012). Novel attempts to optimize vagus nerve stimulation parameters on serotonin neuronal firing activity in the rat brain. *Brain Stimulation, 5*(3), 422–429. https://doi.org/10.1016/j.brs.2011.04.005

Manunta, Y., & Edeline, J.-M. (2004). Noradrenergic induction of selective plasticity in the frequency tuning of auditory cortex neurons. *Journal of Neurophysiology, 92*(3), 1445–1463.

Markovic, J., Anderson, A. K., & Todd, R. M. (2014). Tuning to the significant: Neural and genetic processes underlying affective enhancement of visual perception and memory. *Behavioural Brain Research, 259*, 229–241.

Mather, M., & Harley, C. W. (2016). The locus coeruleus: Essential for maintaining cognitive function and the aging brain. *Trends in Cognitive Sciences, 20*(3), 214–226. https://doi.org/10.1016/j.tics.2016.01.001

Mather, M., & Sutherland, M. R. (2011). Arousal-biased competition in perception and memory. *Perspectives on Psychological Science, 6*(2), 114–133. https://doi.org/10.1177/1745691611400234

Mather, M., Clewett, D., Sakaki, M., & Harley, C. W. (2015). Norepinephrine ignites local hot spots of neuronal excitation: How arousal amplifies selectivity in perception and memory. *Behavioral and Brain Sciences, 39*, 1–100.

Mather, M., Huang, R., Clewett, D., Nielsen, S. E., Velasco, R., Tu, K., Han, S., & Kennedy, B. L. (2020). Isometric exercise facilitates attention to salient events in women via the noradrenergic system. *NeuroImage, 210*, 116560. https://doi.org/10.1016/j.neuroimage.2020.116560

Mathôt, S., Siebold, A., Donk, M., & Vitu, F. (2015). Large pupils predict goal-driven eye movements. *Journal of Experimental Psychology: General, 144*(3), 513–521. https://doi.org/10.1037/a0039168

McBurney-Lin, J., Lu, J., Zuo, Y., & Yang, H. (2019). Locus coeruleus-norepinephrine modulation of sensory processing and perception: A focused review. *Neuroscience & Biobehavioral Reviews, 105*, 190–199. https://doi.org/10.1016/j.neubiorev.2019.06.009

McEwen, B. S. (2007). Physiology and neurobiology of stress and adaptation: Central role of the brain. *Physiological Reviews, 87*, 873–904.

McGaugh, J. L. (2000). Memory: A century of consolidation. *Science, 287*, 248–251.

McGaugh, J. L., McIntyre, C. K., & Power, A. E. (2002). Amygdala modulation of memory consolidation: Interaction with other brain systems. *Neurobiology of Learning and Memory, 78*(3), 539–552. https://doi.org/10.1006/nlme.2002.4082

McGinley, M. J., David, S. V., & McCormick, D. A. (2015). Cortical membrane potential signature of optimal states for sensory signal detection. *Neuron, 87*(1), 179–192. https://doi.org/10.1016/j.neuron.2015.05.038

Mckinnon, A. I., Gray, N. S., & Snowden, R. J. (2020). Enhanced emotional response to both negative and positive images in post-traumatic stress disorder: Evidence from pupillometry. *Biological Psychology, 154*, 107922. https://doi.org/10.1016/j.biopsycho.2020.107922

Menon, V., & Uddin, L. Q. (2010). Saliency, switching, attention and control: A network model of insula function. *Brain Structure and Function, 214*(5–6), 655–667. https://doi.org/10.1007/s00429-010-0262-0

Mergenthaler, P., Lindauer, U., Dienel, G. A., & Meisel, A. (2013). Sugar for the brain: The role of glucose in physiological and pathological brain function. *Trends in Neurosciences, 36*(10), 587–597. https://doi.org/10.1016/j.tins.2013.07.001

Minzenberg, M. J., Watrous, A. J., Yoon, J. H., Ursu, S., & Carter, C. S. (2008). Modafinil shifts human locus coeruleus to low-tonic, high-phasic activity during functional MRI. *Science, 322*(5908), 1700–1702. https://doi.org/10.1126/science.1164908

Mittner, M., Hawkins, G. E., Boekel, W., & Forstmann, B. U. (2016). A neural model of mind wandering. *Trends in Cognitive Sciences, 20*(8), 570–578. https://doi.org/10.1016/j.tics.2016.06.004

Morad, Y., Lemberg, H., Yofe, N., & Dagan, Y. (2000). Pupillography as an objective indicator of fatigue. *Current Eye Research, 21*(1), 535–542. https://doi.org/10.1076/0271-3683(200007)2111-ZFT535

Morris, S. K., Granholm, E., Sarkin, A. J., & Jeste, D. V. (1997). Effects of schizophrenia and aging on pupillographic measures of working memory. *Schizophrenia Research, 27*(2), 119–128. https://doi.org/10.1016/S0920-9964(97)00065-0

Moscovitch, M., & Winocur, G. (1995). Frontal lobes, memory, and aging. *Annals of the New York Academy of Sciences, 769*, 119–150.

Mridha, Z., de Gee, J. W., Shi, Y., Alkashgari, R., Williams, J., Suminski, A., Ward, M. P., Zhang, W., & McGinley, M. J. (2021). Graded recruitment of pupil-linked neuromodulation by parametric stimulation of the vagus nerve. *Nature Communications, 12*(1), 1539. https://doi.org/10.1038/s41467-021-21730-2

Munn, B. R., Müller, E. J., Wainstein, G., & Shine, J. M. (2021). The ascending arousal system shapes neural dynamics to mediate awareness of cognitive states. *Nature Communications, 12*(1), 6016. https://doi.org/10.1038/s41467-021-26268-x

Murphy, P. R., Robertson, I. H., Balsters, J. H., & O'Connell, R. G. (2011). Pupillometry and P3 index the locus coeruleus–noradrenergic arousal function in humans. *Psychophysiology, 48*(11), 1532–1543. https://doi.org/10.1111/j.1469-8986.2011.01226.x

Murphy, P. R., O'Connell, R. G., O'Sullivan, M., Robertson, I. H., & Balsters, J. H. (2014). Pupil diameter covaries with BOLD activity in human locus coeruleus. *Human Brain Mapping, 35*(8), 4140–4154. https://doi.org/10.1002/hbm.22466

Naber, M., Frassle, S., Rutishauser, U., & Einhauser, W. (2013). Pupil size signals novelty and predicts later retrieval success for declarative memories of natural scenes. *Journal of Vision, 13*(2), 11–11. https://doi.org/10.1167/13.2.11

Nassar, M. R., Rumsey, K. M., Wilson, R. C., Parikh, K., Heasly, B., & Gold, J. I. (2012). Rational regulation of learning dynamics by pupil-linked arousal systems. *Nature Neuroscience, 15*(7), 1040–1046. https://doi.org/10.1038/nn.3130

Newman, L. A., Korol, D. L., & Gold, P. E. (2011). Lactate produced by glycogenolysis in astrocytes regulates memory processing. *PLoS One, 6*(12), e28427. https://doi.org/10.1371/journal.pone.0028427

Nielsen, S. E., Barber, S. J., Chai, A., Clewett, D. V., & Mather, M. (2015). Sympathetic arousal increases a negative memory bias in young women with low sex hormone levels. *Psychoneuroendocrinology, 62*, 96–106. https://doi.org/10.1016/j.psyneuen.2015.08.001

Nieuwenhuis, S., Gilzenrat, M. S., Holmes, B. D., & Cohen, J. D. (2005). The role of the locus coeruleus in mediating the attentional blink: A neurocomputational theory. *Journal of Experimental Psychology: General, 134*(3), 291–307. https://doi.org/10.1037/0096-3445.134.3.291

Nieuwenhuis, S., De Geus, E. J., & Aston-Jones, G. (2011). The anatomical and functional relationship between the P3 and autonomic components of the orienting response. *Psychophysiology, 48*(2), 162–175. https://doi.org/10.1111/j.1469-8986.2010.01057.x

Nobukawa, S., Shirama, A., Takahashi, T., Takeda, T., Ohta, H., Kikuchi, M., Iwanami, A., Kato, N., & Toda, S. (2021). Identification of attention-deficit hyperactivity disorder based on the complexity and symmetricity of pupil diameter. *Scientific Reports, 11*(1), 8439. https://doi.org/10.1038/s41598-021-88191-x

Papesh, M. H., Goldinger, S. D., & Hout, M. C. (2012). Memory strength and specificity revealed by pupillometry. *International Journal of Psychophysiology, 83*(1), 56–64. https://doi.org/10.1016/j.ijpsycho.2011.10.002

Partala, T., & Surakka, V. (2003). Pupil size variation as an indication of affective processing. *International Journal of Human-Computer Studies, 59*(1), 185–198. https://doi.org/10.1016/S1071-5819(03)00017-X

Pavlenko, V. B., & Kulichenko, A. M. (2003). Self-initiated motor behavioral act-related neuronal activity in the cat locus coeruleus. *Neurophysiology, 35*(1), 29–37. https://doi.org/10.1023/A:1023994205918

Petrides, M. (2005). Lateral prefrontal cortex: Architectonic and functional organization. *Philosophical Transactions of the Royal Society B-Biological Sciences, 360*(1456), 781–795. https://doi.org/10.1098/rstb.2005.1631

Phillips, M. A., Szabadi, E., & Bradshaw, C. M. (2000). Comparison of the effects of clonidine and yohimbine on pupillary diameter at different illumination levels: Short report. *British Journal of Clinical Pharmacology, 50*(1), 65–68. https://doi.org/10.1046/j.1365-2125.2000.00225.x

Piquado, T., Isaacowitz, D., & Wingfield, A. (2010). Pupillometry as a measure of cognitive effort in younger and older adults. *Psychophysiology, 47*(3), 560–569. https://doi.org/10.1111/j.1469-8986.2009.00947.x

Poe, G. R., Foote, S., Eschenko, O., Johansen, J. P., Bouret, S., Aston-Jones, G., Harley, C. W., Manahan-Vaughan, D., Weinshenker, D., Valentino, R., Berridge, C., Chandler, D. J., Waterhouse, B., & Sara, S. J. (2020). Locus coeruleus: A new look at the blue spot. *Nature Reviews Neuroscience, 21*(11), 644–659. https://doi.org/10.1038/s41583-020-0360-9

Preuschoff, K. (2011). Pupil dilation signals surprise: Evidence for noradrenaline's role in decision making. *Frontiers in Neuroscience, 5*, 115. https://doi.org/10.3389/fnins.2011.00115

Pribram, K. H., & McGuinness, D. (1975). Arousal, activation, and effort in the control of attention. *Psychological Review, 82*(2), 116–149. https://doi.org/10.1037/h0076780

Price, J. L., & Amaral, D. G. (1981). An autoradiographic study of the projections of the central nucleus of the monkey amygdala. *The Journal of Neuroscience, 1*(11), 1242–1259.

Qin, S., Hermans, E. J., van Marle, H. J. F., & Fernández, G. (2012). Understanding low reliability of memories for neutral information encoded under stress: Alterations in memory-related activation in the hippocampus and midbrain. *Journal of Neuroscience, 32*(12), 4032–4041. https://doi.org/10.1523/JNEUROSCI.3101-11.2012

Raizada, R. D. S., & Poldrack, R. A. (2008). Challenge-driven attention: Interacting frontal and brainstem systems. *Frontiers in Human Neuroscience, 2*, 44. https://doi.org/10.3389/neuro.09.003.2007

Rajkowski, J. (1993). Correlations between locus coeruleus (LC) neural activity, pupil diameter and behavior in monkey support a role of LC in attention. *Society for Neuroscience, 93*(4), 1263–1270.

Rajkowski, J., Kubiak, P., & Aston-Jones, G. (1994). Locus coeruleus activity in monkey: Phasic and tonic changes are associated with altered vigilance. *Brain Research Bulletin, 35*(5–6), 607–616. https://doi.org/10.1016/0361-9230(94)90175-9

Rajkowski, J., Majczynski, H., Clayton, E., & Aston-Jones, G. (2004). Activation of monkey locus coeruleus neurons varies with difficulty and performance in a target detection task. *Journal of Neurophysiology, 92*(1), 361–371. https://doi.org/10.1152/jn.00673.2003

Rasch, B., Spalek, K., Buholzer, S., Luechinger, R., Boesiger, P., Papassotiropoulos, A., & de Quervain, D.-F. (2009). A genetic variation of the noradrenergic system is related to differential amygdala activation during encoding of emotional memories. *Proceedings of the National Academy of Sciences, 106*(45), 19191–19196.

Reimer, J., McGinley, M. J., Liu, Y., Rodenkirch, C., Wang, Q., McCormick, D. A., & Tolias, A. S. (2016). Pupil fluctuations track rapid changes in adrenergic and cholinergic activity in cortex. *Nature Communications, 7*, 13289.

Reinhard, G., & Lachnit, H. (2002). Differential conditioning of anticipatory pupillary dilation responses in humans. *Biological Psychology, 60*(1), 51–68. https://doi.org/10.1016/S0301-0511(02)00011-X

Reznikov, L. R., Grillo, C. A., Piroli, G. G., Pasumarthi, R. K., Reagan, L. P., & Fadel, J. (2007). Acute stress-mediated increases in extracellular glutamate levels in the rat amygdala: Differential effects of antidepressant treatment. *European Journal of Neuroscience, 25*(10), 3109–3114.

Richter, M., Gendolla, G. H. E., & Wright, R. A. (2016). Three decades of research on motivational intensity theory: What we have learned about effort and what we still don't know. In A. J. Elliot (Ed.), *Advances in motivation science* (Vol. 3, pp. 149–186). Elsevier. https://doi.org/10.1016/bs.adms.2016.02.001

Robbins, T. W. (2000). From arousal to cognition: The integrative position of the prefrontal cortex. In *Progress in brain research* (Vol. 126, pp. 469–483). Elsevier. https://doi.org/10.1016/S0079-6123(00)26030-5

Robbins, T. W., & Arnsten, A. F. T. (2009). The neuropsychopharmacology of fronto-executive function: Monoaminergic modulation. *Annual Review of Neuroscience, 32*, 267–287. https://doi.org/10.1146/annurev.neuro.051508.135535

Robertson, I. H. (2013). A noradrenergic theory of cognitive reserve: Implications for Alzheimer's disease. *Neurobiology of Aging, 34*(1), 298–308.

Robertson, I. H. (2014). A right hemisphere role in cognitive reserve. *Neurobiology of Aging, 35*(6), 1375–1385.

Robison, M. K., & Unsworth, N. (2017). Individual differences in working memory capacity predict learned control over attentional capture. *Journal of Experimental Psychology: Human Perception and Performance, 43*(11), 1912–1924. https://doi.org/10.1037/xhp0000419

Robison, M. K., Coyne, J. T., Sibley, C., Brown, N. L., Neilson, B., & Foroughi, C. (2022). An examination of relations between baseline pupil measures and cognitive abilities. *Psychophysiology, 56*, e14124. https://doi.org/10.1111/psyp.14124

Roozendaal, B., Castello, N. A., Vedana, G., Barsegyan, A., & McGaugh, J. L. (2008). Noradrenergic activation of the basolateral amygdala modulates consolidation of object recognition memory. *Neurobiology of Learning and Memory, 90*(3), 576–579. https://doi.org/10.1016/j.nlm.2008.06.010

Roozendaal, B., McEwen, B. S., & Chattarji, S. (2009). Stress, memory and the amygdala. *Nature Reviews Neuroscience, 10*(6), 423–433. https://doi.org/10.1038/nrn2651

Rued, H. A., Hilmert, C. J., Strahm, A. M., & Thomas, L. E. (2019). The influence of stress on attentional bias to threat: An angry face and a noisy crowd. *Psychonomic Bulletin & Review, 26*(3), 943–950. https://doi.org/10.3758/s13423-018-1538-2

Sadaghiani, S., & D'Esposito, M. (2014). Functional characterization of the cingulo-opercular network in the maintenance of tonic alertness. *Cerebral Cortex, 25*(9), 2763–2773.

Sakaki, M., Fryer, K., & Mather, M. (2014). Emotion strengthens high priority memory traces but weakens low priority memory traces. *Psychological Science, 25*, 387–395. http://pss.sagepub.com/content/25/2/387.abstract

Salgado, H., Kohr, G., & Trevino, M. (2012). Noradrenergic "tone" determines dichotomous control of cortical spike-timing-dependent plasticity. *Scientific Reports, 2*, 7. https://doi.org/10.1038/srep00417

Sara, S. J. (2009). The locus coeruleus and noradrenergic modulation of cognition. *Nature Reviews Neuroscience, 10*(3), 211–223. https://doi.org/10.1038/nrn2573

Sara, S. J. (2015). Locus coeruleus in time with the making of memories. *Current Opinion in Neurobiology, 35*, 87–94.

Sara, S. J., & Bouret, S. (2012). Orienting and reorienting: The locus coeruleus mediates cognition through arousal. *Neuron, 76*(1), 130–141. https://doi.org/10.1016/j.neuron.2012.09.011

Sara, S. J., & Segal, M. (1991). Plasticity of sensory responses of locus coeruleus neurons in the behaving rat: Implications for cognition. *Progress in Brain Research, 88*, 571–585.

Satpute, A. B., Kragel, P. A., Barrett, L. F., Wager, T. D., & Bianciardi, M. (2019). Deconstructing arousal into wakeful, autonomic and affective varieties. *Neuroscience Letters, 693*, 19–28. https://doi.org/10.1016/j.neulet.2018.01.042

Schneider, M., Hathway, P., Leuchs, L., Sämann, P. G., Czisch, M., & Spoormaker, V. I. (2016). Spontaneous pupil dilations during the resting state are associated with activation of the salience network. *NeuroImage, 139*, 189–201. https://doi.org/10.1016/j.neuroimage.2016.06.011

Schneider-Garces, N. J., Gordon, B. A., Brumback-Peltz, C. R., Shin, E., Lee, Y., Sutton, B. P., Maclin, E. L., Gratton, G., & Fabiani, M. (2010). Span, crunch, and beyond: Working memory capacity and the aging brain. *Journal of Cognitive Neuroscience, 22*(4), 655–669. https://doi.org/10.1162/jocn.2009.21230

Schwabe, L., Hoffken, O., Tegenthoff, M., & Wolf, O. T. (2011). Preventing the stress-induced shift from goal-directed to habit action with a – Adrenergic antagonist. *Journal of Neuroscience, 31*(47), 17317–17325. https://doi.org/10.1523/JNEUROSCI.3304-11.2011

Schwabe, L., Tegenthoff, M., Höffken, O., & Wolf, O. T. (2012). Simultaneous glucocorticoid and noradrenergic activity disrupts the neural basis of goal-directed action in the human brain. *The Journal of Neuroscience, 32*(30), 10146–10155.

Schwarz, L. A., & Luo, L. (2015). Organization of the Locus coeruleus-Norepinephrine System. *Current Biology, 25*(21), R1051–R1056. https://doi.org/10.1016/j.cub.2015.09.039

Seeley, W. W., Menon, V., Schatzberg, A. F., Keller, J., Glover, G. H., Kenna, H., Reiss, A. L., & Greicius, M. D. (2007). Dissociable intrinsic connectivity networks for salience processing and executive control. *The Journal of Neuroscience, 27*(9), 2349–2356.

Servan-Schreiber, D., Printz, H., & Cohen, J. D. (1990). A network model of catecholamine effects: Gain, signal-to-noise ratio, and behavior. *Science, 249*(4971), 892–895. https://doi.org/10.1126/science.2392679

Sharon, O., Fahoum, F., & Nir, Y. (2021). Transcutaneous vagus nerve stimulation in humans induces pupil dilation and attenuates alpha oscillations. *The Journal of Neuroscience, 41*(2), 320–330. https://doi.org/10.1523/JNEUROSCI.1361-20.2020

Shenhav, A., Musslick, S., Lieder, F., Kool, W., Griffiths, T. L., Cohen, J. D., & Botvinick, M. M. (2017). Toward a rational and mechanistic account of mental effort. *Annual Review of Neuroscience, 40*(1), 99–124. https://doi.org/10.1146/annurev-neuro-072116-031526

Shine, J. M. (2019). Neuromodulatory influences on integration and segregation in the brain. *Trends in Cognitive Sciences, 23*(7), 572–583. https://doi.org/10.1016/j.tics.2019.04.002

Shine, J. M., Bissett, P. G., Bell, P. T., Koyejo, O., Balsters, J. H., Gorgolewski, K. J., Moodie, C. A., & Poldrack, R. A. (2016). The dynamics of functional brain networks: Integrated network states during cognitive task performance. *Neuron, 92*(2), 544–554. https://doi.org/10.1016/j.neuron.2016.09.018

Shine, J. M., Aburn, M. J., Breakspear, M., & Poldrack, R. A. (2018a). The modulation of neural gain facilitates a transition between functional segregation and integration in the brain. *eLife, 7*, e31130. https://doi.org/10.7554/eLife.31130

Shine, J. M., van den Brink, R. L., Hernaus, D., Nieuwenhuis, S., & Poldrack, R. A. (2018b). Catecholaminergic manipulation alters dynamic network topology across cognitive states. *Network Neuroscience, 2*(3), 381–396. https://doi.org/10.1162/netn_a_00042

Siegle, G. J., Granholm, E., Ingram, R. E., & Matt, G. E. (2001). Pupillary and reaction time measures of sustained processing of negative information in depression. *Biological Psychiatry, 49*(7), 624–636. https://doi.org/10.1016/S0006-3223(00)01024-6

Simpson, H. M., & Hale, S. M. (1969). Pupillary changes during a decision-making task. *Perceptual and Motor Skills, 29*(2), 495–498. https://doi.org/10.2466/pms.1969.29.2.495

Sorg, O., & Magistretti, P. J. (1991). Characterization of the glycogenolysis elicited by vasoactive intestinal peptide, noradrenaline and adenosine in primary cultures of mouse cerebral cortical astrocytes. *Brain Research, 563*(1), 227–233.

Steinhauer, S. R., & Hakerem, G. (1992). The pupillary response in cognitive psychophysiology and schizophrenia. *Annals of the New York Academy of Sciences, 658*(1), 182–204.

Steinhauer, S. R., Siegle, G. J., Condray, R., & Pless, M. (2004). Sympathetic and parasympathetic innervation of pupillary dilation during sustained processing. *International Journal of Psychophysiology, 52*(1), 77–86.

Steinhauer, S. R., Bradley, M. M., Siegle, G. J., Roecklein, K. A., & Dix, A. (2022). Publication guidelines and recommendations for pupillary measurement in psychophysiological studies. *Psychophysiology, 59*(4), e14035. https://doi.org/10.1111/psyp.14035

Sterpenich, V., D'Argembeau, A., Desseilles, M., Balteau, E., Albouy, G., Vandewalle, G., Degueldre, C., Luxen, A., Collette, F., & Maquet, P. (2006). The locus ceruleus is involved in the successful retrieval of emotional memories in humans. *Journal of Neuroscience, 26*(28), 7416–7423. https://doi.org/10.1523/JNEUROSCI.1001-06.2006

Strange, B. A., & Dolan, R. J. (2004). Beta-adrenergic modulation of emotional memory-evoked human amygdala and hippocampal responses. *Proceedings of the National Academy of Sciences of the United States of America, 101*(31), 11454–11458.

Sutherland, M. R., & Mather, M. (2012). Negative arousal amplifies the effects of saliency in short-term memory. *Emotion, 12*, 1367–1372. https://doi.org/10.1037/a0027860

Szabadi, E. (2018). Functional organization of the sympathetic pathways controlling the pupil: Light-inhibited and light-stimulated pathways. *Frontiers in Neurology, 9*, 1069. https://doi.org/10.3389/fneur.2018.01069

Tervo, D. G. R., Proskurin, M., Manakov, M., Kabra, M., Vollmer, A., Branson, K., & Karpova, A. Y. (2014). Behavioral variability through stochastic choice and its gating by anterior cingulate cortex. *Cell, 159*(1), 21–32. https://doi.org/10.1016/j.cell.2014.08.037

Thayer, R. E. (1978). Toward a psychological theory of multidimensional activation (arousal). *Motivation and Emotion, 2*(1), 1–34. https://doi.org/10.1007/BF00992729

Todd, R. M., Schmitz, T. W., Susskind, J., & Anderson, A. K. (2013). Shared neural substrates of emotionally enhanced perceptual and mnemonic vividness. *Frontiers in Behavioral Neuroscience, 7*, 25. https://doi.org/10.3389/fnbeh.2013.00040

Totah, N. K., Neves, R. M., Panzeri, S., Logothetis, N. K., & Eschenko, O. (2018). The locus coeruleus is a complex and differentiated neuromodulatory system. *Neuron, 99*(5), 1055–1068.e6. https://doi.org/10.1016/j.neuron.2018.07.037

Tromp, J., Nieuwenhuis, S., & Murphy, P. (2022). The effects of neural gain on reactive cognitive control. *Computational Brain & Behavior., 5*(3), 422–433. https://doi.org/10.1007/s42113-022-00140-7

Tsukahara, J. S., & Engle, R. W. (2021). Is baseline pupil size related to cognitive ability? Yes (under proper lighting conditions). *Cognition, 211*, 104643. https://doi.org/10.1016/j.cognition.2021.104643

Tsukahara, J. S., Harrison, T. L., & Engle, R. W. (2016). The relationship between baseline pupil size and intelligence. *Cognitive Psychology, 91*, 109–123. https://doi.org/10.1016/j.cogpsych.2016.10.001

Tully, K., & Bolshakov, V. Y. (2010). Emotional enhancement of memory: How norepinephrine enables synaptic plasticity. *Molecular Brain, 3*(1), 15.

Uematsu, A., Tan, B. Z., Ycu, E. A., Cuevas, J. S., Koivumaa, J., Junyent, F., Kremer, E. J., Witten, I. B., Deisseroth, K., & Johansen, J. P. (2017). Modular organization of the brainstem noradrenaline system coordinates opposing learning states. *Nature Neuroscience, 20*(11), 1602–1611. https://doi.org/10.1038/nn.4642

Unsworth, N., & Miller, A. L. (2021). Individual differences in the intensity and consistency of attention. *Current Directions in Psychological Science, 30*(5), 391–400. https://doi.org/10.1177/09637214211030266

Unsworth, N., & Robison, M. K. (2016). Pupillary correlates of lapses of sustained attention. *Cognitive, Affective, & Behavioral Neuroscience, 16*(4), 601–615. https://doi.org/10.3758/s13415-016-0417-4

Unsworth, N., & Robison, M. K. (2017a). A locus coeruleus-norepinephrine account of individual differences in working memory capacity and attention control. *Psychonomic Bulletin & Review, 24*(4), 1282–1311. https://doi.org/10.3758/s13423-016-1220-5

Unsworth, N., & Robison, M. K. (2017b). The importance of arousal for variation in working memory capacity and attention control: A latent variable pupillometry study. *Journal of*

Experimental Psychology: Learning, Memory, and Cognition, 43(12), 1962–1987. https://doi.org/10.1037/xlm0000421

Unsworth, N., & Robison, M. K. (2020). Working memory capacity and sustained attention: A cognitive-energetic perspective. *Journal of Experimental Psychology: Learning, Memory, and Cognition, 46*(1), 77–103. https://doi.org/10.1037/xlm0000712

Unsworth, N., Robison, M. K., & Miller, A. L. (2019). Individual differences in baseline oculometrics: Examining variation in baseline pupil diameter, spontaneous eye blink rate, and fixation stability. *Cognitive, Affective, & Behavioral Neuroscience, 19*(4), 1074–1093. https://doi.org/10.3758/s13415-019-00709-z

Urai, A. E., Braun, A., & Donner, T. H. (2017). Pupil-linked arousal is driven by decision uncertainty and alters serial choice bias. *Nature Communications, 8*(1), 14637. https://doi.org/10.1038/ncomms14637

Usher, M., Cohen, J. D., Servan-Schreiber, D., Rajkowski, J., & Aston-Jones, G. (1999). The role of locus coeruleus in the regulation of cognitive performance. *Science, 283*(5401), 549–554. https://doi.org/10.1126/science.283.5401.549

Vaishnavi, S. N., Vlassenko, A. G., Rundle, M. M., Snyder, A. Z., Mintun, M. A., & Raichle, M. E. (2010). Regional aerobic glycolysis in the human brain. *Proceedings of the National Academy of Sciences, 107*(41), 17757–17762. https://doi.org/10.1073/pnas.1010459107

Valentino, R. J., & Van Bockstaele, E. (2008). Convergent regulation of locus coeruleus activity as an adaptive response to stress. *European Journal of Pharmacology, 583*(2–3), 194–203. https://doi.org/10.1016/j.ejphar.2007.11.062

Valentino, R. J., Curtis, A. L., Page, M. E., Pavcovich, L. A., & Florin-Lechner, S. M. (1998). Activation of the locus ceruleus brain noradrenergic system during stress: Circuitry, consequences, and regulation. *Advances in Pharmacology, 42*, 781–784. https://doi.org/10.1016/s1054-3589(08)60863-7

Van Den Brink, R. L., Murphy, P. R., & Nieuwenhuis, S. (2016). Pupil diameter tracks lapses of attention. *PLoS One, 11*(10), e0165274. https://doi.org/10.1371/journal.pone.0165274

van der Wel, P., & van Steenbergen, H. (2018). Pupil dilation as an index of effort in cognitive control tasks: A review. *Psychonomic Bulletin & Review, 25*(6), 2005–2015. https://doi.org/10.3758/s13423-018-1432-y

van Marle, H. J. F., Hermans, E. J., Qin, S. Z., & Fernandez, G. (2009). From specificity to sensitivity: How acute stress affects amygdala processing of biologically salient stimuli. *Biological Psychiatry, 66*(7), 649–655. https://doi.org/10.1016/j.biopsych.2009.05.014

van Marle, H. J. F., Hermans, E. J., Qin, S., & Fernández, G. (2010). Enhanced resting-state connectivity of amygdala in the immediate aftermath of acute psychological stress. *NeuroImage, 53*(1), 348–354. https://doi.org/10.1016/j.neuroimage.2010.05.070

van Stegeren, A. H., Roozendaal, B., Kindt, M., Wolf, O. T., & Joëls, M. (2010). Interacting noradrenergic and corticosteroid systems shift human brain activation patterns during encoding. *Neurobiology of Learning and Memory, 93*(1), 56–65. https://doi.org/10.1016/j.nlm.2009.08.004

Varazzani, C., San-Galli, A., Gilardeau, S., & Bouret, S. (2015). Noradrenaline and dopamine neurons in the reward/effort trade-off: A direct electrophysiological comparison in behaving monkeys. *Journal of Neuroscience, 35*(20), 7866–7877. https://doi.org/10.1523/JNEUROSCI.0454-15.2015

Vazey, E. M., Moorman, D. E., & Aston-Jones, G. (2018). Phasic locus coeruleus activity regulates cortical encoding of salience information. *Proceedings of the National Academy of Sciences, 115*(40), E9439–E9448. https://doi.org/10.1073/pnas.1803716115

Verney, S. P., Granholm, E., & Marshall, S. P. (2004). Pupillary responses on the visual backward masking task reflect general cognitive ability. *International Journal of Psychophysiology, 52*(1), 23–36.

Võ, M. L.-H., Jacobs, A. M., Kuchinke, L., Hofmann, M., Conrad, M., Schacht, A., & Hutzler, F. (2007). The coupling of emotion and cognition in the eye: Introducing the pupil old/new effect. *Psychophysiology, 45*, 071003012229007. https://doi.org/10.1111/j.1469-8986.2007.00606.x

Wachtel, P. L. (1967). Conceptions of broad and narrow attention. *Psychological Bulletin, 68*(6), 417–429. https://doi.org/10.1037/h0025186

Wainstein, G., Rojas-Líbano, D., Crossley, N. A., Carrasco, X., Aboitiz, F., & Ossandón, T. (2017). Pupil size tracks attentional performance in attention-deficit/hyperactivity disorder. *Scientific Reports, 7*(1), 8228. https://doi.org/10.1038/s41598-017-08246-w

Wainstein, G., Müller, E. J., Taylor, N., Munn, B., & Shine, J. M. (2022). The role of the locus coeruleus in shaping adaptive cortical melodies. *Trends in Cognitive Sciences, 26*(6), 527–538. https://doi.org/10.1016/j.tics.2022.03.006

Warren, C. M., Eldar, E., van den Brink, R. L., Tona, K.-D., van der Wee, N. J., Giltay, E. J., van Noorden, M. S., Bosch, J. A., Wilson, R. C., Cohen, J. D., & Nieuwenhuis, S. (2016). Catecholamine-mediated increases in gain enhance the precision of cortical representations. *Journal of Neuroscience, 36*(21), 5699–5708. https://doi.org/10.1523/JNEUROSCI.3475-15.2016

Waterhouse, B. D., & Navarra, R. L. (2019). The locus coeruleus-norepinephrine system and sensory signal processing: A historical review and current perspectives. *Brain Research, 1709*, 1–15. https://doi.org/10.1016/j.brainres.2018.08.032

Waterhouse, B. D., Moises, H. C., & Woodward, D. J. (1980). Noradrenergic modulation of somatosensory cortical neuronal responses to lontophoretically applied putative neurotransmitters. *Experimental Neurology, 69*(1), 30–49. https://doi.org/10.1016/0014-4886(80)90141-7

Wetzel, N., Buttelmann, D., Schieler, A., & Widmann, A. (2016). Infant and adult pupil dilation in response to unexpected sounds. *Developmental Psychobiology, 58*(3), 382–392.

Weymar, M., Schwabe, L., Löw, A., & Hamm, A. O. (2012). Stress sensitizes the brain: Increased processing of unpleasant pictures after exposure to acute stress. *Journal of Cognitive Neuroscience, 24*(7), 1511–1518. https://doi.org/10.1162/jocn_a_00174

Widmann, A., Schröger, E., & Wetzel, N. (2018). Emotion lies in the eye of the listener: Emotional arousal to novel sounds is reflected in the sympathetic contribution to the pupil dilation response and the P3. *Biological Psychology, 133*, 10–17.

Wilson, R. S., Nag, S., Boyle, P. A., Hizel, L. P., Yu, L., Buchman, A. S., Schneider, J. A., & Bennett, D. A. (2013). Neural reserve, neuronal density in the locus ceruleus, and cognitive decline. *Neurology, 80*(13), 1202–1208.

Woodhead, M. M. (1964). Searching a visual display in intermittent noise. *Journal of Sound and Vibration, 1*(2), 157–161. https://doi.org/10.1016/0022-460X(64)90077-X

Yebra, M., Galarza-Vallejo, A., Soto-Leon, V., Gonzalez-Rosa, J. J., de Berker, A. O., Bestmann, S., Oliviero, A., Kroes, M. C. W., & Strange, B. A. (2019). Action boosts episodic memory encoding in humans via engagement of a noradrenergic system. *Nature Communications, 10*(1), 3534. https://doi.org/10.1038/s41467-019-11358-8

Yerkes, R. M., & Dodson, J. D. (1908). The relation of strength of stimulus to rapidity of habit-formation. *Journal of Comparative Neurology and Psychology, 18*(5), 459–482. https://doi.org/10.1002/cne.920180503

Young, C. B., Raz, G., Everaerd, D., Beckmann, C. F., Tendolkar, I., Hendler, T., Fernández, G., & Hermans, E. J. (2017). Dynamic shifts in large-scale brain network balance as a function of arousal. *Journal of Neuroscience, 37*(2), 281–290. https://doi.org/10.1523/JNEUROSCI.1759-16.2016

Zekveld, A. A., Koelewijn, T., & Kramer, S. E. (2018). The pupil dilation response to auditory stimuli: Current state of knowledge. *Trends in Hearing, 22*, 2331216518777174. https://doi.org/10.1177/2331216518777174

Zénon, A. (2019). Eye pupil signals information gain. *Proceedings of the Royal Society B: Biological Sciences, 286*(1911), 20191593. https://doi.org/10.1098/rspb.2019.1593

Zerbi, V., Floriou-Servou, A., Markicevic, M., Vermeiren, Y., Sturman, O., Privitera, M., von Ziegler, L., Ferrari, K. D., Weber, B., De Deyn, P. P., Wenderoth, N., & Bohacek, J. (2019). Rapid reconfiguration of the functional connectome after chemogenetic locus coeruleus activation. *Neuron, 103*(4), 702–718.e5. https://doi.org/10.1016/j.neuron.2019.05.034

Zhao, S., Chait, M., Dick, F., Dayan, P., Furukawa, S., & Liao, H.-I. (2019). Pupil-linked phasic arousal evoked by violation but not emergence of regularity within rapid sound sequences. *Nature Communications, 10*(1), 4030. https://doi.org/10.1038/s41467-019-12048-1

Zoladz, P. R., Kalchik, A. E., Hoffman, M. M., Aufdenkampe, R. L., Lyle, S. M., Peters, D. M., Brown, C. M., Cadle, C. E., Scharf, A. R., Dailey, A. M., Wolters, N. E., Talbot, J. N., & Rorabaugh, B. R. (2014). ADRA2B deletion variant selectively predicts stress-induced enhancement of long-term memory in females. *Psychoneuroendocrinology, 48*, 111–122. https://doi.org/10.1016/j.psyneuen.2014.06.012

Zuend, M., Saab, A. S., Wyss, M. T., Ferrari, K. D., Hösli, L., Looser, Z. J., Stobart, J. L., Duran, J., Guinovart, J. J., Barros, L. F., & Weber, B. (2020). Arousal-induced cortical activity triggers lactate release from astrocytes. *Nature Metabolism, 2*(2), 179–191. https://doi.org/10.1038/s42255-020-0170-4

Linking the Superior Colliculus to Pupil Modulation

Chin-An Wang and Douglas P. Munoz

1 Introduction

Pupil size is controlled by the balance activity between the parasympathetic and sympathetic systems (Loewenfeld, 1999; McDougal & Gamlin, 2015; May et al., 2019) and is increasingly popular to be used to index cognitive and neural processing (Einhäuser, 2017; Mathôt, 2018; Ebitz & Moore, 2019). The superior colliculus (SC) is a sensorimotor structure that is causally involved in shifts of gaze and attention (Gandhi & Katnani, 2011; Krauzlis et al., 2013). In addition to luminance and arousal signals modulating the pupil (Fig. 1), mounting evidence has demonstrated that the SC also plays a key role in the control of pupil size (Wang & Munoz, 2015). The SC is an important hub for visual orienting that includes saccades and pupil dilation (Corneil & Munoz, 2014). The orienting response serves to heighten perception to prepare the body for appropriate action required for survival (Sokolov, 1963a; Lynn, 1966). It is therefore essential for the brain to consider and integrate various signals to properly guide visual orienting. Correspondingly, the SC receives multisensory, cognitive, and arousal signals from various cortical and subcortical regions. These signals are integrated into the SC to initiate coordinated orienting responses. Here, we review the evidence linking the SC to multisensory-, cognitive-, and arousal-related pupillary responses, focusing on our recent work in monkeys and humans, and argue that the SC provides a neural substrate for pupillary responses to multisensory, cognitive, and arousal processes.

C.-A. Wang (✉)
Department of Anesthesiology, School of Medicine, College of Medicine, Taipei Medical University, Taipei City, Taiwan

Department of Anesthesiology, Shuang Ho Hospital, Taipei Medical University, New Taipei City, Taiwan
e-mail: josh.wang@tmu.edu.tw

D. P. Munoz
Centre for Neuroscience Studies, Queen's University, Kingston, ON, Canada

M. H. Papesh, S. D. Goldinger (eds.), *Modern Pupillometry*,
https://doi.org/10.1007/978-3-031-54896-3_2

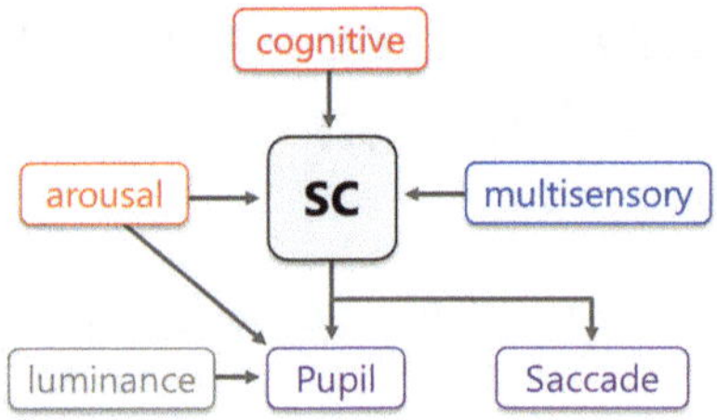

Fig. 1 Schematic depiction of important pupil control signals and the role of the superior colliculus (SC) in coordinating saccade and pupil behavior

1.1 Linking the SC to Pupil Size

The SC is a midbrain structure with neurons organized into a retinotopically coded map of contralateral visual and saccade space. The SC is functionally and anatomically differentiated into superficial layers (SCs) that receive direct inputs from the retina and visual cortex including V1, V2, V3, V4, and middle temporal area, and intermediate/deep layers (SCi) that receive convergent multisensory, cognitive, and arousal inputs from various cortical and subcortical areas including the frontal eye field (FEF), dorsolateral prefrontal cortex, supplementary eye field, lateral intraparietal area (or parietal eye field in humans), anterior cingulate cortex, basal ganglia, and the locus coeruleus (LC) (Edwards et al., 1979; White & Munoz, 2011). Moreover, the SCi projects directly to the brainstem premotor circuit (Sparks, 2002; Horn, 2006; May, 2006) to execute orienting responses including eye/head/body movement, attention shifts, and pupil responses (Hess et al., 1946; Akert, 1949; Sokolov, 1963b; Krauzlis et al., 2013; Corneil & Munoz, 2014).

The SC also connects to the pupil control pathway to provide the necessary connections to change the pupil size. The pretectal olivary nucleus (PON) receives inputs from the retina and the SCs (Harting et al., 1980). Furthermore, the central mesencephalic reticular formation (cMRF or nucleus cuneiformis in humans) is one of the collicular targets involved in horizontal saccade control (Cohen & Büttner-Ennever, 1984), and research has revealed that cMRF also projects directly to the Edinger–Westphal nucleus (May et al., 2016), a critical structure in the parasympathetic pupil control system (Corneil & Munoz, 2014; Wang & Munoz, 2015). In addition, the SCi projects directly and indirectly via the cMRF to the medullary reticular formation (Perkins et al., 2009) and so may influence preganglionic sympathetic neurons via this route.

The relationship between the SC and pupil size has been demonstrated in recent studies (Wang et al., 2012; Joshi et al., 2016; Wang & Munoz, 2021a). To maintain the accuracy of pupil size measurement, subthreshold current was used to electrically stimulate the SCi without evoking saccades (Wang et al., 2012). As shown in Fig. 2a, even with subthreshold (for saccades) stimulation of the SCi (caudal SC), pupil dilation was transiently evoked. To examine the effects of suprathreshold stimulation, the rostral SC was stimulated to evoke very small amplitude saccades (<1.5°) without diminishing the accuracy of pupil size measurement during and after microstimulation (Wang et al., 2012). Similar pupil dilation was observed after

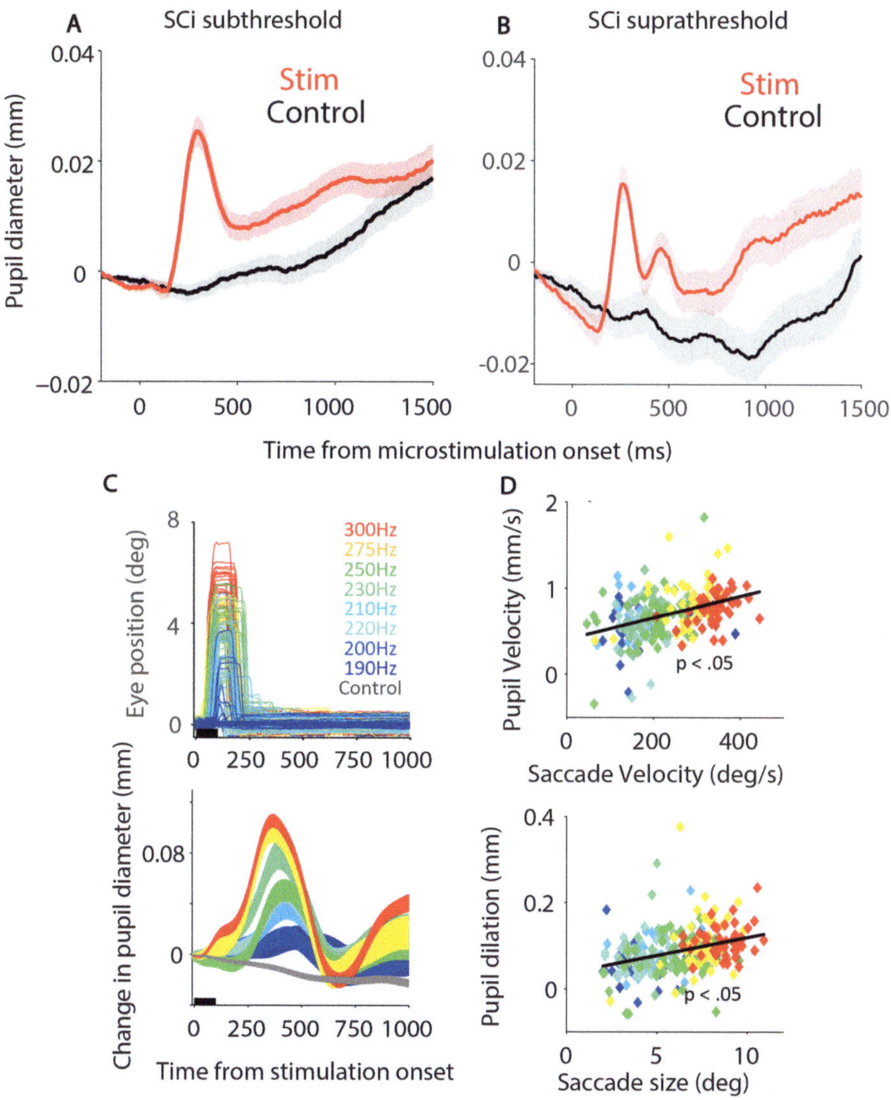

Fig. 2 Effect of SCi microstimulation on saccade and pupil responses. (**a**) Pupil responses following SCi subthreshold microstimulation. (Adapted with permission Wang et al., 2012). (**b**) Pupil responses following SCi suprathreshold microstimulation. (**c**) Effects of stimulation frequency on saccade and pupil responses. (Adapted with permission Wang & Munoz, 2021a). (**d**) Trial-by-trial correlation between normalized pupil and saccade responses in velocity and magnitude. In **a**–**c**, the shaded colored regions surrounding the pupil response represent ±standard error range for different conditions. In **c**, the gray bar on the x-axis indicates the timeline of stimulation (100 ms). Control: control (no stimulation). Stim: microstimulation

microstimulation of the rostral SCi (Fig. 2b). These results together suggest that both the rostral and caudal SC can influence pupil size.

Although the role of the SC on saccade and pupil dilation has been established, whether and how these responses are coordinated remain unknown. More specifically, whether the same (or different) efferent projections from the SCi drive both saccades and pupil dilation (referred to as the common drive hypothesis) is yet to be determined. If the mechanisms of the initiation of saccade and pupil dilation are shared through the SCi, then the comparable modulation of saccade and pupillary responses should be produced from SCi microstimulation. That is, if there are stronger output signals, these signals should evoke larger saccade and pupillary responses. Furthermore, pupil dilation on trials with an evoked saccade should be larger than that on trials without an evoked saccade because signals must surpass the threshold to evoke saccades (Scudder et al., 2002). Using SCi microstimulation with a systematic variation of stimulation frequency (Wang & Munoz, 2021a), it was shown that stimulation frequency modulated the evoked saccade as well as the pupillary responses, with larger saccade and pupillary responses observed for higher stimulation frequencies (Fig. 2c). Trials with saccades were associated with larger pupil velocities and greater dilation, and trial-by-trial correlations were obtained between saccade and pupillary responses, with larger saccade responses also having larger pupillary responses (Fig. 2d). Furthermore, microstimulation of SCi sites, which produced larger amplitude saccades, also evoked larger pupil dilation. These results together demonstrate saccade and pupil movements coordinated through the SCi, likely via a shared common output signal derived from the SCi.

1.2 Bottom-Up Saliency Modulation Between the SC and Pupil Size

An increasing number of studies have suggested that the SCi encodes both stimulus salience and relevance to coordinate various components of orienting (Fecteau & Munoz, 2006; Boehnke & Munoz, 2008; Mysore & Knudsen, 2013), including not only saccades and attention shifts but also pupil dilation (Corneil & Munoz, 2014; Netser et al., 2014; Wang & Munoz, 2015; Wang et al., 2020). Visual contrast, one of the most primitive components of visual salience (Koch & Ullman, 1985; Itti et al., 1998; Itti & Koch, 2001), has pronounced effects on visual responses in the SC and saccade behavior in behaving monkeys (Marino et al., 2012, 2015). Recent studies that investigated the effects of contrast-based saliency on saccade and pupillary responses have further linked these two responses (Wang & Munoz, 2014; Wang et al., 2014). The visual contrast of the stimulus was manipulated while maintaining luminance across contrast levels (Wang & Munoz, 2014). Subjects were required to make a saccade toward a visual target after the disappearance of the central fixation point in the saccade condition but maintained central fixation in the fixation condition (target presented briefly or no stimulus presented in the control

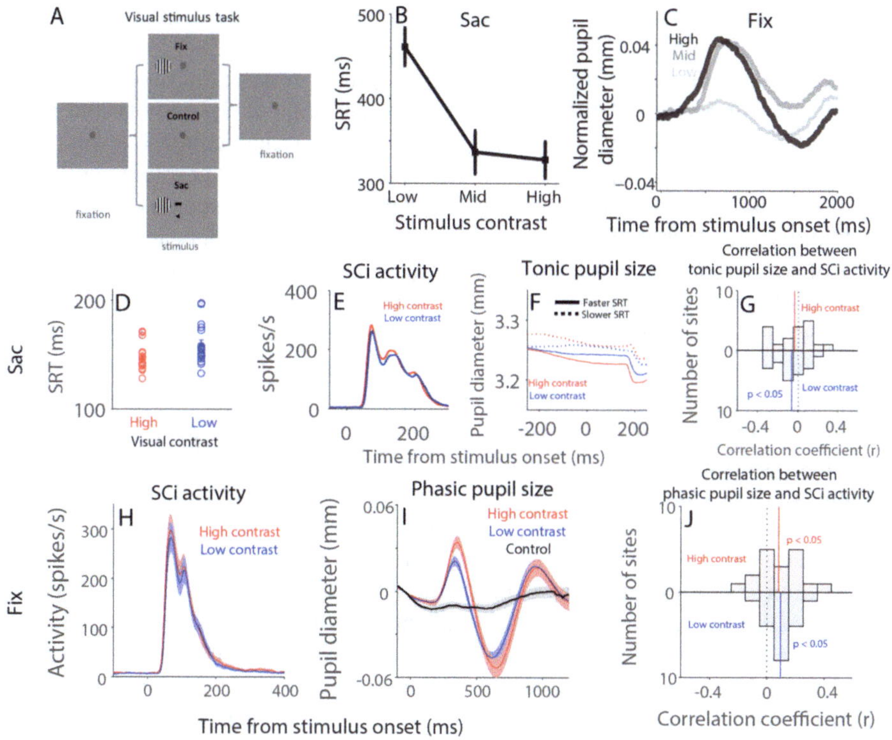

Fig. 3 Effect of bottom-up saliency on saccade, pupil, and SCi responses. (**a**) Schematic of the paradigm (see text for details). (**b**) Effects of stimulus contrast on SRT. (**c**) Effects of stimulus contrast on pupil responses. (Adapted with permission Wang & Munoz, 2014). (**d**, **e**) Effects of stimulus contrast on monkey (**d**) SRT and (**e**) SCi activity. (**f**) Effects of SRT on tonic pupil size. (**g**) Distribution of correlation coefficients for the relationship between tonic pupil size (−50 to 50 ms of stimulus onset) and phasic SCi activity (55–85 ms) in all recording sites. (**h**, **i**) Effects of stimulus contrast on (**h**) SCi activity and (**i**) pupil responses. (**j**) Distribution of correlation coefficients for the relationship between SCi activity (55–85 ms) and phasic pupil responses (305–350 ms) in all recording sites. In **b**, **c**, the error bar represents ±standard error across participants. In **d**, the dark color dots and circles represent the average and individual recording session values of measurements, respectively. In **g**, **j**, the vertical black dotted and colored solid line represent a zero and mean value of the correlation coefficient, respectively. In **h**, **i**, the shaded colored regions surrounding the response represent ±standard error range for different conditions. Fix: fixation condition. Sac: saccade condition. Control: no stimulus fixation condition. SRT: saccade reaction times

condition; Fig. 3a). The modulation of stimulus contrast in humans was manifested not only in saccade reaction times (SRTs) in the saccade condition (Fig. 3b) but also in evoked pupil dilation in the fixation condition (Fig. 3c), with faster and larger evoked pupillary responses elicited by higher contrast stimuli. In accordance with the saliency model (Koch & Ullman, 1985; Itti et al., 1998; Itti & Koch, 2001), the saliency effect on pupil size has been extended to using a pop-out paradigm or varying set size (Wang et al., 2020; Oster et al., 2022). These results demonstrate similar saliency modulation between saccade and pupillary responses evoked by visual stimuli.

The relationship between SC activity and pupillary responses evoked by salient stimuli was further revealed by simultaneous recordings of SC activity and pupil size in behaving monkeys. Stimulus saliency (two visual contrast levels) was manipulated using the same paradigm. The tonic (pre-stimulus absolute pupil size) and phasic (stimulus-evoked pupillary responses) pupillary responses were examined in the saccade and fixation conditions, respectively, to understand the relation between SC activity and pupil size. Similar to previous studies (Marino et al., 2012; Wang & Munoz, 2014; Wang et al., 2014), faster SRTs and larger SCi activity were observed using higher contrast stimuli in the saccade condition (Fig. 3d, e). Interestingly, smaller tonic pupil responses were observed in trials with faster SRTs (Fig. 3f). To explore whether these tonic pupil responses also affected SC visual activity, we examined trial-by-trial correlations between tonic pupil size (around stimulus onset) and phasic SCi activity. Consistent with the adaptive gain theory (Aston-Jones & Cohen, 2005), negative correlations between tonic pupil size and phasic SCi activity were observed, showing smaller tonic pupil sizes linking larger SCi activity, particularly in the low visual contrast condition (Fig. 3g). Moreover, in the fixation condition, stimulus contrast modulated neuronal (Fig. 3h) as well as pupillary (Fig. 3i) responses evoked by salient stimuli, with larger SCi activity and pupillary responses observed in the high contrast condition. More importantly, correlational analysis between phasic SCi and pupillary responses demonstrated positive trial-by-trial correlations, that is, trials with larger SCi activity had larger pupil dilation (Fig. 3j). These results implicate that tonic pupil sizes modulated SC visual activity, and phasic pupillary responses evoked by salient stimuli are mediated by the SC and further support the common-drive hypothesis, that is, the same efferent from the SCi drives both saccade and pupillary responses.

1.3 Cognitive Modulation Between the SC and Pupil Size

1.3.1 Role of the SC and FEF in Pupil Size Modulations Associated with Executive Control

Executive control is a central mechanism that modulates various aspects of cognition. An interleaved pro- and anti-saccade task, in which subjects are instructed, prior to target stimulus appearance, to either automatically look at the peripheral target stimulus (pro-saccade) or to suppress the automatic response and voluntarily look in the opposite direction from the target stimulus (anti-saccade), has been greatly used to examine executive function (Munoz & Everling, 2004). Research has shown that distinct neural preparatory activity, particularly in the SC and FEF, is required before the appearance of the target to successfully generate voluntary or automatic behavior in the pro- and anti-saccade tasks, as the SC, together with the FEF, is causally involved in the control of eye movements (Robinson & Fuchs, 1969; Schiller et al., 1979, 1980; Bruce & Goldberg, 1985; Bruce et al., 1985;

Goldberg & Bruce, 1990). Neuronal recordings in behaving monkeys have shown higher inhibition-related fixation activity in preparation for anti- compared to pro-saccades (Fig. 4a) (Everling et al., 1999; Everling & Munoz, 2000). Moreover, the level of preparatory activity related to motor preparation (saccade-related activity) was higher during pro-saccade preparation and negatively correlated with SRTs (Fig. 4b). Similarly, in human fMRI studies, there was higher FEF blood-oxygen-level-dependent (BOLD) activation during preparation for anti-saccades compared to pro-saccades (Fig. 4c) (Connolly et al., 2002; DeSouza et al., 2003; Manoach et al., 2007), and preparatory FEF activity negatively correlated with SRTs (Alahyane et al., 2014; Connolly et al., 2002).

Because pupil dilation is evoked by microstimulation of both rostral (Fig. 2b) and caudal (Fig. 2a) SC in monkeys, and pupil dilation responses have been obtained with FEF microstimulation (Lehmann & Corneil, 2016; Ebitz & Moore, 2017), pupil size should reflect both types of preparatory activity (Wang et al., 2012). Among human subjects performing pro- and anti-saccades, pupil size was larger in preparation for anti-saccades compared to pro-saccades (Fig. 4d) (Wang et al., 2015). Furthermore, larger pupil dilation prior to stimulus appearance accompanied saccades with faster reaction times. Consistently, this correlation was disrupted with continuous theta-burst stimulation over the FEF in humans (Hsu et al., 2021), demonstrating the role of the FEF in this modulation. Moreover, comparable modulations between pupil dilation and visual stimulus contrast were obtained on saccade latencies and metrics (Wang et al., 2021). As SC activity is systematically modulated by stimulus contrast and correlates with changes in saccade latency and metrics (Marino et al., 2012, 2015), these pupil results reveal that pupil dilation correlates with SC activity. Overall, these results are consistent with human fMRI activation patterns in the frontal areas such as FEF and dorsolateral prefrontal cortex (Cameron et al., 2012; Fernandez-Ruiz et al., 2018) and suggest that pupil size is an effective proxy of neural activity related to executive functioning in the preparation on pro- and anti-saccade generation.

Moreover, these saccadic and pupillary results have been extended to patients with Parkinson's disease (PD) (Wang et al., 2016; Perkins et al., 2021). PD is characterized by motor symptoms attributed to the loss of dopaminergic neurons in the pars compacta of the substantia nigra (Greenfield & Bosanquet, 1953). Converging evidence has found cognitive impairment in patients with PD at or soon after diagnosis (Muslimovic et al., 2005; Rodriguez-Oroz et al., 2009; Leh et al., 2010; Antoniades et al., 2015). Consistently, in a human fMRI study (Cameron et al., 2012), greater FEF activation during anti-saccade than pro-saccade preparation was observed in age-matched controls but not in Parkinson's disease. As shown in Fig. 4e, controls, but not patients with Parkinson's disease, revealed enhanced activation in FEF for correct compared to erroneous anti-saccade trials. Consistent with human fMRI results, pupil dilation during anti-saccade preparation was reduced in patients with Parkinson's disease (Fig. 4f) compared with controls, and enhanced pupil dilation in correct anti-saccade trials was only observed in controls (Wang et al., 2016). These results further link pupil size related to executive functioning within clinical populations.

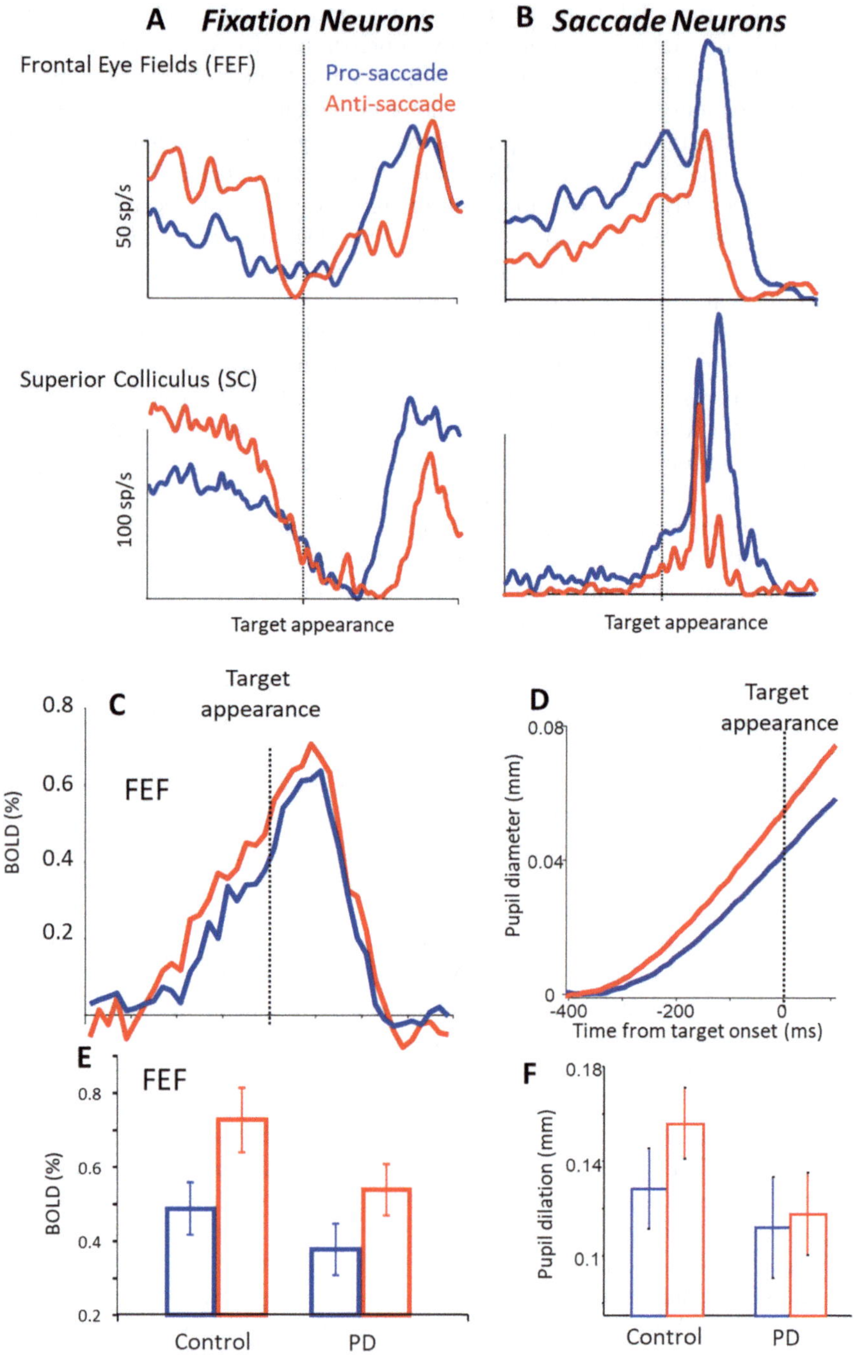

Fig. 4 Effect of pro- and anti-saccade preparation on SC and FEF activity, FEF BOLD activation, and pupil responses. (**a**, **b**) Effects of saccade preparation on (**a**) fixation and (**b**) neurons in the

1.3.2 Role of the SC in Attention-Mediated Local Luminance Pupillary Responses

Attention is an important cognitive process, and it has recently been shown that the pupillary light reflex is also modulated by attention (Binda & Murray, 2015; Mathôt & Van der Stigchel, 2015). Participants, when confronted with bright and dark patch stimuli presented in the periphery, produce a smaller pupil size when their locus of attention overlaps with the bright stimulus compared to the dark stimulus even though the level of global luminance remains unchanged (referred to as the local luminance modulation) (Binda et al., 2013; Mathot et al., 2013; Naber et al., 2013). These results demonstrate that pupil size can be modulated by the luminance level at the location selected by spatial attention, independent of global luminance. Saccade planning and working memory are interdependent with attention (Awh & Jonides, 2001; Awh et al., 2006; Bisley, 2011; Kowler, 2011; Melcher & Piazza, 2011; Hanning et al., 2016; Ohl & Rolfs, 2017), and similar neural networks are recruited during these cognitive operations (Ikkai & Curtis, 2011; Jerde et al., 2012). Studies using pupil local luminance paradigms have revealed that the observed local luminance modulation via attention (Binda et al., 2013, 2014; Mathot et al., 2013, 2014) is similar to that obtained through saccade planning or working memory (Mathôt et al., 2015; Fabius et al., 2017; Unsworth & Robison, 2017, 2018). Moreover, direct comparison between the effects resulting from saccade planning and working memory has demonstrated that pupil size following patch presentation is smaller when the bright patch, compared to the dark patch, is spatially aligned with the target location (Fig. 5a), together suggesting the mediation of attentional mechanisms on this behavior (Wang et al., 2018b). Recent research has further revealed the role of the SCi on the local luminance modulation in behaving monkeys (Wang & Munoz, 2018). Pupil size was altered according to the local luminance level at the spatial location corresponding to the microstimulated location in the SCi map of monkeys, with smaller pupil size when the bright, compared to dark, patch was spatially aligned to the stimulated location (Fig. 5b) (similar results in FEF: Ebitz & Moore, 2017 in monkeys; Hsu et al., 2022 in humans). Moreover, inactivation of the SCi reversed this local luminance modulation (Fig. 5c). These findings implicate a causal role of the SCi in coordinating the local luminance modulation and highlight preparatory responses of optimizing pupil size for visual processing before the next fixation. In summary, these results reveal a central role of the SC in mediating some cognitive modulations observed in pupil size.

Fig 4 (continued) FEF and SC. (Adapted with permission, Munoz & Everling, 2004). (**c**) Effects of saccade preparation on fMRI FEF activation. (Adapted with permission, Connolly et al., 2002). (**d**) Effects of saccade preparation on pupil responses. (Adapted with permission, Wang et al., 2015). (**e**) Effects of saccade preparation on fMRI FEF activation (mean beta weight) in Parkinson's disease. (Adapted with permission, Cameron et al., 2012). (**f**) Effects of saccade preparation on the pupil response in Parkinson's disease. (Adapted with permission, Wang et al., 2016). In **e**, **f**, the error bar represents ±standard error across participants. Control: age-matched control. PD: Parkinson's disease

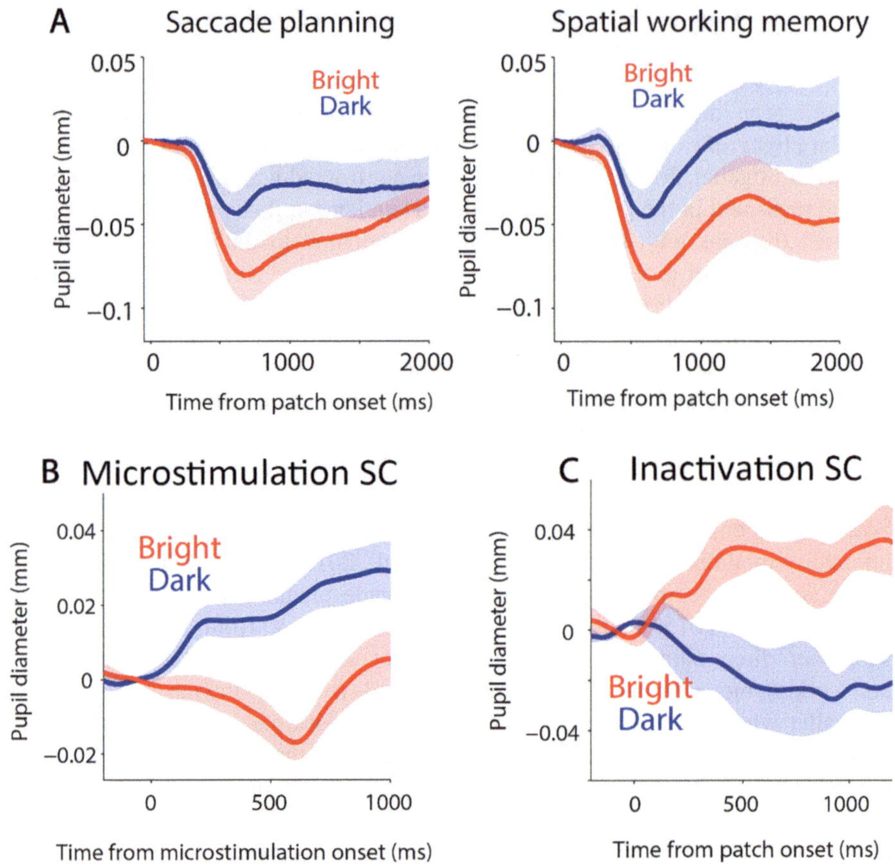

Fig. 5 Effect of saccade planning, working memory, and SCi activation on the local luminance pupil response. (**a**) Change in pupil diameter following the presentation of two patch stimuli when the bright (or dark) patch stimulus spatially aligned with the location for saccade planning or spatial working memory. (Adapted with permission, Wang et al., 2018). (**b**, **c**) Change in pupil diameter following the presentation of patch stimuli when the bright (or dark) patch stimulus spatially aligned with the affected site via (**b**) electrical microstimulation and (**c**) lidocaine microinjection. (Adapted with permission, Wang & Munoz, 2018). The shaded colored regions surrounding the response represent ±standard error range for different conditions

1.4 Arousal Effects on Saccades and Pupil Size

Pupil size has long been linked to arousal (Bradshaw, 1967), and pupil dilation is consistently observed when subjects are aroused (e.g., emotional stimulus presentation: Wang et al., 2018a; Cherng et al., 2020; spectral density: Nguyen et al., 2022). Studies have used pupil size to index the level of arousal (Aston-Jones & Cohen, 2005; Nassar et al., 2012; Eldar et al., 2013; de Gee et al., 2014; Murphy et al., 2014; Ebitz & Platt, 2015; Urai et al., 2017). Research in behaving animals further suggests that the rates of pupil size (i.e., pupil velocity) represent an effective index of arousal fluctuation (Reimer et al., 2014; McGinley et al., 2015; Vinck et al., 2015; Joshi et al., 2016). Research has used pupil velocity to examine the arousal modulation between human saccade and pupillary responses evoked by sensory stimuli (Wang et al., 2017). As shown in Fig. 6a, temporal dynamics of correlation

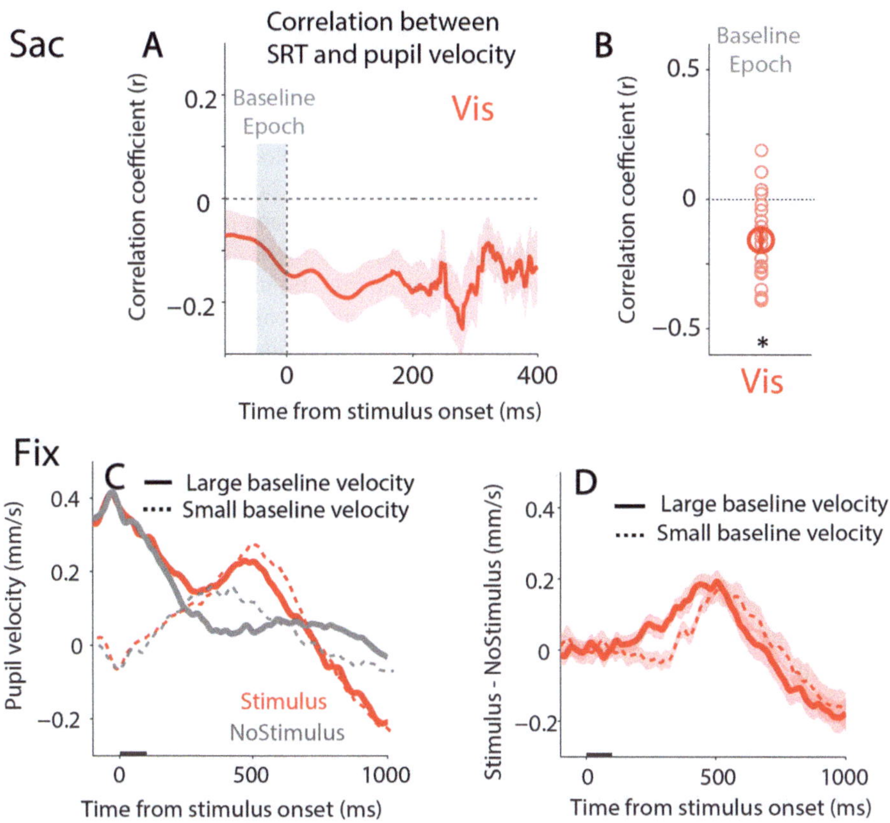

Fig. 6 Effect of baseline pupil velocity (arousal signals) on saccade and pupil responses. (Adapted with permission Wang et al., 2017). (**a**) Correlation coefficients between SRT and pupil velocity dynamics following the presentation of visual stimuli. (**b**) Distribution of correlation coefficients for the relationship between SRT and pupil velocity in the baseline epoch ($N = 20$). (**c**) Baseline pupil effects with larger and smaller baseline pupil velocity (50 ms to stimulus onset). (**d**) Normalized pupil velocity (stimulus minus no-stimulus condition) between larger and smaller baseline pupil velocity following visual stimulus presentation. In **a**, **d**, the shaded colored regions surrounding the response represent ±standard error range for different conditions. In **b**, the vertical colored bold circle represents the mean value of the correlation coefficient and the error bar represents ±standard error across participants. The colored small circle represents the correlation coefficient for each participant. The horizontal dotted line represents a zero value of correlation coefficient ($r = 0$). * indicates that differences are statistically significant. In **c**, **d**, the gray bar on the x-axis indicates the timeline of stimulus presentation

coefficients between SRT and pupil velocity demonstrated a negative correlation between SRT and pupil velocity on a trial-by-trial basis. Moreover, the negative correlation was present at the baseline epoch (Fig. 6b), showing trials with higher pupil velocities (i.e., higher arousal level) before stimulus onset having faster SRTs. Similar results were obtained in pupillary responses evoked by visual stimuli in the fixation condition. Trials were separated according to pupil velocity (higher or lower) during the baseline pupil epoch. As shown in Fig. 6c, the dynamics of pupil velocity were different between the higher and lower baseline pupil velocity conditions in the control (no-stimulus) condition (gray lines). Contrasted pupillary responses between stimulus and no-stimulus conditions (stimulus minus no-stimulus) were used to normalize the data. Consistent with arousal modulations on SRTs, stronger pupillary responses in the higher pupil velocity condition, compared to the lower pupil velocity condition, were obtained (Fig. 6d). Together, these results suggest the concerted modulation of arousal on saccade and pupillary responses.

To investigate the neural mechanisms underlying these pupil velocity modulations, we further examined the relation between pre-stimulation pupil velocity and saccades evoked by SCi microsimulation in behaving monkeys (Wang et al., 2022). As shown in Fig. 7, pre-stimulation pupil velocities correlated with saccade behavior driven by SCi microstimulation. Trials were separated into the two groups: with or without saccades induced by SCi microstimulation (Fig. 7a). Higher pupil velocities were obtained on trials without evoked saccades compared to trials with evoked saccades (Fig. 7b, c). For trials with evoked saccades by SCi microstimulation, saccade latencies were also associated with pre-stimulation pupil velocities. Trials were separated according to pre-stimulation pupil velocities (Fig. 7d), and saccades induced by SCi microstimulation between the higher and lower pupil velocity conditions were significantly different (Fig. 7e). Correlational analysis between pre-stimulation pupil velocity and saccade latencies on a trial-by-trail basis across sampled sites showed that faster saccade latencies were obtained in trials with lower pre-stimulation pupil velocities (Fig. 7f). These results reveal that the arousal fluctuation indexed by pupil velocity correlated with saccade behavior coordinated by the SCi. Notably, behavioral research generally shows that arousing stimuli can enhance saccade responses, producing faster saccade reaction times and larger saccade peak velocities (Crommelinck & Roucoux, 1976; van Steenbergen et al., 2011; Di Stasi et al., 2013; Simola et al., 2015; DiGirolamo et al., 2016; McSorley & Morriss, 2017; Wang et al., 2017). However, here, we found that lower pupil velocities (lower arousal levels) correlated with higher saccade probabilities and faster saccade latencies induced by SCi microstimulation, seemingly conflicting with other findings. It is important to note that, in this experiment, monkeys were required to maintain central fixation, and rostral SCi activity induced by the presence of a central fixation point (Munoz & Wurtz, 1993; Dorris & Munoz, 1995) should be enhanced during higher pupil velocities, leading to reduced caudal activity, correspondingly resulting in lower evoked saccade probabilities and longer evoked saccade latencies induced by SCi microstimulation. Future studies that systematically manipulate rostral and caudate SCi activity in the context of arousal effects are needed to directly test this possibility. Note that the performance is often optimal at

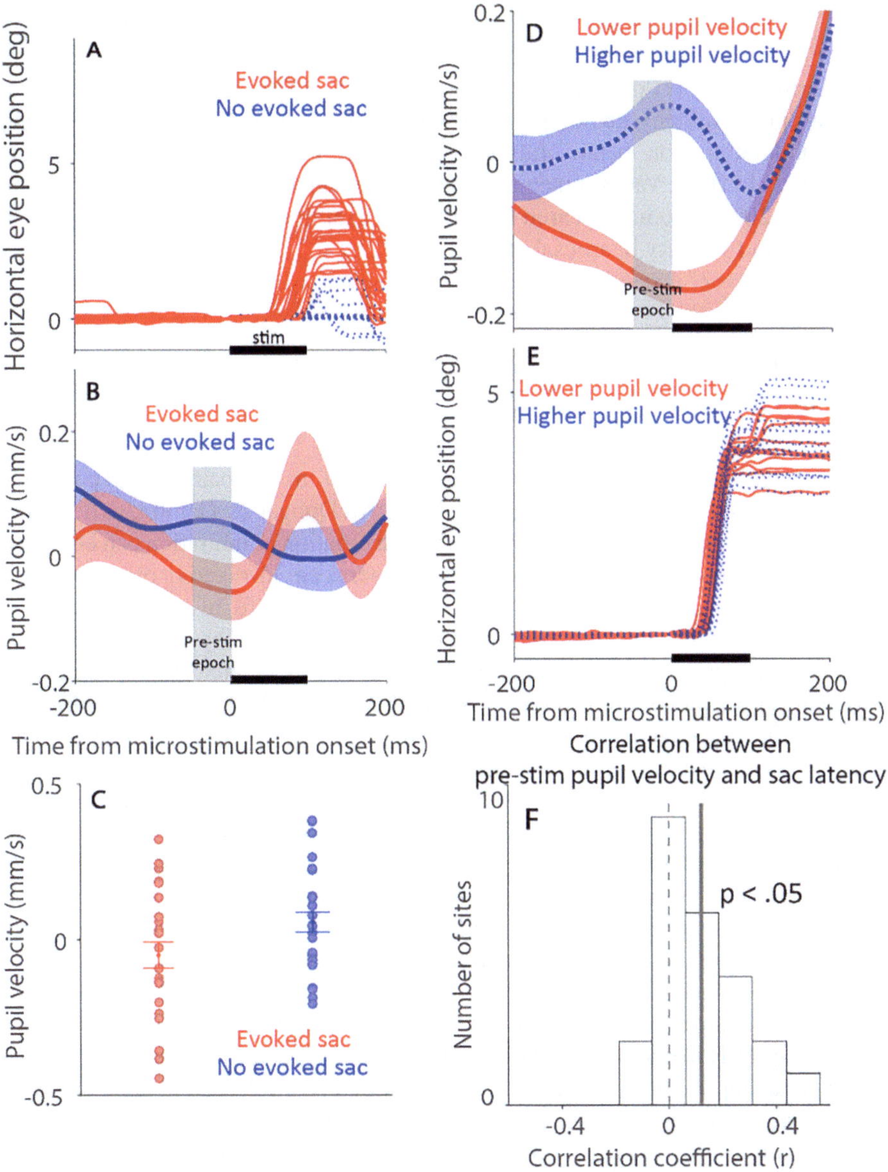

Fig. 7 Effect of pre-stimulation pupil velocity on saccades evoked by SCi microstimulation. (Adapted with permission Wang et al., 2022). (**a**) Horizontal eye position following microstimulation in trials with or without an evoked saccade. (**b**) Dynamics of pupil velocity in trials with or without an evoked saccade. (**c**) Pupil velocity in the pre-stim epoch between trials with or without an evoked saccade. (**d**) Dynamics of pupil velocity following microstimulation in trials with higher or lower pre-stim pupil velocity. (**e**) Horizontal eye position following microstimulation in trials with higher or lower pre-stim pupil velocity. (**f**) Distribution of correlation coefficients for the relationship between pre-stim pupil velocity and saccade latency in all stimulation sites. In **a**, **b**, **d**, **e**, the black bar on the x-axis indicates the time line of microstimulation. In **b**, **d**, the shaded colored regions surrounding the pupillary response represent ±standard error range for different conditions. In **c**, the error bar represents ±standard error across trials. The colored small circle represents the mean value for each condition, and the large circle represents the value for each trial. In **f**, the vertical thick line represents the mean correlation coefficients. The vertical dotted line represents a zero value of correlation coefficient (r = 0). stim: microstimulation. pre-stim: pre-microstimulation

the intermediate level of arousal (McCormick et al., 2020). As both absolute pupil size and pupil velocity could associate with fluctuations in the arousal level, and their modulations on behavioral and electrophysiological levels are different (Reimer et al., 2016; Van Kempen et al., 2019; Yamagishi & Furukawa, 2020; Wang & Munoz, 2021b), future research is required to examine the roles of absolute pupil size and pupil velocity in arousal effects.

1.5 Parasympathetic and Sympathetic Contributions to Observed Pupillary Responses

As pupil size is controlled by the balanced activity of the sympathetic and parasympathetic nervous systems (Loewenfeld, 1999; McDougal & Gamlin, 2015; May et al., 2019), sympathetic and parasympathetic excitation or inhibition can both lead to changes in pupil size. Although these two systems are involved in observed pupil modulations, the degree of involvement could be different. To investigate the contribution of these two systems on transient pupil dilation evoked by salient stimuli, we compared the pupil response onset latency (PROL) associated with orienting or with pupillary light or darkness reflex (Wang & Munoz, 2014). As shown in Fig. 8a, b, the PROL associated with orienting and the light reflex were relatively similar, and they were much earlier than the darkness reflex PROL. As the transient response of the light reflex is primarily driven by activation of the parasympathetic system, and that of the darkness reflex is mostly modulated by activation from the sympathetic pathway (Bitsios et al., 1996; Clarke et al., 2003; Loewenfeld, 1999; Nisida et al., 1960), these results suggest that inhibition of parasympathetic pathways initiates the fastest dilation of the pupil, while activation of sympathetic pathways likely facilitates a slower increase in pupil size.

Varying background luminance is another means to understand the relative contributions of sympathetic and parasympathetic modulation on the pupillary response (Steinhauer & Hakerem, 1992). Higher luminance levels lead to greater activation of the parasympathetic system, while lower luminance levels lead to greater activation of the sympathetic system. If the pupil dilation is mainly mediated by parasympathetic inhibition, PROL and dilation size should be slower and reduced with low compared to high background luminance due to minimal parasympathetic activation in the dark. As shown in Fig. 8c (Wang et al., 2012), pupil dilation was reliably evoked by SCi microstimulation in two levels of background luminance in the stimulation condition compared to the no-stimulation condition. Pupil size was then normalized (stimulation minus no-stimulation condition), PROLs in the two background luminance conditions were similar but fast (~150 ms after stimulation), and larger pupil dilation was obtained with lower luminance background (Fig. 8d). Although these results seem to suggest that pupil dilation evoked by SCi microstimulation is mainly mediated by sympathetic excitation, it is very unlikely that the sympathetic system can be recruited within 150 ms after stimulation. Thus, it is

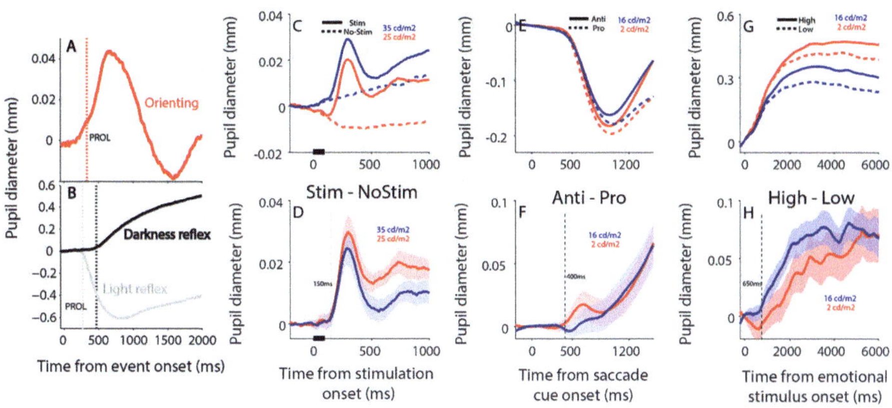

Fig. 8 Effect of background luminance on pupil response onset latencies. (**a**) Pupil response following the presentation of salient visual stimuli. (Adapted with permission Wang & Munoz, 2014). (**b**) Pupil response following the change of background luminance (light reflex: increase 10 cd/m^2; darkness reflex: decrease 10 cd/m^2). (**c**) Monkey pupil dynamics following SCi microstimulation with dim and bright backgrounds (25 and 35 cd/m^2). (Adapted with permission Wang et al., 2012). (**d**) Normalized pupil dynamics (stim minus no-stim pupil size) following SCi microstimulation. (**e**) Pupil dynamics following instructed saccade preparation cue with dim and bright backgrounds (2 and 16 cd/m^2). (Adapted with permission Cherng et al., 2020). (**f**) Normalized pupil dynamics (anti-saccade minus pro-saccade pupil size) following instructed saccade preparation cue. (**g**) Pupil dynamics following acoustic emotional arousing sounds with dim and bright backgrounds (2 and 16 cd/m^2). (**h**) Normalized pupil dynamics (high-arousal minus low-arousal pupil size) following acoustic emotional arousing sounds. In **c**, **d**, the black bar on the x-axis indicates the time line of microstimulation. In **d**, **f**, **h**, the shaded colored regions surrounding the pupillary response represent ±standard error range for different conditions. The dashed lines indicate the estimated PROL. Stim: microstimulation. No-stim: no microstimulation. PROL: pupil response onset latency

more reasonable to assume that both systems are important to pupil dilation induced by SC microstimulation.

To understand the contributions of the sympathetic and parasympathetic nervous systems to pupillary responses associated with emotion and saccade preparation, we examine the pupil size related to emotional sounds and saccade preparation with varying background luminance (Cherng et al., 2020). Larger pupil dilation responses were observed with a brighter background (Fig. 8e), suggesting an involvement of the parasympathetic system. However, the normalized pupillary responses (anti-saccade minus pro-saccade condition) showed earlier dilation around 400 ms after instructed saccade cue in the darker background condition (Fig. 8f), noting the sympathetic contribution. Therefore, these two systems likely play an important role in mediating pupil dilation responses related to saccade preparation. As shown in Fig. 8g, larger pupil dilation was evoked by higher arousing auditory stimuli and by lower background luminance, and the normalized pupillary responses (high- minus low-arousal condition) showed similar PROLs (Fig. 8h), together suggesting the primary involvement of the sympathetic system in pupil dilation associated with emotional arousal.

In summary, we argue that the parasympathetic and sympathetic systems are important to observed pupillary responses. Notably, although varying background luminance level is useful, the results are often mixed and less clear-cut, probably due to a limited number of background luminance levels and the influence of baseline pupil size (e.g., the law of initial values: Lacey, 1956; Chen et al., 2022) driven by the global luminance level. The pharmacological approach that respectively blocks the parasympathetic sphincter or sympathetic dilator muscle (e.g., Steinhauer et al., 2004) is thus suggested as a complementary approach to address individual contribution of the parasympathetic or sympathetic system to various pupil modulations.

2 Conclusions

Pupil size is modulated by sensory, cognitive, and affective processing and is an effective proxy of neural activity related to those processes. Here, we show that the similar modulation between saccade and pupillary responses by bottom-up saliency, cognitive, and arousal processes and the SC is implicated in underlying these modulations (Fig. 1). The SC, a structure underlying the orienting response, receives important control signals from the areas such as the FEF and LC, providing a fast route that enables subjects to quickly respond to environmental threats for survival. Therefore, this pathway likely mediates some pupillary responses particularly under time constraints. Future studies are needed to explore the involvement of the SC in the pupil modulations in tasks with or without time constraints.

Acknowledgments This work was supported by the Taiwan Ministry of Science and Technology Grant (109-2636-H-038-005, 110-2636-H-038-005, and 111-2628-H-008-003) to CW and by the Canadian Institutes of Health Research Grant (MOP-FDN-148418) and the Canada Research Chair Program to DPM.

References

Akert, K. (1949). Der visuelle Greifreflex. *Helvetica Physiologica et Pharmacologica Acta, 7*, 112–134.

Alahyane, N., Brien, D. C., Coe, B. C., Stroman, P. W., & Munoz, D. P. (2014). Developmental improvements in voluntary control of behavior: Effect of preparation in the fronto-parietal network? *Neuroimage, 98*, 103–117.

Antoniades, C. A., Demeyere, N., Kennard, C., Humphreys, G. W., & Hu, M. T. (2015). Antisaccades and executive dysfunction in early drug-naive Parkinson's disease: The discovery study. *Movement Disorders, 30*, 843–847.

Aston-Jones, G., & Cohen, J. D. (2005). An integrative theory of locus coeruleus-norepinephrine function: Adaptive gain and optimal performance. *Annual Review of Neuroscience, 28*, 403–450.

Awh, E., & Jonides, J. (2001). Overlapping mechanisms of attention and spatial working memory. *Trends in Cognitive Sciences, 5*, 119–126.

Awh, E., Armstrong, K. M., & Moore, T. (2006). Visual and oculomotor selection: Links, causes and implications for spatial attention. *Trends in Cognitive Sciences, 10*, 124–130.

Binda, P., & Murray, S. O. (2015). Keeping a large-pupilled eye on high-level visual processing. *Trends in Cognitive Sciences, 19*, 1–3.

Binda, P., Pereverzeva, M., & Murray, S. O. (2013). Attention to bright surfaces enhances the pupillary light reflex. *The Journal of Neuroscience, 33*, 2199–2204.

Binda, P., Pereverzeva, M., & Murray, S. O. (2014). Pupil size reflects the focus of feature-based attention. *Journal of Neurophysiology, 112*, 3046–3052.

Bisley, J. W. (2011). The neural basis of visual attention. *The Journal of Physiology, 589*, 49–57.

Bitsios, P., Prettyman, R. & Szabadi, E. (1996). Changes in autonomic function with age: A study of pupillary kinetics in healthy young and old people. *Age Ageing, 25*, 432–438.

Boehnke, S. E., & Munoz, D. P. (2008). On the importance of the transient visual response in the superior colliculus. *Current Opinion in Neurobiology, 18*, 544–551.

Bradshaw, J. (1967). Pupil size as a measure of arousal during information processing. *Nature, 216*, 515–516.

Bruce, C. J., & Goldberg, M. E. (1985). Primate frontal eye fields. I. Single neurons discharging before saccades. *Journal of Neurophysiology, 53*, 603–635.

Bruce, C. J., Goldberg, M. E., Bushnell, M. C., & Stanton, G. B. (1985). Primate frontal eye fields. II. Physiological and anatomical correlates of electrically evoked eye movements. *Journal of Neurophysiology, 54*, 714–734.

Cameron, I. G. M., Pari, G., Alahyane, N., Brien, D. C., Coe, B. C., Stroman, P. W., & Munoz, D. P. (2012). Impaired executive function signals in motor brain regions in Parkinson's disease. *NeuroImage, 60*, 1156–1170.

Chen, J.-T., Kuo, Y.-C., Hsu, T.-Y., & Wang, C.-A. (2022). Fatigue and arousal modulations revealed by saccade and pupil dynamics. *International Journal of Environmental Research and Public Health, 19*, 9234.

Cherng, Y.-G., Baird, T., Chen, J.-T., & Wang, C.-A. (2020). Background luminance effects on pupil size associated with emotion and saccade preparation. *Scientific Reports, 10*, 15718.

Clarke, R. J., Zhang, H. & Gamlin, P. D. (2003). Characteristics of the pupillary light reflex in the alert rhesus monkey. *Journal of Neurophysiology, 89*, 3179–3189.

Cohen, B., & Büttner-Ennever, J. A. (1984). Projections from the superior colliculus to a region of the central mesencephalic reticular formation (cMRF) associated with horizontal saccadic eye movements. *Experimental Brain Research, 57*, 167–176.

Connolly, J. D., Goodale, M. A., Menon, R. S., & Munoz, D. P. (2002). Human fMRI evidence for the neural correlates of preparatory set. *Nature Neuroscience, 5*, 1345–1352.

Corneil, B. D., & Munoz, D. P. (2014). Overt responses during covert orienting. *Neuron, 82*, 1230–1243.

Crommelinck, M., & Roucoux, A. (1976). Characteristics of cat's eye saccades in different states of alertness. *Brain Research, 103*, 574–578.

de Gee, J. W., Knapen, T., & Donner, T. H. (2014). Decision-related pupil dilation reflects upcoming choice and individual bias. *Proceedings of the National Academy of Sciences of the United States of America, 111*, E618–E625.

DeSouza, J. F., Menon, R. S., & Everling, S. (2003). Preparatory set associated with pro-saccades and anti-saccades in humans investigated with event-related FMRI. *Journal of Neurophysiology, 89*,1016–1023.

Di Stasi, L. L., Catena, A., Cañas, J. J., Macknik, S. L., & Martinez-Conde, S. (2013). Saccadic velocity as an arousal index in naturalistic tasks. *Neuroscience and Biobehavioral Reviews, 37*, 968–975.

DiGirolamo, G. J., Patel, N., & Blaukopf, C. L. (2016). Arousal facilitates involuntary eye movements. *Experimental Brain Research, 234*, 1967–1976.

Dorris, M. C., & Munoz, D. P. (1995). A neural correlate for the gap effect on saccadic reaction times in monkey. *Journal of Neurophysiology, 73*, 2558–2562.
Ebitz, R. B., & Moore, T. (2017). Selective modulation of the pupil light reflex by microstimulation of prefrontal cortex. *The Journal of Neuroscience, 37*, 5008–5018.
Ebitz, R. B., & Moore, T. (2019). Both a gauge and a filter: Cognitive modulations of pupil size. *Frontiers in Neurology, 9*, 1190.
Ebitz, R. B., & Platt, M. L. (2015). Neuronal activity in primate dorsal anterior cingulate cortex signals task conflict and predicts adjustments in pupil-linked arousal. *Neuron, 85*, 628–640.
Edwards, S. B., Ginsburgh, C. L., Henkel, C. K., & Stein, B. E. (1979). Sources of subcortical projections to the superior colliculus in the cat. *The Journal of Comparative Neurology, 184*, 309–329.
Einhäuser, W. (2017). The pupil as marker of cognitive processes. In *Cognitive science and technology* (pp. 141–169). Springer International Publishing.
Eldar, E., Cohen, J. D., & Niv, Y. (2013). The effects of neural gain on attention and learning. *Nature Neuroscience, 16*, 1146–1153.
Everling, S., & Munoz, D. P. (2000). Neuronal correlates for preparatory set associated with pro-saccades and anti-saccades in the primate frontal eye field. *The Journal of Neuroscience, 20*, 387–400.
Everling, S., Dorris, M. C., Klein, R. M., & Munoz, D. P. (1999). Role of primate superior colliculus in preparation and execution of anti-saccades and pro-saccades. *The Journal of Neuroscience, 19*, 2740–2754.
Fabius, J. H., Mathôt, S., Schut, M. J., Nijboer, T. C. W., & Van der Stigchel, S. (2017). Focus of spatial attention during spatial working memory maintenance: Evidence from pupillary light response. *Visual Cognition, 25*, 10–20.
Fecteau, J. H., & Munoz, D. P. (2006). Salience, relevance, and firing: A priority map for target selection. *Trends in Cognitive Sciences, 10*, 382–390.
Fernandez-Ruiz, J., Peltsch, A., Alahyane, N., Brien, D. C., Coe, B. C., Garcia, A., & Munoz, D. P. (2018). Age related prefrontal compensatory mechanisms for inhibitory control in the antisaccade task. *NeuroImage, 165*, 92–101.
Gandhi, N. J., & Katnani, H. A. (2011). Motor functions of the superior colliculus. *Annual Review of Neuroscience, 34*, 205–231.
Goldberg, M. E., & Bruce, C. J. (1990). Primate frontal eye fields. III. Maintenance of a spatially accurate saccade signal. *Journal of Neurophysiology, 64*, 489–508.
Greenfield, J. G., & Bosanquet, F. D. (1953). The brain-stem lesions in Parkinsonism. *Journal of Neurology, Neurosurgery, and Psychiatry, 16*, 213–226.
Hanning, N. M., Jonikaitis, D., Deubel, H., & Szinte, M. (2016). Oculomotor selection underlies feature retention in visual working memory. *Journal of Neurophysiology, 115*, 1071–1076.
Harting, J. K., Huerta, M. F., Frankfurter, A. J., Strominger, N. L., & Royce, G. J. (1980). Ascending pathways from the monkey superior colliculus: An autoradiographic analysis. *The Journal of Comparative Neurology, 192*, 853–882.
Hess, W. R., Buergi, S., & Bucher, V. (1946). Motorische Funktionen des Tektal – und Tegmentalgebietes (motor functions of tectal and tegmental areas). *Monatsschrift für Psychiatrie und Neurologie, 112*, 1–52.
Horn, A. K. E. (2006). The reticular formation. In *Progress in brain research* (pp. 127–155). Elsevier.
Hsu, T. -Y., Hsu, Y. -F., Wang, H. -Y., & Wang, C.-A. (2021). Role of the frontal eye field in human pupil and saccade orienting responses. *The European Journal of Neuroscience, 54*, 4283. https://doi.org/10.1111/ejn.15253
Hsu, T.-Y., Wang, H.-Y., Chen, J.-T., & Wang, C.-A. (2022). Investigating the role of human frontal eye field in the pupil light reflex modulation by saccade planning and working memory. *Frontiers in Human Neuroscience, 16*, 793.
Ikkai, A., & Curtis, C. E. (2011). Common neural mechanisms supporting spatial working memory, attention and motor intention. *Neuropsychologia, 49*, 1428–1434.

Itti, L., & Koch, C. (2001). Computational modelling of visual attention. *Nature Reviews. Neuroscience, 2*, 194–203.
Itti, L., Koch, C., & Niebur, E. (1998). A model of saliency-based visual attention for rapid scene analysis. *IEEE Transactions on Pattern Analysis and Machine Intelligence, 20*, 1254–1259.
Jerde, T. A., Merriam, E. P., Riggall, A. C., Hedges, J. H., & Curtis, C. E. (2012). Prioritized maps of space in human frontoparietal cortex. *The Journal of Neuroscience, 32*, 17382–17390.
Joshi, S., Li, Y., Kalwani, R. M., & Gold, J. I. (2016). Relationships between pupil diameter and neuronal activity in the locus coeruleus, colliculi, and cingulate cortex. *Neuron, 89*, 221–234.
Koch, C., & Ullman, S. (1985). Shifts in selective visual attention: Towards the underlying neural circuitry. *Human Neurobiology, 4*, 219–227.
Kowler, E. (2011). Eye movements: The past 25 years. *Vision Research, 51*, 1457–1483.
Krauzlis, R. J., Lovejoy, L. P., & Zenon, A. (2013). Superior colliculus and visual spatial attention. *Annual Review of Neuroscience, 36*, 165–182.
Lacey, J. I. (1956). *The evaluation of autonomic responses: toward a general solution* (p. 67). Ann.
Leh, S. E., Petrides, M., & Strafella, A. P. (2010). The neural circuitry of executive functions in healthy subjects and parkinson's disease. *Neuropsychopharmacology, 35*, 70–85.
Lehmann, S. J., & Corneil, B. D. (2016). Transient pupil dilation after subsaccadic microstimulation of primate frontal eye fields. *The Journal of Neuroscience, 36*, 3765–3776.
Loewenfeld, I. E. (1999). *The pupil: Anatomy, physiology, and clinical applications.* Butterworth- Heinemann.
Lynn, R. (1966). *Attention, arousal and the orientation reaction*. Pergamon Press.
Manoach, D. S., Thakkar, K. N., Cain, M. S., Polli, F. E., Edelman, J. A., Fischl, B., & Barton, J. J. (2007). Neural activity is modulated by trial history: A functional magnetic resonance imaging study of the effects of a previous antisaccade. *Journal of Neuroscience, 27*, 1791–1798.
Marino, R. A., Levy, R., Boehnke, S., White, B. J., Itti, L., & Munoz, D. P. (2012). Linking visual response properties in the superior colliculus to saccade behavior. *The European Journal of Neuroscience, 35*, 1738–1752.
Marino, R. A., Levy, R., & Munoz, D. P. (2015). Linking express saccade occurance to stimulus properties and sensorimotor integration in the superior colliculus. *Journal of Neurophysiology, 114*, 879–892.
Mathôt, S. (2018). Pupillometry: Psychology, physiology, and function. *Journal of Cognition, 1*.
Mathôt, S., & Van der Stigchel, S. (2015). New light on the mind's eye: The pupillary light response as active vision. *Current Directions in Psychological Science, 24*, 374–378.
Mathot, S., van der Linden, L., Grainger, J., Vitu, F., Mathôt, S., van der Linden, L., Grainger, J., & Vitu, F. (2013). The pupillary light response reveals the focus of covert visual attention. *PLoS One, 8*, e78168.
Mathôt, S., Dalmaijer, E., Grainger, J., & Van der Stigchel, S. (2014). The pupillary light response reflects exogenous attention and inhibition of return. *Journal of Vision, 14*, 7.
Mathôt, S., van der Linden, L., Grainger, J., & Vitu, F. (2015). The pupillary light response reflects eye-movement preparation. *Journal of Experimental Psychology. Human Perception and Performance, 41*, 28–35.
May, P. J. (2006). The mammalian superior colliculus: laminar structure and connections. *Progress in Brain Research, 151*, 321–378.
May, P. J., Warren, S., Bohlen, M. O., Barnerssoi, M., & Horn, A. K. E. (2016). A central mesencephalic reticular formation projection to the Edinger-Westphal nuclei. *Brain Structure & Function, 221*, 4073–4089.
May, P. J., Reiner, A., Gamlin, P. D., May, P. J., Reiner, A., & Gamlin, P. D. (2019). Autonomic regulation of the eye. In *Oxford research encyclopedia of neuroscience* (pp. 1–27).
McCormick, D. A., Nestvogel, D. B., & He, B. J. (2020). Neuromodulation of brain state and behavior. *Annual Review of Neuroscience, 43*, 391–415.
McDougal, D. H., & Gamlin, P. D. (2015). Autonomic control of the eye. *Comprehensive Physiology, 5*, 439–473.

McGinley, M. J., Vinck, M., Reimer, J., Batista-Brito, R., Zagha, E., Cadwell, C. R., Tolias, A. S., Cardin, J. A., & McCormick, D. A. (2015). Waking state: Rapid variations modulate neural and behavioral responses. *Neuron, 87*, 1143–1161.

McSorley, E., & Morriss, J. (2017). What you see is what you want to see: Motivationally relevant stimuli can interrupt current resource allocation. *Cognition and Emotion, 31*, 168–174.

Melcher, D., & Piazza, M. (2011). The role of attentional priority and saliency in determining capacity limits in enumeration and visual working memory. *PLoS One, 6*, e29296.

Munoz, D. P., & Everling, S. (2004). Look away: The anti-saccade task and the voluntary control of eye movement. *Nature Reviews. Neuroscience, 5*, 218–228.

Munoz, D. P., & Wurtz, R. H. (1993). Fixation cells in monkey superior colliculus. I. Characteristics of cell discharge. *Journal of Neurophysiology, 70*, 559–575.

Murphy, P. R., Vandekerckhove, J., & Nieuwenhuis, S. (2014). Pupil-linked arousal determines variability in perceptual decision making. *PLoS Computational Biology, 10*, e1003854.

Muslimovic, D., Post, B., Speelman, J. D., & Schmand, B. (2005). Cognitive profile of patients with newly diagnosed Parkinson disease. *Neurology, 65*, 1239–1245.

Mysore, S. P., & Knudsen, E. I. (2013). A shared inhibitory circuit for both exogenous and endogenous control of stimulus selection. *Nature Neuroscience, 16*, 473–478.

Naber, M., Alvarez, G. A., & Nakayama, K. (2013). Tracking the allocation of attention using human pupillary oscillations. *Frontier in Psychology, 4*, 919.

Nassar, M. R., Rumsey, K. M., Wilson, R. C., Parikh, K., Heasly, B., & Gold, J. I. (2012). Rational regulation of learning dynamics by pupil-linked arousal systems. *Nature Neuroscience, 15*, 1040–1046.

Netser, S., Dutta, A., & Gutfreund, Y. (2014). Ongoing activity in the optic tectum is correlated on a trial-by-trial basis with the pupil dilation response. *Journal of Neurophysiology, 111*, 918–929.

Nguyen, K. T., Liang, W. K., Juan, C. H., & Wang, C. A. (2022). Time-frequency analysis of pupil size modulated by global luminance, arousal, and saccade preparation signals using Hilbert-Huang transform. *International Journal of Psychophysiology, 176*, 89–99.

Nisida, I., Okada, H. & Nakano, O. (1960). The activity of the ciliospinal centers and their inhibition in pupillary light reflex. *Japanese Journal of Physiology, 10*, 73–84.

Ohl, S., & Rolfs, M. (2017). Saccadic eye movements impose a natural bottleneck on visual short-term memory. *Journal of Experimental Psychology. Learning, Memory, and Cognition, 43*, 736–748.

Oster, J., Huang, J., White, B. J., Radach, R., Itti, L., Munoz, D. P., & Wang, C. A. (2022). Pupillary responses to differences in luminance, color and set size. *Experimental Brain Research, 2022*(1), 1–13.

Perkins, E., Warren, S., & May, P. J. (2009). The mesencephalic reticular formation as a conduit for primate collicular gaze control: Tectal inputs to neurons targeting the spinal cord and medulla. *The Anatomical Record: Advances in Integrative Anatomy and Evolutionary Biology, 292*, 1162–1181.

Perkins, J. E., Janzen, A., Bernhard, F. P., Wilhelm, K., Brien, D. C., Huang, J., Coe, B. C., Vadasz, D., Mayer, G., Munoz, D. P., & Oertel, W. H. (2021). Saccade, pupil, and blink responses in rapid eye movement sleep behavior disorder. *Movement Disorders, 36*, 1720–1726.

Reimer, J., Froudarakis, E., Cadwell, C. R., Yatsenko, D., Denfield, G. H., & Tolias, A. S. (2014). Pupil fluctuations track fast switching of cortical states during quiet wakefulness. *Neuron, 84*, 355–362.

Reimer, J., McGinley, M. J., Liu, Y., Rodenkirch, C., Wang, Q., McCormick, D. A., & Tolias, A. S. (2016). Pupil fluctuations track rapid changes in adrenergic and cholinergic activity in cortex. *Nature Communications, 7*, 13289.

Robinson, D. A., & Fuchs, A. F. (1969). Eye movements evoked by stimulation of frontal eye fields. *Journal of Neurophysiology, 32*, 637–648.

Rodriguez-Oroz, M. C., Lage, P. M., Sanchez-Mut, J., Lamet, I., Pagonabarraga, J., Toledo, J. B., García-Garcia, D., Clavero, P., Samaranch, L., Irurzun, C., Matsubara, J. M., Irigoien, J., Bescos, E., Kulisevsky, J., Pérez-Tur, J., & Obeso, J. A. (2009). Homocysteine and cog-

nitive impairment in Parkinson's disease: A biochemical, neuroimaging, and genetic study. *Movement Disorders, 24*, 1437–1444.
Schiller, P. H., True, S. D., & Conway, J. L. (1979). Effects of frontal eye field and superior colliculus ablations on eye movements. *Science, 206*, 590–592.
Schiller, P. H., True, S. D., & Conway, J. L. (1980). Deficits in eye movements following frontal eye-field and superior colliculus ablations. *Journal of Neurophysiology, 44*, 1175–1189.
Scudder, C. A., Kaneko, C. R., & Fuchs, A. F. (2002). The brainstem burst generator for saccadic eye movements: A modern synthesis. *Experimental Brain Research, 142*, 439–462.
Simola, J., Le Fevre, K., Torniainen, J., & Baccino, T. (2015). Affective processing in natural scene viewing: Valence and arousal interactions in eye-fixation-related potentials. *NeuroImage, 106*, 21–33.
Sokolov, E. N. (1963a). *Perception and the conditioned reflex*. Macmillan.
Sokolov, E. N. (1963b). Higher nervous functions: The orienting reflex. *Annual Review of Physiology, 25*, 545–580.
Sparks, D. L. (2002). The brainstem control of saccadic eye movements. *Nature Reviews. Neuroscience, 3*, 952–964.
Steinhauer, S. R., & Hakerem, G. (1992). The pupillary response in cognitive psychophysiology and schizophrenia. *Annals of the New York Academy of Sciences, 658*, 182–204.
Steinhauer, S. R., Siegle, G. J., Condray, R., & Pless, M. (2004). Sympathetic and parasympathetic innervation of pupillary dilation during sustained processing. *International Journal of Psychophysiology, 52*, 77–86.
Unsworth, N., & Robison, M. K. (2017). Pupillary correlates of covert shifts of attention during working memory maintenance. *Attention, Perception, & Psychophysics, 79*, 782–795.
Unsworth, N., & Robison, M. K. (2018). Tracking working memory maintenance with pupillometry. *Attention, Perception, & Psychophysics, 80*, 461–484.
Urai, A. E., Braun, A., & Donner, T. H. (2017). Pupil-linked arousal is driven by decision uncertainty and alters serial choice bias. *Nature Communications, 8*, 14637.
Van Kempen, J., Loughnane, G. M., Newman, D. P., Kelly, S. P., Thiele, A., O'Connell, R. G., & Bellgrove, M. A. (2019). Behavioural and neural signatures of perceptual decision-making are modulated by pupil-linked arousal. *eLife, 8*, e42541.
van Steenbergen, H., Band, G. P. H., & Hommel, B. (2011). Threat but not arousal narrows attention: Evidence from pupil dilation and saccade control. *Frontiers in Psychology, 2*, 281.
Vinck, M., Batista-Brito, R., Knoblich, U., & Cardin, J. A. (2015). Arousal and locomotion make distinct contributions to cortical activity patterns and visual encoding. *Neuron, 86*, 740–754.
Wang, C.-A., & Munoz, D. P. (2014). Modulation of stimulus contrast on the human pupil orienting response. *The European Journal of Neuroscience, 40*, 2822–2832.
Wang, C.-A., & Munoz, D. P. (2015). A circuit for pupil orienting responses: Implications for cognitive modulation of pupil size. *Current Opinion in Neurobiology, 33*, 134–140.
Wang, C.-A., & Munoz, D. P. (2018). Neural basis of location-specific pupil luminance modulation. *Proceeding of the National Academy of Sciences*, 201809668.
Wang, C.-A., & Munoz, D. P. (2021a). Coordination of pupil and saccade responses by the superior colliculus. *Journal of Cognitive Neuroscience, 33*, 919–932.
Wang, C. A., & Munoz, D. P. (2021b). Differentiating global luminance, arousal and cognitive signals on pupil size and microsaccades. *The European Journal of Neuroscience, 54*, 7560–7574.
Wang, C.-A., Boehnke, S. E., White, B. J., & Munoz, D. P. (2012). Microstimulation of the monkey superior colliculus induces pupil dilation without evoking saccades. *The Journal of Neuroscience, 32*, 3629–3636.
Wang, C.-A., Boehnke, S. E., Itti, L., & Munoz, D. P. (2014). Transient pupil response is modulated by contrast-based saliency. *The Journal of Neuroscience, 34*(2), 408–417.
Wang, C.-A., Brien, D. C., & Munoz, D. P. (2015). Pupil size reveals preparatory processes in the generation of pro-saccades and anti-saccades. *The European Journal of Neuroscience, 41*, 1102–1110.

Wang, C.-A., McInnis, H., Brien, D. C., Pari, G., & Munoz, D. P. (2016). Disruption of pupil size modulation correlates with voluntary motor preparation deficits in Parkinson's disease. *Neuropsychologia, 80*, 176–184.
Wang, C.-A., Blohm, G., Huang, J., Boehnke, S. E., & Munoz, D. P. (2017). Multisensory integration in orienting behavior: Pupil size, microsaccades, and saccades. *Biological Psychology, 129*, 36–44.
Wang, C.-A., Baird, T., Huang, J., Coutinho, J. D., Brien, D. C., & Munoz, D. P. (2018a). Arousal effects on pupil size, heart rate, and skin conductance in an emotional face task. *Frontiers in Neurology, 9*, 1–13.
Wang, C.-A., Huang, J., Yep, R., & Munoz, D. P. (2018b). Comparing pupil light response modulation between saccade planning and working memory. *Journal of Cognition, 1*, 33.
Wang, C.-A., Huang, J., Brien, D. C., & Munoz, D. P. (2020). Saliency and priority modulation in a pop-out paradigm: Pupil size and microsaccades. *Biological Psychology, 153*, 107901.
Wang, C.-A., Nguyen, K. T., & Juan, C.-H. (2021). Linking pupil size modulated by global luminance and motor preparation to saccade behavior. *Neuroscience, 476*, 90–101.
Wang, C.-A., White, B., & Munoz, D. P. (2022). Pupil-linked arousal signals in the midbrain superior colliculus. *Journal of Cognitive Neuroscience, 34*, 1–15.
White, B. J., & Munoz, D. P. (2011). The superior colliculus. In S. Liversedge, I. Gilchrist, & S. Everling (Eds.), *Oxford handbook of eye movements* (pp. 195–213). Oxford University Press.
Yamagishi, S., & Furukawa, S. (2020). Factors influencing saccadic reaction time: Effect of task modality, stimulus saliency, spatial congruency of stimuli, and pupil size. *Frontiers in Human Neuroscience, 14*, 571893.

Part II
Psychological Science

Pupillometry in Developmental Psychology

Robert Hepach

1 Introduction

Eye tracking machines capture pupil size, but appropriate experimental designs are required to capture psychological data through the systematic study of *changes* in pupil size. The origins of studying psychological processes in young children through pupil dilation date back to the mid-twentieth century, which marked the advent of modern infant cognition. At the time, Robert Fantz and colleagues studied infants' visual attention to stimuli of varying visual complexities (Fantz, 1963, 1964). Their method of visual habituation and measuring infants' gaze direction and duration laid the groundwork for a research stream that revolutionized the study of infant cognitive development (Cohen, 2009; Cohen & Cashon, 2007). At the core of Fantz's work lays the idea of an orienting response: an organism habituates as a stimulus becomes familiar, but the subject enters a state of heightened alertness when novel stimuli are presented, as a way of taking in new information to adjust their behaviour. The greater the novelty and motivational significance of the stimulus, the greater the orienting response (Nieuwenhuis et al., 2011; Sara & Bouret, 2012; Sokolov, 1963). The 'symptoms' of such a response can include increased visual attention, heart rate deceleration, postural freezing, and increased pupil dilation. During the same time period in the 1960s, but less well known or at least less frequently cited, Fitzgerald and colleagues studied the changes in infants' pupil diameter while presenting social and non-social stimuli (Fitzgerald, 1968; Fitzgerald et al., 1967; see Sirois & Brisson, 2014, for a review). The work by Fitzgerald and colleagues extended an emerging line of research in adults using pupil dilation to measure arousal (Bradshaw, 1967; Hess & Polt, 1960) and cognitive effort (Kahneman et al., 1969; see Goldwater, 1972; Loewenfeld, 1993, for reviews). While adult research on pupil dilation progressed in subsequent decades (for recent

R. Hepach (✉)
Department of Experimental Psychology, University of Oxford, Oxford, UK
e-mail: robert.hepach@psy.ox.ac.uk

M. H. Papesh, S. D. Goldinger (eds.), *Modern Pupillometry*,
https://doi.org/10.1007/978-3-031-54896-3_3

reviews, see Eckstein et al., 2017; Laeng et al., 2012; Sirois & Brisson, 2014), the method lay dormant in developmental psychology. This changed with the introduction of corneal-reflection eye trackers to developmental psychological laboratories in the 1990s and early 2000s (Aslin & McMurray, 2004; Gredebäck et al., 2009; Oakes, 2012).

The seminal works by Jackson and Sirois (2009) and Gredebäck and Melinder (2010) reintroduced developmental psychology to the technique of measuring infants' pupil dilation to capture various aspects of psychological engagement (Gredebäck & Melinder, 2010; Jackson & Sirois, 2009; for reviews, see Hepach & Westermann, 2016; Zhang & Emberson, 2020). In the domain of physical cognition, infants at 7 months of age showed greater pupil dilation when viewing impossible physical events, i.e. outcomes that violated their expectations (Jackson & Sirois, 2009). The authors embedded measures of pupil dilation within the so-called violation of expectation (VOE) paradigms and argued that measuring pupil dilation provided distinct advantages over measuring more established measures of cumulative looking time (Sirois & Jackson, 2012). In the domain of social cognition, 12- but not 8-month-old infants showed greater pupil dilation to irrational social interactions, i.e. outcomes that violated their social expectations (Gredebäck & Melinder, 2010). These initial studies highlighted the potential of pupil dilation beyond VOE paradigms to capture individual differences not only in social but also in physical cognition. Both studies, above and beyond pioneering the method, included analysis techniques to deal with luminance artefacts and to embrace signal analysis modelling techniques (Jackson & Sirois, 2009; see also Nyström et al., 2016). Experimental paradigms were developed to rule out that the observed psychological changes in pupil dilation were a mere consequence of luminance or movement artefacts. Given the temperamental volatility of infants and young children, this required and continues to require creativity, skill, and patience on the researcher's part.

In any pupillometry study, the developmental psychologist's typical infant or toddler attention span and temperament can be the experimental psychologist's nightmare. The human pupils continuously oscillate, and, consequently, changes in pupil size are a dynamic signal to analyse. At any given point in time, pupil dilatations are a result of (1) luminance adaptation to the environment and the visual field in focus (Loewenfeld, 1993), (2) eye and head movement (Brisson et al., 2013), (3) changes in tonic arousal state (Anderson & Colombo, 2009; Granholm & Steinhauer, 2004), and (4) psychological engagement (Eckstein et al., 2017; Goldwater, 1972; Laeng et al., 2012). It is for those reasons that changes in pupil dilation are best measured in dark, luminance-controlled settings where the stimulus is presented auditorily to a (near) motionless participant to avoid any movement artefacts and light adaptation responses. However, young children (and some adults) struggle to spend prolonged periods of time in a state of bodily inertia (in dark environments). This brings with it unique challenges to developmental psychologists studying physical and social cognition in early ontogeny. For example, in the studies reviewed for this chapter, the majority used visual stimuli to measure pupil dilation as a way of focussing participants' attention on a computer screen. This already constitutes a non-ideal scenario because psychological changes in pupil dilation need to be separated from luminance effects as children are visually scanning the stimulus presented to them. Across the studies reviewed here, several suggestions have been made to address such issues. For pupillometry research with adult subjects, methods' reviews and reflections are published in regular intervals (Eckstein et al., 2017; Goldwater, 1972; Laeng et al., 2012; Loewenfeld, 1993; Mathôt, 2018). The present chapter is a modest attempt to providing such a reflection on methods and the state of the art of pupillometry in developmental science (see also Hepach & Westermann, 2016).

This chapter pursues two broad goals. The first is to underscore the utility and potential of measuring pupil dilation (henceforth short for *psychologically induced* pupil dilation) in infants and young children. We document the tremendous progress that has been made to establish that psychologically induced changes can be reliably measured in the research domains of (1) surprise and violation of expectation paradigms, (2) task-evoked pupil dilation and neural activity, and (3) arousal state and motivation. The second goal is to document and reflect on the heterogeneity in how pupil dilation data are analysed. This chapter is based on a review of all published studies on pupillometry from 2007 to date (i.e. early 2022) focussed on early and middle childhood (2 months to 16 years based on a total of $N > 5700$ participants). The chapter is structured as follows:

- The first part lays out the main theoretical concepts and introduces three levels of analysis to provide a frame of reference for reviewing the studies: Capturing and processing pupil dilation data (*Level 1*). Separating psychologically induced pupil dilation from luminance effects on pupil size (*Level 2*). Inferring a psychological process from changes in pupil dilation (*Level 3*).
- The second part reviews the extant body of literature in three areas of research: (1) surprise and violation of expectation paradigms, (2) task-evoked pupil dilation and neural activity, and (3) arousal state and motivation.
- The final part of the chapter provides a summary of current challenges in developmental pupillometry with an outlook to future directions before presenting a final conclusion.

For terminology, we refer to *pupillometry* as the science of studying systematic changes in pupil size. *Pupil size* (estimated in pixels) and *pupil diameter* (estimated in mm) essentially provide the same information as outputted by the eye tracking system per sample taken. *Pupillary oscillations* refer to continuous changes in pupil size, i.e. increases and decreases over time. *Pupil dilatations* reflect the recurrent increases in pupil size (over time), whereas *pupil dilation* refers to a single event increase in pupil size. These pupil dilations can either be time-locked changes, the so-called phasic *task-evoked pupillary responses* (*TEPR*s) measured in violation of expectation paradigms or to index neural activity (Beatty & Lucero-Wagoner, 2000), or they can be tonic *state-evoked pupillary responses* (*SEPR*s) to measure changes in arousal and motivation. *Pupil constriction* refers to the decrease in pupil size, and the *pupillary light reflex* is the involuntary decrease in pupil size in response to a bright stimulus (in visual focus).

2 Theoretical and Conceptual Foundations of Developmental Pupillometry

Modern-day eye tracking systems capture pupil size with remarkable accuracy, and although individual commercial systems differ in their specifications, each involves some element of separating out the shape of the pupil from the rest of the eye (Mathôt, 2018). It is beyond the remits of this chapter and the expertise of the author to evaluate these techniques. The majority of studies reviewed here used systems by Tobii (https://www.tobii.com/), Eyelink (https://www.sr-research.com/), ASL (https://est-kl.com/), or SensoMotoric Instruments, but more recent start-up solutions to mobile eye tracking are also available (https://pupil-labs.com/). In principle, it is also not impossible to build one's own eye tracker (https://eyewriter.org/diy/). In fact, even without an eye tracker and using images collected via a webcam, researchers have tapped into systematic changes in infants' pupil size (Bainbridge et al., 2021). Here, we focus on three distinct problems that developmental researchers face when designing a pupillometry study, even if the most state-of-the-art commercial eye tracking system is used:

- How to capture (in terms of time windows) and process (analysis pipelines) pupil size data? (*Level 1: The Signal*)
- How to separate psychologically induced pupil dilation from luminance effects on pupil size? (*Level 2: Evidence of Psychosensory Stimulation*)
- How to infer a psychological process from changes in pupil dilation? (*Level 3: The Psychological Mechanism*)

2.1 *Level 1: The Signal*

For an experimental psychologist, it is relatively easy to capture pupil size using an eye tracker, but that does not mean they have collected data. Much like video footage from a behavioural study that needs to be coded, the streams of eye tracking information on pupil size need to be processed. Given the versatility of modern eye trackers and the software that is used to run them, researchers are often confronted with this first level of analysis towards the end of their study, after data collection is completed. At this point, the information from the eye must be extracted from the eye tracking system. As with any physiological variable, data need to be filtered and pre-processed to remove artefacts. This is the so-called pre-processing of eye tracking data before any statistical analyses can be conducted. One example of a highly intuitive and illustrative software to guide researchers is TimeStudio, which presents a graphical user interface to carry out these steps (Nyström et al., 2016; see also Forbes, 2020). In its 'rawest' form, pupil size is captured per sample, e.g. for eye trackers that sample at 120 Hz, this would result in 120 samples per second (on average). Each sample provides information above and beyond pupil size, and eye tracking systems differ in the type of additional information provided. Based on this information on pupil size across the duration of a study session, the goal in the subsequent processing steps is to transpose the data to a format where each row presents the data collected in response to a stimulus, per subject per trial. At the same time, the exported data need to be filtered to remove outliers. The resulting data per trial can be aggregated, e.g. averaged across trials, and filtered again, e.g. dropping trials with too few data points or which exceed the sample mean by a certain standard deviation (see Carter & Luke, 2020; Kret & Sjak-Shie, 2019, for best practices advice).

2.2 *Level 2: Evidence of Psychosensory Stimulation*

At the point of successful data extraction and pre-processing, researchers have to calculate the dependent measure per unit, i.e. per individual experimental trial of their experimental design for each subject. Consequently, the question arises whether pupil dilation changes are a mere consequence of luminance effects. The pupils regulate the influx of light into the eyes, and looking at bright and dark surfaces will constrict and dilate pupils, respectively. On this second level of analysis, following the pre-processing stage, researchers are expected to plausibly rule out any luminance artefacts in their data. Given the temperamental volatility of infants and young children, it is easy to see how children gazing away from the computer screen can alter their pupil size. This may be confounded with the experimental manipulation, e.g. if one condition is less engaging and participants consequently look away more often from the screen. In addition, the changing brightness of a visual scence which children focus on as they move their eyes across the illuminated screen surface increases fluctuations in pupil size (Brisson et al., 2013) or can result in confounded variance along the scale or the levels of predictor variables, e.g. if one experimental condition uses brighter stimuli than another condition. The crux then is to demonstrate that psychosensory stimulation systematically affected changes in pupil dilation above and beyond luminance effects. From the present review of published studies, the following are the most common steps taken in the post-processing stage:

- Baseline correction of pupil data to control for the initial light adaption to the stimulus.
- Measuring the luminosity, or brightness, of the stimuli onto which pupil dilation is measured and using this information as a covariate in statistical analyses.
- Using an identical visual image across conditions during the crucial measurement of pupil dilation.

2.3 Level 3: The Psychological Mechanism

Successful pre- and post-processing of the pupillary data establishes that artefacts were removed (pre-processing) and luminance effects were ruled out or at least controlled for (post-processing). This raises the question what psychological mechanism(s) caused the observed changes in pupil dilation. Although this is the third level of analysis, it is often the initial level on which studies are based and designed, i.e. when articulating the psychological phenomenon under investigation. It is important to bear in mind that there is no one-to-one correspondence between a subject's pupil size and the psychological mechanism involved. Knowing about a person's pupil size does not reveal the specific psychological mechanism involved. At the same time and based on comparative work with non-human primates, the psychophysiological mechanisms underlying changes in pupil size are well-understood (Joshi et al., 2016; Sara & Bouret, 2012). The two broad categories of changes in pupil size, assuming that luminance is controlled for, are as follows (Granholm & Steinhauer, 2004):

- Tonic pupil size reflects a subject's arousal state much like heart rate or skin conductance. This is because the mechanism dilating the pupils results from sympathetic nervous system activity in similar ways as mechanisms driving heart rate and skin conductance change. The intuition underlying pupil dilation changes is the same as for other physiological variables, but measuring the pupil size does not require participants to wear electrodes or devices that make pupillometry particularly appealing to researchers working with infants and young children.
- Phasic changes in pupil size are a result of central nervous system activity originating in the locus coeruleus (LC), the brain's norepinephrine (NE) hub, and pathways that innervate the muscles constricting and dilating the pupils (Poe et al., 2020). Comparative work with monkeys has documented that phasic changes in LC activity are mirrored in phasic changes in pupil dilation (Joshi et al., 2016; Sara & Bouret, 2012). An increase in attention towards (motivationally) significant stimuli increases pupil dilation. NE is released during shifts in attention, and changes in pupil dilation are symptoms of such attentional shifts (Aston-Jones & Cohen, 2005; Nieuwenhuis et al., 2011). Furthermore, selectively blocking these pathways reduces psychologically induced pupil dilation.

3 Surprise and Violation of Expectation Paradigms

Viewing unexpected events triggers an attention response and results in increased pupil dilation. In adult research, the earliest applications of pupil dilation saw subjects play guessing games in which pupil size increased in periods of increased uncertainty (Friedman et al., 1973). In research with young children and, particularly, with infants, measures of pupil dilation are used in much the same way as the more historically established measures of cumulative looking time. In these violation of expectation (VOE) paradigms, participants are familiarized with a sequence of events, which tap into children's social or physical cognition. The general logic is that *if* children understand something about the social or physical principles underlying the observed events, *then* surprising outcomes, i.e. an unexpected continuation of the event, should result in increased pupil dilation. The experimental design typically revolves around a two (outcome)-by-two (context) factorial design where one of the two outcomes is unexpected for one level of the context factor (but expected for the second level of context) and vice versa. The underlying logic of the pupillary surprise response thus mirrors the intuition underlying measures of looking time: novel, unexpected events should draw children's attention more than familiar, expected ones.

The first application of pupillometry to study young children's physical cognition was pioneered by Jackson and Sirois (2009). The authors familiarized 8-month-old infants with a self-propelled train on a circular track. On the top of the track, the train would temporarily move behind an occluder on the left-hand side and exit shortly thereafter on the right-hand side (from the perspective of the child). The infants showed increased pupil dilation if the train exited from behind the occluder with a colour that was different from the one it had before it moved behind the occluder. To the infants, who at that age understand object permanence, this was surprising given that they would expect the same-coloured train to exit from behind the occluder. To validate and anchor their pupil dilation measure, the authors also measured changes in children's cumulative looking time. The above-mentioned unexpected event triggered both greater looking time and pupil dilation in children, but the authors found pupil dilation to be the more sensitive measure, i.e. less susceptible to trial effects than looking time. Importantly, the authors introduced functional data analysis techniques to model the pupil responses via the so-called spline fitting. This allowed for continuous data analysis to pinpoint the moment at which the difference between experimental conditions was the strongest, i.e. when specifically in the unfolding of the scene, the surprise response was the strongest. In subsequent work, the general pattern of results was replicated in set-ups that used the so-called drawbridge paradigm (Sirois & Jackson, 2012), and yet other work documented a developmental progression of infants' object permanence from 10 to 18 months of age (Pätzold & Liszkowski, 2020).

Building on these findings that validated the measure of pupil dilation using paradigms previously established in the context of assessing measures of cumulative looking time, other research focussed on studying children's understanding of communicative cues. In one study, 12-month-old infants expected an adult to point to a location where an interesting object was hidden, and infants showed greater pupil dilation if the container the agent communicatively pointed towards turned out to be empty. Interestingly, this was not the case if an empty container was revealed without there having been a prior point by an adult. In addition, merely highlighting a container with a visual attention getter did not induce an expectation in infants (Pätzold & Liszkowski, 2019; see also Gustafsson et al., 2015, 2016 for research on 12-month-olds' perception of social interactions and joint attention). In a further broadening of the scope of using pupillometry to study children's social cognition, other research focussed on children's understanding of others' mental states including desires and (false) beliefs. In one study, 24-month-old children viewed an agent emoting either positively or negatively towards two novel objects, respectively. In the 'positive' condition, the actress expressed their liking of the object, while in the 'negative' condition, she expressed disgust for an object. From the child's perspective, one object elicited a positive emotion while the other object elicited a negative emotion in the adult. In a test phase, in which the adult would reach for one of the two objects, children showed greater pupil dilation if the adult reached for the negatively valenced object instead of the positively valenced object. This suggested that children could cognitively represent another's desire and form an expectation based on that desire as to what the agent would subsequently preferentially reach for (Vaish et al., 2018). In the context of representing others' mental states, Dörrenberg and colleagues found evidence that 24-month-old children showed greater pupil dilation (and longer looking time) when agents showed belief-incongruent behaviour, and this pattern depended on the order in which experimental conditions were presented to the participants. The authors went to great lengths to validate the measures of pupil dilation against the measures of looking time and anticipatory looking, and indeed in some of the experimental conditions, the authors found correlations between the dependent measures (Dörrenberg et al., 2018).

To study the development of children's surprise responses to unexpected events, Krüger and colleagues studied 1- to 6-year-old children's pupillary surprise response to animal sounds that did not match the picture of the animal that the children were looking at (Krüger et al., 2020; see also Wetzel et al., 2016 for an earlier demonstration of this effect in 14-month-old infants). The design was particularly elegant because the authors used an identical visual image, e.g. a cat, and paired it with an expected (cat) or unexpected (cow) sound (see also Chen & Westermann, 2018). In this way, the authors could control the luminance of the stimuli with the only difference being the auditory cue given. This surprise response to unexpected events is not limited to early childhood. Nine-year-old children show a pupillary surprise response when their prediction of events reveals to be incorrect. Furthermore, there are individual differences such that the so-called cognitive conflict does not increase children's learning, but learning success is rather contingent on children's inhibitory control (Brod et al., 2020). This application of pupillometry to study children's memory and encoding of information mirrors findings with much younger participants. In one study, 17-month-old infants showed greater pupil dilation to mislabelled events, but children's individual surprise response did not predict the encoding of novel information (Csink et al., 2021).

In summary, measuring pupil dilation in VOE paradigms provides a viable alternative to measuring cumulative looking time although the benefits and drawbacks are still contested (Aslin, 2012). Pupillometry offers more flexibility in testing age groups (8 months to 12 years in the current review) and provides a means to time-lock children's responses in ways that cumulative looking time measures in VOE paradigms cannot. At the same time, it is noteworthy that of all the studies reviewed here, only 12% (11 out of 93) were based on VOE paradigms and several of those studies used pupil dilation to study individual differences above and beyond a factorial VOE experimental design. The sample sizes ranged from 17 to 239 (average $n = 57$, total $N = 629$). In the majority of the studies, and similar to studies measuring cumulative looking time, the statistical inference technique was based on analyses of variance (ANOVAs) or t-tests with the exception of Csink and colleagues who used a permutation analysis (Csink et al., 2021). To control for luminance artefacts, many VOE studies matched their stimuli's luminance to rule out that systematic changes across experimental condition could be a result of luminance artefacts (Jackson & Sirois, 2009; Pätzold & Liszkowski, 2019; Sirois & Jackson, 2012). In other cases, an identical image was presented while pupil dilation was measured (Krüger et al., 2020; Pätzold & Liszkowski, 2020; Vaish et al., 2018). Almost all VOE studies referred to cognitive processes when interpreting pupil size changes with the most common ones being surprise (Brod et al., 2020; Krüger et al., 2020; Vaish et al., 2018) and attention or cognitive load (Dörrenberg et al., 2018; Gustafsson et al., 2016; Pätzold & Liszkowski, 2019).

4 Task-Evoked Pupil Dilation and Neural Activity

States of increased attention result in greater pupil dilation. The connection between neural activity and pupil dilation is well-documented in adult and comparative research and captured by the term phasic task-evoked pupillary responses (Beatty & Lucero-Wagoner, 2000; see also Goldinger & Papesh, 2012). Much like event-related potentials analyses of electroencephalography, changes in pupil size that are time-locked to stimulus onset often show a characteristic peak dilation following events that trigger an attention response. Studies using VOE paradigms, as reviewed in the previous section, are thus a particular version of measuring surges in attention mirroring designs used in looking time studies. However, and importantly, the use of pupillometry is not limited to such factorial designs, and, in fact, in adult research, it is much more common to interpret changes in pupil dilation as 'markers' of neural activity without necessarily invoking a specific psychological process such as surprise. Many developmental studies with infants and young children adopted this framework of measuring pupil dilation in what we call here phasic *task-evoked pupillary dilations* (*TEPD*s) to specify that the pupillary response is a peak dilation rather than a constriction of the pupil.

Studying *TEPD*s broadens the scope of pupillometry. It allows researchers to use dynamic stimuli, e.g. videos, to measure the time-locked point-in-time changes in pupil dilation in response to a critical event. In one of the first applications of this experimental logic in developmental science, Gredebäck and Melinder (2010) presented 6- and 12-month-old infants with videos in which two (female) protagonists were modelling a feeding action: one actress moved a spoon (with food) either towards the second adult's mouth or head. This made the resolutions of the scene either rational or irrational. The authors found that the irrational resolution of moving the spoon to the recipient's head increased pupil dilation more than viewing the rational action. This was the case for 12- but not for 6-month-old infants (Gredebäck & Melinder, 2010; see also Gredebäck & Melinder 2011 for an example of employing the same paradigm with 4-month-old infants). It is important to stress that this line of work extends the scope of pupillometry beyond experimental designs, to study individual differences, e.g., in children's action perception (see also Verschoor et al., 2013, 2015). In one extension of the 'spoon-feeding paradigm', 6- and 10-month-old infants' pupillary responses were contingent on their prior anticipation of the correct outcome of the feeding action. The more infants expected the spoon to go towards the recipient's mouth, the greater their pupil dilation if the spoon went instead to a bowl (Gredebäck et al., 2018; see also Marciszko et al., 2020). Similarly, in a longitudinal study, infants' *TEPD*s in response to seeing an irrational event (one agent asking for an object by extending their arm but another actor placing the object on their head) at 9 months predicted how much children themselves would respond to a request for help at 2 years of age. This positive relation between *TEPD*s and active helping behaviour mirrored the relation between children's event-related potentials (ERP) responses and their helping behaviour (Juvrud et al., 2019). Children's pupil dilation responses when attending to others' needs may be rooted in 5-month-olds' responses to helping and hindering events (Tan & Hamlin, 2022) and, as a response, persists across early and middle childhood (Hepach & Herrmann, 2019). The scope of studying individual differences in social cognition further opens pupillometry to study clinical populations, e.g. children diagnosed with autism (DiCriscio & Troiani, 2017; Zivan et al., 2021). This includes comparing children with autism spectrum disorder (ASD) to matched controls in how they perceive faces (Anderson et al., 2006; Falck-Ytter, 2008) or others' gaze (Nuske et al., 2015) and how groups differ with regard to general visual attention (Blaser et al., 2014; DiCriscio et al., 2019) as well as attentional control (Murofushi et al., 2021).

A further experimental variation of studying *TEPD*s in pupillometry is to present stimuli in the so-called oddball paradigms that provide a window into studying children's language perception and, more specifically, their processing of speech. Hochmann and Papeo (2014) studied 3- and 6-month-old infants' *TEPD*s in response to infrequent and frequent auditory stimuli. More specifically, the authors presented infants with a 'standard' condition in which syllables of a consonant and a vowel were repeated four times. In the 'deviant' condition, the fourth repetition introduced a novel consonant–vowel syllable. Six- but not 3-month-old infants showed greater pupil dilation in response to the final syllable when it was presented in the deviant compared to the standard condition. This study extended the age range in which to study pupil dilation to include infants as young as 3 months (Hochmann & Papeo, 2014). Similar match–mismatch paradigms have subsequently been used to study *TEPD*s in 4-month-old (Marno et al., 2015), 30-month-old (Tamási et al., 2017, 2019), as well as in longitudinal studies including 18-, 21-, and 24-month-old infants (Arias-Trejo et al., 2021). In addition to studying infants' and children's pupillary responses in oddball-like paradigms, other research further broadened the scope of pupillometry to study individual differences in language processing, e.g. by comparing bilingual to monolingual children at the age of

20 months (Byers-Heinlein et al., 2017) and 31 months (Kuipers & Thierry, 2013). Furthermore, pupillometry has been used to study the priming effects in 24-month-olds (Angulo-Chavira & Arias-Trejo, 2021), speech processing in 5-month-olds (Fló, 2021) as well as in 6- to 8-month-old infants (Calignano et al., 2021), gaze following in 4-month-old infants (Michel et al., 2021), comprehension of communicative cues in 3- and 5-year-old children (Schulze & Buttelmann, 2021), and sentence comprehension in 6-year-old children with specific learning impairments (Lum et al., 2017).

Children's social cognition is not only limited to evaluating and understanding others' interactions and language perception but also encompasses studying infants' and children's processing of others' emotions. Jessen and colleagues demonstrated that viewing happy compared to fearful facial expressions triggered greater pupil dilation in 7-month-old infants – even if the stimuli were presented subliminally (Jessen et al., 2016). This underscores the value of pupillometry to index subconscious or preconscious processes as they have been documented in research with adults (Laeng & Endestad, 2012; see Laeng et al., 2012 for a review). Infants and young children show different degrees of pupil dilation in response to different facial expressions. This has been documented in the context of fear perception in 6-month-old (Hoehl et al., 2017) and 14-month-old infants (Gredebäck et al., 2012) as well as, more broadly, in the perception of basic emotions in 14-month-old infants (Aktar et al., 2018).

One suggested mechanism to explain children's sensitivity to others' expressions of emotions, even when cues are subtle and presented subconsciously, is empathic contagion and, more specifically, pupillary mimicry. Infants as young as 4, 6, and 9 months showed increased pupil dilation that was proportional to the size of the pupil in an image that they were viewing. Crucially, and based on elegant control conditions, the mimicry effects could not be explained by the luminance effects of viewing larger vs. smaller dark images in focal view (Aktar et al., 2021; Fawcett et al., 2016, 2017; Haynes et al., 2021). In addition, children's own pupillary responses may synchronize not only with the images they view but also with their caregiver's pupillary response. Aktar and colleagues studied 222 infant–parent dyads across infant age groups of 6, 12, and 18 months in a paradigm in which the authors captured both the parent's and the child's pupillary response to negative vs. neutral facial expression. Strikingly, children's pupil dilation to dynamic expression of emotions was positively related to their parent's pupil dilation to viewing the same set of stimuli, an effect that was not contingent on a specific age group (Aktar et al., 2021). Infants not only respond to different emotional expressions but also use others' expressions to predict their subsequent behaviour. Fourteen- but not 10-month-old infants showed greater pupil dilation if an actor expressing a happy emotion subsequently (angrily) thumped a toy and if an angry actor subsequently (happily) patted a toy – compared to conditions in which the emotional valence of the expression and the actions matched (Hepach & Westermann, 2013). This reduced sensitivity in 10-month-old infants may be rooted in a more general premature perception of others' actions given that 9- to 12-month-old infants do not show the same increase in pupil dilation to viewing impossible body movements as adults do (Morita et al., 2012).

In the broader context of perceiving normativity, 3- and 4-year-old children showed increased pupil dilation if an action constituted a moral transgression (i.e. harming someone) compared to when the identical action was merely a conventional transgression (Yucel et al., 2020). Similarly, and in the domain of children's emotion processing, research has expanded to include psychopathology, e.g. the underlying mechanisms of social anxiety disorders (Keil et al., 2018) and of autism spectrum conditions. Based on studying systematic changes in *TEPD*s, Nuske and colleagues found that 2- to 5-year-old children with ASD, in comparison to typically developing children, showed reduced emotional reactivity in response to fearful expressions when these were presented subliminally, but there were no group differences when stimuli were presented supraliminally (Nuske et al., 2014b). In light of the above-mentioned findings by Jessen et al., this suggests that pupillometry can not only tap into preconscious processes but it also possible to detect individual differences in children's responses to subliminal stimuli. In other work, children's pupillary responses to viewing emotional expressions were mediated by whether the individuals expressing the emotions were familiar to children, e.g. their teachers and caretakers (Nuske et al., 2014b), and differences between children with ASD and typically developing children depend on the specific emotion that children are viewing (Sepeta et al., 2012).

In the earliest application of pupillometry in adult research, participants were confronted with cognitively taxing tasks, which revealed a proportional increase in pupil dilation with increased task complexity (Kahneman et al., 1969; Kahneman & Beatty, 1966). In developmental science, this provides a way to monitor the level of children's cognitive engagement when confronted with novel and challenging tasks to provide a window into the underlying mechanisms of learning (Breitwieser & Brod, 2021; Chen & Westermann, 2018; Wu et al., 2021). In one study, Ackermann and colleagues studied 24-month-old children's curiosity in certain types of object categories and how this predicted their learning success of novel objects (from those categories). In this paradigm, children's intrinsic interest in objects from four categories including vehicles, clothes, animals, and drinks was measured by how much their pupils dilated when viewing objects of each category (with the accompanying label). Children's curiosity in novel objects was moderated by their individual interest in each category: Children learned novel objects better from categories that they were inherently interested in (Ackermann et al., 2020). Pupillometry thus provides a measure of tapping into intrinsic motivation, i.e. children's reward mechanisms, at ages as young as 7 months (Tummeltshammer et al., 2019). More broadly, changes in children's cognitive load have been studied in the domains of episodic memory at 4 to 7 months (Hellmer et al., 2018); scalar errors at 4, 6, 10, and 14 months (Addyman et al., 2014); prediction errors and implicit learning at 6 months (Zhang et al., 2019); visual attention at 8 months (Pérez et al., 2020) as well as at 10, 14, and 18 months (Achermann et al., 2021); logical reasoning inferences at 10 months (Ekramnia et al., 2021); visual working memory at 13 months (Cheng et al., 2019); predictive processing at 18 and 24 months (Kayhan et al., 2019); cognitive control at 3 and 8 years (Chatham et al., 2009) as well as 9 years (Chevalier, 2018); and executive functions in the context of bilingualism at 8 years (Enke et al., 2022).

In summary, the study of phasic, point-in-time task-evoked pupillary dilations (*TEPD*s) moved pupillometry beyond the scope of VOE paradigms to more fully embrace the study of individual differences. This research emphasis is also reflected in the fact that the majority of studies with developmental populations published to date were based on measures of *TEPD*s (67%, 62 out of 93). Of these 62 studies, 22 used an experimental design based on two experimental conditions in which consistent, matching, rational, familiar, appropriate, congruent, expected, normal, old, possible, probable, or standard events were compared with inconsistent, non-matching, irrational, novel, inappropriate, incongruent, unexpected, backward, new, impossible, improbable, or deviant events, respectively. The sample sizes ranged from 12 to 249 (average $n = 56$, total $N = 3520$) and the ages studied ranged from 3 months to 16 years. The study of *TEPD*s further documents the great diversification in the statistical modelling techniques. While the majority of studies (61%, 38 out of 62) used an ANOVA- or t-test-based analysis approach, 10 studies used a mixed-modelling approach and 6 used linear regression, including hierarchical models, to estimate the effects on pupil dilation. This shift is a result of *TEPD*s being used to study individual differences beyond studying group-level effects in factorial designs.

5 Arousal State and Motivation

Pupil dilation, similar to other physiological variables, reflects autonomous nervous system activity (Bradley et al., 2008; Kahneman et al., 1969; Libby Jr et al., 1973) even in children (Wass et al., 2015). The psychological states corresponding to states of increased arousal can be described as subjects being 'switched on'. In other words, a subject's tonic arousal level, as measured in pupil size over multiple seconds or even minutes, is higher – compared to a baseline period – in states of heightened psychological engagement. It is worth bearing in mind that measuring children's arousal state does not tap into fundamentally different psychological processes than measuring phasic task-evoked pupillary dilations. Rather, the studies reviewed below addressed questions that did not require researchers to pinpoint the onset of a pupillary response. Instead, measuring the so-called tonic state-evoked pupil dilations (*SEPD*s) provides a means to study individual differences in tonic baseline arousal states (e.g. identifying hyperactive individuals), physiological responses to emotional stimuli, and motivation.

The pursuit of psychological goals is accompanied by corresponding physiological changes. Whilst the goal, or need, remains unfulfilled, subjects are in states of increased psychological tension until the goal is attained (Hornstein, 1972). This psychological tension can be measured via changes in pupil dilation through *SEPD*s. In the context of studying 18-month-old infants' to 5-year-old children's motivation to help others, children's change in pupil dilation in response to seeing others needing help was related to whether and how fast they subsequently helped others. The greater the change in pupil dilation, the more likely and faster children were to help. This positive relation was not observed for children's baseline pupil dilation, and, in fact, the relation persisted if baseline levels of tonic arousal state were controlled for (Hepach et al., 2013, 2016, 2019a). In addition to measuring children's response to pupil dilation and building on the logic underlying psychological tension, it is further possible to measure pupil dilation after the resolution of the helping situation. For example, 2- and 3-year-old children who saw an adult needing help showed similar levels of resolution pupil dilation after they themselves helped and after seeing someone else provided help (Hepach et al., 2012, 2017a, 2019b). This was not the case if children themselves were the cause for the adult needing help in which case pupil dilation was greater after someone else helped compared to when children could remedy the situation (Hepach et al., 2017a). In addition, and similarly, if the adult needing help had previously helped the child, children's pupil dilation remained elevated if they merely saw someone else provide the help, but pupil dilation was lower if *they* could in return help their previous benefactor (Hepach et al., 2019b). This overall pattern of resolution pupil dilation across studies provided a means to infer whether the initial motivation (to help) was resolved in which case pupil dilation should return to baseline levels. In guilt- and gratitude-like situations, children's motivation to help was to provide the help themselves, possibly to remedy the situation or to return a previously received help (see also Hepach, 2017).

Children's *SEPD*s in response to viewing others' emotional displays reflect changes in their own emotional arousal. Geangu and colleagues presented 6- and 12-month-old infants with videos of another infant expressing a neutral, positive, or negative emotional expression. Participants' pupil dilation was measured across a time window of 25 s as they watched the videos. Looking time patterns revealed that children focussed most on the face, and pupil dilation patterns showed that, in particular, the expressed negative emotion resulted in greater pupil dilation (Geangu et al., 2011). In this application of the method, it is noteworthy that the authors did not only document phasic *TEPD*s but rather children's tonic arousal state, i.e. *SEPD*s, remained elevated while watching negative emotions expressed on video. Similar work found that infants' individual pupil dilation to viewing videos of positive and negative emotions in other infants was related to their parents' self-reported altruism and perspective taking (Upshaw et al., 2015). This raises the possibility that children's pupil dilation responses are not only a biophysiological reflex to viewing others' emotions but also shaped by their cultural upbringing from early in development. Prunty and colleagues presented 6-, 9-, and 12-month-old infants with both static and dynamic expressions of basic emotions and found that dynamic stimuli elicited greater *SEPD*s than static images (Prunty et al., 2022). Based on critical luminance controls, such as including stimulus luminosity in their statistical modelling, the authors could rule out that this effect was driven by luminosity differences between the two types of stimulus categories. In contrast to prior work, the authors found *SEPD*s in response to expressions of anger and happiness but not for any of the other basic emotions. This raises the question whether children's *SEPD*s are influenced by how much of the scene they are viewing. In videos that show the face, effects may be different compared to videos that show the face, body, and part of the surroundings. In addition, children's *SEPD*s in response to viewing emotional facial expressions may be further moderated by individual differences in infants' temperament and parents' emotionality (Aktar et al., 2016).

The study of *SEPD*s provides a means to capture not only individual- but also group-level differences in tonic arousal states. In a study comparing autistic and typically developing children at the age of 4 years, Nuske and colleagues found that tonic baseline pupil diameter did not differ between groups (Nuske et al., 2014a). However, in addition to studying resting-state physiology, the so-called hypersensitivity may be a marker or developmental precursor to autism. Nyström and colleagues found that 10-month-old infants with an older autistic sibling showed a greater pupillary constriction, i.e. a greater reduction in response to light, in comparison to a matched group of infants who did not have an autistic sibling. There was no difference in tonic pupil diameter across groups, and the authors concluded that autism may be accompanied by a hypersensitivity to light instead of an overall elevated arousal state (Nyström et al., 2015). In a crucial follow-up to their initial findings, Nyström and colleagues found that 3-year-old children who received an autism diagnosis had greater pupillary constriction at age 9 to 10 months than a matched control group (Nyström et al., 2018). Other work investigating the underlying mechanisms of autism found differences across groups in children's emotional reactivity to emotionally loaded objects. For example, in a social-referencing paradigm, children watched videos of an adult emoting positively or negatively towards a box, but children never saw the content of the box. However, typically developing 2- to 5-year-old children showed increased *SEPD*s, as an indicator of emotional reactivity, in response to viewing a close-up of the box, whereas no such pattern was found for same-age children diagnosed with autism (Nuske et al., 2016). In contexts that involved children viewing others needing help, 2- to 5-year-old children, both with autism and typically developing, showed greater pupil dilation when an adult's need remained unfulfilled, whereas no such increase was found if a non-social control remained unresolved (Hepach et al., 2020). Overall, the results from these studies point to context specificity and atypical emotional reactivity in children with autism, and more research is needed to further explore this phenomenon.

Overall, pupillometry studies that measure *SEPD*s further broaden researchers' experimental design toolkits. In many of the studies reviewed above, the stimulus onto which pupil dilation was measured was not itself part of the experimental manipulation and therefore remained identical across experimental conditions or across groups. This is because in studying *SEPD*s, researchers are interested in individual differences in response to the identical visual stimulus or they are interested in the effects of an experimental manipulation that occurs between the presentations of a neutral stimulus onto which pupil dilation is measured. In other work, the versatility of measuring *SEPD*s has been demonstrated in studies on children's memory and imitation (Sonne et al., 2016), visual attention (Wass & Smith, 2014), metacognition (Paulus et al., 2013), and music perception (Bainbridge et al., 2021).

In summary, the study of tonic state-evoked pupil dilations (*SEPDs*) is the third most common experimental paradigm used in developmental pupillometry; 20 out of the 93 studies (22%) reviewed here focussed their analyses on measuring the tonic arousal state or shifts in the tonic arousal state. The sample sizes ranged from 12 to 211 (average n = 79, total N = 1583), and the ages studied ranged from 2 months to 9 years. The majority of studies (15 out of 20) used an ANOVA- or t-test-based analysis approach, while other studies used multiple linear regressions or linear mixed models.

6 Current Challenges and Future Directions

Developmental pupillometry is an emerging, fast expanding method, but it is also being applied in an area of research that is still evolving. This is in part due to individual research groups within their specific field of study having turned to measuring pupil dilation because it provided a practical and/or conceptual solution to a problem they were facing in addressing a research question. While each research team worked through the three levels of analysis (see *Theoretical and Conceptual Foundations of Developmental Pupillometry*), there is nevertheless a considerable, and impressive, degree of overlap across studies in how practical problems of pre- and post-processing pupil data were resolved. The high level of congruence notwithstanding, no standard protocol on how to process pupil data exists to date. Moving forward, the field may limit itself if opportunities for consolidation and coordination are missed. In the following, we turn to specific points that stood out from reviewing the extant literature on developmental pupillometry.

6.1 Studies Using Pupil Dilation Require Their Own Design

There is no free lunch when it comes to collecting pupil size information in any given eye tracking study. The idea that one can ‘go back’ to a completed eye-tracking study and perform post hoc analyses on pupil diameter, even if those were preregistered, is misleading. There are numerous sources of luminance artefacts that could potentially confound any experimental manipulation. A study intending to analyse changes in pupil dilation, as a dependent variable, should be designed as a pupillometry study from the beginning, especially with regard to the age range under investigation, the number of participants, the duration of the baseline phase, the duration of the analysis window, and the kind of luminance control. The following are the most common design decisions for each of the three research areas in which pupillometry has been used:

- *Surprise and violation of expectation paradigms (TEPDs)*
  - Participants’ ages ranged from 8 months to 12 years.

  - Number of participants in individual studies ranged from 17 to 239.
  - The length of the baseline chosen ranged from 117 ms to 2 s (*Median* = 533 ms).
  - The length of the window in which the effect was measured ranged from 1 s to 27 s (*Median* = 6 s).
  - Common luminance controls included using an identical visual image across experimental conditions, matching the luminance of stimuli, and investigating whether changes in looking time could statistically explain changes in pupil dilation.

- *Task-evoked pupil dilation and neural activity* (*TEPDs*)

  - Participants' ages ranged from 3 months to 16 years.
  - Number of participants in individual studies ranged from 12 to 249.
  - The length of baselines chosen ranged from 100 ms to 10 s (*Median* = 500 ms).
  - The length of the window in which the effect was measured ranged from 900 ms to 64 s (*Median* = 3.4 s).
  - Common luminance controls included using an identical visual image across experimental conditions, matching the luminance of stimuli, and comparing the luminance (as a dependent measure) across conditions.

- *Arousal state and motivation* (*SEPD*s).

  - Participants' ages ranged from 2 months to 9 years.
  - Number of participants in individual studies ranged from 12 to 211.
  - The length of baselines chosen ranged from 100 ms to 60 s (*Median* = 7 s).
  - The length of the window in which the effect was measured ranged from 1.2 s to 20 mins (*Median* = 10 s).
  - Common luminance controls included using an identical visual image across experimental conditions, using identical video used across conditions, and matching the luminance of stimuli.

Considerations of the type of design to use go hand in hand with the decision of how to best pre-process the pupil size information and how to calculate the dependent measure per participant. The following offers a selection of questions for which individual studies needed (and need) to provide answers to:

- *Should data for the left eye, the right eye, or both eyes be included?* The majority of the studies reviewed here calculated the average pupil size available for both eyes given the high correlation between left and right pupil information. However, it is worth considering whether to include samples for which data from only one eye could be tracked. These may represent cases with low overall measuring accuracy, and including such samples may increase signal noise.
- *Do we use pupil dilation from every individual sample, or do we exclude samples that were **not** mapped to be part of a fixation?* Most eye tracking studies, especially those in which fixation patterns were analysed, will exclude samples that were part of a saccade or could not be classified as either a fixation or a saccade. Applying a fixation filter to pupil dilation will reduce the number of data points but could increase the stability of the signal. One consequence of this approach

is that a separate decision needs to be made with regard to how a fixation is defined (Wass et al., 2013).

- *How do we filter pupil data?* The vast majority of pupillometry studies apply some kind of filter given that the inaccuracy of the eye tracking system can result in data points that are physiologically impossible, e.g. a doubling in pupil size from one sample to another. Several examples exist such as moving average filters (Gredebäck & Melinder, 2010), low-pass digital filters (Jackson & Sirois, 2009), quantile-cut-off filters (Hepach et al., 2012), etc.
- *Do we interpolate gaps in pupil dilation data?* This is not a trivial question. The size and the duration of gaps between eye tracking samples vary. One needs to consider the costs and benefits of interpolating across large gaps in data, especially if those were the result of the participant shifting their attention away from the computer screen.

In the age of open science, researchers will seek to preregister their pupillometry designs prior to data collection. It is noteworthy that of all the studies reviewed here, only five studies (just over 5%) included a preregistration. To date, there exist no guidelines as to what pre-processing steps need to be included in data analysis. One promising avenue is illustrated by a recent call for researchers to take part in a multi-laboratory collaborative data analysis project. The key idea is for laboratories to apply their established processing pipelines to identical data set to answer the same question (Sirois et al., 2023; Calignano et al., 2023). Similar undertakings, though outside the scope of pupillometry, have resulted in somewhat heterogeneous analysis strategies with in part conflicting conclusions (Silberzahn et al., 2018).

6.2 How Many and Which Psychological Processes Increase Pupil Dilation?

There are many psychological processes that increase pupil dilation. From the studies reviewed here, the most common processes were arousal (30%), cognitive load (19%), attention (11%), and surprise (7%). In total, the 93 studies reviewed revealed 25 different psychological processes, even after grouping studies with nuanced differences in the processes they suggested. This heterogeneity is a consequence of the diverse psychological phenomena studied, i.e., the research questions that gave rise to the respective study conducted. The five most common phenomena were emotion processing (8%), language and speech processing (6%), pupil mimicry (4%), helping behaviour (4%), and autism and emotion processing (4%). A total of 66 phenomena were investigated. This breadth in the spectrum of applications of developmental pupillometry is a strength of the method. It illustrates its versatility to address diverse research questions. At the same time, as the number of (published) pupillometry studies in development nears 100, it seems appropriate to consider whether there are indeed fewer broad processes underlying the currently documented psychologically mediated effects on children's pupil size.

One step towards a more unified understanding of the psychological mechanisms involved is to appreciate the three levels of analysis introduced in this chapter: the signal, the psychosensory stimulation, and the psychological processes. When reporting and interpreting pupil dilation data, it is worth asking oneself whether this could be done descriptively, i.e. stating 'We found greater pupil dilation [...].' as opposed to 'We found greater mental effort [...].', especially in the Results section of a study. Given that changes in pupil dilation reflect so many psychological processes, it would be misleading to write of increases in pupil dilation reflecting a singular activity in mental effort, surprise, attention, or cognitive load. It is difficult, if not impossible, to – with a sufficient degree of certainty – identify a singular psychological process or motivation from changes in pupil dilation documented in a single study. One example: in the line of research measuring young children's motivation to help others, as reviewed under the *Arousal and Motivation* section above, children showed increased *SEPD*s if others were not helped at all or not helped in ways to fulfil the child's initial motivation. In these scenarios, the specific increase in pupil dilation, in a given experimental condition, could be the result of several – in part – competing motivations: to see if others helped, to interact with others, to avoid guilt, or to seek social approval (see Pletti et al., 2017). Therefore, the psychological process, or motivation, needs to be inferred across experimental conditions or better across studies and methods. In this specific example, Hepach and colleagues built on their initial findings that 2-year-old children were motivated to see others helped (Hepach et al., 2012) through (1) replicating the pattern of results in separate studies (Hepach et al., 2017a, 2019b), (2) documenting a negative relation between changes in pupil dilation and the latency to help (Hepach et al., 2013, 2019a), and (3) using posture analyses methods showing increased upper-body posture effects in 2-year-olds if others were helped, thus corroborating earlier findings using pupillometry (Hepach et al., 2017b, 2022).

A further step forward may be to build on the unified approach currently pursued in pre-processing pupil dilation data and to apply the same approach to the identification of psychological mechanisms. As it stands, the same psychological mechanism (e.g. 'cognitive load') is implicated in nine different areas of research including studies on communication, behavioural inhibition, face perception in autism, cognitive control, executive functions, language processing, logical reasoning, memory, and scalar errors. Conversely, and focussing on language processing as an example, different processes have been implicated in addition to 'cognitive load'. These include pupil dilation as a marker of neural activity, arousal, cognitive resources, or attention. Following on from the argument above, changes in pupil dilation are not diagnostic of a single psychological mechanism. Rather, they provide an invaluable means to measure the degree of arousal, cognitive load, etc. That is, once a more parsimonious ascription of psychological processes to underlying psychological phenomena is adopted, one can, for example, begin to compare the extent to which memory tasks elicit greater pupil dilation than language processing tasks or the degree to which viewing negative emotions results in greater pupil dilation than viewing positive emotions. This will pave the way for a systematic review of the effect sizes reported in published studies to provide an answer to the question of what effects can be considered small, medium, or large in pupillometry work with developmental populations.

7 Conclusions

The application of pupillometry in the three main areas of research reviewed here can be described by a popular US folk rhyme: *something old, something borrowed, something new, and something blue*. The 'old' element is the introduction of studying task-evoked pupil dilations (*TEPD*s) to measure neural activity in infants and young children. Furthermore, pupillometry 'borrowed' the VOE paradigms that traditionally analysed changes in participants' looking time. The 'new' element is represented by the studies on arousal and motivation (*SEPD*s), an application of the method that is now more numerous in developmental populations compared to research with adults. Finally, 'blue' may reflect the sentiment that some researchers feel when the application of pupillometry proves to be challenging given the volatility of the pupillary signal. The present chapter sought to demonstrate that systematically studying psychologically induced changes in pupil size has allowed and will continue to allow researchers to map infants' and toddlers' internal psychological engagement onto external events. After all, developmental psychologists are generally, though not singularly, interested in finding ways to look inside a child's mind. Pupillometry provides such a way.

References

Achermann, S., Falck-Ytter, T., Bölte, S., & Nyström, P. (2021). Updating expectations about unexpected object motion in infants later diagnosed with autism spectrum disorder. *Journal of Autism and Developmental Disorders, 51*(11), 4186–4198.

Ackermann, L., Hepach, R., & Mani, N. (2020). Children learn words easier when they are interested in the category to which the word belongs. *Developmental Science, 23*(3), e12915.

Addyman, C., Rocha, S., & Mareschal, D. (2014). Mapping the origins of time: Scalar errors in infant time estimation. *Developmental Psychology, 50*(8), 2030.

Aktar, E., Mandell, D. J., de Vente, W., Majdandžić, M., Raijmakers, M. E., & Bögels, S. M. (2016). Infants' temperament and mothers', and fathers' depression predict infants' attention to objects paired with emotional faces. *Journal of Abnormal Child Psychology, 44*(5), 975–990.

Aktar, E., Mandell, D. J., de Vente, W., Majdandžić, M., Oort, F. J., van Renswoude, D. R., Raijmakers, M. E., & Bögels, S. M. (2018). Parental negative emotions are related to behavioral and pupillary correlates of infants' attention to facial expressions of emotion. *Infant Behavior and Development, 53*, 101–111.

Aktar, E., Nimphy, C. A., Kret, M. E., Pérez-Edgar, K., Bögels, S. M., & Raijmakers, M. E. (2021). Pupil responses to dynamic negative facial expressions of emotion in infants and parents. *Developmental Psychobiology, 63*(7), e22190.

Anderson, C. J., & Colombo, J. (2009). Larger tonic pupil size in young children with autism spectrum disorder. *Developmental Psychobiology: The Journal of the International Society for Developmental Psychobiology, 51*(2), 207–211.

Anderson, C. J., Colombo, J., & Jill Shaddy, D. (2006). Visual scanning and pupillary responses in young children with autism spectrum disorder. *Journal of Clinical and Experimental Neuropsychology, 28*(7), 1238–1256.

Angulo-Chavira, A. Q., & Arias-Trejo, N. (2021). Mediated semantic priming interference in toddlers as seen through pupil dynamics. *Journal of Experimental Child Psychology, 208*, 105146.

Arias-Trejo, N., Angulo-Chavira, A. Q., Avila-Varela, D. S., Chua-Rodriguez, F., & Mani, N. (2021). Developmental changes in phonological and semantic priming effects in Spanish-speaking toddlers. *Developmental Psychology, 58*, 236.

Aslin, R. N. (2012). Infant eyes: A window on cognitive development. *Infancy, 17*(1), 126–140.

Aslin, R. N., & McMurray, B. (2004). Automated corneal-reflection eye tracking in infancy: Methodological developments and applications to cognition. In *Infancy* (Vol. 6, Issue 2, pp. 155–163). Wiley Online Library.

Aston-Jones, G., & Cohen, J. D. (2005). An integrative theory of locus coeruleus-norepinephrine function: Adaptive gain and optimal performance. *Annual Review of Neuroscience, 28*, 403–450.

Bainbridge, C. M., Bertolo, M., Youngers, J., Atwood, S., Yurdum, L., Simson, J., Lopez, K., Xing, F., Martin, A., & Mehr, S. A. (2021). Infants relax in response to unfamiliar foreign lullabies. *Nature Human Behaviour, 5*(2), 256–264.

Beatty, J., & Lucero-Wagoner, B. (2000). The pupillary system. In *Handbook of psychophysiology* (Vol. 2, pp. 142–162).

Blaser, E., Eglington, L., Carter, A. S., & Kaldy, Z. (2014). Pupillometry reveals a mechanism for the Autism Spectrum Disorder (ASD) advantage in visual tasks. *Scientific Reports, 4*(1), 1–5.

Bradley, M. M., Miccoli, L., Escrig, M. A., & Lang, P. J. (2008). The pupil as a measure of emotional arousal and autonomic activation. *Psychophysiology, 45*(4), 602–607.

Bradshaw, J. (1967). Pupil size as a measure of arousal during information processing. *Nature, 216*(5114), 515–516.

Breitwieser, J., & Brod, G. (2021). Cognitive prerequisites for generative learning: Why some learning strategies are more effective than others. *Child Development, 92*(1), 258–272.

Brisson, J., Mainville, M., Mailloux, D., Beaulieu, C., Serres, J., & Sirois, S. (2013). Pupil diameter measurement errors as a function of gaze direction in corneal reflection eyetrackers. *Behavior Research Methods, 45*(4), 1322–1331.

Brod, G., Breitwieser, J., Hasselhorn, M., & Bunge, S. A. (2020). Being proven wrong elicits learning in children–but only in those with higher executive function skills. *Developmental Science, 23*(3), e12916.

Byers-Heinlein, K., Morin-Lessard, E., & Lew-Williams, C. (2017). Bilingual infants control their languages as they listen. *Proceedings of the National Academy of Sciences, 114*(34), 9032–9037.

Calignano, G., Girardi, P., & Altoè, G. (2023). First steps into the pupillometry multiverse of developmental science. *Behavior Research Methods*, 1–20.

Calignano, G., Dispaldro, M., Russo, S., & Valenza, E. (2021). Attentional engagement during syllable discrimination: The role of salient prosodic cues in 6-to 8-month-old infants. *Infant Behavior and Development, 62*, 101504.

Carter, B. T., & Luke, S. G. (2020). Best practices in eye tracking research. *International Journal of Psychophysiology, 155*, 49–62.

Chatham, C. H., Frank, M. J., & Munakata, Y. (2009). Pupillometric and behavioral markers of a developmental shift in the temporal dynamics of cognitive control. *Proceedings of the National Academy of Sciences, 106*(14), 5529–5533.

Chen, Y.-C., & Westermann, G. (2018). Different novelties revealed by infants' pupillary responses. *Scientific Reports, 8*(1), 1–8.

Cheng, C., Kaldy, Z., & Blaser, E. (2019). Focused attention predicts visual working memory performance in 13-month-old infants: A pupillometric study. *Developmental Cognitive Neuroscience, 36*, 100616.

Chevalier, N. (2018). Willing to think hard? The subjective value of cognitive effort in children. *Child Development, 89*(4), 1283–1295.

Cohen, L. B. (2009). The evolution of infant cognition: A personal account. *Infancy, 14*(4), 403–413.
Cohen, L. B., & Cashon, C. H. (2007). Infant cognition. In *Handbook of child psychology* (Vol. 2). https://doi.org/10.1002/9780470147658.chpsy0205
Csink, V., Mareschal, D., & Gliga, T. (2021). Does surprise enhance infant memory? Assessing the impact of the encoding context on subsequent object recognition. *Infancy, 26*(2), 303–318.
DiCriscio, A. S., & Troiani, V. (2017). Pupil adaptation corresponds to quantitative measures of autism traits in children. *Scientific Reports, 7*(1), 1–9.
DiCriscio, A. S., Hu, Y., & Troiani, V. (2019). Brief report: Visual perception, task-induced pupil response trajectories and ASD features in children. *Journal of Autism and Developmental Disorders, 49*(7), 3016–3030.
Dörrenberg, S., Rakoczy, H., & Liszkowski, U. (2018). How (not) to measure infant theory of mind: Testing the replicability and validity of four non-verbal measures. *Cognitive Development, 46*, 12–30.
Eckstein, M. K., Guerra-Carrillo, B., Singley, A. T. M., & Bunge, S. A. (2017). Beyond eye gaze: What else can eyetracking reveal about cognition and cognitive development? *Developmental Cognitive Neuroscience, 25*, 69–91.
Ekramnia, M., Mehler, J., & Dehaene-Lambertz, G. (2021). Disjunctive inference in preverbal infants. *Iscience, 24*(10), 103203.
Enke, S., Gunzenhauser, C., Hepach, R., Karbach, J., & Saalbach, H. (2022). Differences in cognitive processing? The role of verbal processes and mental effort in bilingual and monolingual children's planning performance. *Journal of Experimental Child Psychology, 213*, 105255.
Falck-Ytter, T. (2008). Face inversion effects in autism: A combined looking time and pupillometric study. *Autism Research, 1*(5), 297–306.
Fantz, R. L. (1963). Pattern vision in newborn infants. *Science, 140*(3564), 296–297.
Fantz, R. L. (1964). Visual experience in infants: Decreased attention to familiar patterns relative to novel ones. *Science, 146*(3644), 668–670.
Fawcett, C., Wesevich, V., & Gredebäck, G. (2016). Pupillary contagion in infancy: Evidence for spontaneous transfer of arousal. *Psychological Science, 27*(7), 997–1003.
Fawcett, C., Arslan, M., Falck-Ytter, T., Roeyers, H., & Gredebäck, G. (2017). Human eyes with dilated pupils induce pupillary contagion in infants. *Scientific Reports, 7*(1), 1–7.
Fitzgerald, H. E. (1968). Autonomic pupillary reflex activity during early infancy and its relation to social and nonsocial visual stimuli. *Journal of Experimental Child Psychology, 6*(3), 470–482.
Fitzgerald, H. E., Lintz, L. M., Brackbill, Y., & Adams, G. (1967). Time perception and conditioning an autonomic response in human infants. *Perceptual and Motor Skills, 24*(2), 479–486.
Fló, A. (2021). Evidence of ordinal position encoding of sequences extracted from continuous speech. *Cognition, 213*, 104646.
Forbes, S. H. (2020). PupillometryR: An R package for preparing and analysing pupillometry data. *Journal of Open Source Software, 5*(50), 2285.
Friedman, D., Hakerem, G., Sutton, S., & Fleiss, J. L. (1973). Effect of stimulus uncertainty on the pupillary dilation response and the vertex evoked potential. *Electroencephalography and Clinical Neurophysiology, 34*(5), 475–484.
Geangu, E., Hauf, P., Bhardwaj, R., & Bentz, W. (2011). Infant pupil diameter changes in response to others' positive and negative emotions. *PLoS One, 6*(11), e27132.
Goldinger, S. D., & Papesh, M. H. (2012). Pupil dilation reflects the creation and retrieval of memories. *Current Directions in Psychological Science, 21*(2), 90–95.
Goldwater, B. C. (1972). Psychological significance of pupillary movements. *Psychological Bulletin, 77*(5), 340.
Granholm, E. E., & Steinhauer, S. R. (2004). Pupillometric measures of cognitive and emotional processes. *International Journal of Psychophysiology, 52*(1), 1–6. https://doi.org/10.1016/j.ijpsycho.2003.12.001
Gredebäck, G., & Melinder, A. (2010). Infants' understanding of everyday social interactions: A dual process account. *Cognition, 114*(2), 197–206.

Gredebäck, G., & Melinder, A. (2011). Teleological reasoning in 4-month-old infants: Pupil dilations and contextual constraints. *PLoS One, 6*(10), e26487.

Gredebäck, G., Johnson, S., & von Hofsten, C. (2009). Eye tracking in infancy research. *Developmental Neuropsychology, 35*(1), 1–19.

Gredebäck, G., Eriksson, M., Schmitow, C., Laeng, B., & Stenberg, G. (2012). Individual differences in face processing: Infants' scanning patterns and pupil dilations are influenced by the distribution of parental leave. *Infancy, 17*(1), 79–101.

Gredebäck, G., Lindskog, M., Juvrud, J. C., Green, D., & Marciszko, C. (2018). Action prediction allows hypothesis testing via internal forward models at 6 months of age. *Frontiers in Psychology, 9*, 290.

Gustafsson, E., Brisson, J., Beaulieu, C., Mainville, M., Mailloux, D., & Sirois, S. (2015). How do infants recognize joint attention? *Infant Behavior and Development, 40*, 64–72.

Gustafsson, E., Brisson, J., Mailloux, D., Mainville, M., Beaulieu, C., & Sirois, S. (2016). Do infants recognize engagement in social interactions? The case of face-to-face conversation. *Infancy, 21*(5), 685–696.

Haynes, K. T., Kelsey, C. M., & Grossmann, T. (2021). Probing infants' sensitivity to pupil size when viewing eyes. *Infancy, 26*(2), 291–302.

Hellmer, K., Söderlund, H., & Gredebäck, G. (2018). The eye of the retriever: Developing episodic memory mechanisms in preverbal infants assessed through pupil dilation. *Developmental Science, 21*(2), e12520.

Hepach, R. (2017). Prosocial arousal in children. *Child Development Perspectives, 11*(1), 50–55.

Hepach, R., & Herrmann, E. (2019). The development of prosocial attention across two cultures. *Frontiers in Psychology, 10*, 138.

Hepach, R., & Westermann, G. (2013). Infants' sensitivity to the congruence of others' emotions and actions. *Journal of Experimental Child Psychology, 115*(1), 16–29.

Hepach, R., & Westermann, G. (2016). Pupillometry in infancy research. *Journal of Cognition and Development, 17*(3), 359–377.

Hepach, R., Vaish, A., & Tomasello, M. (2012). Young children are intrinsically motivated to see others helped. *Psychological Science, 23*(9), 967–972.

Hepach, R., Vaish, A., & Tomasello, M. (2013). A new look at children's prosocial motivation. *Infancy, 18*(1), 67–90.

Hepach, R., Vaish, A., Grossmann, T., & Tomasello, M. (2016). Young children want to see others get the help they need. *Child Development, 87*(6), 1703–1714.

Hepach, R., Vaish, A., & Tomasello, M. (2017a). Children's intrinsic motivation to provide help themselves after accidentally harming others. *Child Development, 88*(4), 1251–1264.

Hepach, R., Vaish, A., & Tomasello, M. (2017b). The fulfillment of others' needs elevates children's body posture. *Developmental Psychology, 53*(1), 100.

Hepach, R., Vaish, A., Müller, K., & Tomasello, M. (2019a). The relation between young children's physiological arousal and their motivation to help others. *Neuropsychologia, 126*, 113–119.

Hepach, R., Vaish, A., Müller, K., & Tomasello, M. (2019b). Toddlers' intrinsic motivation to return help to their benefactor. *Journal of Experimental Child Psychology, 188*, 104658.

Hepach, R., Hedley, D., & Nuske, H. J. (2020). Prosocial attention in children with and without autism spectrum disorder: Dissociation between anticipatory gaze and internal arousal. *Journal of Abnormal Child Psychology, 48*(4), 589–605.

Hepach, R., Engelmann, J. M., Herrmann, E., Gerdemann, S. C., & Tomasello, M. (2022). Evidence for a developmental shift in the motivation underlying helping in early childhood. *Developmental Science*, e13253.

Hess, E. H., & Polt, J. M. (1960). Pupil size as related to interest value of visual stimuli. *Science, 132*(3423), 349–350.

Hochmann, J.-R., & Papeo, L. (2014). The invariance problem in infancy: A pupillometry study. *Psychological Science, 25*(11), 2038–2046.

Hoehl, S., Hellmer, K., Johansson, M., & Gredebäck, G. (2017). Itsy bitsy spider…: Infants react with increased arousal to spiders and snakes. *Frontiers in Psychology, 8*, 1710.

Hornstein, H. A. (1972). Promotive tension: The basis of prosocial behavior from a lewinian perspective1. *Journal of Social Issues, 28*(3), 191–218.

Jackson, I., & Sirois, S. (2009). Infant cognition: Going full factorial with pupil dilation. *Developmental Science, 12*(4), 670–679.

Jessen, S., Altvater-Mackensen, N., & Grossmann, T. (2016). Pupillary responses reveal infants' discrimination of facial emotions independent of conscious perception. *Cognition, 150*, 163–169.

Joshi, S., Li, Y., Kalwani, R. M., & Gold, J. I. (2016). Relationships between pupil diameter and neuronal activity in the locus coeruleus, colliculi, and cingulate cortex. *Neuron, 89*(1), 221–234.

Juvrud, J., Bakker, M., Kaduk, K., DeValk, J. M., Gredebäck, G., & Kenward, B. (2019). Longitudinal continuity in understanding and production of giving-related behavior from infancy to childhood. *Child Development, 90*(2), e182–e191.

Kahneman, D., & Beatty, J. (1966). Pupil diameter and load on memory. *Science, 154*(3756), 1583–1585.

Kahneman, D., Tursky, B., Shapiro, D., & Crider, A. (1969). Pupillary, heart rate, and skin resistance changes during a mental task. *Journal of Experimental Psychology, 79*(1p1), 164.

Kayhan, E., Heil, L., Kwisthout, J., van Rooij, I., Hunnius, S., & Bekkering, H. (2019). Young children integrate current observations, priors and agent information to predict others' actions. *PLoS One, 14*(5), e0200976.

Keil, V., Hepach, R., Vierrath, S., Caffier, D., Tuschen-Caffier, B., Klein, C., & Schmitz, J. (2018). Children with social anxiety disorder show blunted pupillary reactivity and altered eye contact processing in response to emotional faces: Insights from pupillometry and eye movements. *Journal of Anxiety Disorders, 58*, 61–69.

Kret, M. E., & Sjak-Shie, E. E. (2019). Preprocessing pupil size data: Guidelines and code. *Behavior Research Methods, 51*(3), 1336–1342.

Krüger, M., Bartels, W., & Krist, H. (2020). Illuminating the dark ages: Pupil dilation as a measure of expectancy violation across the life span. *Child Development, 91*(6), 2221–2236.

Kuipers, J.-R., & Thierry, G. (2013). ERP-pupil size correlations reveal how bilingualism enhances cognitive flexibility. *Cortex, 49*(10), 2853–2860.

Laeng, B., & Endestad, T. (2012). Bright illusions reduce the eye's pupil. *Proceedings of the National Academy of Sciences, 109*(6), 2162–2167.

Laeng, B., Sirois, S., & Gredebäck, G. (2012). Pupillometry: A window to the preconscious? *Perspectives on Psychological Science, 7*(1), 18–27.

Libby, W. L., Jr., Lacey, B. C., & Lacey, J. I. (1973). Pupillary and cardiac activity during visual attention. *Psychophysiology, 10*(3), 270–294.

Loewenfeld, I. E. (1993). *The pupil: Anatomy, physiology, and clinical applications* (Vol. 2). Iowa State University Press.

Lum, J. A., Youssef, G. J., & Clark, G. M. (2017). Using pupillometry to investigate sentence comprehension in children with and without specific language impairment. *Journal of Speech, Language, and Hearing Research, 60*(6), 1648–1660.

Marno, H., Farroni, T., Vidal Dos Santos, Y., Ekramnia, M., Nespor, M., & Mehler, J. (2015). Can you see what I am talking about? Human speech triggers referential expectation in four-month-old infants. *Scientific Reports, 5*(1), 1–10.

Marciszko, C., Forssman, L., Kenward, B., Lindskog, M., Fransson, M., & Gredebäck, G. (2020). The social foundation of executive function. *Developmental Science, 23*(3), e12924. Chicago.

Mathôt, S. (2018). Pupillometry: Psychology, physiology, and function. *Journal of Cognition, 1*(1).

Michel, C., Kayhan, E., Pauen, S., & Hoehl, S. (2021). Effects of reinforcement learning on gaze following of gaze and head direction in early infancy: An interactive eye-tracking study. *Child Development, 92*(4), e364–e382.

Morita, T., Slaughter, V., Katayama, N., Kitazaki, M., Kakigi, R., & Itakura, S. (2012). Infant and adult perceptions of possible and impossible body movements: An eye-tracking study. *Journal of Experimental Child Psychology, 113*(3), 401–414.

Murofushi, Y., Kubota, M., Akira, I., Hayakawa, I., Ozawa, H., & Abe, Y. (2021). Atypical and non-spontaneous attentional control in "just Look" tasks for face recognition in ASD children revealed by gaze tracking pupillometry. https://doi.org/10.21203/rs.3.rs-1039951/v1

Nieuwenhuis, S., De Geus, E. J., & Aston-Jones, G. (2011). The anatomical and functional relationship between the P3 and autonomic components of the orienting response. *Psychophysiology, 48*(2), 162–175.
Nuske, H. J., Vivanti, G., & Dissanayake, C. (2014a). Brief report: Evidence for normative resting-state physiology in autism. *Journal of Autism and Developmental Disorders, 44*(8), 2057–2063.
Nuske, H. J., Vivanti, G., & Dissanayake, C. (2014b). Reactivity to fearful expressions of familiar and unfamiliar people in children with autism: An eye-tracking pupillometry study. *Journal of Neurodevelopmental Disorders, 6*(1), 1–16.
Nuske, H. J., Vivanti, G., & Dissanayake, C. (2015). No evidence of emotional dysregulation or aversion to mutual gaze in preschoolers with autism spectrum disorder: An eye-tracking pupillometry study. *Journal of Autism and Developmental Disorders, 45*(11), 3433–3445.
Nuske, H. J., Vivanti, G., & Dissanayake, C. (2016). Others' emotions teach, but not in autism: An eye-tracking pupillometry study. *Molecular Autism, 7*(1), 1–13.
Nyström, P., Gredebäck, G., Bölte, S., & Falck-Ytter, T. (2015). Hypersensitive pupillary light reflex in infants at risk for autism. *Molecular Autism, 6*(1), 1–6.
Nyström, P., Falck-Ytter, T., & Gredebäck, G. (2016). The TimeStudio project: An open source scientific workflow system for the behavioral and brain sciences. *Behavior Research Methods, 48*(2), 542–552.
Nyström, P., Gliga, T., Nilsson Jobs, E., Gredebäck, G., Charman, T., Johnson, M. H., Bölte, S., & Falck-Ytter, T. (2018). Enhanced pupillary light reflex in infancy is associated with autism diagnosis in toddlerhood. *Nature Communications, 9*(1), 1–5.
Oakes, L. M. (2012). Advances in eye tracking in infancy research. *Infancy, 17*(1), 1–8. https://doi.org/10.1111/j.1532-7078.2011.00101.x
Pätzold, W., & Liszkowski, U. (2019). Pupillometry reveals communication-induced object expectations in 12-but not 8-month-old infants. *Developmental Science, 22*(6), e12832.
Pätzold, W., & Liszkowski, U. (2020). Pupillometric VoE paradigm reveals that 18-but not 10-month-olds spontaneously represent occluded objects (but not empty sets). *PLoS One, 15*(4), e0230913.
Paulus, M., Proust, J., & Sodian, B. (2013). Examining implicit metacognition in 3.5-year-old children: An eye-tracking and pupillometric study. *Frontiers in Psychology, 4*, 145.
Pérez, D. L., Ramotowska, S., Malinowska-Korczak, A., Haman, M., & Tomalski, P. (2020). Working together to orient faster: The combined effects of alerting and orienting networks on pupillary responses at 8 months of age. *Developmental Cognitive Neuroscience, 42*, 100763.
Pletti, C., Scheel, A., & Paulus, M. (2017). Intrinsic altruism or social motivation—What does pupil dilation tell us about children's helping behavior? *Frontiers in Psychology, 8*, 2089.
Poe, G. R., Foote, S., Eschenko, O., Johansen, J. P., Bouret, S., Aston-Jones, G., Harley, C. W., Manahan-Vaughan, D., Weinshenker, D., & Valentino, R. (2020). Locus coeruleus: A new look at the blue spot. *Nature Reviews Neuroscience, 21*(11), 644–659.
Prunty, J. E., Keemink, J. R., & Kelly, D. J. (2022). Infants show pupil dilatory responses to happy and angry facial expressions. *Developmental Science, 25*(2), e13182.
Sara, S. J., & Bouret, S. (2012). Orienting and reorienting: The locus coeruleus mediates cognition through arousal. *Neuron, 76*(1), 130–141.
Schulze, C., & Buttelmann, D. (2021). Children understand communication intuitively, but indirect communication makes them think twice—Evidence from pupillometry and looking patterns. *Journal of Experimental Child Psychology, 206*, 105105.
Sepeta, L., Tsuchiya, N., Davies, M. S., Sigman, M., Bookheimer, S. Y., & Dapretto, M. (2012). Abnormal social reward processing in autism as indexed by pupillary responses to happy faces. *Journal of Neurodevelopmental Disorders, 4*(1), 1–9.
Silberzahn, R., Uhlmann, E. L., Martin, D. P., Anselmi, P., Aust, F., Awtrey, E., Bahník, Š., Bai, F., Bannard, C., & Bonnier, E. (2018). Many analysts, one data set: Making transparent how variations in analytic choices affect results. *Advances in Methods and Practices in Psychological Science, 1*(3), 337–356.
Sirois, S., & Brisson, J. (2014). Pupillometry. *Wiley Interdisciplinary Reviews: Cognitive Science, 5*(6), 679–692.

Sirois, S., & Jackson, I. R. (2012). Pupil dilation and object permanence in infants. *Infancy, 17*(1), 61–78.

Sirois, S., Brisson, J., Blaser, E., Calignano, G., Donenfeld, J., Hepach, R., ... & Valenza, E. (2023). The pupil collaboration: A multi-lab, multi-method analysis of goal attribution in infants. *Infant Behavior and Development, 73*, 101890.

Sokolov, E. N. (1963). Perception and the conditioned reflex. *International Journal of Psychophysiology, 35*(2–3), 197–217.

Sonne, T., Kingo, O. S., & Krøjgaard, P. (2016). Empty looks or paying attention? exploring infants' visual behavior during encoding of an elicited imitation task. *Infancy, 21*(6), 728–750.

Tamási, K., McKean, C., Gafos, A., Fritzsche, T., & Höhle, B. (2017). Pupillometry registers toddlers' sensitivity to degrees of mispronunciation. *Journal of Experimental Child Psychology, 153*, 140–148.

Tamási, K., McKean, C., Gafos, A., & Höhle, B. (2019). Children's gradient sensitivity to phonological mismatch: Considering the dynamics of looking behavior and pupil dilation. *Journal of Child Language, 46*(1), 1–23.

Tan, E., & Hamlin, J. K. (2022). Mechanisms of social evaluation in infancy: A preregistered exploration of infants' eye-movement and pupillary responses to prosocial and antisocial events. *Infancy, 27*(2), 255–276.

Tummeltshammer, K., Feldman, E. C., & Amso, D. (2019). Using pupil dilation, eye-blink rate, and the value of mother to investigate reward learning mechanisms in infancy. *Developmental Cognitive Neuroscience, 36*, 100608.

Upshaw, M. B., Kaiser, C. R., & Sommerville, J. A. (2015). Parents' empathic perspective taking and altruistic behavior predicts infants' arousal to others' emotions. *Frontiers in Psychology, 6*, 360.

Vaish, A., Hepach, R., & Grossmann, T. (2018). Desire understanding in 2-year-old children: An eye-tracking study. *Infant Behavior and Development, 52*, 22–31.

Verschoor, S. A., Spapé, M., Biro, S., & Hommel, B. (2013). From outcome prediction to action selection: Developmental change in the role of action–effect bindings. *Developmental Science, 16*(6), 801–814.

Verschoor, S. A., Paulus, M., Spapé, M., Biro, S., & Hommel, B. (2015). The developing cognitive substrate of sequential action control in 9-to 12-month-olds: Evidence for concurrent activation models. *Cognition, 138*, 64–78.

Wass, S. V., Smith, T. J., & Johnson, M. H. (2013). Parsing eye-tracking data of variable quality to provide accurate fixation duration estimates in infants and adults. *Behavior Research Methods, 45*, 229–250. Chicago.

Wass, S. V., & Smith, T. J. (2014). Individual differences in infant oculomotor behavior during the viewing of complex naturalistic scenes. *Infancy, 19*(4), 352–384.

Wass, S. V., De Barbaro, K., & Clackson, K. (2015). Tonic and phasic co-variation of peripheral arousal indices in infants. *Biological Psychology, 111*, 26–39.

Wetzel, N., Buttelmann, D., Schieler, A., & Widmann, A. (2016). Infant and adult pupil dilation in response to unexpected sounds. *Developmental Psychobiology, 58*(3), 382–392.

Wu, C.-J., Liu, C.-Y., Yang, C.-H., & Jian, Y.-C. (2021). Eye-movements reveal children's deliberative thinking and predict performance on arithmetic word problems. *European Journal of Psychology of Education, 36*(1), 91–108.

Yucel, M., Hepach, R., & Vaish, A. (2020). Young children and adults show differential arousal to moral and conventional transgressions. *Frontiers in Psychology, 11*, 548.

Zhang, F., & Emberson, L. L. (2020). Using pupillometry to investigate predictive processes in infancy. *Infancy, 25*(6), 758–780.

Zhang, F., Jaffe-Dax, S., Wilson, R. C., & Emberson, L. L. (2019). Prediction in infants and adults: A pupillometry study. *Developmental Science, 22*(4), e12780.

Zivan, M., Morag, I., Yarmolovsky, J., & Geva, R. (2021). Hyper-reactivity to salience limits social interaction among infants born pre-term and infant siblings of children with ASD. *Frontiers in Psychiatry, 12*, 650.

Pupillometry, Attention Control, and Working Memory

Matthew Robison

The seminal pupillometric study in cognitive psychology was published by Hess and Polt (1964). They gave participants five mental multiplication problems of increasing difficulty, and they took snapshots of the participants' eyes at a rate of two pictures per second while they solved the problems. Hess and Polt reported pupillary dilations that scaled with problem difficulty. Pupils were maximally dilated immediately before participants gave their response and constricted after the response. Hess and Polt thus concluded, "It seems evident then, that in this situation the pupil response is a direct reflection of neurological activity and is associated with specific areas of the brain and nervous system" (p. 1191). Such a finding seems as though it would have sparked a generation of cognitive psychologists interested in using pupillometry to test theories about cognitive and neural mechanisms. But surprisingly, and perhaps because of how difficult it was to image and measure pupil diameter, cognitive psychologists left pupillometry relatively untouched for several decades. Over the past 20 or so years, as pupillometry has become almost trivially easy with state-of-the-art eye-tracking equipment, there has been a surge of studies leveraging the technique to investigate rather specific research questions in the field of attention. Here, I will focus on four areas of research: (1) conflict resolution/cognitive control, (2) sustained attention, (3) working memory, and (4) individual differences. Pupillometry has proved a fruitful tool for addressing important theoretical questions in each of these four domains. I will review a selection of studies published in each of these fields, and I will present previously published and novel analyses of data from my own research that demonstrate findings in each area.

M. Robison (✉)
Department of Psychology, University of Texas at Arlington, Arlington, TX, USA
e-mail: matthew.robison@uta.edu

M. H. Papesh, S. D. Goldinger (eds.), *Modern Pupillometry*,
https://doi.org/10.1007/978-3-031-54896-3_4

1 Conflict Resolution and Cognitive Control

One of the most important areas of attention research, and indeed still one of the most hotly debated, is the area of cognitive control. The research in this field asks important questions like, how does the human cognitive system optimize in a world that has both predictable and unpredictable environments? How does the attentional system bias behavior? And how can these biases be overcome when they need to be? Take, for example, the Stroop task (Stroop, 1935). It is one of the oldest and most popular tasks in all cognitive psychology. The task goal is simple, yet rather difficult to execute. When a colored word (e.g., **RED**) appears, report the font color of the word, not the word itself. Because our cognitive system has developed a tendency to read the words, there is an existing stimulus-response binding that must be overridden in situations where that automaticity conflicts with the demands of the task. If, for example, the word **RED** is presented, there is an incongruency between the automatic response (red) and the correct response (blue). This conflict is apparent in two ways: people make more errors on incongruent trials compared to congruent trials, and correct responses to incongruent trials are typically delivered more slowly than correct responses to congruent trials. Researchers have spent decades trying to understand how the cognitive system resolves such conflict, where in the processing stream (e.g., perception and response production) the conflict occurs, and how this might manifest an important developmental or individual difference (Jensen & Rowher, 1966; McLeod, 1991). Perhaps unsurprisingly, then, researchers have leveraged pupillometry to answer these and other questions.

1.1 Stroop Task

Brown et al. (1999) published one of the first studies to show a pupillary Stroop effect. They had participants report the color of a stimulus that was either a colored word (congruent and incongruent trials) or colored Xs (neutral trials). They measured pupil diameter during the task, and they reported greater pupil diameter when participants saw the colored words vs. the colored Xs. Here, the conflict was only present in the word trials compared to X trials. Thus, there was a pupillary conflict effect in this version of the Stroop task. Follow-up experiments showed similar effects, with responses to incongruent color-word stimuli eliciting greater pupillary responses than Xs (Siegle et al., 2004, 2008).

Laeng et al. (2011) pointed out that these earlier studies did not control for lexical processing and that the larger pupillary responses for incongruent trials may have been due to additional processing of the word over the Xs. Therefore, they presented participants with congruent colored words, incongruent colored words, colored nonwords, and colored words that were not colors. Again, they demonstrated a pupillary Stroop effect, with congruent color words producing the smallest pupillary responses, noncolor words producing larger pupillary responses, and incongruent color words producing still larger pupillary responses. Laeng et al. (2011) interpreted these findings to mean that there is both semantic processing and conflict resolution occurring in resolving Stroop interference and that pupil responses can differentiate these two processes. Therefore, in addition to being used to study the cognitive processing associated with conflict resolution, pupillometry may also be used to understand semantic processing as well.

Figure 1 shows a plot of the pupillary Stroop effect from a recent study. These data were from a larger individual-differences examination of attention control, and part of the task battery was a Stroop task with pupillometry (Unsworth & Robison, 2017c). We gave participants a classic Stroop task with congruent (e.g., **RED**) and incongruent (e.g., **RED**) color-word stimuli. Two-thirds of the trials were congruent, and the remaining trials were incongruent. As is clearly visible in Fig. 1, incongruent trials elicited greater pupillary responses than incongruent trials. These data demonstrate the general pattern found in the literature (Brown et al., 1999, Siegle et al., 2004, 2008; Laeng et al., 2011; Rondeel et al., 2015). The pupillary response is clearly sensitive to Stroop interference and can be used as a metric of the additional attentional outlay required to resolve such conflict.

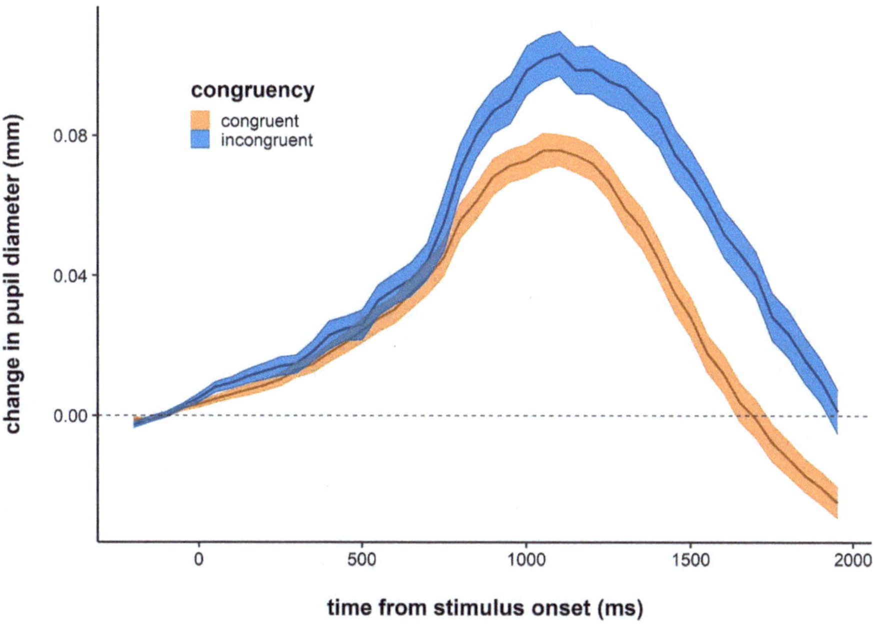

Fig. 1 Pupillary responses to congruent and incongruent trial in the Stroop color-word task. Congruent trials (blue waveform) elicited smaller pupillary responses than incongruent trials (orange waveform). Shaded regions around each line indicate +/− one standard error of the man. (Data come from Unsworth and Robison (2017c))

1.2 Flanker Tasks

Flanker tasks have been used much the same way that the Stroop task has been to study conflict resolution processes. In the traditional arrow flanker task (Eriksen & Eriksen, 1974), participants must report the direction of a central arrow while ignoring the direction of flanking errors. On congruent trials, the central arrow is flanked by arrows pointing in the same direction (<<<<<). On incongruent trials, the central arrow is flanked by arrows pointing in the opposite direction (<<><<). Incongruent trials typically produce more errors and longer response times on correct trials. Just like the Stroop task, researchers have started to leverage pupillometry to understand flanker effects. For example, Geva et al. (2013) gave participants the Attention Network Test (ANT; Petersen & Posner, 2012). Embedded in the ANT are congruent, incongruent, and neutral arrow flanker trials. Geva et al. (2013) demonstrated a pupillary flanker effect, such that pupillary responses were larger for incongruent trials compared to congruent and neutral trials. This effect was subsequently replicated by van Bochove et al. (2013). Wendt et al. (2014) leveraged a flanker task to demonstrate a slightly different phenomenon. They showed that participants' pupils were significantly greater following a high-conflict trial. Wendt et al. argued these data suggest that participants were exerting more effort following high-conflict trials, which corroborated reaction time data from such trials. So, like the Stroop task, pupillometry can be leveraged in combination with flanker tasks to probe how people overcome stimulus–stimulus conflict.

1.3 Antisaccade Task

In the antisaccade task, participants are instructed to look away from a stimulus that flashes in their periphery. The dominant behavioral response is to look toward the flashing stimulus, so participants must override this response to execute the correct behavior. The ability to do so appears to be a rather reliable indicator of an individual's attention control ability (Kane et al., 2001; Meier et al., 2018; Unsworth et al., 2004). Wang et al. (2015) examined the pupillary correlates of preparation for an antisaccade trial. They randomly interleaved antisaccade trials with prosaccade trials (look toward the cue) and recorded both eye movements and pupil diameter. Wang et al. reported greater prestimulus pupil dilation in preparation for antisaccade trials compared to prosaccade trials. Dalmaso et al. (2020) subsequently replicated this effect but showed that it only arose when antisaccade and prosaccade trials were randomly interleaved, not when they were presented in a blocked manner. Hutchison et al. (2020) also replicated this effect with interleaved prosaccade and antisaccade trials. Thus, it appears pupil dilation in the preparatory interval before antisaccade trials indexes the additional effort people must exert to instantiate the nondominant goal of looking away from the flashing cue, especially when that goal is not being reinforced by the task. In their study, Hutchison et al. also included thought probes – screens asking participants to report their attentional state in the moments immediately preceding the probe. They demonstrated that when participants reported mind-wandering on a thought probe, the effect was not observed. These data suggest that preparatory instantiation of the task goal (look away) is indeed reflected in the pupil, and only when participants were adequately engaged with the task.

1.4 Continuous Performance Tasks

Continuous performance tasks (CPTs) are another common set of tasks used to measure cognitive control. Tasks like the AX-CPT can be used to study modes of control like proactive and reactive control and, thus, can be used to examine clinical, developmental, and individual differences in cognitive control. In the AX-CPT, participants are told to respond in one way to a specific sequence of cue-probe pairs (As followed by Xs) and to withhold or provide a different response to any other pair of stimuli. Importantly, AX trials comprise 70% of all trials, making this response dominant. The remaining 30% of trials comprises an equal proportion of AY trials (as followed by any other nontarget stimulus), BX trials (Xs preceded by any nontarget stimulus), and BY trials (a combination of two nontarget stimuli). From patterns of reaction times and error rates, the data can reveal whether a participant is being proactive or reactive in their response (see Braver, 2012, for review).

Recently, cognitive psychologists have embedded pupillometry into well-described tasks like the AX-CPT to examine physiological correlates of proactive and reactive control processes. For example, Chiew and Braver (2013) had participants complete a version of the AX-CPT with a motivational manipulation. On a baseline run of the task, participants were told to ignore a pretrial cue (green square or green dollar sign). In a motivated run, they were told that this cue indicated whether a trial was incentivized (dollar sign) or non-incentivized (square). Participants were told that they could earn up to an additional $5 for fast and accurate performance on the incentive trials. The pupillary data showed a complex and theoretically informative pattern of responses. In general, the behavioral data suggested a near-complete shift to a proactive mode of control during the motivated run. This shift was accompanied by an increase in pupil diameter. Importantly, there was a *decrease* in evoked pupillary responses for non-incentive trials in the reward block. These findings highlight the difference between what sustained (tonic) and transient (phasic) changes in pupil diameter can tell researchers about a given group, experimental, developmental, or developmental difference.

Chatham et al. (2009) provide an illustrative example of how pupillometry can be leveraged to understand developmental differences in cognitive control using a CPT. They had groups of 3.5- and 8-year-old children complete a cartoon version of AX-CPT while they recorded pupil diameter. The reaction time data suggested a shift toward proactive control among the 8-year-olds. Importantly, the pupil data offered an explanation as to why. Examining the waveforms for the age groups revealed greater pupil dilations among 8-year-olds during the delay period (after the cue and before the probe), and 3.5-year-olds showed greater pupillary responses during the probe period (after the probe appeared). Thus, the pupil data clearly showed that the 8-year-olds were being proactive in engaging effort to maintain the cue context in memory, whereas 3.5-year-olds were being reactive and retrieving context information after the probe appeared (Chatham et al., 2009). Thus, pupillometry may be a powerful and relatively practical method of measuring developmental differences in cognitive abilities, even among quite young cohorts.

1.5 Summary

Several commonly used tasks that measure attentional processes like interference resolution, inhibition of prepotent responses, and proactive/reactive control have been used in combination with pupillometry to understand how attention is exerted in these contexts, how attentional states affect the implementation of such processes, and how these processes change across development. Some of the basic findings (e.g., pupillary Stroop effects, pupillary flanker effects) can be leveraged in future research to better understand how the attentional system implements complex cognitive operations and how these processes might differ with regard to clinical, individual, and developmental differences.

2 Spatial Attention

The pupil's primary role in the visual system is to regulate the amount of light that is allowed through the eye to hit the retina. When there is a lot of light in the environment, the pupil constricts. When it is relatively dark, the pupil dilates. Importantly, it has been demonstrated that the pupil is also sensitive to changes in *perceived* luminance (Laeng & Endestad, 2012). But recently, several studies have shown that the pupil can indicate where in space someone is attending, even when they are not directly looking at that region of space (Binda et al., 2013, 2014; Mathôt et al., 2013). That is, even when participants are not actually looking at an object or a region of space, the pupil dilates/constricts in response to the luminance of the object or spatial region where they are covertly attending.

Specifically, Binda et al. (2013) presented participants with a screen that had a black disc on one side, a white disc on the other side, and a central fixation point. Within each disk was a colored point that participants used for the behavioral task (counting changes in color). At the beginning of each trial, the task cued participants to either the right or left hemifield. Importantly, in some conditions, they instructed participants to keep their eyes focused on the central fixation point but to shift their attention to the cued disc. They demonstrated that these covert shifts of attention produced differences in pupil size. When the participants attended the white disc, the pupil showed a relative constriction compared to when participants attended the black disc. Thus, the pupil showed a light reflex even though the actual amount of environmental luminance was identical across the attend-white attend-black trials. These effects were subsequently replicated by Binda et al. (2014), showing that the pupil shows a relative dilation when people attend bright objects vs. dark objects centrally (i.e., without shifting their attention spatially). These findings have been leveraged to make inferences about the role of covert spatial attention and the top-down exertion of cognitive processes like attention on bottom-up feature perception like brightness.

3 Sustained Attention

In many tasks, people show a systematic decline in performance as a function of time. Historically, this phenomenon has been referred to as the *vigilance decrement*. It is still under debate as to what psychological or physiological phenomenon causes this decrement. One rather dominant explanation is resource theory (Smit et al., 2004; Warm et al., 2008), which argues that attention-demanding tasks require cognitive resources, which are limited. The longer someone spends on a task, the smaller that pool of resources becomes, and task performance deteriorates as a result. However, recently, researchers have started to challenge the notion of a physical resource limitation, instead positing that vigilance decrements are a manifestation of shifts in the allocation of resources in response to an unrewarding situation, mental cost/benefit tradeoff assessments, and changes in motivation (Kurzban et al., 2013; Thomson et al., 2015). A careful adjudication of these theories and their competing predictions is beyond the scope of this chapter. However, researchers have started to use pupillometry to understand the behavioral and physiological correlates of vigilance decrements and have started to leverage pupillometry to test competing predictions made by theories of sustained attention.

3.1 *Oddball Tasks*

Visual and auditory oddball tasks are among the most common tasks for electrophysiological research because they produce clear and consistent event-related potentials. As it turns out, oddball tasks also produce large and consistent pupillary effects. In a typical oddball task, participants see or hear a stream of stimuli, and one category of stimuli is labeled as targets, and another category as nontargets (or distractors). In active tasks, participants must make a response to infrequent target stimuli and make a different response (or withhold their response) to frequent nontarget stimuli. In passive tasks, participants do not make responses but simply listen to (or watch) the stream of stimuli.

Friedman et al. (1973) were the first to report a pupillary oddball effect. They reported a pupillary dilation in response to target stimuli, which scaled with target probability. The more infrequently targets were presented, the larger the pupillary response was. This effect has subsequently been replicated several times in both the visual and auditory domains, and the consistency of the effect has been used as leverage to test theories about other cognitive and neural processes (Kuchinke et al., 2007; Gilzenrat et al., 2010; Murphy et al., 2011; Preuschoff et al., 2011).

For example, Kamp and Donchin (2015) presented words to participants visually and asked them to categorize them in one of three different ways (edible vs. inedible, living vs. nonliving, and smaller or larger than a shoebox). One category was always frequently presented, and one was always infrequently presented. In some blocks, they also presented images of frequent category items rather than words. The pupil dilated more in response to infrequent items as opposed to frequent words and pictures, and there was no difference in the pupillary response to frequent words and pictures. This is different than what occurred with the event-related potentials. Pictures elicited the largest P3 responses in the electrophysiological data. Thus, there was a dissociation between the information provided by the pupil data and the EEG data. The pupillary response also predicted subsequent memory for items in a surprise recognition test. Items that were accurately recognized as presented during the oddball task (both frequent and infrequent) were associated with larger evoked pupillary responses during the encoding (oddball) portion of the task.

As another example of how these findings have been leveraged to test theories about cognitive and neural processes, Murphy et al. (2011) had participants complete an auditory oddball task while they recorded both pupil diameter and EEG. They demonstrated a quadratic relationship between pretrial pupil diameter and reaction time variability, suggesting optimal task performance was associated with moderate pretrial pupil diameter. They also showed that the largest pupillary responses came immediately after trials with the smallest pretrial pupil diameters, indicating the pupillary response may reflect, to a certain degree, a re-engagement with the task following moments of low arousal or disengagement. They argued that these results are consistent with the Adaptive Gain theory of locus coeruleus functioning (the Individual Differences section of this chapter discusses this theory in more detail).

Finally, a recent set of experiments by Strauch et al. (2020) manipulated decision-making requirements, target probabilities, and motor response requirements within participants. Their design allowed them to disambiguate contributions of decision-making, target probability, and motor responses to pupillary responses. They found that targets elicited greater responses, even when they were frequent, compared to nontargets. When items were presented at different probabilities, without a distinction being made between targets and nontargets, the pupillary responses did not differ. Thus, targets only elicited larger responses when there was a decision to be made. Thus, Strauch et al. (2020) used these data to argue that decision-making is a two-stage process that encompasses determining whether a stimulus is goal-relevant or not and then categorizing it as a target or nontarget (or assessing its value).

3.2 N-*Back Tasks*

Although *n*-back tasks are typically used as measures of working memory, several studies have used *n*-back tasks to induce a vigilance decrement. For example, Hopstaken et al. (2015b) had participants complete a 2-h session of a two-back task. At the end of each of the seven task blocks, Hopstaken et al. asked participants to report their current level of task engagement and mental fatigue on a continuous scale. They collected pupillary data in addition to behavioral data. In accompaniment with behavioral performance (d') decreasing, subjective fatigue increasing, and task engagement decreasing, stimulus-evoked pupil dilations systematically declined across blocks. Crucially, at the end of the sixth block, they told participants that their remaining time in the experiment would depend on their performance on the immediately ensuing block, relative to their performance on the preceding six blocks, and that it would range anywhere from an additional 5 min to an additional 40 min. In response to this instruction, participants' behavioral accuracy, stimulus-evoked pupil diameter, and subjective measures of fatigue and task engagement all rebounded to levels at or even above the first block. Hopstaken et al. observed similar effects in two similar studies with *n*-back tasks (Hopstaken et al., 2015a, 2016).

In addition to the group-level time-on-task effects, Hopstaken et al. (2015a, b, 2016) reported significant correlations among changes in task performance, task engagement, fatigue, baseline pupil diameter, and evoked pupillary responses. That is, participants who showed steeper declines in task performance and engagement and steeper increases in fatigue across blocks showed steeper decreases in baseline pupil diameter and evoked pupillary responses.

These data suggest that pupillary measures of effort (i.e., evoked responses) corroborate subjective measures of task engagement and fatigue and measures of task performance across blocks and that they are additionally sensitive to manipulations that affect such measures. Hopstaken et al. argue that these data are inconsistent with resource theories of the vigilance decrement because objective measures of performance, subjective measures of fatigue/engagement, and physiological measures of alertness and effort can all be "restored," even without rest, with a powerful extrinsic motivator. Again, experiments like this show how pupillometry can be leveraged to test important theoretical predictions regarding attentional processes.

3.3 *Psychomotor Vigilance*

For decades, the psychomotor vigilance task (PVT) has primarily been used to understand the consequences of sleep loss on cognitive functioning and indeed still is (Basner & Dinges, 2011; Dinges et al., 1987; Drummond et al., 2005; Lim & Dinges, 2008). But recently, cognitive psychologists have begun to use the PVT as a measure of sustained attention in basic experimentation and individual differences designs (Massar et al., 2016; Unsworth et al., 2010; Unsworth & McMillan, 2014;

Unsworth & Robison, 2016). In recent instantiations of the task, researchers typically show a "timer" that starts at 0. After a random amount of time, usually between 2 and 10 s, the timer begins counting like a stopwatch. The participant's task is simple: press a key as quickly as possible when they notice the timer begin counting. The PVT produces several stereotypical patterns of data. First, participants' reaction times tend to slow over time, demonstrating a vigilance decrement even over short intervals (~ 10 min). Second, individual differences in reaction times, especially in the slowest reaction times, correlate well with other measures of attention, measures of working memory capacity, and measures of fluid intelligence (Unsworth et al., 2010; Unsworth & McMillan, 2014). Third, the task produces a stereotypical pattern of pupillary responses. Therefore, it has been a useful task for the application of pupillometry to sustained attention research.

As an example, Massar et al. (2016) investigated the effect of incentives on sustained attention using a combination of PVT and pupillometry. In three runs of the task, Massar et al. offered trial-level incentives for speed. In an unincentivized (baseline) block, participants completed the standard PVT. In a low-incentive block, participants received 1 cent for every trial where their reaction time was faster than their median reaction time in the baseline run. In a high-incentive block, participants received 10 cents for every reaction time meeting that same criterion. Massar et al. found that incentives speeded reaction times and increased pupil diameter, with the high-incentive block having the largest effects.[1] Thus, Massar et al. (2016) showed that both behavior and pupil diameter are sensitive to motivational manipulations in the PVT. Similar to Hopstaken et al. (2015a, b, 2016), Massar et al. used these data to argue that resource accounts of vigilance are limited since the motivational manipulations affected both behavioral and physiological indices of sustained attention.

As another example, Unsworth and Robison (2016) gave participants a 30-min version of the PVT while they recorded pupil diameter. They focused on a few findings. First, they demonstrated a systematic decrease in both pretrial (baseline) pupil diameter and evoked pupillary responses with time on task. This latter effect is replotted in Fig. 2. Thus, as participants spent more time on the task, their decrements in task performance were accompanied by a similar decrement in evoked pupillary responses. These findings replicate what Hopstaken et al. (2015a, b, 2016) observed in the two-back task.

In addition to the time-on-task analyses, Unsworth and Robison (2016) also embedded thought probes at various points throughout the task to measure momentary attentional states. They showed that compared to when participants reported being off-task (e.g., mind-wandering), their reaction times were slower, and their pupillary responses were smaller in the preceding trial compared to when they reported being on-task (see Fig. 3). Importantly, this was a within-subjects analysis.

[1] Massar et al. (2016) reported changes in pupil diameter overall but did not specifically compare the magnitude of the pupillary responses after stimulus onsets.

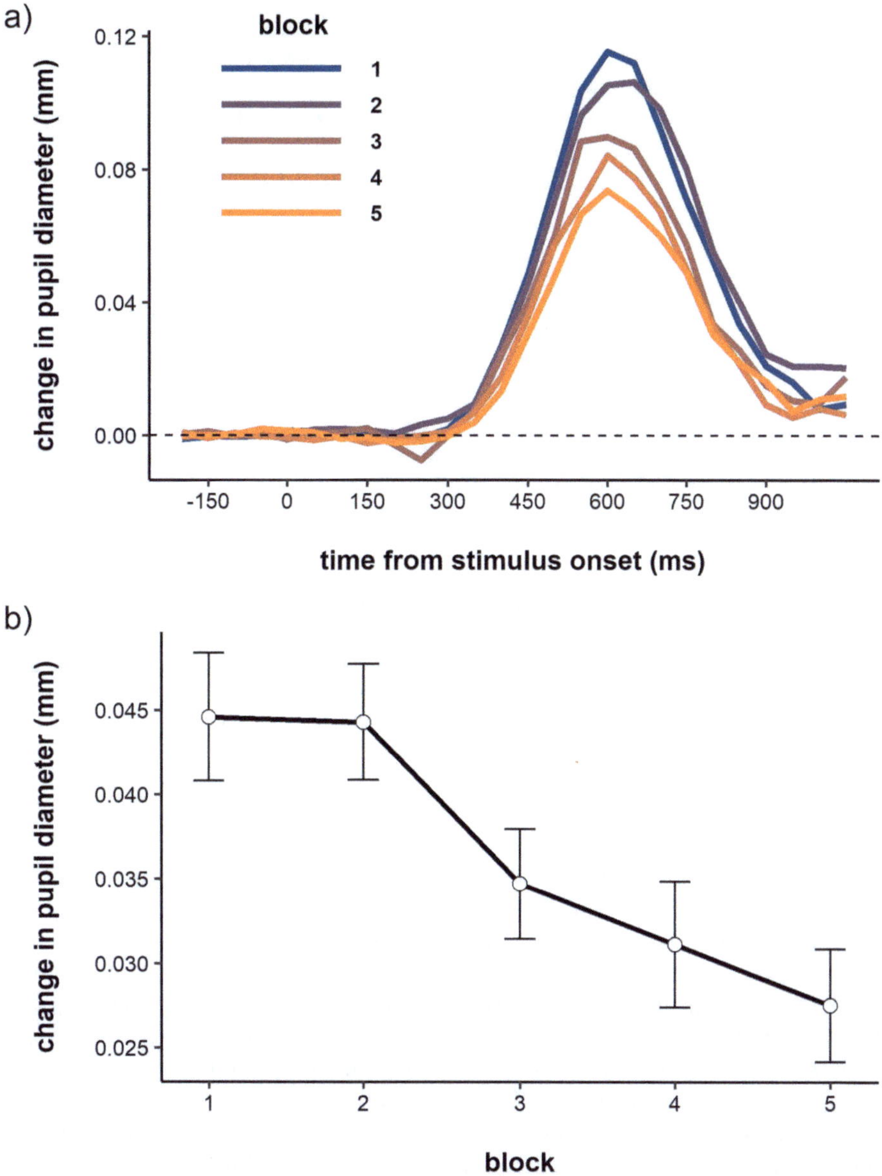

Fig. 2 (**a**) Waveforms of evoked pupillary responses as a function of block and (**b**) average evoked pupillary responses as a function of block in the psychomotor vigilance task (PVT). Error bars represent +/– one standard error of the mean. (Data are replotted from Unsworth and Robison (2016))

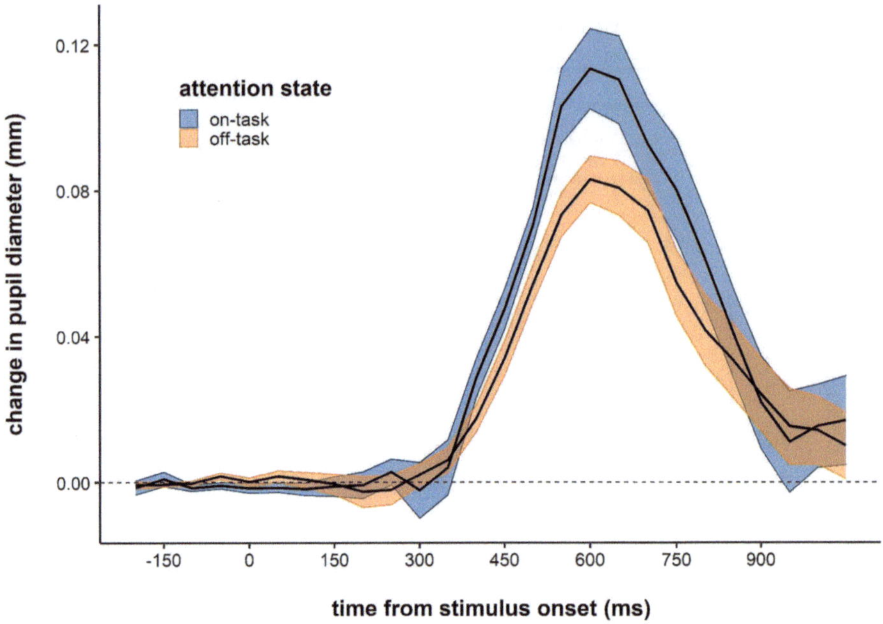

Fig. 3 Evoked pupillary responses preceding reports of being on-task vs. reports of being in an off-task state (e.g., mind-wandering) during the psychomotor vigilance task. Shaded error bars represent +/− one standard error of the mean. (Data are replotted from Unsworth and Robison (2016))

Therefore, the pupillary response was sensitive to momentary fluctuations of the attentional state. Replicated this effect in several follow-up experiments.

In a subsequent study, Unsworth et al. (2018) examined pupillary dynamics during the intertrial interval preceding the onset of the stimulus and found meaningful differences in how the pupil changed preceding fast and slow responses. Specifically, they showed a "ramping up" of pupil diameter in the seconds preceding stimulus onset for fast trials, indicating that participants exerted a degree of preparatory control (or "intrinsic alertness") to respond quickly. They also showed that evoked pupillary responses were larger for the fastest trials compared to the slowest trials.

In a re-analysis of the data of Unsworth and Robison (2016) for this chapter, I examined whether this effect emerged in that dataset, as well. Specifically, I rank-ordered each participant's reaction times from fastest to slowest and binned them into five quintiles. Then, I examined evoked pupillary responses on trials as a function of quintile. The resulting waveforms are plotted in Fig. 4. From inspection of the waveforms, it appears there are differences in both the magnitude of the waveforms *and* the latency of the peaks of the waveforms.

To examine this statistically, I computed the peak of the task-evoked response per quintile for each participant. Then, borrowing a technique from the event-related potential literature (Luck, 2014; Kiesel et al., 2008). I computed the fractional

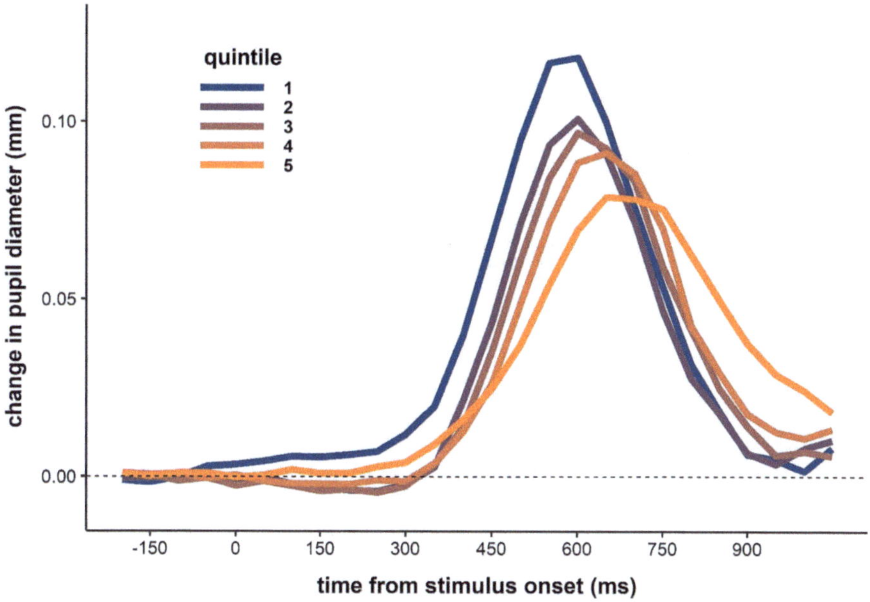

Fig. 4 Evoked pupillary response waveforms by reaction time (RT) quintile. RTs are ranked from fastest to slowest and binned into quintiles (quintile 1 = fastest RTs, quintile 5 = slowest RTs). (Data are from Unsworth and Robison (2016))

(50%) peak latency of the response per quintile for each participant. This measure gives an indication of the point at which the waveform has reached 50% of its peak. A linear mixed effect model with quintile (1–5) as a continuous fixed effect and subject as a random effect examined these measures statistically. In the first model, the magnitude of the response was the dependent variable. In the second model, the fractional peak latency was the dependent variable. The models indicated significant effects of the RT quintile on both the magnitude ($b = -0.004$ mm, SE = 0.001, $t = -4.04$, $p < 0.001$) and the latency of the pupillary response ($b = 16.28$ ms, SE = 5.52, $t = 3.95$, $p < 0.001$). Thus, faster RTs in the PVT were associated with larger pupillary responses that onset earlier. These effects are plotted in Fig. 5.

Pupillary responses typically peak well after the response has been given. Thus, changes in pupil diameter are a sluggish signal. However, pupil diameter is a rather clean signal in the sense that relatively few trials are required to estimate group-level and even individual effects reliably. Furthermore, the data above show that in addition to using the magnitude of pupillary response, the latency of pupillary responses can also be informative. However, it is worth noting that because the pupillary response usually happens *after* a response has been made, it cannot be used as a reasonable proxy of the speed/onset of neural processes per se. But even with this limitation, the fact that the latency of pupillary responses lengthened with reaction time in the analysis above may mean that the latency of pupillary responses could be used as a measure of experimental, developmental, and/or individual differences in the future. This is an area that certainly requires additional research.

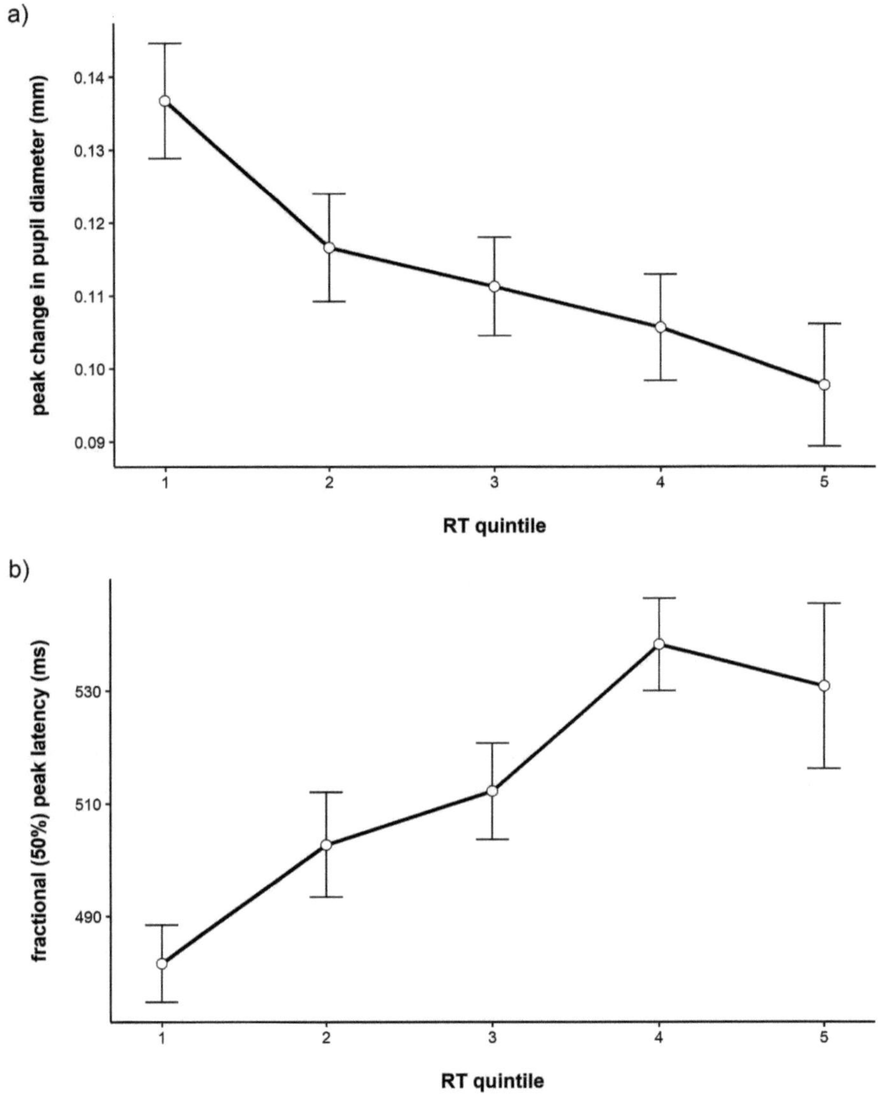

Fig. 5 Effects of reaction time (RT) quintile on the (**a**) magnitude and (**b**) peak latency of evoked pupillary responses in the psychomotor vigilance task. Faster RTs led to larger and earlier-peaking pupillary responses than slower RTs. Error bars represent +/− one standard error of the mean. (Data are from Unsworth and Robison (2016))

4 Working Memory

Working memory is often defined as the capacity-limited ability to actively maintain and manipulate goal-relevant information in the service of ongoing cognitive operations. As such, it is a core cognitive construct, and it shares conceptual and theoretical overlap with attention control (e.g., Engle & Kane, 2004). Some of the earliest applications of pupillometry in cognitive psychology were concerned with how pupillometry could be used to study the short-term memory system. For example, following up on Hess and Polt's (1964) seminal study, Kahneman and Beatty (1966) gave participants a short-term memory task while they took images of the eyes to measure pupil diameter. In interleaved trials, participants heard and then immediately recalled strings of single digits (set size varying from three to seven), strings of four monosyllabic words, or strings of four digits to which participants had to add one to each digit. List items were read aloud in synchrony with the clicking of the camera (1/s). At the end of the list, there was a 2-s pause, and then participants reported the items in serial order at a rate of one item per second, again in synchrony with the camera. Kahneman and Beatty (1966) reported several important findings. First, the magnitude of the pupillary response scaled with list length in the digit-recall condition. Second, participants' pupils dilated to a greater extent when they were warned that a long list was upcoming, reflecting a degree of preparatory attention. Third, the pupil returned to a baseline level after participants reported the items, which they argued reflected an off-loading of information from memory. Fourth, they compared the digit recall (four digits), word recall, and digit transformation tasks and found that word recall produced greater pupillary dilations than digit recall, but digit transformation produced still larger dilations. Finally, Kahneman and Beatty reported a time-on-task effect in their control experiment. That is, pupillary responses were shallower in the final block of trials compared to an initial block, reminiscent of the time-on-task effects reported more recently during sustained attention tasks (Hopstaken et al., 2015a, 2016; Unsworth & Robison, 2016).

In a follow-up to this study, Peavler (1974) gave participants lists of digits that were either 5, 9, or 13 digits in length. In a control block, participants were told that they did not need to remember or report the digits. Then, in a recall block, they were instructed they would need to report the digits strings as accurately as possible. Peavler reported a dilation of the pupil only when participants were tasked with recalling the digits. Additionally, the pupil showed a dilation with increasing digit presentation for all five items of the five-item lists but a rounding of the dilation for the longer list lengths. So just as working memory is capacity-limited, it appears that there is an upper bound on the magnitude of pupillary dilation as people maintain information in working memory.

In more recent instantiations of these designs, Alnæs et al. (2014) demonstrated a monotonic increase in pupil dilation as participants tracked 2, 3, 4, and 5 moving objects (see Aminihajibashi et al., 2019, 2020; Wahn et al., 2016, for similar findings). In visual working memory tasks, Unsworth and Robison (2015, 2018b) showed an increase in pupillary responses during a delay from set sizes 1 to 5, but no increase in pupil dilation as set size grew to 6, 7, and 8 items. Again, these findings indicate that the magnitude of pupillary dilations is maximal at or near the theoretical capacity limits of the working memory system.

More specific applications of pupillometry have been investigated in working memory research. For example, Unsworth and Robison (2018b) examined the pupillary correlates of various working memory-related phenomena. For example, in one experiment, Unsworth and Robison examined the maintenance and removal of information in working memory. On each trial, they presented participants with an array of items (colored squares, set sizes 1–8) to remember over a delay. On half of the trials, halfway through the 5-s delay period, an auditory cue told participants they needed to "hold" the items for the test, or they could "drop" the items because no test would ensue. The pupil dilated in both conditions until the hold/drop cue, after which the pupil constricted for the drop trials and continued to stay dilated for the hold trials. This indicated that when participants received the drop cue, they no longer effortfully maintained the items in working memory, which was reflected by the pupillary constriction. In another experiment, Unsworth and Robison (2018b) presented participants with red and blue oriented bars. The participant's task was to remember the orientation of the bars over a delay interval for a subsequent test. Importantly, participants were told that only the red bars would ever be tested. Thus, the blue bars should have been "filtered" and not allowed to gain access to working memory. Previously, it had been demonstrated such a filtering ability to do so is one important feature of individual differences in working memory capacity (e.g., Vogel et al., 2005). Pupillary responses were maximal when participants saw either four red bars with two blue bars or four red bars alone and minimal when participants saw just two red bars. In trials where participants saw two red bars and two blue bars, the pupillary response was greater than when participants saw just two red bars but less than when participants saw just four red bars. This indicated that, on average, participants did not completely filter the irrelevant information from working memory, especially when the total amount of information did not exceed typical capacity limits.

As another example, Robison and Unsworth investigated "lapses" in working memory maintenance – trials where participants accurately report only a few (or sometimes zero) items correctly. Using a discrete whole-report paradigm with a fixed set size of six items, Adam et al. (2015) showed that participants regularly encounter these instances, and the frequency with which they do so is an important individual difference. Using this same paradigm while measuring pupil diameter, Robison and Unsworth (2019) showed that pupil dilations during the maintenance interval differ for successful vs. unsuccessful trials (see Fig. 6).

As a final example of how pupillometry has been leveraged to understand working memory, two recent studies have taken advantage of the pupillary light reflexes in response to covert shifts of attention discussed earlier. Unsworth and Robison (2017b) presented participants with one grey-oriented bar on the right and left hemifields of the screen. The right side of the screen was white, the left was black, and the center region was grey. Before each trial, participants were cued to one side of

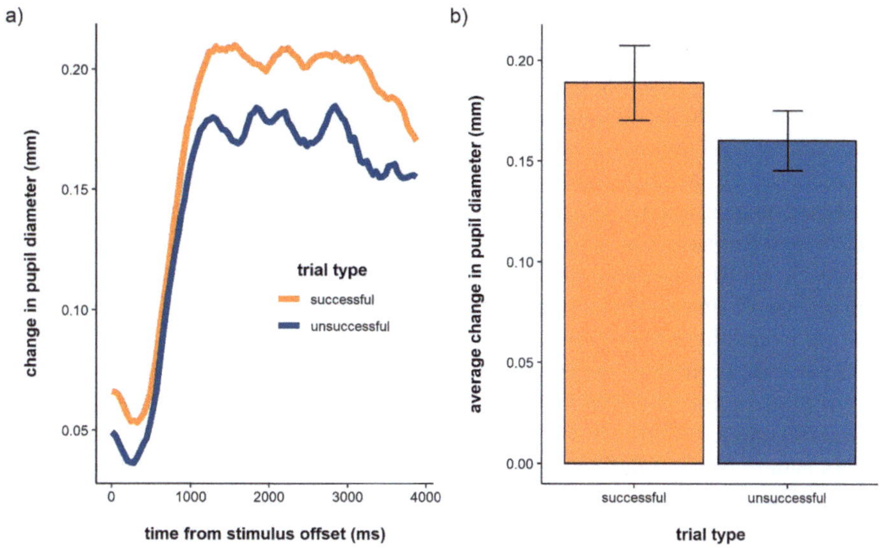

Fig. 6 (**a**) Evoked pupillary response for successful trials (recall >3 items) vs. unsuccessful trials (recall <3 trials) in a visual working memory task. (**b**) Average change in pupil diameter across the delay. Error bars represent +/− one standard error of the mean. (Data are from Robison and Unsworth (2019, Experiment 1))

the screen by a central left- or right-pointing arrow. A grey-oriented bar then appeared in each hemifield, and the participant's task was to remember the orientation of the bar on the cued side of the screen over a 4-s delay. Pupillary dilations were larger when participants attended to the black side of the screen vs. the white side of the screen, indicating that participants were covertly attending to the cued side of the screen during the delay to maintain the item in working memory. Unsworth and Robison (2017b) argued that these results support the theory that people use spatial attention as a means of keeping working memory representations active (e.g., Awh & Jonides, 2001).

5 Individual Differences

Most of the research I have discussed in this chapter has used within- or between-subject experimental manipulations to examine how a given independent variable can affect baseline pupil diameter and/or evoked pupillary responses within an attention control or working memory task. As it turns out, baseline pupil diameter, pretrial pupillary dynamics, and evoked pupillary responses can all provide valuable information regarding individual differences in attention control and working memory capacity.

To illustrate such effects, I re-examined a dataset in which participants completed a 30-min version of the PVT (Robison, 2018). Thought probes were embedded in the task to measure task-related and task-unrelated thoughts (TUTs). At the end of each of the five blocks, participants rated their subjective motivational state and state of alertness on a seven-point scale. This allowed us to examine intra- and interindividual patterns of motivation and alertness within the task. The task was designed as a between-subjects experiment, and the results are reported in Robison (2018) and Robison et al. (2020). However, the combined sample size across experiments ($N = 140$) allows for an examination of interindividual correlations. The following novel analyses are exploratory in nature and beg replication. However, several interesting correlations emerged.

First, and perhaps unsurprisingly, participants who reported higher subjective motivation performed better on the task in the form of faster ($r(136) = -0.38$, $p < 0.01$) and less variable RTs ($r(136) = -0.42$, $p = 0.01$). But interestingly, there was a pattern of interrelations among evoked responses, motivation, and task performance, as well. Participants who reported higher motivation demonstrated greater average evoked pupillary responses ($r(137) = 0.18$, $p = 0.03$), and participants who showed greater evoked pupillary responses showed faster average RTs ($r(134) = -0.24$, $p = 0.01$) and reported fewer TUTs ($r(137) = -0.24$, $p < 0.01$). Furthermore, participants who showed more trial-to-trial variability in pretrial pupil diameter reported more TUTs ($r(136) = 0.20$, $p = 0.02$) and slower ($r(133) = 0.32$, $p < 0.01$) and more variable RTs ($r(134) = 0.43$, $p < 0.01$). Figure 7 shows scatterplots of some of these correlations. These correlations were found in exploratory analyses, and they are quite modest in magnitude. Therefore, they deserve follow-up work. However, these preliminary findings data suggest that evoked pupillary responses and baseline pupil measures are not only affected by group-level differences induced by motivational manipulations (e.g., Massar et al., 2016; Hopstaken et al., 2015a, b, 2016) but that they are also sensitive to individual differences in motivation and indicative of task performance.

As another example, Unsworth et al. (2020) examined pupillary dynamics in the "waiting" period of the PVT. As discussed in the section on the antisaccade task, pupillary activity in anticipation of a stimulus can carry meaningful information. Here, Unsworth et al. examined how the pupil changed across the duration of the waiting period, which ranged from 2 to 10 s. On average, the pupil was relatively flat. However, the patterns of anticipatory activity diverged when examining trial-level and individual differences. The pupil showed a gradual dilation in anticipation of the stimulus on fast trials, whereas on slow trials, the pupil showed a relative constriction. Unsworth et al. (2020) argued that this preparatory dilation/constriction of the pupil is a physiological indicator of *intrinsic alertness* – the individual increasing and focusing their attention in preparation for the onset of the stimulus in order to react quickly. Intrinsic alertness can be differentiated from phasic alertness – a short-term readiness following a cue or warning signal that dissipates quickly – and tonic alertness – long-term, slow-changing readiness that is determined by an individual's general level of arousal/wakefulness (Langner et al., 2012;

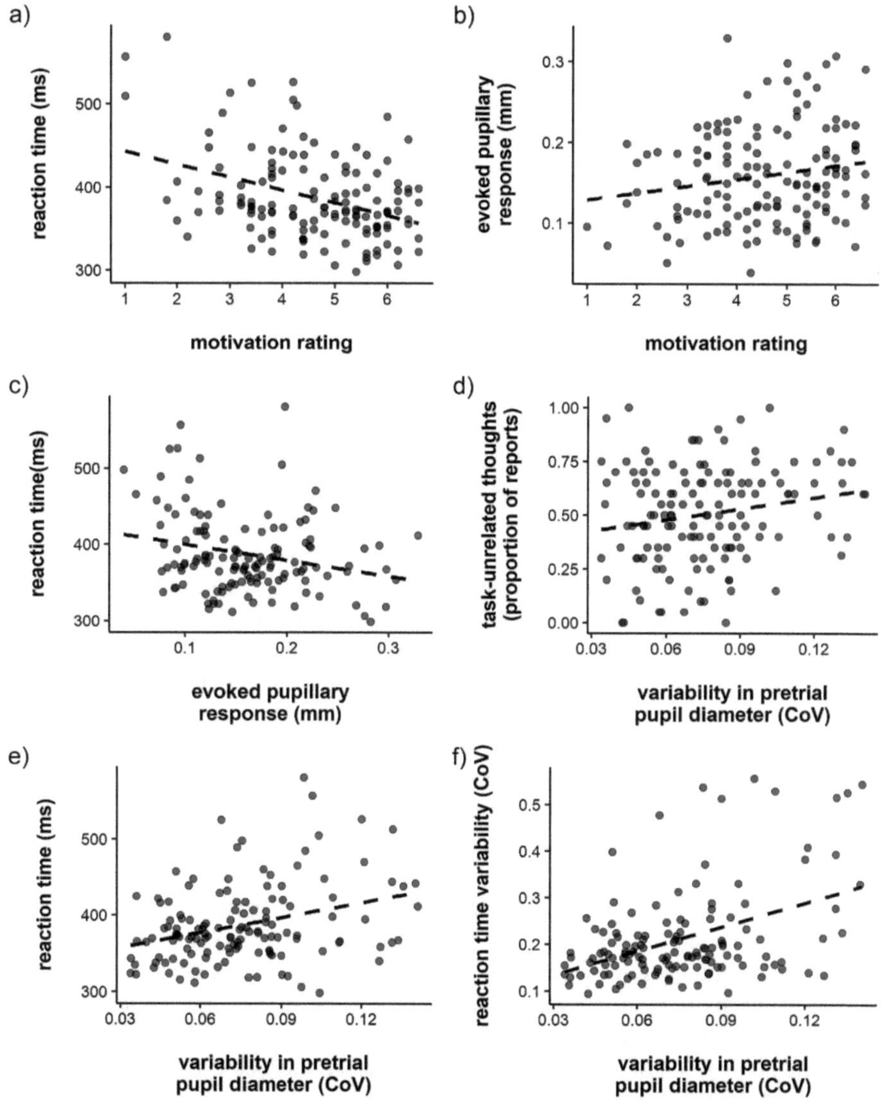

Fig. 7 Scatterplots of behavioral, subjective, and pupillary individual differences in the psychomotor vigilance task. (Data are from Experiment 2 in Robison (2018). See Robison (2018) for more) between (**a**) average reaction time and motivation, (**b**) average evoked pupillary response and motivation, (**c**) average evoked pupillary response and average reaction time, (**d**) intraindividual variability in pretrial pupil diameter and task-unrelated thougths, (**e**) intraindividual variability in pretrial pupil diameter and average reaction time, and (**f**) intraindividual variability in pretrial pupil diameter and intrainvidual variability in reaction time

Sadaghiani & D'Espositio, 2015, Sturm & Willmes, 2001; Unsworth & Robison, 2020; van Zomeran & Brouwer, 1994).

At the level of individual differences, Unsworth et al. binned reaction times by rank-ordering them from fastest to slowest and grouping them into five quintiles. They used the slowest bin of trials, an individual-difference metric. This measure has been regularly used in factor-analytic analyses of attention control (e.g., Unsworth et al., 2010; Unsworth & Spillers, 2010). Participants who showed greater increases in intrinsic alertness, measured by pupillary dilation during the wait period, had faster reaction times, especially in this slowest bin of trials. These results indicate that intrinsic alertness reduced the occurrence and severity of attentional lapses. Thus, Unsworth et al. (2020) offered one explanation for why individuals differ in their tendency to experience attentional lapses – intrinsic alertness – and they leveraged pupillometry to bolster this argument.

To offer an explanation as to what these pupillary measures might indicate, Nash Unsworth and I recently proposed a theory regarding individual differences in attention control, leveraging pupillometry to propose structural and/or functional differences in brain networks that are important for the implementation of attention (Unsworth & Robison, 2017a). Testing the theory rests on several assumptions. First, it relies on the idea that pupil diameter can be used as an indirect measure of both tonic and phasic activity of neurons in the locus coeruleus (LC; see Chap. 1 of this volume). Second, it relies on the hypothesis that there are several large-scale brain networks that interact to produce the patterns of behavior we observe in attention-control tasks. To the extent that these assumptions are true, we argue that we can leverage pupillometry to test theories as to why individuals differ in their ability to control and sustain their attention. We specifically discuss the fronto-parietal control network (FPN), the default-mode network (DMN), and the salience network (SN) as important networks in attention control. The FPN is a collection of brain regions in the frontal and parietal cortices that are particularly important for the implementation of attention control (Corbetta & Shulman, 2002; Miller & Cohen, 2001; Peterson & Posner, 2012; Vincent et al., 2008). The DMN is a collection of brain regions that seem to underlie internally directed cognition, whether it be task-irrelevant (e.g., mind-wandering) or task-relevant (i.e., goal-directed) cognition (Andrews-Hanna et al., 2014; Raichle et al., 2001). The DMN is particularly active during periods of task-irrelevant thought when the task is externally directed, like the sorts of tasks discussed in the present chapter. Thus, in many situations, the DMN and FPN work antagonistically. That is, DMN activity must be suppressed by the FPN for effective attentional control. However, they do work together when the task involves an internally directed process (e.g., autobiographical memory; Spreng et al., 2009). The SN is a collection of brain regions that seem particularly important for processing errors and adjusting behavior accordingly (Menon & Uddin, 2010; Seeley et al., 2007).

Our theory argues that one crucial reason why individuals differ in attention control is their ability to regulate arousal during goal-directed activity. Because the LC is a nucleus that regulates the release of NE and thus arousal (Berridge & Waterhouse, 2003), we posit dysregulation of the LC-NE system as a driver of arousal dysregulation. The LC projects into all areas of the cortex, and it has particularly dense projections into the areas that comprise the large-scale networks mentioned above (FPN, DMN, and SN). The theory builds on ideas posed by Aston-Jones and Cohen in their influential Adaptive-Gain theory of LC function (2005). The Adaptive-Gain theory posits that tonic LC activity lies on a spectrum and that there is a functional link between tonic LC activity and phasic activity. When tonic LC activity is very low, the organism will be drowsy and relatively unresponsive. When tonic LC activity is very high, the organism will be hyperaroused, stressed, and respond indiscriminately. But when tonic LC activity is moderate, phasic activity will be target- and goal-oriented, responding to relevant events and not responding to irrelevant events.

In a recent latent variable analysis, we had participants complete a battery of working memory and attention control tasks. Pupil diameter was recorded during the attention control tasks (antisaccade, PVT, and Stroop).[2] Thought probes were also embedded in the attention control tasks to allow us to measure instances of task-unrelated thought (TUT). We specified latent factors for working memory capacity, attention control, baseline pupil diameter, task-evoked pupillary responses, intraindividual variability in baseline pupil diameter, intraindividual variability in task-evoked pupillary responses, and TUTs. The strongest predictors of TUTs, working memory capacity, and attention control abilities were the two variability pupil factors. That is, participants who experienced more variability in baseline pupil diameter (tonic LC activity) and evoked pupillary responses (phasic LC activity) exhibited lower attention control and more TUTs.

Subsequent studies have observed negative correlations between variability in resting state pupil diameter and working memory capacity (Aminihajibashi et al., 2019), negative correlations between variability in pretrial pupil diameter and working memory capacity (Aminihajibashi et al., 2020; Robison & Unsworth, 2019; Robison & Brewer, 2020; Unsworth & Robison, 2015), and a positive correlation between pretrial pupil diameter and the tendency to experience TUTs (Robison & Brewer, 2020). Although some studies have observed a positive correlation between mean resting state pupil diameter and working memory capacity (Heitz et al., 2008; Tsukahara et al., 2016), this finding is not consistently observed (see Unsworth et al., 2019, for a recent meta-analysis of the relation). Finally, some studies have observed a positive correlation between mean resting state pupil diameter and measures of fluid intelligence (Tsukahara et al., 2016; Tsukahara & Engle, 2020), but a recent study did not replicate this finding (Aminihajibashi et al., 2020).

Collectively, we interpret the relationships observed between measures of pupil diameter, especially the measures of *variability* in pupil diameter to mean that some individuals have difficulty regulating their arousal. Precisely what causes this trait remains to be determined. However, arousal dysregulation does appear to be an underlying feature of individuals with poor attention control and low working

[2] Unfortunately, due to eye-tracker malfunction, valid pupil data were unavailable for the majority of participants in the antisaccade task, so we were only able to examine pupil data from the Stroop and PVT.

memory capacity. Our theory posits that periodic breakdowns of the appropriate and consistent delivery of NE to the FPN allow the DMN to become active, causing subjective attentional lapses, such as mind-wandering. Such attentional lapses can produce slow and/or inappropriate responses and poor maintenance of goals and goal-relevant information in working memory, especially in attention-demanding situations. Importantly, pupillometry was heavily leveraged to build and test this theory.

6 Conclusion

Although pupillometry was first introduced as a method to cognitive psychologists six decades ago, the surge of studies employing it has only happened in the last 20 years. But across a wide variety of domains including cognitive control, sustained attention, spatial attention, and working memory, pupillometry has been used to examine important theoretical questions and even test competing predictions made by models of human cognitive processes. Even with this surge of research, it feels as though we have only begun to scratch the surface of pupillometry's potential to provide insight into the psychological and neural mechanisms of human cognition.

References

Adam, K. C. S., Mance, I., Fukuda, K., & Vogel, E. K. (2015). The contribution of attentional lapses to individual differences in visual working memory capacity. *Journal of Cognitive Neuroscience, 27*, 1601–1616.

Alnæs, D., Sneve, M. H., Espeseth, T., Endestad, T., van de Pavert, S. H., & Laeng, B. (2014). Pupil size signals mental effort deployed during multiple object tracking and predicts brain activity in the dorsal attention network and the locus coeruleus. *Journal of Vision, 14*, 1–20.

Aminihajibashi, S., Hagen, T., Foldal, M. D., Laeng, B., & Espeseth, T. (2019). Individual differences in resting-state pupil size: Evidence for association between working memory capacity and pupil size variability. *International Journal of Psychophysiology, 140*, 1–7.

Aminihajibashi, S., Hagen, T., Andreassen, O. A., Laeng, B., & Espeseth, T. (2020). The effects of cognitive abilities and task demands on tonic and phasic pupil sizes. *Biological Psychology, 156*, 107945.

Andrews-Hanna, J. R., Smallwood, J., & Spreng, R. N. (2014). The default network and self-generated thought: Component processes, dynamic control, and clinical relevance. *Annals of the New York Academy of Sciences, 1316*, 29–52.

Aston-Jones, G., & Cohen, J. D. (2005). An integrative theory of locus coeruleus-norepinephrine function: Adaptive gain and optimal performance. *Annual Reviews of Neuroscience, 28*, 403–450.

Awh, E., & Jonides, J. (2001). Overlapping mechanisms of attention and spatial working memory. *Trends in Cognitive Sciences, 5*, 119–126.

Basner, M., & Dinges, D. F. (2011). Maximizing sensitivity of the psychomotor vigilance test (PVT) to sleep loss. *Sleep, 34*, 581–591.

Berridge, C. W., & Waterhouse, B. D. (2003). The locus coeruleus–noradrenergic system: modulation of behavioral state and state-dependent cognitive processes. *Brain Research Reviews, 42*, 33–84.

Binda, P., Pereverzeva, M., & Murray, S. O. (2013). Attention to bright surfaces enhances the pupillary light reflex. *Journal of Neuroscience, 5*, 2199–2204.

Binda, P., Pereverzeva, M., & Murray, S. O. (2014). Pupil size reflects the focus of feature-based attention. *Journal of Neurophysiology, 12*, 3046–3052.

Braver, T. S. (2012). The variable nature of cognitive control: A dual mechanisms framework. *Trends in Cognitive Sciences, 16*, 106–113.

Brown, G. G., Kindermann, S. S., Siegle, G. J., Granholm, E., Wong, E. C., & Buxton, R. B. (1999). Brain activation and pupil response during covert performance of the Stroop Color Word task. *Journal of the International Neuropsychological Society, 5*, 308–319.

Chatham, C. H., Frank, M. J., & Munakata, Y. (2009). Pupillometric and behavioral markers of a developmental shift in the temporal dynamics of cognitive control. *Proceedings of the National Academy of Sciences, 106*(14), 5529–5533.

Chiew, K. S., & Braver, T. S. (2013). Temporal dynamics of motivation-cognitive control interactions revealed by high-resolution pupillometry. *Frontiers in Psychology, 4*, 15.

Corbetta, M., & Shulman, G. L. (2002). Control of goal-directed and stimulus-driven attention in the brain. *Nature Reviews Neuroscience, 3*, 215–229.

Dalmaso, M., Castelli, L., & Galfano, G. (2020). Microsaccadic rate and pupil size dynamics in pro−/anti-saccade preparation: The impact of intermixed vs. blocked trial administration. *Psychological Research, 84*, 1320–1332.

Dinges, D. F., Orne, M. T., Whitehouse, W. G., & Orne, E. C. (1987). Temporal placement of a nap for alertness: Contributions of circadian phase and prior wakefulness. *Sleep, 10*, 313–329.

Drummond, S. P., Bischoff-Grethe, A., Dinges, D. F., Ayalon, L., Mednick, S. C., & Meloy, M. J. (2005). The neural basis of the psychomotor vigilance task. *Sleep, 28*, 1059–1068.

Eriksen, B. A., & Eriksen, C. W. (1974). Effects of noise letters upon the identification of a target letter in a non-search task. *Perception & Psychophysics, 16*, 143–149.

Engle, R. W., & Kane, M. J. (2004). Executive attention, working memory capacity, and a two-factor theory of cognitive control. In B. H. Ross (Ed.), *The Psychology of Learning and Motivation*, (Vol. 44), pp. 145–199. Elsevier.

Friedman, D., Hakerem, G., Sutton, S., & Fleiss, J. L. (1973). Effect of stimulus uncertainty on the pupillary dilation response and the vertex evoked potential. *Electroencephalography and Clinical Neurophysiology, 34*, 475–484.

Geva, R., Zivan, M., Warsha, A., & Olchik, D. (2013). Alerting, orienting or executive attention networks: Differential patterns of pupil dilations. *Frontiers in Behavioral Neuroscience, 7*, 145.

Gilzenrat, M. S., Nieuwenhuis, S., Jepma, M., & Cohen, J. D. (2010). Pupil diameter tracks changes in control state predicted by the adaptive gain theory of locus coeruleus function. *Cognitive, Affective, & Behavioral Neuroscience, 10*, 252–269.

Heitz, R. P., Schrock, J. C., Payne, T. W., & Engle, R. W. (2008). Effects of incentive on working memory capacity: Behavioral and pupillometric data. *Psychophysiology, 45*, 119–129.

Hess, E. H., & Polt, J. M. (1964). Pupil size in relation to mental activity during simple problem-solving. *Science, 143*, 1190–1192.

Hopstaken, J. F., Van Der Linden, D., Bakker, A. B., & Kompier, M. A. (2015a). A multifaceted investigation of the link between mental fatigue and task disengagement. *Psychophysiology, 52*, 305–315.

Hopstaken, J. F., van der Linden, D., Bakker, A. B., & Kompier, M. A. (2015b). The window of my eyes: Task disengagement and mental fatigue covary with pupil dynamics. *Biological Psychology, 110*, 100–106.

Hopstaken, J. F., van der Linden, D., Bakker, A. B., Kompier, M. A., & Leung, Y. K. (2016). Shifts in attention during mental fatigue: Evidence from subjective, behavioral, physiological, and eye-tracking data. *Journal of Experimental Psychology: Human Perception and Performance, 42*, 878–889.

Hutchison, K. A., Moffitt, C. C., Hart, K., Hood, A. V., Watson, J. M., & Marchak, F. M. (2020). Measuring task set preparation versus mind wandering using pupillometry. *Journal of Experimental Psychology: Learning, Memory, and Cognition, 46*, 280–295.

Jensen, A. R., & Rohwer, W. D., Jr. (1966). The Stroop color-word test: A review. *Acta Psychologica, 25*, 36–93.

Kahneman, D., & Beatty, J. (1966). Pupil diameter and load on memory. *Science, 154*, 1583–1585.

Kamp, S. M., & Donchin, E. (2015). ERP and pupil responses to deviance in an oddball paradigm. *Psychophysiology, 52*, 460–471.

Kane, M. J., Bleckley, M. K., Conway, A. R., & Engle, R. W. (2001). A controlled-attention view of working-memory capacity. *Journal of Experimental Psychology: General, 130*, 169–183.

Kiesel, A., Miller, J., Jolicœur, P., & Brisson, B. (2008). Measurement of ERP latency differences: A comparison of single-participant and jackknife-based scoring methods. *Psychophysiology, 45*, 250–274.

Kuchinke, L., Võ, M. L. H., Hofmann, M., & Jacobs, A. M. (2007). Pupillary responses during lexical decisions vary with word frequency but not emotional valence. *International Journal of Psychophysiology, 65*, 132–140.

Kurzban, R., Duckworth, A., Kable, J. W., & Myers, J. (2013). An opportunity cost model of subjective effort and task performance. *Behavioral and Brain Sciences, 36*, 661–679.

Laeng, B., & Endestad, T. (2012). Bright illusions reduce the eye's pupil. *Proceedings of the National Academy of Sciences, 109*, 2162–2167.

Laeng, B., Ørbo, M., Holmlund, T., & Miozzo, M. (2011). Pupillary stroop effects. *Cognitive Processing, 12*, 13–21.

Langner, R., Kellermann, T., Eickhoff, S. B., Boers, F., Chatterjee, A., Willmes, K., & Sturm, W. (2012). Staying responsive to the world: Modality-specific and -nonspecific contributions to speeded auditory, tactile, and visual stimulus detection. *Human Brain Mapping, 33*, 398–418.

Lim, J., & Dinges, D. (2008). Sleep deprivation and vigilant attention. *Annals of the New York Academy of Sciences, 1129*, 305–322.

Luck, S. J. (2014). *An introduction to the event-related potential technique*. MIT press.

MacLeod, C. M. (1991). Half a century of research on the Stroop effect: An integrative review. *Psychological Bulletin, 109*, 163–203.

Massar, S. A., Lim, J., Sasmita, K., & Chee, M. W. (2016). Rewards boost sustained attention through higher effort: A value-based decision making approach. *Biological Psychology, 120*, 21–27.

Mathôt, S., Van der Linden, L., Grainger, J., & Vitu, F. (2013). The pupillary light response reveals the focus of covert visual attention. *PLoS One, 8*, e78168.

Meier, M. E., Smeekens, B. A., Silvia, P. J., Kwapil, T. R., & Kane, M. J. (2018). Working memory capacity and the antisaccade task: A microanalytic–macroanalytic investigation of individual differences in goal activation and maintenance. *Journal of Experimental Psychology: Learning, Memory, and Cognition, 44*, 68–84.

Menon, V., & Uddin, L. Q. (2010). Saliency, switching, attention and control: A network model of insula function. *Brain Structure and Function, 214*, 655–667.

Miller, E. K., & Cohen, J. D. (2001). An integrative theory of prefrontal cortex functioning. *Annual Review of Neuroscience, 24*, 167–202.

Murphy, P. R., Robertson, I. H., Balsters, J. H., & O'Connell, R. G. (2011). Pupillometry and P3 index the locus coeruleus–noradrenergic arousal function in humans. *Psychophysiology, 48*, 1532–1543.

Peavler, W. S. (1974). Pupil size, information overload, and performance differences. *Psychophysiology, 11*, 559–566.

Petersen, S. E., & Posner, M. I. (2012). The attention system of the human brain: 20 years after. *Annual Review of Neuroscience, 35*, 73–89.

Preuschoff, K., t Hart, B. M., & Einhauser, W. (2011). Pupil dilation signals surprise: Evidence for noradrenaline's role in decision making. *Frontiers in Neuroscience, 5*, 115.

Raichle, M. E., MacLeod, A. M., Snyder, A. Z., Powers, W. J., Gusnard, D. A., & Shulman, G. L. (2001). A default mode of brain function. *Proceedings of the National Academy of Sciences, 98*, 676–682.

Robison, M. K. (2018). Regulating mind-wandering and sustained attention with goal-setting, feedback, and incentives [Doctor Dissertation].

Robison, M. K., & Brewer, G. A. (2020). Individual differences in working memory capacity and the regulation of arousal. *Attention, Perception, & Psychophysics, 82*, 3273–3290.

Robison, M. K., & Unsworth, N. (2019). Pupillometry tracts fluctuations in working memory performance. *Attention, Perception, & Psychophysics, 81*, 407–419.

Robison, M. K., Unsworth, N., & Brewer, G. A. (2020). Examining the effects of goal-setting, feedback, and incentives on sustained attention. *Journal of Experimental Psychology: Human Perception and Performance, 47*, 869.

Rondeel, E., Van Steenbergen, H., Holland, R., & van Knippenberg, A. (2015). A closer look at cognitive control: Differences in resource allocation during updating, inhibition and switching as revealed by pupillometry. *Frontiers in Human Neuroscience, 9*, 494.

Sadaghiani, S., & D'Esposito, M. (2015). Functional characterization of the cingulo-opercular network in the maintenance of tonic alertness. *Cerebral Cortex, 25*, 2763–2773.

Seeley, W. W., Menon, V., Schatzberg, A. F., Keller, J., Glover, G. H., Kenna, H., et al. (2007). Dissociable intrinsic connectivity networks for salience processing and executive control. *Journal of Neuroscience, 27*, 2349–2356.

Siegle, G. J., Steinhauer, S. R., & Thase, M. E. (2004). Pupillary assessment and computational modeling of the Stroop task in depression. *International Journal of Psychophysiology, 52*, 63–76.

Siegle, G. J., Ichikawa, N., & Steinhauer, S. (2008). Blink before and after you think: Blinks occur prior to and following cognitive load indexed by pupillary responses. *Psychophysiology, 45*, 679–687.

Smit, A. S., Eling, P. A., & Coenen, A. M. (2004). Mental effort causes vigilance decrease due to resource depletion. *Acta Psychologica, 115*, 35–42.

Spreng, R. N., Mar, R. A., & Kim, A. S. (2009). The common neural basis of autobiographical memory, prospection, navigation, theory of mind, and the default mode: A quantitative meta-analysis. *Journal of Cognitive Neuroscience, 21*, 489–510.

Strauch, C., Koniakowsky, I., & Huckauf, A. (2020). Decision making and oddball effects on pupil size: Evidence for a sequential process. *Journal of Cognition, 3*, 7.

Stroop, J. R. (1935). Studies of interference in serial verbal reactions. *Journal of Experimental Psychology, 18*, 643–662.

Sturm, W., & Willmes, K. (2001). On the functional neuroanatomy of intrinsic and phasic alertness. *NeuroImage, 14*, S76–S84.

Thomson, D. R., Besner, D., & Smilek, D. (2015). A resource-control account of sustained attention: Evidence from mind-wandering and vigilance paradigms. *Perspectives on Psychological Science, 10*, 82–96.

Tsukahara, J. S., & Engle, R. W. (2020). *The locus coeruleus-norepinephrine system and fluid intelligence*. PsyArXiv.

Tsukahara, J. S., & Engle, R. W. (2021). Fluid intelligence and the locus coeruleus–norepinephrine system. *Proceedings of the National Academy of Sciences, 118*(46), e2110630118.

Tsukahara, J. S., Harrison, T. L., & Engle, R. W. (2016). The relationship between baseline pupil size and intelligence. *Cognitive Psychology, 91*, 109–123.

Unsworth, N., & McMillan, B. D. (2014). Similarities and differences between mind-wandering and external distraction: A latent variable analysis of lapses of attention and their relation to cognitive abilities. *Acta Psychologica, 150*, 14–25.

Unsworth, N., & Robison, M. K. (2015). Individual differences in the allocation of attention to items in working memory: Evidence from pupillometry. *Psychonomic Bulletin & Review, 22*, 757–765.

Unsworth, N., & Robison, M. K. (2016). Pupillary correlates of lapses of sustained attention. *Cognitive, Affective, & Behavioral Neuroscience, 16*, 601–615.

Unsworth, N., & Robison, M. K. (2017a). A locus coeruleus-norepinephrine account of individual differences in working memory capacity and attention control. *Psychonomic Bulletin & Review, 24*, 1282–1311.

Unsworth, N., & Robison, M. K. (2017b). Pupillary correlates of covert shifts of attention during working memory maintenance. *Attention, Perception, & Psychophysics, 79*, 782–795.

Unsworth, N., & Robison, M. K. (2017c). The importance of arousal for variation in working memory capacity and attention control: A latent variable pupillometry study. *Journal of Experimental Psychology: Learning, Memory, & Cognition, 43*, 1962–1987.

Unsworth, N., & Robison, M. K. (2018b). Tracking working memory maintenance with pupillometry. *Attention, Perception, & Psychophysics, 80*, 461–484.

Unsworth, N., & Robison, M. K. (2020). Working memory capacity and sustained attention: A cognitive-energetic perspective. *Journal of Experimental Psychology: Learning, Memory, and Cognition, 46*, 77–103.

Unsworth, N., & Spillers, G. J. (2010). Working memory capacity: Attention, memory, or both? A direct test of the dual-component model. *Journal of Memory and Language, 62*, 392–406.

Unsworth, N., Schrock, J. C., & Engle, R. W. (2004). Working memory capacity and the antisaccade task: Individual differences in voluntary saccade control. *Journal of Experimental Psychology: Learning, Memory, and Cognition, 30*, 1302–1321.

Unsworth, N., Redick, T. S., Lakey, C. E., & Young, D. L. (2010). Lapses in sustained attention and their relation to executive control and fluid abilities: An individual differences investigation. *Intelligence, 38*, 111–122.

Unsworth, N., Robison, M. K., & Miller, A. L. (2018). Pupillary correlates of fluctuations in sustained attention. *Journal of Cognitive Neuroscience, 30*, 1241–1253.

Unsworth, N., Miller, A. L., & Robison, M. K. (2020). Individual differences in lapses of sustained attention: Oculometric indicators of intrinsic alertness. *Journal of Experimental Psychology: Human Perception and Performance, 46*, 569–592.

Unsworth, M., Miller, A. L., & Robison, M. K. (2019). Is working memory capacity related to baseline pupil diameter? *Psychonomic Bulletin & Review, 28*, 228–237.

van Bochove, M. E., Van der Haegen, L., Notebaert, W., & Verguts, T. (2013). Blinking predicts enhanced cognitive control. *Cognitive, Affective, & Behavioral Neuroscience, 13*, 346–354.

van Zomeren, A. H., & Brouwer, W. H. (1994). *Clinical neuropsychology of attention*. Oxford Press.

Vincent, J. L., Kahn, I., Snyder, A. Z., Raichle, M. E., & Buckner, R. L. (2008). Evidence for a frontoparietal control system revealed by intrinsic functional connectivity. *Journal of Neurophysiology, 100*, 3328–3342.

Vogel, E. K., McCollough, A. W., & Machizawa, M. G. (2005). Neural measures reveal individual differences in controlling access to working memory. *Nature, 438*, 500–503.

Wahn, B., Ferris, D. P., Hairston, W. D., & König, P. (2016). Pupil sizes scale with attentional load and task experience in a multiple object tracking task. *PLoS One, 11*, e0168087.

Wang, C. A., Brien, D. C., & Munoz, D. P. (2015). Pupil size reveals preparatory processes in the generation of pro-saccades and anti-saccades. *European Journal of Neuroscience, 41*, 1102–1110.

Warm, J. S., Parasuraman, R., & Matthews, G. (2008). Vigilance requires hard mental work and is stressful. *Human Factors, 50*, 433–441.

Wendt, M., Kiesel, A., Geringswald, F., Purmann, S., & Fischer, R. (2014). Attentional adjustment to conflict strength. *Experimental Psychology, 61*, 55–67.

Utility of Pupillometric Methods for Studying Visual Search

Juan D. Guevara Pinto and Michael C. Hout

1 Introduction

One of the most common (and perhaps most important) tasks that we engage in during our everyday lives is searching for objects in our environments. From hunter-gatherers foraging for nutritious food to modern-day office workers searching for an important file on their computer desktops, humans have consistently been tasked with finding task-relevant "targets" amidst a backdrop awash in visual stimuli. This process of "visual search" is ubiquitous in daily life, but it is also quite common in a variety of societally important tasks (e.g., airport security screening), and thus (not surprisingly), it has been one of the most extensively studied topics in cognitive science. At the time of writing this, a Google Scholar search for the phrase "visual search" yielded over 14,000 entries just for the year 2022. Basic science researchers have focused a great deal of attention on understanding the cognitive processes underlying search, and applied researchers have directed efforts at ways to improve search performance. Because of this, a wide variety of different research methodologies have been adopted and modified over the years.

In its most basic form, researchers study visual search in the laboratory by having participants search for a target picture (usually on a computer screen) hidden among a visual display of (often similar looking) nontarget items referred to as "distractors." This is somewhat akin to the kids' game "Where is Waldo?" as participants are tasked with finding a target item as quickly as possible. Many search paradigms also involve "target absent" displays wherein the observer must search exhaustively to correctly determine that their target is not in the display. Such

J. D. Guevara Pinto (✉)
Department of Psychology, Rollins College, Winter Park, FL, USA
e-mail: jguevarapinto@rollins.edu

M. C. Hout
Department of Psychology, New Mexico State University, Las Cruces, NM, USA

M. H. Papesh, S. D. Goldinger (eds.), *Modern Pupillometry*,
https://doi.org/10.1007/978-3-031-54896-3_5

emphasis on the *speed* of visual searches has obvious implications for real-world behaviors (e.g., someone might want to find their keys as quickly as possible so that they can leave their home and get to work), but its methodological importance likely originates from earlier work in visual attention and perception. Some of the earliest visual search paradigms (still in use today) were developed by Anne Treisman to understand how different visual features, such as color, shape, or orientation, are perceived "preattentively" (Tversky, 2020). For example, when observers search through a display of red *X*s and green *T*s, they typically are much slower at finding a red *T* than a blue one, as the blue "singleton" target tends to "pop out" given that it is presented in a different color than the rest of the items in the display (see Fig. 1). In fact, during such "pop-out searches," response times (RTs) remain relatively stable regardless of how many distractors are in the display, suggesting that some features are indeed processed preattentively. By contrast, when searching for a red *T* (among red *X*s and green *T*s), RTs increase proportionally with the number of items in the display because directed attention is required to "bind" features together into an integrated whole. This "feature integration" is what allows observers to determine that an item is not the target either because it is red but not a *T*, or because it is a *T* but not a red one. This key finding is at the core of Treisman's *Feature Integration Theory* (FIT; Treisman & Gelade, 1980), but it is the examination of RTs as a function of the set size of items in the display, or "search slopes," that is most germane here.

As a result of Treisman's seminal work on feature integration, search slopes became the standard measurement of choice in visual search research, with some studies examining search slopes across hundreds of thousands of visual search trials (e.g., Wolfe, 1998a). Such attempts to quantify search performance have also informed modern frameworks of visual search. For instance, Wolfe's *Guided Search* model (e.g., Wolfe, 2007; Wolfe et al., 1989) expands upon (and in some ways

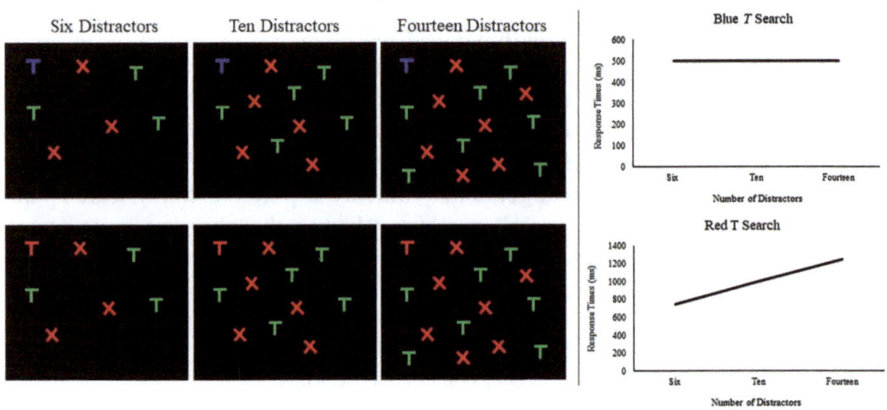

Fig. 1 Examples of singleton ("pop-out") searches (upper left panels) and conjunction searches (lower left panels). Right panels reflect changes in response times (RTs) as a function of set size for both types of search

detracts from) FIT's core ideas by suggesting that features processed preattentively are also matched to a "target template." This template is a mental representation of what the target looks, and it "guides" the search by selectively directing attention to areas of the display containing matching features. In essence, the more areas in the display that contain a good match to the target, the steeper the search slope becomes.

Search slopes have provided crucial insights into various aspects of visual search performance. Indeed, some early views on visual search (which preferred the use of search slopes) explicitly stated that "eye movements are probably not the determining factor in visual searches" (Wolfe, 1998b, p. 14). However, search slopes are a relatively crude tool for studying behavior because analyzing response time alone does not allow for a more nuanced exploration of attentional distributions (e.g., where did the observer look or fail to look), object recognition timing (e.g., how long were objects examined and how frequently), the particular source of search errors (e.g., was a target missed because it was not viewed or because it was viewed but not recognized), among other factors. This preference for search slopes over eye movement recordings was, of course, at least partially due to the scarcity of eye movement recording devices available more than two decades ago. But because of improvements in eye-tracking technology and the broader accessibility of such devices, it is no surprise that eye movement measures have recently been rising in popularity among visual search researchers (Godwin et al., 2021). Oculomotor metrics now allow researchers not just to estimate the processing of individual items based on RTs and the set size of the display but to *directly* measure how many items are directly fixated (and for how long), as well as what areas of the display attract attention the most (or least). However, as with many changes in scientific paradigms, it takes time to adopt new approaches. Thus, the debate regarding whether search slopes are still informative in light of eye movement recordings is still ongoing (e.g., Eimer, 2017; Hulleman & Olivers, 2017).

Despite this relatively slow evolution of experimental methodologies in the visual search literature, some researchers have already embraced novel techniques such as *pupillometry* to better understand search processes. As indicated by Fig. 2, when searching for both "visual search" and "pupillometry," we see an exponential increase in entries in the last 10 years alone, suggesting that more and more researchers are starting to see the value in adopting this technique to fully understand search processes. We believe that the field of visual search is one that can certainly benefit from this technique, particularly given that pupillometry is a well-established, physiological index of attention allocation (e.g., Hess & Polt, 1964; Kahneman & Beatty, 1966; Laeng et al., 2012). When we search for information in our environment, we cannot focus on every object, person, or area in a scene to find a given target. Importantly, when we select a particular item for closer examination, we tend to become (to varying degrees) unaware of the other items in our field of view. This reflects the limits of our attentional capacity (e.g., Cowan et al., 2005; Huang & Pashler, 2005; Kristjánsson & Draschkow, 2021), and it results from the interaction of the bottom-up properties of the environment (e.g., clutter, distractor homogeneity), our top-down goals (e.g., what does the target look like; Hout & Goldinger, 2014; see also Wolfe & Horowitz, 2017, for list of factors used to guide attention),

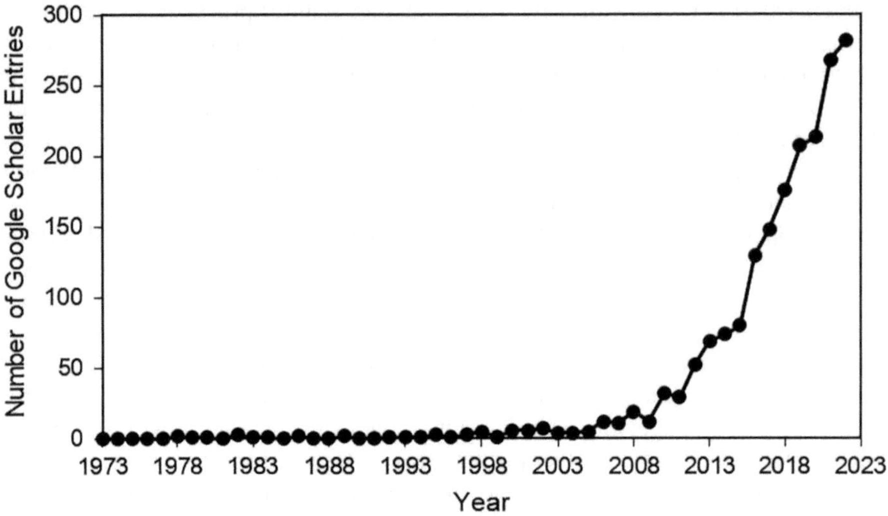

Fig. 2 Number of Google Scholar entries for searches including "visual search" and "pupillometry" across a 50-year span

and our selection "history" (e.g., what items or location have we previously attended).[1] As described by Kahneman (1973), attention is not an unlimited "resource," and it is consumed as we exert *effort* to accomplish multiple cognitive tasks (e.g., determining whether a selected item is the target or not). Although operationalizing terms such as "attention" and "effort" is not easy to do (see Bruya & Tang, 2018 for alternative perspectives on Kahneman's ideas), it is well accepted that devoting attention to processing one item in our environment generally reduces our ability to process others. If you are trying to find your seat at a crowded baseball stadium, you might miss part of the game or the face of an old friend standing nearby while your attention is focused on finding the right combination of letters and numbers that indicate that you have found your seat.

One of the earliest demonstrations of researchers examining the pupillary response during visual search comes from Porter et al. (2007), who were interested in examining the attentional load imposed by traditional search tasks. Pupillary reflexes clearly indexed the mental load imposed by search arrays. Relative to easy searches, challenging searches (e.g., those with larger set sizes and higher distractor homogeneity) yielded larger *tonic*[2] pupil dilations. But perhaps more interestingly, the moment-to-moment dilations provided insights into the dynamics of mental

[1] Please see Shomstein et al. (2023) for a recent proposal on how taxonomies of attention selection need to be revised to accommodate *all* attentional phenomena.

[2] Pupillary responses are divided between *tonic* and *phasic* pupil dilation. Tonic pupil dilation reflects the sustained component of the pupil response across a prolonged period of time and is generally modulated by arousal, environmental lighting, and overall task difficulty. Phasic pupil dilation reflects that transient or event-related component of the pupil response, and it is modulated

effort during search: The data suggested that observers engaged in greater mental effort at the later stages of the search (i.e., when a person is close to making a response) relative to the earlier stages (i.e., during exploratory search). Additionally, when participants were required to search for two targets in the same trial, the additional memory load of remembering where the first target has yielded an increase in pupil size relative to single-target searches. This indicates that increasing spatial memory requirements during visual search directly impacts tonic pupillary responses, suggesting that multiple target searches involve some (albeit limited) memory for item locations.

Porter and colleagues' findings are important as they call into question an implicit assumption made by most traditional visual search frameworks, which is that the amount of attention allocated to the search process is rigid. By this, we mean that it is assumed that attention allocation is determined at early search stages through the activation of priority maps (e.g., Treisman & Gelade, 1980; Duncan & Humphreys, 1989; Wolfe et al., 1989) and that allocation remains relatively unaltered through the remainder of the search until the target is found. The results from Porter et al. suggest otherwise, highlighting the utility of methods like pupillometry for providing novel insights about long-held assumptions. In fact, one of the primary advantages of pupillometric methods for visual search research lies in quantifying the unfolding of mental events as they occur *during* the search process. These are, in and of themselves, different than the advantages conferred by more traditional eye-tracking measures such as target fixation times and fixation durations, which can help delineate the stages of the search (e.g., Castelhano et al., 2008; Malcolm & Henderson, 2009) but do not provide much information regarding mental events happening within each stage. For example, knowing that an observer fixated on the target but did not make an overt target-present response (via a button press) indicates that the selected item did not reach a target-detection criterion (e.g., Wolfe & van Wert, 2010) and thus resulted in a target miss not because it was not examined but because it was not recognized. However, this fact provides little evidence about the extent of item processing itself, such as whether the target came closer to reaching the recognition criterion relative to other distractor items.

To test whether targets are processed more extensively than distractors even when an overt target-present response was not made, Privitera et al. (2010) asked participants to search for categorical targets (e.g., icons of musical instruments or notations) presented within a continuous stream of rapidly presented real-world object icons (700 ms per icon). This technique is known as Rapid Serial Visual Presentation (RSVP), and the process involves presenting images in sequential order, one-by-one, for a brief (usually for less than a second) amount of time (see Fig. 3). Importantly, Privatera et al.'s participants were instructed to either make an overt target-present response as soon as a target was presented in the continuous RSVP stream or to count all targets presented throughout the stream and then report

---

by the onset and offset of task-relevant stimuli (commonly referred to as task-evoked pupillary responses; TEPRs).

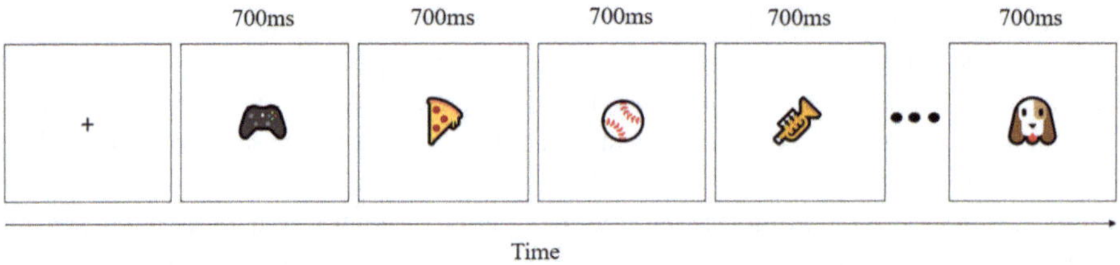

Fig. 3 Example of a Rapid Visual Serial Presentation (RSVP) procedure. Variants of this procedure include presentation rate (i.e., timing) for each image, whether responses are required at the moment a target(s) appears or occur at the end of the trial, and the length (i.e., number of images) of the RSVP stream

the total number at the end of the display. The results showed that although target-detection registered via overt responses yielded larger early *phasic* pupil dilation 500–700 ms after the onset of the target, targets detected without an overt response still yielded significantly larger dilation than distractor items later, following approximately 1400–1600 ms post-target onset. This suggests that targets are indeed processed more extensively than distractors and that subthreshold target detection can be inferred from phasic pupillary dilation even when an overt response is not made by the observer.

While these and other (e.g., Backs & Walrath, 1992; Takeuchi et al., 2011) early studies showed the vast potential of pupillometry in visual search research, more recent work has been increasingly systematic in exploiting the temporal dynamics of pupillary responses to uncover important aspects of how visual search is conducted. In the next sections, we will discuss the cases of the *low-prevalence effect (LPE)* and *domain-specific expertise*, two distinct areas of research that have received high degrees of attention in the past decade given their importance to applied professions (e.g., airport security, medical image screening, and sports refereeing). We are just starting to see the emergence of pupillometric methods in both fields, but the results from the few studies that employ these methods are as fascinating as they are critical to improving search procedures in applied domains.

1.1 Case of the **Low-Prevalence Effect**

One of the most frequently discussed applications of visual search research is its use in professional settings such as airport security screening. This profession includes searching (relatively quickly) for illegal/prohibited (e.g., drugs) and/or potentially dangerous (e.g., weapons, bombs) objects concealed within passengers' luggage. The time pressure of moving the passenger line along may impact performance, but there are other internal cognitive factors that might impose greater challenges. For example, Biggs and Mitroff (2014a) reviewed the adverse effects of several task-specific demands during this type of security screening task, including low target visibility due to clutter and/or the intent-to-hide objects by the passengers themselves, the lack of precise knowledge regarding what the target(s) may look like, and *low-target prevalence*. This latter issue reflects the statistical regularity with which targets are encountered by airport security officers, which is likely to be very low.[3] This produces a degree of bias stemming from the fact that any given piece of luggage is unlikely to contain an illegal or dangerous item. When an illegal item is indeed present, security officers are, therefore, more likely to *miss* it (relative to the same item had it shown up more regularly). This is known as the *low-prevalence effect* (LPE; Wolfe et al., 2007), and it has been studied extensively across various literatures in cognitive science; it is the extremely robust finding that as the likelihood of encountering a target on any given trial decreases, the likelihood of missing that target when it is presented increases (for a review, see Horowitz, 2017).

In visual search, the LPE has been traditionally identified by an elevated miss rate during low-prevalence conditions (e.g., 10% target-present or less) relative to high-prevalence conditions (e.g., 50–90%). Such effects have been replicated numerous times across various levels of prevalence (e.g., Rich et al., 2008; Evans et al., 2013; Wolfe et al., 2013). Beyond elevated miss rates, behavioral measures of the LPE include faster responses during target-absent trials during low (relative to high) prevalence conditions (Wolfe et al., 2005; Rich et al., 2008; Godwin et al., 2015), consistent with a preemptive search termination decision prior to the target being observed. Importantly, the increase in miss rates and the hastened target-absent responses are not a result of a speed-accuracy tradeoff, as rare targets are still missed at disproportionate rates even when observers are asked to slow down their searches (e.g., Wolfe et al., 2007). Instead, it is believed that the elevated miss rates associated with the LPE reflect a shift to a more conservative target identification criterion rather than a decrease in sensitivity. Wolfe and van Wert (2010) proposed the Multiple-Decision Model (MDM) of visual search to account for prevalence-related effects observed during search (see also Hout et al., 2022, for a similar model of "interactive" searches). According to this model, observers select individual items for processing in a serial fashion, based on preattentively defined activation maps of the search display (see Wolfe, 1994, for a description of how these maps are produced). Once an item has been selected for processing, the observer then completes an internal two-alternative forced choice (2-AFC) decision, determining whether the selected item is (or is not) the target by comparing the selected item to a target template (Hout & Goldinger, 2014). If the observer determines that the item is the target, then a target-present response is made and the trial ends. However, if the observer determines that the attended item is not the target, then the item with

[3] Unfortunately, the precise prevalence rates for illegal items (such as firearms) at airport security points is not publicly available, but it is possible to estimate it from other measures reported by the Transportation Security Administration (TSA) in the United States. For instance, in 2018, the TSA reported detecting approximately 4432 firearms (avg. of 11 detected daily) across airports in the United States (tsa.gov/news/press/factsheets/tsa-numbers, 2021). At the same time, the TSA reports screening about 5.5 million carry-on items on a daily basis, indicating the daily probability of detecting a firearm among any of the screened items is 0.000002%.

the next highest activation is selected for processing and the 2-AFC decision is made anew.

A second process accumulates information simultaneously as the observer selects items for processing and performing the corresponding 2-AFC decisions. This second process reflects the observer's quitting threshold, and it is expressed as a drift-diffusion process where evidence from the search display is accumulated to determine whether the search should be terminated. If the accumulated evidence reaches the quitting threshold before the target is selected for processing, then the observer terminates the search and a target-absent response is made. Thus, under this model, a target-present response can only be made if the target is selected for processing before the quitting threshold is reached (and when the selected item is assessed by the observer to have been a close enough match to their target template). Moreover, although both the internal 2-AFC decision and the quitting threshold occur in parallel, they are independent of one another, with the internal decision process being modeled using signal detection theory (SDT) and the quitting threshold being modeled by a drift-diffusion process.

The independence of the two processes outlined in the MDM indicates that miss rates during low-prevalence search can potentially be explained individually by each process. More precisely, each process allows for a specific type of search error. Miss errors attributable to an early search termination (i.e., selection errors wherein the target is never looked at) are reflected by a decrease (over time) in the observer's quitting threshold during low-prevalence search. This explains not only the higher miss rates for rare targets but also the decrease in RTs during target-absent trials, suggesting that under low-prevalence conditions, observers do not search exhaustively; they terminate the search before selecting every item in the search display for inspection. Regarding the recognition decision, low target-prevalence involves a conservative-shift of the target-present criterion, whereby the observer requires additional evidence for any selected item to be perceived as the target before a target-present decision can be made. This, in turn, makes it harder for observers to identify the target item as such after it has been selected for processing (i.e., identification errors, whereby a target is examined but not identified as the to-be-located object).

Traditional behavioral measures (i.e., miss rates and RTs) have provided ample evidence that low-target prevalence reduces observers' quitting thresholds. As discussed earlier, one of the hallmarks of the LPE is that observers make faster target-absent responses when targets are rare, relative to when they are frequent (e.g., Wolfe et al., 2005; Rich et al., 2008). This has been further supported by more recent evidence from studies tracking observers' eye movements during the search, where it has been found that low-prevalence searches yield fewer item fixations during target-absent trials, in comparison to high-prevalence conditions (e.g., Godwin et al., 2015; Schwark et al., 2012). Quite often, observers make target-absent responses even before they fixate directly on the target during low-prevalence conditions (Peltier & Becker, 2016). Furthermore, the rising popularity of eye-tracking research in the past decade has also provided evidence for the prevalence-induced shift in the internal target identification criterion suggested by the MDM.

For instance, Hout et al. (2015) found that observers repeatedly failed to detect target objects despite looking directly at them. Observers in this study were asked to search for two target objects from different categories simultaneously, one of which appeared more frequently than the other (this is referred to as a "relative prevalence" manipulation). However, only one target (or none) was ever present in the display. Despite the overall target prevalence of the search task being high (50% target-present), Hout et al. (2015) observed that viewers were more likely to miss the rare target relative to the comparatively more common one. Even when a correct target-present response was ultimately made, observers took longer to render those decisions when fixating on a rare target relative to a common one. Hout et al. (2015) also examined the LPE during a more "passive" type of visual search task by using an RSVP procedure (see Fig. 3), which involved presenting each item in a search array one-by-one, each for a brief amount of time. Participants here were not allowed to make target present/absent responses until the end of the RSVP stream of items and were thus forced to view every item in the display before making their decision. By removing the need to select potential targets across space, the RSVP procedure eliminated the possibility of observers making selection errors, and early termination errors were eliminated by only allowing search decisions at the end of the stream. Despite, therefore, being forced to look at every object in the stream, Hout et al. (2015) found that observers still missed disproportionately more rare targets than common ones, indicating that identification errors are a significant contribution to the LPE as proposed by the MDM.

Given the extensive research on the LPE in visual search, it may be tempting to think that methodologies such as pupillometry may not have much left to offer in terms of novel insights for understanding the impact of target prevalence on cognitive processes. However, the MDM makes clear predictions regarding the target identification process, some of which are yet unanswered and may not be easily answered by only relying on behavioral and eye movement measures alone. For instance, according to the MDM, when targets are selected for processing, a conservative shift in the target identification criterion occurs during low-prevalence conditions, thus demanding greater evidence for a selected item to be judged as the target. That implies that there is a greater cognitive demand when processing rare targets, which results in the elevated miss rates that are the hallmark of the LPE. However, if this is indeed the case, one might ask, what are the consequences of *correctly* identifying a rare target as such? Given that correct target identification (i.e., hits) is the end-product of successful target processing, that would suggest that hits for rare targets might be more "costly" in terms of processing resources than hits for common targets. Importantly, because cognitive processing does not occur in a vacuum, we must next ask what would happen if an observer needed to perform a subsequent cognitive operation after identifying a rare target. Would the increased cognitive demands of correctly identifying a rare target impair subsequent processing to a greater extent relative to identifying less-demanding common targets?

Such questions are theoretically fascinating but have not received much attention in the LPE literature until recently. Perhaps, this is due to the important practical considerations involved in attempting to eliminate the LPE in applied situations.

That is, reducing *miss rates* to near zero (i.e., increasing *hit rates* to perfect performance) might be the primary goal of a researcher interested in the LPE from a practical point of view. Once observers consistently and accurately identify rare targets, there might be little interest left (from an applied perspective) in examining the cognitive consequences of such idealistic levels of performance. On the other hand, it is also possible that these questions have not been studied extensively due to the constraints in research methodologies utilized in the LPE literature, which historically involved mostly behavioral approaches (e.g., miss rates and RTs) and only the fairly recent adoption of oculomotor measures (e.g., fixation counts and durations). Neither approach, however, is especially useful in indexing the cognitive demands imposed by a task, and thus, questions about the cognitive consequences of completing a demanding target identification go unasked. This is where the utility of novel approaches such as pupillometry starts to become evident: By relying on established physiological measures of cognitive processing, we can begin asking questions that look beyond our traditional conceptualization of "performance."

Papesh and Guevara Pinto (2019) used pupillometry to assess the broader impact of target prevalence on cognition after a target identification is made. To do so, they relied on the attentional blink (AB) paradigm, an experimental task that tracks not only the processing demands of target identifications in the temporal dimension (e.g., Raymond et al., 1992; see Martens & Wyble, 2010, for a review) but also one in which the source of errors has been attributed to the locus–coeruleus norepinephrine (LC–NE) system[4] (Niewuenhius et al., 2005; Warren et al., 2009). In the standard AB paradigm, observers are presented with a sequence of stimuli (e.g., digits or symbols) in an RSVP stream and are asked to identify two different targets (e.g., two letters) embedded in the stream. Importantly, researchers using this paradigm can manipulate the temporal distance, or "lag," between the two targets by varying the number of intervening nontarget stimuli. The typical pattern of results is such that observers tend to miss the second target (T2) following the correct identification of the first target (T1), particularly when T2 is presented shortly after T1 (150–400 ms post-T1). One potential explanation for this robust finding is that the observer is *still* processing T1 when T2 is presented, and thus T2 is missed despite being directly viewed; it is as if the observer's attention "blinked" for the brief moment, T2 was presented. In their study, Papesh and Guevara Pinto (2019) examined whether correctly identifying a rare target produced a stronger attentional blink effect than identifying common targets and whether phasic pupillary responses could be used to track and distinguish the processing demands involved with identifying targets of different prevalence levels. Observers were presented with the standard AB paradigm described above, where they passively searched for two letters among a rapid stream of digits and symbols (each item presented for 15 ms, with a 75 ms interstimulus interval). Importantly, T1 could be any of three letters: The letter *M* (presented on 88% of trials; a high-prevalence target) or the letters *A*

[4] As discussed in several chapters, the LC-NE system is implicated in phasic pupillary responses to task-relevant stimuli.

and *R* (presented on 6% of trials each; low-prevalence targets). The second target (T2) was randomly drawn from a pool of different letters, with an equal presentation rate across the experiment.

Papesh and Guevara Pinto's (2019) results conceptually replicated previous work on the LPE: Low-prevalence T1 targets yielded a higher miss rate than the high-prevalence T1 targets (e.g., Hout et al., 2015; Peltier & Becker, 2016). Furthermore, their predictions regarding the AB effect were supported, as low-prevalence T1 targets yielded larger phasic pupillary dilation than the high-prevalence T1 target, suggesting that processing rare targets is indeed more cognitively demanding than processing common ones, even when a correct target identification response is made. However, because even phasic pupillary responses to cognitive events unfold over an extended period of time, simply analyzing the pupil size at the end of the trial period is likely to yield a null effect, as transient pupillary changes may be unobservable. For example, Fig. 4 illustrates baseline-corrected pupil size after the onset of T1. The different pupil size between high and low T1 targets clearly emerges only after approximately 700 ms following target presentation for trials in which an AB was observed (i.e., missing T2 following the correct identification of T1). This time-course for prevalence-induced pupillary responses is, in fact, consistent with the time-course of stimulus-driven neural activity in the locus coeruleus (LC). When a task-relevant stimulus is processed, the LC enters a brief period of phasic activity approximately 150 ms after the onset of the target, facilitating the

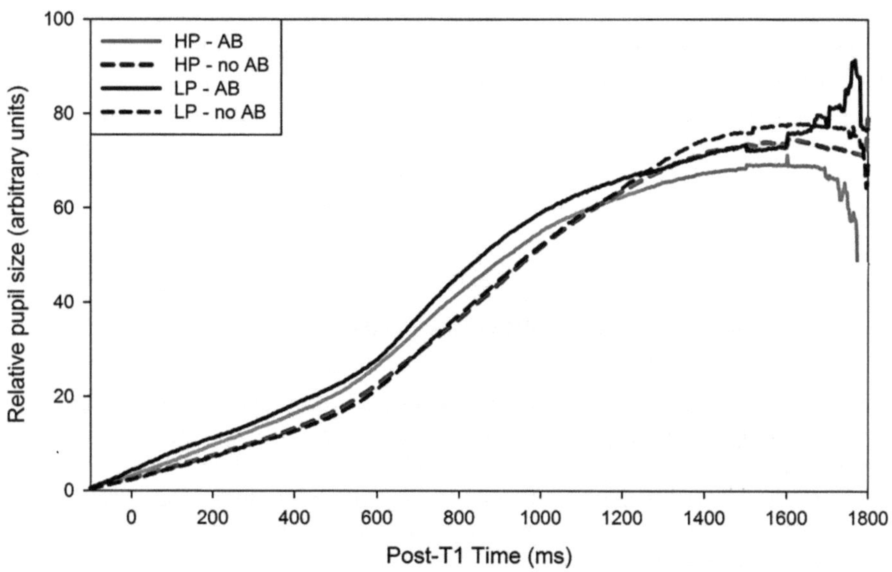

Fig. 4 Average baseline-corrected pupil size in arbitrary units. Time 0 denotes T1 onset. Solid lines reflect trials where an attentional blink (AB) was observed and dashed lines reflect trials where no AB was observed. Low- (LP) and high-(HP) prevalence T1 targets are reflected by gray and black colors, respectively. (Figure reproduced from Papesh & Guevara Pinto, 2019)

processing of any other external stimuli (e.g., Aston-Jones & Bloom, 1981; Joshi et al., 2016). However, 50–100 ms following this phasic phase, the LC becomes autoinhibitory, resulting in a refractory period (between 200 and 400 ms after the onset of the target) during which stimulus-driven LC activity is not possible (Aston-Jones et al., 1994; Usher, 1999). It is during this refractory period that the AB effect takes place (Nieuwenhuis et al., 2005), positioning the dynamics of the LC-NE system as the neurological source for the AB. Furthermore, as shown in Fig. 4, the results from Papesh and Guevara Pinto (2019) suggest that the cognitive "costs" of correctly identifying low-prevalence targets may also be attributable to the LC-NE system, as prevalence-induced pupillary effects closely map onto the time-course of the LC refractory period.[5] In other words, correctly identifying rare targets "prolongs" the LC refractory phase and thus limits the processing of any new stimuli, suggesting that target prevalence impacts cognition in broader ways than simply making targets harder to detect.

In summary, as predicted by the multiple-decision model of visual search (Wolfe & van Wert, 2010), the processing of rare or infrequent targets imposes a greater demand on cognition due to a conservative shift in the target identification criterion. That is, rare targets require more evidence to be recognized as "targets." By adopting pupillometry as an innovative approach to examine the LPE, Papesh and Guevara Pinto (2019) provided evidence showing that relative to high-prevalence targets, the processing of low-prevalence targets indeed requires more extensive effort and this has, in turn, "downstream" consequences for any additional cognitive processing (i.e., the identification of a second target). Furthermore, such effects can be assessed by measuring phasic pupillary responses during passive search tasks, with a time course consistent with that of activity in the LC-NE system. This is the *first* advantage that pupillometry has to offer to us: It allows us to study cognition not at a single point in time, but as a more dynamic, complex sequence of events that unfold over time. This, in turn, lets researchers examine more nuanced predictions from established models of visual search (e.g., the MDM) that otherwise would not be easily examined using traditional measures of search performance.

1.2 Case of Domain-Specific Expertise

When it comes to conducting searches in applied settings, the low-prevalence effect is not the only challenge faced by observers. At a much more foundational level, observers need to be trained in the specialized area that they will be searching through. Importantly, the length and quality of the training may vary across

[5] Note that due to an approximate 300–400 ms delay between LC activity and phasic pupillary responses, the LC refractory period in the experimental paradigm used by Papesh and Guevara Pinto (2019) would *begin* to occur about 500–700 ms following the onset of T1. This is consistent with the time course illustrated in Fig. 4, where the differences in pupil size between HP-AB and LP-AB trials begins to emerge around 700 ms post-T1 onset.

professions. For example, new officers for the Transportation Security Agency (TSA) in the United States are expected to complete only about 80 h of training before being tasked with the responsibility of keeping travelers safe (jobs.tsa.gov/transport-security-officer), while medical students train for multiple years before they begin work as radiologists. Thankfully, multiple studies have demonstrated that experienced (or expert) searchers tend to outperform novice searchers in their domain of visual search expertise (e.g., Biggs et al., 2013; Manning et al., 2006). While it is certainly safe to assume that the more practice and experience one gets with a task, the easier the task becomes, researchers are still working to understand not only how long it takes to develop expert-level experience within a visual search domain but also exactly what the precise *cognitive benefits* are that experience provides to observers.

Since the inception of the earliest visual search paradigms, it has been observed that one of the primary benefits of repetitive exposure to a search task, regardless of domain, is that searches tend to be conducted faster and more accurately (e.g., Hout & Goldinger, 2010, 2012). This is evident by observing significant declines in RTs (and often shallower search slopes) after repeated exposure to distractor identities, spatial configurations, or some combination of the two. Furthermore, the effects of learning on oculomotor behaviors following repetitive visual search are also quite interesting. For instance, experts tend to make fewer eye-movements relative to novices in a wide range of visual tasks, including chess (Charness et al., 2001; Reingold et al., 2001), medical image perception (Drew et al., 2013; Kundel et al., 2007; Manning et al., 2006), and baggage screening (Biggs et al., 2013; Biggs & Mitroff, 2014b). This reduction in the number of eye-movements implies that experience with the task allows expert observers to extract more information from a visual display in a single fixation relative to novice observers.

This was recently demonstrated by Papesh et al. (2021) who measured progressive changes in the oculomotor behaviors of novice observers as they gained experience in a visual search task across fourteen training sessions. Observers memorized a list of 20 target categories (e.g., *bottle*, *toy train*) before completing a "hybrid search" task (i.e., search through memory and a visual display simultaneously) in which they had to look for objects from any of the relevant target categories among distractors drawn from other real-world object categories (c.f., Cunningham & Wolfe, 2014). Each trial contained between 0 and 3 targets. Papesh et al. (2021) documented the development of visual search expertise by examining how various standard dependent measures (e.g., RTs, number of fixations) changed over sessions, and they also calculated the size of the *functional viewing field* (FVF). The FVF (often referred to more generally as the "attentional window") is the area of a visual display that is directly processed by the observer within any given fixation, with items falling within this area being processed at a higher resolution than those falling outside it (i.e., in parafoveal or peripheral vision; Sanders, 1970). In line with previous work on expertise development, the results showed that the size of the FVF *increased* with experience (see Fig. 5), allowing observers to find the target(s) much faster by making fewer fixations (see Brams et al., 2019, for a review on search behaviors and expertise).

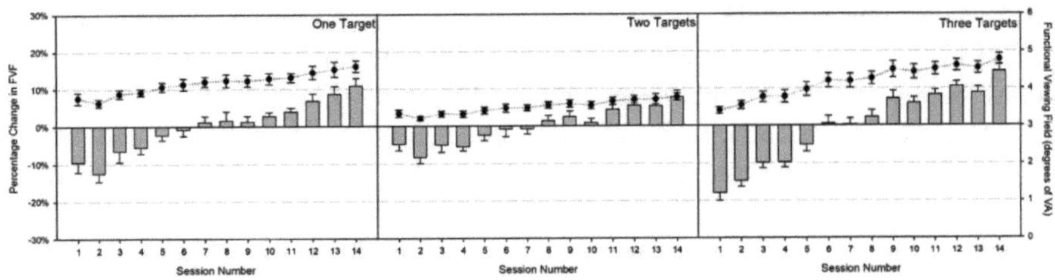

Fig. 5 Functional viewing field estimates in degrees of visual angle (circles) and percent change in functional viewing field (bars) across fourteen training sessions. The left panel shows trials with one target, the middle panel shows trials with two targets, and the right panel shows trials with three targets. Error bars reflect ±1 standard error. (Figure reproduced from Papesh et al., 2021)

Several studies have established that the FVF narrows in size as visual search tasks become more difficult (e.g., Drew et al., 2017; Guevara Pinto & Papesh, 2019; Hulleman et al., 2020; Young & Hulleman, 2013) because a narrow focus of attention seems to allow observers to closely examine each potential target while reducing processing interference from nearby distractors. Because observers tend to make fewer fixations (i.e., they adopt a broader FVF) during visual search after a training period, it is assumed that the benefit of expertise is that it makes challenging visual searches less cognitively demanding in general. Thus, in comparison to novice searchers, experts seem to have more attentional resources available to process multiple items in the display simultaneously from a single fixation. In other words, when search is less demanding of cognitive resources (because of acquired expertise), observers can gather more information from the display without having to overtly move their eyes as often or as far.

While the assumption that experience makes visual search less cognitively demanding is a reasonable one, it still requires empirical support to be validated. This is the *second* advantage that pupillometry offers visual search researchers: It provides a unique tool to test our untested assumptions. In this instance, if visual search is more demanding for novice searchers than for experts, then tonic pupillary responses should be able to be used to quantify this increase in cognitive demand. Evidence supporting this assumption was recently provided by Castner et al. (2020) who examined pupillary responses between expert and novice searchers during anomaly searches in orthopantomograms (OPTs; or dental radiographs, see Fig. 6). Experts consisted of professional dentists from a university clinic, while novices consisted of sixth-semester dental students. Both groups searched among 15–20 OPTs, each of which could contain more than one anomaly (i.e., search target), and anomalies were categorized based on their difficulty of detection as easy, medium, difficult, or ambiguous. Importantly, the level of difficulty reflected the "diagnostic" difficulty of detecting the anomaly and not differences in perceptual clarity. Baseline pupil size was measured for 2 s prior to the onset of each OPT, after which the search progressed in two phases: First, participants were given a free-viewing exploratory period to search for areas indicative of any pathologies in need of an intervention. Novice searchers were given 90 s to explore the OPT, while experts

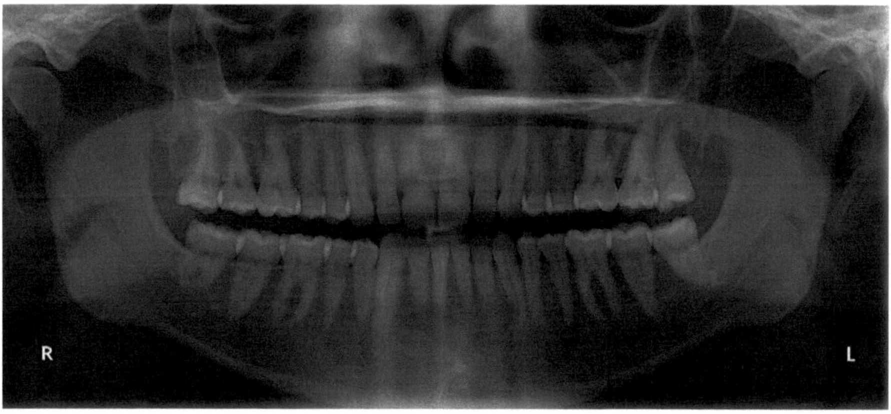

Fig. 6 Example of a normal orthopantomography (OPT). Image downloaded from Wikimedia Commons (https://commons.wikimedia.org/wiki/File:X-ray_of_all_32_human_teeth.jpg)

were given only 45 s. Following the exploration period, both groups of searchers were given unlimited time to use an on-screen drawing tool and mark any anomalies they had found. The results showed that relative to experts, novice searchers showed larger increases from baseline pupil size after detecting an anomaly during the exploratory phase. However, these changes in novices' pupil size were consistent in magnitude regardless of the level of difficulty for any particular anomaly, suggesting that *all* anomalies were equally demanding. On the other hand, changes in tonic pupil size for expert searchers were largest for the difficult anomalies and smallest for the easy anomalies, reflecting a *gradual* modulation of visual search difficulty on pupillary responses. Because this was not the case for novice searchers, Castner et al. (2020) concluded that novices incur a high cognitive load across all levels of difficulty due to their lack of knowledge and experience in identifying anomalies. This is in line with the assumption that experience reduces the cognitive demand associated with visual search, as cognitive resources were less taxed when experts encountered an easy or moderately difficult anomaly.

It is important to note that, while it has been demonstrated that prior knowledge reduces the cognitive load during visual search (e.g., Biggs et al., 2013; Johnson et al., 2002), the difficulty-modulated pupillary responses of experts observed by Castner et al. (2020) could also reflect differences in the overall arousal of the observer, and/or the difficulty of making a medical diagnostic decision rather than difficulty in the search process itself. Dentists and other medical practitioners might also engage in alternative search strategies in a more naturalistic setting relative to strategies employed during a laboratory experiment. For instance, when viewing a 2-D medical image in a clinical practice, doctors can make adjustments to the image in terms of position, brightness, contrast, and perhaps most importantly, the scale or "depth" of the image. That is, most medical practitioners are able to "zoom" in and out when considering a potential anomaly, and there are likely important individual differences in search strategies across practitioners. For instance, radiologists have been noted to adopt one of two general search strategies when searching through a computed tomography (CT) scan: So-called "drillers" tend to focus on a small area of the image by zooming in and out to closely examine a potential anomaly before moving on to a different area, whereas "scanners" tend to scan an entire level of the image first before zooming in to get a closer look at different "slices." Importantly, drillers have been reported to outperform scanners in detecting nodules when searching through CT scans (Drew et al., 2013), perhaps because they tend to "exploit" each area by focusing attention on a narrow region of space rather than "exploring" the image by diffusing attention throughout all areas. Indeed, tonic neural activity in the locus coeruleus (LC) has been associated with an explorative attentional control state, while phasic LC activity has been associated with shifts to more focused, exploitative attention to task-relevant stimuli (Aston-Jones & Cohen, 2005).

Recently, Brunyé et al. (2023) used pupillometry to demonstrate how zooming behavior during visual search may reflect changes in attentional control states. Pathologists searched through 14 high-resolution images of breast tissue biopsies (digital whole slide images), which included a full spectrum of diagnoses ranging from benign to invasive cancers. During their search for anomalies, observers were able to zoom in and out before having to provide a diagnosis and rate the image on a 1–7 difficulty scale. Importantly, the researchers measured pupil dilations at each level of zoom (e.g., low magnification vs. high magnification) during the entire image viewing period (i.e., tonic pupillary responses). Results showed that, at low magnification levels, tonic pupil diameter increased as the observers' subjective ratings of difficulty increased, consistent with the notion that tonic pupillary responses of expert searchers can index increases in cognitive demand during visual search. Similar results have also been replicated elsewhere (Drew et al., 2023), suggesting that pupillometry can potentially be used in medical diagnosis to signal the need for additional training in medical students and/or the need for a second opinion. Furthermore, Brunyé et al. (2023) also found that tonic pupillary dilation increased in size at each higher level of zoom (i.e., at higher magnification levels). They attributed this finding to an overall increase in *perceptual* difficulty the deeper the pathologist searched, likely as a result of the amount of low-level cellular features in the images increasing at higher magnification levels. In other words, as pathologists adopted higher levels of magnification, a whole new, detail-rich scene may have emerged, much like observing your backyard at a distance and then getting closer to it to attend to the small patch of grass near your feet (allowing you better notice the stem of each blade of grass, the texture of the soil, and so forth). This suggests that zooming in increased the amount of information available for processing (e.g., visual clutter), thus making the search, overall, more challenging. Most important to our discussion here, it also suggests that pupillometry is sensitive to changes in search difficulty that occur *during* the search process.

Although tonic pupil size can index the overall change in search difficulty after increasing the magnification level, it does not capture any potential attention shifts at the *moment* of zooming in. This raises the question: do the pupils also reveal a shift from an explorative state of attention to a more exploitative state once a potential target region is viewed in closer detail? To test this, Brunyé et al. measured pupil size 5000 ms before and after each discrete zoom-in and zoom-out "event" (i.e., phasic pupillary responses). Interestingly, the results showed that pupil size actually constricted about 500 ms prior to the onset of zoom-in events, and this was attributed to the pupil's response to the upcoming onset of motion during zoom rather than an attentional shift to an exploitative state of attention control. However, phasic pupillary responses to zooming-out events did provide evidence for a shift from exploitation *back* to exploration, as pupil size significantly increased immediately prior to the onset of zoom-out events. Brunyé and colleagues interpreted this effect as the culmination in evidence accumulation from a focused (i.e., exploitative) state of attention once a decision was reached. That is, pathologists could confidently zoom-out back to lower levels of magnification and terminate their search if an anomaly was detected or move on to explore other areas of interest.

In summary, the cognitive benefits of experience in visual tasks involve not only faster and more accurate performance but also the ability to extract more information from the display in a single fixation (relative to less-experienced observers; Charness et al., 2001; Manning et al., 2006; Biggs et al., 2013). It has been hypothesized that this occurs because prior experience makes the search task less demanding, freeing attention resources to process a greater number of items in parallel, relative to novice searchers who seemingly adopt a more sequential or serial approach to their search by processing one item at a time. While these ideas are based on the findings whereby experts make fewer fixations during search tasks in their domain of expertise (and/or adopt a broader FVF following extensive search training; e.g., Papesh et al., 2021), it has not been directly tested until recently. By measuring tonic pupillary responses during anomaly searches of orthopantomographies, Castner et al. (2020) provided the first (to our knowledge) empirical support suggesting that novice searchers experience a greater cognitive demand during search than experienced searchers. This reflects the utility of pupillometry in testing implied assumptions regarding the cognitive effort required during applied search tasks. Additionally, it is also possible to use pupillometry to assess distinct behaviors (i.e., scanning or drilling strategy) engaged in naturally by expert searchers during visual search (e.g., exploration and exploitation) and to relate such behaviors to specific attentional control states (Brunyé et al., 2023). This suggests that pupil dilation could be a promising window into the study of specific search strategies adopted by observers (and perhaps how to *improve* them!).

1.3 What Other Aspects of Visual Search Could Benefit from Pupillometry?

So far, we have discussed the utility of pupillometry in visual search research by highlighting one of the challenges faced by searchers in societally important search tasks (e.g., the broader impacts on cognition of the *low-prevalence effect*), as well as the cognitive benefits associated with being an expert searcher (e.g., greater availability of attention resources to process more items without the need to make additional eye movements). This is important given that the objective of much laboratory visual search research is to understand the underlying mechanisms of how we conduct searches so that we can improve performance in the real world. We want to, for instance, make it easier for security officers to detect threats at airport baggage screening checkpoints as well as for doctors to detect anomalies in their patients' scans. However, a significant portion of the visual search literature adopts a more theoretical perspective by focusing on studying the mechanisms of visual attention more broadly in order to inform cognitive frameworks and models of visual search. For instance, as mentioned earlier, a prominent debate in the literature considers whether eye movements, and the functional field of view surrounding them, should be incorporated into existing models of visual search (e.g., Hulleman & Olivers, 2017; Wolfe, 2021). The core of this discussion is whether all elements in a visual search display receive equal attention within a single fixation (thus the call to focus on eye movements instead of search slopes and the number of items in the display). This is in fact a relatively old debate, as many others have raised questions in the past about the diffusion or "spread" of attention across foveal (i.e., central) and peripheral vision during cognitive tasks (e.g., Engle, 1977; Belopolsky & Theeuwes, 2010; Eriksen & St. James, 1986). What can pupillometry contribute to such discussions?

Certainly, pupil dilation in cognitive research is most commonly explored as a physiological index of arousal and cognitive effort (e.g., Kahneman & Beatty, 1966; Laeng et al., 2012). However, pupil size is also known to change in response to environmental luminance, with constriction occurring in bright environments and dilation in darker ones. In turn, this pupillary light reflex can be expected to impact the processing of a visual image. Constricted pupils, for example, reduce the scattering of light on the retina by focusing incoming light more directly on the fovea, which, in turn, improves the discrimination of fine details. On the other hand, dilated pupils allow more light to enter the retina, providing a better opportunity to absorb light from the environment, though at the cost of having a lower foveal resolution. This is why, for instance, if you wake up in the middle of the night to get some water, you might only be able to see the general contours and layout of the objects in your room rather than the finer details.

Interestingly, researchers have begun to use this pupillary light reflex as a medium for understanding the processing differences in foveal and peripheral vision as they relate to pupil size. For example, Mathôt and Ivanov (2019) tasked observers with completing either a foveal discrimination task or a peripheral detection task (on different experimental sessions). On each trial, participants fixated on a central dot and waited for a Gabor patch to be briefly presented for 650 ms. In the foveal discrimination task, the Gabor patch was presented at the center of the display, and participants had to indicate whether it was tilted to the right or to the left of the screen (see Fig. 7). By contrast, in the peripheral detection task, the patch appeared only in 50% of the trials, and it was presented at a random point of 7.70° of visual angle in eccentricity surrounding the central dot. Participants in this peripheral detection task had

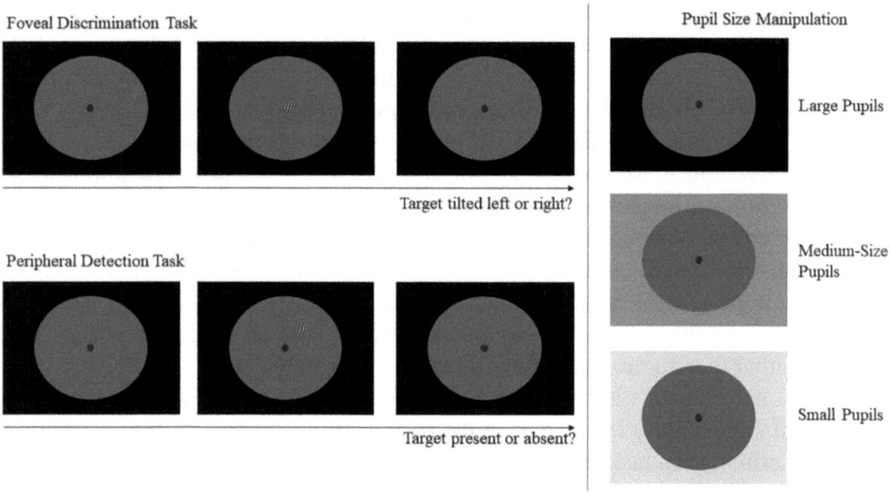

Fig. 7 Examples of the experimental tasks (left panel) and pupil size manipulations (right panel) used by Mathôt and Ivanov (2019). (This figure was recreated from Mathôt & Ivanov, 2019)

to indicate whether the patch appeared or not during the trials. Importantly, the researchers directly manipulated (rather than passively measuring) the observer's pupil size by varying the background luminance in the display across trials, with some trials adopting a bright background luminance (eliciting small pupils) and others a dark background luminance (eliciting comparatively larger pupils). Overall, mean performance was equal between the two tasks, although the peripheral detection task yielded larger pupil sizes relative to the foveal discrimination task. More importantly, the results showed that performance on the peripheral detection task improved during trials with darker backgrounds: There was a strong positive correlation between the measured pupil size and accuracy in detecting Gabor targets in the periphery. The authors replicated this effect in a second experiment in which the Gabor patches were replaced by words (e.g., *apple*), and others have found this same effect using even more faint stimuli (e.g., Eberhardt et al., 2021). As regard the foveal discrimination task, brighter backgrounds did not influence discrimination performance, either with Gabor or word stimuli, arguably because the stimuli were easily perceived from the distance where observers were seated. Indeed, smaller pupils significantly improved foveal discrimination performance in a third experiment wherein the stimuli (upper- or lower-case letters) were presented at a greater distance from the observer (5.5 meters away from the observer), just above the threshold size for perception. Together, these findings suggest that pupil size can be directly manipulated to improve processing in foveal or peripheral vision, simply by manipulating environmental luminance.

In some ways, these studies have cleverly reversed the typical pupillometric approach by directly manipulating observers' pupil size (to examine the impact on cognition) rather than measuring pupil size as a response to the degree of cognitive processing. This is the *third* advantage that we believe pupillometry has to offer visual search research: Pupil size can be manipulated as an independent variable to examine the detection performance of items across a visual display. In fact, it has been recently proposed that the cognitive system has "co-opted" control of the pupillary light reflex as a functional filter for visual information to achieve task-related goals (Ebitz & Moore, 2019). For example, it is well known that pupils dilate as general arousal increases (e.g., Laeng et al., 2012; Mathôt & Van der Stigchel, 2015), which could reflect the top-down goals of an individual, such as when someone is actively searching for any threats in their environment. In such a case, it would be important to detect any unexpected stimuli in the periphery, and thus attention should be spread throughout the field of view to explore a scene in its entirety rather than to more centrally focus on any specific area.

Mathôt and Ivanov (2019) reported being surprised by their finding that pupil sizes were significantly larger during the peripheral detection task than during the foveal discrimination task (regardless of background luminance), even though both tasks were equated in difficulty. They suggested that such an effect may be caused by visual attention being directed to the periphery, rather than to the central area, during the peripheral detection task. Data from Brocher et al. (2018) may support this idea. They presented observers with four shapes (a combination of black and white circles or triangles) contained in a small space (64 × 64 pixels) at both the left and right sides of a central fixation. Observers' task was simply to count the number of white triangles at each side while remaining fixated on a central fixation cross, but the experimenters manipulated the distance away from the center that the stimuli were presented on each trial (e.g., 12.5°, 20°, 27.5°, 35°, or 42.5° degrees away in the periphery). They observed that phasic pupillary responses following the onset of the stimuli increased systematically in size as the stimuli were presented further away in the periphery. Furthermore, they replicated this effect in a second experiment and also showed that the overall pupil size was larger when the task required observers to make a covert shift of attention toward the periphery (i.e., count white triangles) relative to when they were simply required to *detect* whether any peripheral items appeared.

The findings described above indicate a close relationship between pupil size and foveal/peripheral task performance, one which is not mediated by task difficulty (at least when pupil size is manipulated through environmental changes). Specifically, when pupils constrict to focus an overabundance of light in bright environments directly onto the fovea, processing in central vision is incidentally improved. Comparatively, when light is scarce (i.e., dark environments), pupils dilate to allow more light overall into the eye, improving processing in peripheral vision. Brocher et al. (2018) demonstrated that the effect of pupil size on peripheral performance is further facilitated if it is accompanied by an actual shift in covert attention. By "covert" attention, we mean attending somewhere in a display without moving your eyes, like (while driving) when you attend a car passing you on the left while keeping your gaze focused in front of you.

Let us consider again the debate regarding the spread of attention surrounding fixations (e.g., functional viewing field; FVF) made during visual search (e.g.,

Engle, 1977; Belopolsky & Theeuwes, 2010; Eriksen & St. James, 1986; Hulleman & Olivers, 2017). Recently, there has been an increase in work examining the dynamics between foveal and peripheral processing during visual search. For instance, in the latest iteration of the *Guided Search* model, Wolfe (2021) describes the possibility that attention flexibly spreads out and narrows back again (in a back-and-forth "dance") throughout an entire search period. According to Guided Search 6.0, there may be three distinct FVFs that emerge during the search process. The *resolution* FVF reflects the individual's limitation in peripheral processing. This is the traditional FVF, which indicates the limits of foveal and parafoveal vision, with items falling within these limits being processed in high resolution and items falling outside the limits being processed in very poor (if any) detail. In contrast, the *exploratory* FVF is much broader and is used to direct attention overtly via eye movements to new areas of the display that require further inspection (i.e., areas presumably falling outside the resolution FVF). Finally, the *attentional* FVF reflects the limits of covert deployments of attention and is used to *exploit* and process items that can be attended to without having to fixate directly on them. The general idea behind this updated version of the iconic visual search model is that the spread of attention may vary depending on the intentionality and attentional control state of the observer (e.g., explore vs exploit) at each timepoint during the search process and that the three types of attention spread that result from such attention states are *functionally* distinct.

Could pupillometry provide any support for this updated model? We believe it can. For example, the direct manipulation of pupil size (to explore concomitant changes in peripheral processing) through the pupillary light reflex may be useful for informing us about the resolution FVF, as this is thought to reflect the distinction between foveal and peripheral processing (e.g., Eberhardt et al., 2021; Mathôt & Ivanov, 2019). Meanwhile, manipulations of pupil size through covert attention may be informative of the attentional FVF, as larger pupils are associated with more disparate deployments of covert attention into the periphery (Brocher et al., 2018). Simply put, by dilating the pupils, we can open a window to study the spread of attention and how it dynamically changes during visual search. In fact, some have already started to use the pupillary light reflex as an index of the size of the attention window (e.g., Tkacz-Domb & Yeshurun, 2018). Note that the opportunity for examining the existence of functionally distinct FVFs, like the ones proposed by Wolfe (2021), is only possible when pupillometry is reconceptualized not as a dependent measure but as a variable that can be manipulated in an experimental design, opening novel avenues for exciting, new visual search research.

2 Limitations and Conclusion

We believe that there are new insights to be made in visual search from the adoption of emerging methodologies like pupillometry. That said, we acknowledge that like most other methods, pupillometry has its own limitations that may make its adaptability less feasible. The most general and obvious challenge, we believe, is obtaining clean pupil size recordings during data collection. During visual search, particularly during standard "free-viewing" search conditions, collecting useful pupil data can sometimes be a challenge. When observers freely move their eyes across a visual display, an eye-tracking camera gets an inconsistent view of the pupil as the eye rotates and the pupil is "reshaped" with every new fixation depending on gaze position (see Hayes & Petrov, 2016, for a way to control and correct these artifacts). Moreover, eye movements occur at variable moments in time during standard visual search, making it also potentially difficult to examine phasic pupillary responses (as if they were time-locked at the start of a fixation). For these, and perhaps other reasons, "passive" search tasks (i.e., RSVP procedures) are typically preferred over "active" ones (i.e., standard free-viewing conditions) when adopting pupillometric methods. Such methods can control for the timing and positioning of the stimuli in relation to the pupils (cf. see Stotle et al., 2020, for an alternative approach to deal with variable stimuli onsets during standard searches) and allow the observer to keep their eyes fixed on a single, central point throughout the task. These tasks also lend themselves well to the use of simplistic stimuli such as symbols, images, or scenes, which may also be preferred as researchers can easily control for bottom-up features of the stimuli such as brightness and size.

Another potential limitation is that this methodology is still relatively "young," and more evidence is needed to understand how *relative* pupil size may vary in response to different experimental manipulations. For instance, in the studies discussed above, Brocher et al. (2018) noted that *larger* pupil sizes are observed as task-relevant stimuli were presented further away in the periphery (indicative of attentional shifts towards the periphery), but Castner et al. (2020) showed that expert searchers (whose attention have "spread out" towards the periphery due to experience) have *smaller* pupil sizes than novices searchers when processing stimuli in their domain of expertise. Such differences in pupil sizes across these two studies are likely the result of differences in tonic and phasic pupil responses, where tonic neural activity from the locus coeruleus (LC) reported in Caster et al. (2020) represents changes in overall task demand, while the phasic LC activity described by Brocher et al. (2018) represents moment-to-moment shifts in attention towards the periphery. Further research is certainly needed to fully comprehend how pupillary response functions during search when taking into consideration evidence across various cognitive phenomena.

Despite these limitations, visual search research can benefit greatly from the wider adoption of pupillometric methods. Benefits from popularizing pupillometry as a measure in visual search research include (but are not limited to): (1) Allowing researchers to study cognition not in isolation but as a continuous process that unfolds over time (e.g., Papesh & Guevara Pinto, 2019); (2) allowing researchers to assess the benefits of experience on search by testing the assumptions of use-inspired basic search processes (e.g., Brunyé et al., 2023; Castner et al., 2020); and (3) allowing researchers to develop new experimental paradigms that utilize pupil dilations as an independent variable that impacts cognitive processing (e.g., Brocher et al., 2018; Mathôt & Ivanov, 2019).

Finally, the goal of this chapter was not to provide a "recipe" of how to incorporate pupillometry into existing visual search paradigms. Rather, it is to illustrate some of the benefits of adopting pupillometry and to inspire novel insights derived therefrom that will move the field forward. Indeed, there are likely many more benefits and applications that we might not be currently aware of, each one with the potential of bringing new insights to a well-established field. After all, examining novel ideas sometimes requires novel approaches.

References

Aston-Jones, G., & Bloom, F. E. (1981). Norepinephrine-containing locus coeruleus neurons in behaving rats exhibit pronounced responses to non-noxious environmental stimuli. *Journal of Neuroscience, 1*, 887–900.

Aston-Jones, G., & Cohen, J. D. (2005). An integrative theory of locus coeruleus-norepinephrine function: Adaptive gain and optimal performance. *Annual Review of Neuroscience, 28*(1), 403–450.

Aston-Jones, G., Rajkowski, J., Kubiak, P., & Alexinsky, T. (1994). Locus coeruleus neurons in monkey are selectively activated by attended cues in a vigilance task. *Journal of Neuroscience, 14*, 4467–4480.

Backs, R. W., & Walrath, L. C. (1992). Eye movement and pupillary responses indices of mental workload during visual search of symbolic display. *Applied Ergonomics, 23*(4), 243–254.

Belopolsky, A. V., & Theeuwes, J. (2010). No capture outside the attentional window. *Vision Research, 50*, 2543–2550.

Biggs, A. T., & Mitroff, S. R. (2014a). Improving the efficacy of security screening tasks: A review of visual search challenges and ways to mitigate their adverse effects. *Applied Cognitive Psychology, 29*(1), 142–148.

Biggs, A. T., & Mitroff, S. R. (2014b). Different predictors of multiple-target search accuracy between nonprofessional and professional visual searchers. *The Quarterly Journal of Experimental Psychology, 67*, 1335–1348.

Biggs, A. T., Cain, M. S., Clark, K., Darling, E. F., & Mitroff, S. R. (2013). Assessing visual search performance differences between transportation security administration officers and nonprofessional visual searchers. *Visual Cognition, 21*, 330–352.

Brams, S., Ziv, G., Levin, O., Spitz, J., Wagemans, J., Williams, A. M., & Helsen, W. F. (2019). The relationship between gaze behavior, expertise, and performance: A systematic review. *Psychological Bulletin, 145*(10), 980.

Brocher, A., Harbecke, R., Graf, T., Memmert, D., & Hüttermann, S. (2018). Using task effort and pupil size to track covert shifts of visual attention independently of a pupillary light reflex. *Behavior Research Methods, 50*, 2551–2567.

Brunyé, T. T., Drew, T., Kerr, K. F., Shucard, H., Powell, K., Weaver, D. L., & Elmore, J. G. (2023). Zoom behavior during visual search modulates pupil diameter and reflects adaptive control states. *PLoS One, 18*(3), e0282616.

Bruya, B., & Tang, Y. Y. (2018). Is attention really effort? Revisiting Daniel Kahneman's influential 1973 book attention and effort. *Frontiers in Psychology, 9*, 1133.

Castelhano, M. S., Pollatsek, A., & Cave, K. (2008). Typicality aids search for an unspecified target, but only in identification, and not in attentional guidance. *Psychonomic Bulletin & Review, 15*, 795–801.

Castner, N., Appel, T., Eder, T., Richter, J., Scheiter, K., Keutel, C., Hüttig, F., Duchowski, A., & Kasneci, E. (2020). Pupil diameter differentiates expertise in dental radiography visual search. *PLoS One, 15*(5), e0223941.

Charness, N., Reingold, E. M., Pomplun, M., & Stampe, D. M. (2001). The perceptual aspect of skilled performance in chess: Evidence from eye movements. *Memory and Cognition, 29*, 1146–1152.

Cowan, N., Elliott, E. M., Scott Saults, J., Morey, C. C., Mattox, S., Hismjatullina, A., & Conway, A. R. (2005). On the capacity of attention: its estimation and its role in working memory and cognitive aptitudes. *Cognitive Psychology, 51*(1), 42–100.

Cunningham, C. A., & Wolfe, J. M. (2014). The role of object categories in hybrid visual and memory search. *Journal of Experimental Psychology: General, 143*(4), 1585–1599.

Drew, T., Vo, M. L.-H., Olwal, A., Jacobson, F., Seltzer, S. E., & Wolfe, J. M. (2013). Scanners and drillers: Characterizing expert visual search through volumetric images. *Journal of Vision, 13*, 1–13.

Drew, T., Boettcher, S. E. P., & Wolfe, J. M. (2017). One visual search, many memory searches: An eye-tracking investigation of hybrid search. *Journal of Vision, 17*(11), 1–10.

Drew, T., Konold, C. E., Lavelle, M., Brunyé, T. T., Kerr, K. F., Shucard, H., Weaver, D. L., & Elmore, J. G. (2023). Pathologist pupil dilation reflects experience level and difficulty in diagnosing medical images. *Journal of Medical Imaging, 10*(2), e025503.

Duncan, J., & Humphreys, G. W. (1989). Visual search and stimulus similarity. *Psychological Review, 96*(3), 433–458.

Eberhardt, L. V., Strauch, C., Hartmann, T. S., & Huckauf, A. (2021). Increasing pupil size is associated with improved detection in performance in the periphery. *Attention, Perception, & Psychophysics, 84*, 138–149.

Ebitz, R. B., & Moore, T. (2019). Both a gauge and a filter: Cognitive modulation of pupil size. *Frontiers in Neurology, 9*(1190), 1–4.

Eimer, M. (2017). Why the item will remain the unit of attentional selection in visual search. *Behavioral and Brain Sciences, 40*, e137.

Engle, F. L. (1977). Visual conspicuity, visual search and fixation tendencies of the eye. *Vision Research, 17*, 95–108.

Eriksen, C. W., & St. James, J. D. (1986). Visual attention within and around the field of focal attention: A zoom lens model. *Perception & Psychophysics, 40*, 225–240.

Evans, K. K., Birdwell, R. L., & Wolfe, J. M. (2013). If you don't find it often, you often don't find it: Why some cancers are missed in breast cancer screening. *PLoS One, 8*(5), e64366.

Godwin, H. J., Menneer, T., Riggs, C. A., Cave, K. R., Thaibsyah, M., & Donnelly, N. (2015). The effects of increasing target prevalence on information processing during visual search. *Psychonomic Bulletin & Review, 22*, 469–475.

Godwin, H. J., Hout, M. C., Alexdóttir, K. J., Walenchok, S. C., & Barnhart, A. S. (2021). Avoiding potential pitfalls in visual search and eye-movements experiments: A tutorial review. *Attention, Perception, & Psychophysics, 83*(7), 2753–2783.

Guevara Pinto, J. D., & Papesh, M. H. (2019). Incidental memory following rapid object processing: The role of attention allocation strategies. *Journal of Experimental Psychology: Human Perception and Performance, 45*(9), 1174–1190.

Hayes, T. R., & Petrov, A. A. (2016). Mapping and correcting the influence of gaze position on pupil size measurements. *Behavior Research Methods, 48*(2), 510–527.

Hess, E. H., & Polt, J. M. (1964). Pupil size in relation to mental activity during simple problem-solving. *Science, 143*, 1190–1192.

Horowitz, T. S. (2017). Prevalence in visual search: From the clinic to the lab and back again. *Japanese Psychological Research, 59*(2), 65–108.

Hout, M. C., & Goldinger, S. D. (2010). Learning in repeated visual search. *Attention, Perception, & Psychophysics, 72*, 1267–1282.

Hout, M. C., & Goldinger, S. D. (2012). Incidental learning speeds visual search by lowering response thresholds, not by improving efficiency: Evidence from eye movements. *Journal of Experimental Psychology: Human Perception and Performance, 38*(1), 90–112.

Hout, M. C., & Goldinger, S. D. (2014). Target templates: The precision of mental representations affects attentional guidance and decision-making in visual search. *Attention, Perception, & Psychophysics, 77*(1), 128–149.

Hout, M. C., Walenchok, S. C., Goldinger, S. D., & Wolfe, J. M. (2015). Failures of perception in the low-prevalence effect: Evidence from active and passive visual search. *Journal of Experimental Psychology: Human Perception and Performance, 41*, 977–994.

Hout, M. C., White, B., Madrid, J., Godwin, H. J., & Scarince, C. (2022). Examining the effects of passive and active strategy use during interactive search for LEGO® bricks. *Journal of Experimental Psychology: Applied, 28*(1), 35–51.

Huang, L., & Pashler, H. (2005). Attention capacity and task difficulty in visual search. *Cognition, 94*(3), 101–111.

Hulleman, J., & Olivers, C. N. (2017). On the brink: The impending demise of the item in visual search. *Behavioral and Brain Sciences, 40*, 1–69.

Hulleman, J., Lund, K., & Skarratt, P. (2020). Medium versus difficult visual search: How a quantitative change in the functional visual field leads to a qualitative difference in performance. *Attention, Perception, & Psychophysics, 82*, 118–139.

Johnson, D. N., McGrath, A., & McNeil, C. (2002). Cuing interacts with perceptual load in visual search. *Psychological Science, 13*, 284–287.

Joshi, S., Li, Y., Kalwani, R. M., & Gold, J. L. (2016). Relationships between pupil diameter and neural activity in the locus coeruleus, colliculi, and cingulate cortex. *Neuron, 89*(1), 221–234.

Kahneman, D. (1973). *Attention and effort*. Prentice-Hall.

Kahneman, D., & Beatty, J. (1966). Pupil diameter and load on memory. *Science, 154*, 1583–1585.

Kristjánsoon, A., & Draschkow, D. (2021). Keeping it real: Looking beyond capacity limits in visual cognition. *Attention, Perception, & Psychophysics, 83*, 1375–1390.

Kundel, H. L., Nodine, C. F., Conant, E. F., & Weinstein, S. P. (2007). Holistic component of image perception in mammogram interpretation: Gaze-tracking study. *Radiology, 242*, 396–402.

Laeng, B., Sirois, S., & Gredebäck, G. (2012). Pupillometry: A window to the preconscious? *Perspectives in Psychological Science, 7*(1), 18–27.

Malcolm, G. L., & Henderson, J. M. (2009). The effects of target template specificity on visual search in real-world scenes: Evidence from eye movements. *Journal of Vision, 9*, 1–13.

Manning, D., Ethell, S., Donovan, T., & Crawford, T. (2006). How do radiologists do it? The influence of experience and training on searching for chest nodules. *Radiography, 12*, 134–142.

Martens, S., & Wyble, B. (2010). The attentional blink: Past, present, and future of a blind spot of perceptual awareness. *Neuroscience & Biobehavioral Reviews, 34*(6), 947–957.

Mathôt, S., & Ivanov, Y. (2019). The effect of pupil size and peripheral brightness on detection and discrimination performance. *PeerJ, 7*, e8820.

Mathôt, S., & Van der Stigchel, S. (2015). New light on the mind's eye: The pupillary light response as active vision. *Current Directions in Psychological Science, 24*(5), 374–378.

Nieuwenhuis, S., Gilzenrat, M. S., Holmes, B. D., & Cohen, J. D. (2005). The role of the locus coeruleus in mediating the attentional blink: A neurocomputational theory. *Journal of Experimental Psychology: General, 134*(3), 291–307.

Papesh, M. H., & Guevara Pinto, J. D. (2019). Spotting rare items makes the brain "blink" longer: Evidence from pupillometry. *Attention, Perception, & Psychophysics, 81*, 2635–2647.

Papesh, M. H., Hout, M. C., Guevara Pinto, J. D., Robbins, A., & Lopez, A. (2021). Eye movements reflect expertise development in hybrid search. *Cognitive Research: Principles and Implications, 6*(1), 1–20.

Peltier, C., & Becker, M. W. (2016). Decision processes in visual search as a function of target prevalence. *Journal of Experimental Psychology: Human Perception and Performance, 42*, 1466–1476.

Porter, G., Troscianko, T., & Gilchrist, I. D. (2007). Effort during visual search and counting: Insights from pupillometry. *Quarterly Journal of Experimental Psychology, 60*(2), 211–229.

Privitera, C. M., Renninger, L. W., Carney, T., Klein, S., & Aguilar, M. (2010). Pupil dilation during visual target detection. *Journal of Vision, 10*(10), 3–3.

Raymond, J. E., Shapiro, K. L., & Arnell, K. M. (1992). Temporary suppression of visual processing in an RSVP task: An attentional-blink? *Journal of Experimental Psychology: Human Perception and Performance, 18*(3), 849–860.

Reingold, E. M., Charness, N., Pomplun, M., & Stampe, D. M. (2001). Visual span in expert chess players: Evidence from eye movements. *Psychological Science, 12*, 48–55.

Rich, A. N., Kunar, M. A., Van Wert, M. J., Hidalgo-Sotelo, B., Horowitz, T. S., & Wolfe, J. M. (2008). Why do we miss rare targets? Exploring the boundaries of the low prevalence effect. *Journal of Vision, 8*, 1–17.

Sanders, A. F. (1970). Some aspects of the selective process in the functional visual field. *Ergonomics, 13*, 101–117.

Schwark, J., Sandry, J., MacDonald, J., & Dolgov, I. (2012). False feedback increases detection of low-prevalence targets in visual search. *Attention, Perception, & Psychophysics, 74*(8), 1583–1589.

Shomstein, S., Zhang, X., & Dubbelde, D. (2023). Attention and platypuses. *WIRES Cognitive Science, 14*(1), e1600.

Stotle, M., Gollan, B., & Ansorge, U. (2020). Tracking visual search demands and memory load through pupil dilation. *Journal of Vision, 20*(6), 1–19.

Takeuchi, T., Puntous, T., Tuladhar, A., Yoshimoto, S., & Shirama, A. (2011). Estimation of mental effort in learning visual search by measuring pupil dilation. *PLoS One, 6*(7), e21973.

Tkacz-Domb, S., & Yeshurun, Y. (2018). The size of the attentional window when measured by the pupillary response to light. *Scientific Reports, 8*, 11878.

Triesman, A., & Gelade, G. (1980). A feature-integration theory of attention. *Cognitive Psychology, 12*(1), 97–136.

Tversky, B. (2020). Anne Marie Treisman (1937–2018). *American Psychologist, 75*(4), 592–593.

Usher, M. (1999). The role of the locus coeruleus in the regulation of cognitive performance. *Science, 283*(5401), 549–554.

Warren, C. M., Breuer, A. T., Kantner, J., Fiset, D., Blais, C., & Masson, M. E. (2009). Target—distractor interference in the attentional blink implicates the locus coeruleus—norepinephrine system. *Psychonomic Bulletin & Review, 16*(6), 1106–1111.

Wolfe, J. M. (1998a). What can 1 million trials tell us about visual search? *Psychological Science, 9*(1), 33–39.

Wolfe, J. M. (1998b). Visual search. In H. Pashler (Ed.), *Attention* (pp. 13–73). University College London Press.

Wolfe, J. M. (2007). Guided Search 4.0: Current progress with a model of visual search. In W. Gray (Ed.), *Integrated models of cognitive systems* (pp. 99–119). Oxford.

Wolfe, J. M. (2021). Guided search 6.0: An updated model of visual search. *Psychonomic Bulletin & Review*, 1–33. Epub ahead of print. https://doi.org/10.3758/s13423-020-01859-9

Wolfe, J. M., & Horowitz, T. S. (2017). Five factors that guide attention in visual search. *Nature Human Behaviour, 1*(3), 1–8.

Wolfe, J. M., & Van Wert, M. J. (2010). Varying target prevalence reveals two dissociable decision criteria in visual search. *Current Biology, 20*, 121–124.

Wolfe, J. M., Cave, K. R., & Franzel, S. L. (1989). Guided search: An alternative to the feature integration model for visual search. *Journal of Experimental Psychology. Human Perception and Performance, 15*(3), 419–433.

Wolfe, J. M. (1994). Guided search 2.0: A revised model of visual search. *Psychonomic Bulletin & Review, 1*(2), 202–238. https://doi.org/10.3758/BF03200774

Wolfe, J. M., Horowitz, T. S., & Kenner, N. M. (2005). Rare items often missed in visual searches. *Nature, 435*, 439–440.

Wolfe, J. M., Horowitz, T. S., Van Wert, M. J., Kenner, N. M., Place, S. S., & Kibbi, N. (2007). Low target prevalence is a stubborn source of errors in visual search tasks. *Journal of Experimental Psychology: General, 136*(4), 623–638.

Wolfe, J. M., Brunelli, D. N., Rubinstein, J., & Horowitz, T. S. (2013). Prevalence effects in newly trained airport checkpoint screeners: Trained observers miss rare targets, too. *Journal of Vision, 13*(3), 1–9.

Young, A. H., & Hulleman, J. (2013). Eye movements reveal how task difficulty molds visual search. *Journal of Experimental Psychology: Human Perception and Performance, 39*, 168–190.

Pupillometry and Declarative Long-Term Memory

Brianna E. Cairney, Felicia Chaisson, and Megan H. Papesh

> With one singular exception, time's arrow is straight. Unidirectionality of time is one of nature's most fundamental laws. It has relentlessly governed all happenings in the universe—cosmic, geological, physical, biological, psychological—as long as the universe has existed...The singular exception is provided by the human ability to remember past happenings (Tulving, 2002, pp. 1–2).

1 Introduction to LTM

The nature and structure of recognition memory have been debated since Tulving (1983) introduced the term "episodic memory" to describe memory for everyday experiences (for a review, see Squire, 2004). Episodic memories, however, are not always concrete and are often experienced as vague feelings of knowing. For example, when shopping at the grocery store, you may run into work colleagues, old friends, family members, or a vaguely familiar clerk from a different store. You recognize most of these people with ease, accurately accessing your memory of each person and, more specifically, from where you know them. But memory is not always so specific; sometimes, as in Mandler's (1980) famous example, we see our local butcher on the bus and experience a feeling of familiarity that can range from vague ("I think I know that guy") to strong ("Seriously – how do I know that guy?!"). Importantly, this experience is not necessarily resolved with recollected details; you cannot name the butcher but know that you know him. Such apparent memory failures provide important anecdotal evidence about the everyday function of memory;

B. E. Cairney · F. Chaisson
Louisiana State University, Baton Rouge, LA, USA

M. H. Papesh (✉)
University of Massachusetts Lowell, Lowell, MA, USA
e-mail: megan_papesh@uml.edu

M. H. Papesh, S. D. Goldinger (eds.), *Modern Pupillometry*,
https://doi.org/10.1007/978-3-031-54896-3_6

memory can go undetected and under-appreciated until we experience such salient and frustrating failures. An interesting facet of such everyday memory failures, however, is what they reveal about the general function and architecture of the recognition memory system: Memories can be strong, weak, detailed, or partial, yet they are all “memory.”

How do we study such diverse experiences that nevertheless fall under the same categorical umbrella? At their core, memories live in the brain. Decades of research support distinguishing memory systems by the neural pathways and structures that support them. Ever since Patient H.M. had his hippocampi removed[1] and lost the ability to form new episodic memories but retained virtually all other memory types (Scoville & Milner, 1957; Squire, 2009), researchers have investigated the diverse brain regions involved in learning and memory. These diverse brain regions support distinct *types* of learning and, by association, memories. For example, the Multiple Memory Systems (MMS) framework has inspired research for over 40 years (Zola-Morgan & Squire, 1982; see Squire, 2009, for a brief review). According to the MMS framework, declarative memories are primarily served by medial temporal lobe structures, which include the hippocampus, but declarative learning and retrieval also frequently involve other sites, such as frontal structures and the diencephalon. Nondeclarative memories, which emerge from motor, stimulus-response, and/or perceptual learning, are supported by other regions, including the basal ganglia (motor), amygdala and cerebellum (stimulus-response), and neocortical sites (perceptual).

Episodic memories are arguably the quintessential declarative memory. As described by Tulving (2002), they give humans the unique ability to time travel. Of course, in scientific terms, “time travel” merely means activating or reinstating brain states, most frequently in the hippocampus. For example, when participants who complete memory tests while undergoing fMRI experience episodic remembering, there is often more activity in the hippocampus than when they experience vaguer feelings of recollection (e.g., Aggleton et al., 2005; Eldridge et al., 2005; Holdstock et al., 2005; Uncapher & Rugg, 2005; Verfaellie et al., 2008; Yonelinas et al., 2002). Single-cell neural recording confirms that the hippocampus not only encodes episodic memories (Wixted et al., 2018) but also provides a unique episodic signal relative to other brain regions that only provide a generic “old/new” signal (Urgolites et al., 2022). Over time, consolidation processes ensure that some of these episodic memories become semantic, whereby they lose their episodic quality but become stable knowledge (e.g., Smith & Squire, 2009). Importantly, both episodic and semantic memories are declarative, relying on similar or overlapping brain structures.

[1] Note: There is a suggestion that Patient H.M.’s brain damage was more extensive than commonly thought (Dittrich, 2017). For example, postmortem examination showed evidence of atrophy in frontal structures that may have been the result of the original resection surgery.

2 LC-NE and Memory

As mentioned, declarative memories vary in their strength and quality – some memories are strong and vivid, whereas some may be simply strong and others may be weak and vaguely detailed. How can the MMS framework support the creation of such diverse memories within the same medial temporal lobe structures? We suggest that differences in episodic and metacognitive qualities derive from the effects of the locus coeruleus–norepinephrine (LC–NE) system, which mediates arousal and innervates virtually all brain regions involved in learning (except the basal ganglia; several chapters in this volume provide reviews and alternatives). Several theories have been articulated to explain the role of the LC-NE system in processing emotional stimuli (e.g., via projections to the amygdala Strange & Dolan, 2004; Strange et al., 2003), and others focus more on arousal and processing priority (e.g., Bouret & Sara, 2005; Mather et al., 2016). Regardless of the theory one adopts, the LC-NE system's role as a neural gain modulator may explain why some events produce richly detailed memories and others produce vague feelings of knowing.

The LC releases and distributes NE, which binds with three major receptor categories (α_1, α_2, and β-adrenoreceptors) that collectively act as *neural gain amplifiers* (Bickler & Hansen, 1996; Delaney et al., 2007; Egli et al., 2005; Gilsbach & Hein, 2008; Langer, 2008; Ramos & Arnsten, 2007; Rinne et al., 2013; Starke, 2001), capable of enhancing neural signal-to-noise ratios. Although the colloquialism "rush of adrenaline" suggests a global ramping up of neural excitability, NE's dominant effect is to suppress neural activity, allowing selected neural populations (or representations) to take priority (Clewett et al., 2018; Foote et al., 1975; Freedman et al., 1977; Polack et al., 2013; Segal & Bloom, 1976; Waterhouse & Woodward, 1980). The LC is particularly engaged during arousal and orienting, which prepare an organism to act (Fries et al., 2001; Polack et al., 2013). By modulating neural gain, the LC-NE system can increase attentional salience for selected stimuli (or memories), while competing stimuli are perceptually suppressed.

In addition to cortical projections, the LC also projects to subcortical regions implicated in memory, affecting synaptic plasticity in the hippocampus and amygdala (Bergado et al., 2007; Berridge & Waterhouse, 2003; Harley, 1987; Seol et al., 2007). For example, when rats encounter novel objects, phasic firing bursts in LC neurons co-occur (Sara, 2009; Vankov et al., 1995), suggesting that NE helps modulate long-term potentiation. Indeed, LC activity seems to affect all memory stages, including encoding (e.g., Bliss & Collingridge, 1993; Lynch, 2004; Traynelis et al., 2010), consolidation (e.g., Cirelli et al., 1996; Cirelli & Tononi, 2000; Mather et al., 2016; Moncada et al., 2011; Moncada & Viola, 2007), and retrieval (e.g., Devauges & Sara, 1991; Murchison et al., 2004; Sara & Devauges, 1988). Given the LC's role in neural gain modulation, differences in memory quality are likely to emerge based on emotional characteristics, processing priorities, or salience (both bottom-up and top-down).

Measuring LC activity can prove challenging: The structure is located deep in the brainstem and is rarely, if ever, mandated for single-cell recording in human seizure disorder patients. Whereas pharmacological and microelectrode methods are used in nonhuman animals, pupillometry is the primary method used in humans (e.g., Goldinger & Papesh, 2012; Laeng et al., 2012, but see Clewett et al., 2017; Hauser et al., 2019; Sterpenich et al., 2006). In primate studies, single-cell LC recordings are time-locked with pupil fluctuations, verifying that pupil changes directly reflect LC neural firing (Aston-Jones & Cohen, 2005a; Rajkowski et al., 1994, 2004). Joshi et al. (2016) showed that LC activity precedes both spontaneous and stimulus-driven pupillary changes, with dilation occurring ≈310 ms after LC neuron spikes. During passive fixation, pupil size is directly correlated with LC spike rates. Given a startling tone, LC activity transiently increased, directly before pupil dilation. In human studies, there are clear linkages between pupil size and LC activity under tightly controlled conditions (Aston-Jones & Cohen, 2005b; Clewett et al., 2018; Goldinger & Papesh, 2012; Tsukahara et al., 2016). Sterpenich et al. (2006) found that, when participants successfully encoded faces in emotional contexts, they showed larger pupils and enhanced activity in both the LC region and amygdala, suggesting that the LC helps coordinate the fronto-hippocampal network required for memory. More recently, Clewett et al. (2018) used fMRI, melanin-weighted MRI, and pupillometry to reveal tight coupling for arousal, LC engagement, and transient connectivity changes among memory-relevant brain structures. Although pupillometry does not allow researchers to separate the influence of other related neuromodulatory systems (e.g., dopamine), arousal-mediated changes in declarative memory appear to be primarily driven by the noradrenergic system (Hauser et al., 2019). Given the tight correspondence between pupil size and LC-NE activity, pupillometry is an ideal method for studying how the neural systems involved in declarative memory can produce such diverse memories.

3 LTM Pupillometry Studies

3.1 Episodic Memory: Encoding and Retrieval Processes

Inspired by the neurophysiology of memory and arousal, and aided by advances in eye-tracking technology, researchers have used pupillometry to answer questions about long-term memory. For example, what determines whether something will be remembered or forgotten? How are memory decisions made and what sort of details, if any, accompany their retrieval? Models of memory often rely on several key variables from behavioral memory paradigms: The old/new decision, recollective experiences, and participants' subjective confidence. In this section, we will review each of these elements and how pupillometry may provide more nuanced information to supplement the behavioral data.

Old/New Decisions Arguably, the most basic aspect of measuring memory is recording decisions of whether an item or experience is old (previously encountered) or new (novel to the participant). Given the conscious deliberation involved in such declarative memory decisions, the medial temporal lobe system will be critically involved. Although recording old/new decisions will reveal what participants consciously infer following the neural reconstruction (or not) of recent memories, such isolated decisions do not reveal anything about the likelihood of participants' accuracy or the neural processes that were recruited during the reconstruction process. To provide more in-depth information about participants' old/new decisions, researchers have utilized pupillometry to investigate whether pupil size can reveal more about old/new decisions than the raw decision states themselves.

Võ et al. (2008) were the first to discover the "pupil old/new effect," in which the eyes' pupils differentiate between studied and unstudied information. During an encoding phase, participants viewed a list of positive, negative, and neutral words to commit to memory. After a short break, they completed a recognition task by responding to words on a list containing old and new items. Although pupil size did not differentiate successful from unsuccessful learning at encoding, there was a reliable difference in pupil size at retrieval: Old items correctly remembered as old (hits) elicited increased pupillary responses compared to new items correctly classified as new (correct rejections). The authors hypothesized that the need to engage in retrieval and postretrieval recollection processes may underlie this effect. Specifically, memory models that were popular at the time (e.g., dual-process framework, Yonelinas, 2001, 2002) proposed that memory decisions involving the recollection of episodic details would be slower and more cognitively demanding than decisions supported by vague feelings of familiarity or no memory at all. The difference in processes and effort may explain why pupils were larger for correctly remembered items than for correctly rejected items. Notably, the pupil's old/new effect was diminished for positive and negative words compared to neutral words, which was attributed to the less cognitive effort involved in encoding and retrieving emotional, relative to neutral, information.

The introduction of the pupil's old/new effect inspired many follow-up experiments. Similar effects have been reported for items consciously recollected compared to items that participants report as merely familiar, when participants engage in deep or elaborative encoding rather than shallow encoding, and when participants inaccurately report recalling a foil that is semantically related to the target item they previously attempted to encode (Otero et al., 2011). The pupil old/new effect also appears to be an automatic response that is independent of how participants respond to old and new items. For example, Heaver and Hutton (2011) had participants imagine they survived a car crash and then complete an old/new word recognition task. Participants were divided into one of three groups: a standard group that performed the task as accurately as possible, a malingering group that responded as though they had amnesia, and a single response group that reported "new" to all items (whether true or not). Regardless of which response group they were assigned to, all participants had increased pupil sizes for old items compared to new items.

The foregoing studies suggest an alluring possibility: Could pupil size be used to discriminate true from false memories? The answer, unfortunately, appears to be "no." Brocher and Graf (2017) conducted a systematic exploration of the pupil old/

new effect, examining whether effects change based on participants' processing orientation, attention, and accuracy. For example, in Experiment 1, participants encoded words and nonwords before a subsequent "recognition" phase, during which they made lexical decisions instead of recognition responses to old and new items. In this way, participants' processing orientation was not primarily on memory retrieval, despite some items being familiar from the encoding phase. No pupil old/new effects emerged. Subsequent studies revealed that pupil old/new effects emerged when participants made old/new decisions for false alarms (i.e., new words incorrectly judged "old"). These results challenge the notion that pupil size is an automatic and veridical reflection of memory, but nevertheless confirm a close relationship between intentional memory use and the pupillary response.

Although many memory studies examine questions related to what people remember about past events, an equally important consideration is what they know to be new. Without the ability to appreciate novelty, people would go through life constantly experiencing déjà vu, the feeling that something is familiar. Several studies have used pupillometry to better understand the underlying mechanisms of novelty detection. Unlike pupil *dilation* observed in pupil old/new effect research, complex stimuli have also been shown to elicit pupil *constriction*, which may be as informative for memory processes as dilation. In a declarative memory task, Naber et al. (2013) presented participants with pictures of complex natural scenes and then tested them with familiar (i.e., previously viewed) and novel images. During encoding, subsequently remembered images elicited increased pupil constriction relative to subsequently forgotten images. Similarly, during retrieval, both truly novel and forgotten (i.e., familiar scenes judged "novel") scenes were associated with more pupil constriction. Together, these findings suggest that pupil constriction during complex scene viewing signals subjective novelty. Appreciating novelty is important during memory tasks, as other research shows that items perceived as novel are more likely to be remembered, even when novelty is subjective or task irrelevant (e.g., Kishiyama et al., 2004; Knight, 1996).

In addition to measuring the old/new decision, researchers interested in memory processes will also often measure response time. As discussed, different memory processes may unfold across shorter or faster timescales (e.g., recollection is thought to be slow, whereas familiarity is thought to be fast). Response times, however, are not the only temporal element in a memory experiment. Time pressure, whether resulting from increased time spent on a task or having a time limit imposed, may also play a role. The *decision urgency pupil dilation model*, proposed by Murphy et al. (2016), highlights the influence of context and time on pupil size. For example, imagine trying to mentally solve a complex multiplication problem. More time spent on working out the solution corresponds to a gradual increase in pupil size, thought to reflect a self-imposed sense of "mental hurrying" or urgency. When a time constraint is imposed (e.g., you must solve the math problem within 5 s), the sense of urgency increases, resulting in greater pupil dilation responses. This response, referred to as the "urgency signal," may confound pupillometric memory study findings if time pressure effects are independent of task difficulty, but it has been largely overlooked in the literature.

To account for the influence of time pressure on pupillary changes, Gross and Dobbins (2020) dissociated the urgency signal from task difficulty. Participants completed an old/new word recognition task in which levels of encoding (i.e., shallow vs. deep) and trial length were manipulated. There was no difference in recall for probe words during the test, but pupil dilation was greater for speeded than unspeeded trials. Slower individual decisions elicited increased pupil dilation, lending support to the claim that pupil size increases with task urgency. Put differently, as more time passed, the urgency to produce a response increased, thereby eliciting pupil dilation; this effect was steeper when under explicit time constraints. Although recall was better for probe words encoded in the deep processing task, pupil dilation remained constant across all task and timing conditions. Thus, although much of the literature attributes pupil dilation to encoding efficacy or task difficulty, future research may wish to consider "urgency signal" contributions to pupillary changes during retrieval.

Recollective Detail (Remembering Versus Knowing) Memories are more than just a state of oldness or newness. They bring with them different levels of detail and awareness (or reliving) that make them each unique. Tulving (1985) originally proposed the Remember/Know (RK) paradigm to differentiate between states of awareness (e.g., noetic vs. autonoetic) during memory. Since that time, however, the RK paradigm has most often been used inform debates about the underlying processes supporting recognition memory (e.g., whether there are one or two underlying processes). Summarizing this debate is outside the scope of this chapter, but we will consider several studies that have used pupillometry to better understand the memories and processes involved in remembering versus knowing.

Kafkas and Montaldi (2011) used an incidental encoding task and modified RK paradigm to determine whether eye movements and pupil size can differentiate between memories accompanied by recollection and those associated with retrieval. Participants encoded pictures of man-made and natural items without any instructions to commit them to memory while their eye movements and pupil sizes were recorded. At the test, participants saw old and new pictures and rated them according to whether they were new, one of three strengths of familiar, or recollected. Although fixation patterns at encoding differentiated between recollection and familiarity (more fixations were associated with recollection), pupil size at encoding did not differentiate between recollection and familiarity. Instead, the researchers found a clearly linear relationship between pupil size at learning and subsequent feelings of memory: Items that were subsequently forgotten were associated with the largest encoding diameters and those that were later recollected had the smallest encoding diameters, with the three levels of familiarity falling orderly in between. This is consistent with the effects of novelty discussed earlier, but difficult to compare to seemingly inconsistent results from Otero et al. (2011). Otero et al. (2011) observed a larger pupil old/new effect for remembered, relative to known, items, but differences in paradigms render these studies challenging to consider simultaneously. For example, Kafkas and Montaldi (2011) used incidental learning of pictures and the encoding phase as their interest period, whereas Otero et al. (2011)

used intentional learning of words and the retrieval phase as their interest period. Given that pupil old/new effects depend on participants' approach to the memory task (e.g., Brocher & Graf, 2017), it is not surprising that different effects emerge with different stimuli, paradigms, and interest periods.

Subsequent research by Kafkas and Montaldi (2012) further clarified the roles of recollection and familiarity by measuring eye movement variables and brain responses (via fMRI) at retrieval. Using a matching-to-sample incidental learning paradigm and the same modified RK procedure as their 2011 study, Kafkas and Montaldi (2012) found that recollection produced larger pupil diameters at retrieval than both familiarity and "new" decisions. Importantly, they also found that recollection and strength-matched familiarity decisions recruited different brain structures, with recollection recruiting more hippocampal regions. Although some have suggested that the RK paradigm encourages participants to adopt different criteria for R and K judgments (e.g., Dunn, 2004), and that R judgments may not be different types of memories but just stronger memories (e.g., Dunn, 2004, 2008; Rotello & Zeng, 2008; Wixted, 2009; Wixted & Mickes, 2010), these results suggest that there may be some qualitative differences between recollection and familiarity that can be observed through the eyes and in the brain (but see Wais et al., 2010).

Subjective Confidence (Memory Strength) In addition to recording participants' memory decisions and their recollective details, researchers often ask participants to report on their subjective confidence as a means of evaluating memory strength. By itself, subjective confidence estimates may not be a pure reflection of memory strength, as the decisions may be influenced by factors related to fluency or ease of retrieval. For example, van Rijn et al. (2012) examined pupil sizes as participants repeatedly retrieved paired associates from multiple set sizes. With each retrieval repetition, which presumably produces stronger memory traces, participants' pupils dilated *less*, which is consistent with dilation reflecting retrieval effort, rather than confidence, per se. Papesh et al. (2012) collected confidence-based memory decisions in a memory study with intentional learning of words presented in different voices. Across the study and test, words were presented in the same or different voices to produce differences in retrieval accuracy and confidence (participants tend to emphasize the extremes of confidence scales in memory research, see Mickes et al., 2011). Pupillometry data revealed that higher confidence decisions were associated with larger pupil diameters and better memory. Subsequent research by Oliveira et al. (2021) replicated this effect with false alarms in a continuous recognition memory paradigm (false alarms were examined because participants did not use the full confidence scale for hits), but clearly this is an area ripe for future research.

3.2 *Episodic Memory: Factors That Affect Learning*

Old/new decisions, recollective details, and memory strength all depend on various factors present during encoding and/or retrieval (and, potentially, individual differences in working memory capacity, see Miller et al., 2019). Even anecdotally, it is easy to appreciate that some memories are simply more memorable than others. In the following section, we discuss some of the factors that have been shown to affect learning and retrieval success and how pupillometry has been used to provide a fuller appreciation for these effects.

Emotions and Arousal As anyone with a vivid flashbulb memory can attest, emotional or arousing events can produce stronger and seemingly more detailed memories. Even without the extremes that produce flashbulb memories, autobiographical memories tend to have more emotional content than other types of memory (e.g., word learning). Although these memories contain a mixture of episodic and semantic details, they are most often studied for their episodic and emotional qualities. Relative to unemotional tasks (e.g., counting), retrieving autobiographical memories is associated with enlarged pupils (El Haj et al., 2019), which reflects their episodic emotional content. This effect, however, diminishes with age, reflecting age-related changes in the LC-NE system (Hämmerer et al., 2017).

The effects of emotions on memory are said to occur because of tight connections between the amygdala and hippocampus, with the amygdala modulating hippocampal activity during both encoding (e.g., Kilpatrick & Cahill, 2003) and retrieval (e.g., Strange & Dolan, 2004). The amygdala does not operate alone, of course, but has reciprocal connections with other structures, including the LC, that might mediate more centralized arousal (Berntson et al., 2003; Davis & Whalen, 2001). To examine the influence of emotions and arousal on learning and memory, Sterpenich et al. (2006) had participants encode neutral faces that appeared atop negative (e.g., a snake) or neutral backgrounds. At the test, participants saw the neutral faces in isolation and made decisions about their old/new status and the type of background they were originally paired with. Pupil sizes during learning were larger for faces that were subsequently recognized than subsequently forgotten, but only when they appeared with a negative context. At retrieval, the pupil's old/new effect was observed regardless of the learning context. fMRI data, however, revealed that the LC and amygdala interacted during retrieval to restore the central arousal effects that emerged during learning, and that the amygdala modulated the response of medial temporal lobe sites.

Arousal is not necessarily dependent on just emotional stimuli – a state of threat can elicit similar pupillary responses. Clewett et al. (2018) used a threat of monetary punishment to induce arousal in participants. This proved effective, as participants recalled more high-priority items and fewer distracting, low-priority items under threat conditions. Memory effects were accompanied by greater pupil dilations during encoding of high-priority compared to low-priority scenes when under threat. Through fMRI scans during encoding, this study also demonstrated a link between

LC activity and pupil dilation, bolstering the argument that this region of the brain focuses limited attention and enhances encoding for critical information when under arousal, which can be observed exogenously via pupil dilation.

Value-Directed Learning/Memory Rewards and the opportunity to pursue high-value items influence many cognitive processes. Unsurprisingly, the typical result is that people are willing to work harder and for longer to earn higher rewards. A similar effect is observed in long-term memory, such that items associated with higher values tend to be remembered better than items associated with lower values (Castel, 2007). Ariel and Castel (2014) investigated value-directed remembering using pupillometry to quantify resource allocation during learning. They found that participants' pupils dilated more when encoding high-, relative to low-, value words and that high-value items were more likely to be remembered. This result did not emerge because participants ignored the low-value words; fixation data revealed that participants were equally likely to fixate high- and low-value words. The authors also argued that the pupil effect was not driven by the incentives themselves (e.g., pupils dilate to high incentives without ramping up attention to the accompanying item). For example, Kahneman and Peavler (1969) temporally dissociated the incentive cue from the study item. They found that participants' pupils dilated more to study items with high values than low values, and dilation did not begin when the incentive cue was presented; rather, dilation emerged only after the (value-primed) study item appeared.

Subsequent research by Miller et al. (2019) confirms the connection between value and attentional engagement. In their Experiment 2, low and high working memory capacity participants completed a delayed free recall task after encoding items whose serial positions were associated with different values. In some lists, the point values decreased with each serial position, and in others, the point values increased with each serial position. When the high-value items appeared at the beginning of the lists, participants recalled more primacy items, and when the high-value items appeared at the end of the lists, participants recalled more recency items. Pupil size also nicely tracked value: When the lists ascended in points, pupil sizes were largest in the middle and end of the list. That this effect emerged for both low and high working memory capacity individuals was important, as it had previously been suggested that low working memory capacity individuals struggle to sustain attention. Although the low working memory capacity participants did not recruit the same level of attentional resources as the high working memory capacity participants, the fact that they could titrate their attention based on motivations disconfirmed the notion that they are unable to sustain attention across a serial list of study items.

3.3 Semantic Memory

Although all declarative memories begin as episodes, some consolidate over time and become semantic memories (Smith & Squire, 2009). Broadly speaking, semantic memories are declarative knowledge – memories for facts, word meaning, etc. Although semantic memory can be measured through general knowledge assessments or synonym/antonym tests, researchers more commonly measure semantic memory through priming and categorization paradigms. Recently, researchers have also begun to use pupillometry to enhance understanding of the underlying neural mechanisms associated with semantic memory.

Semantic Priming What is the first word you think of when you read DOCTOR? What about when you read DOG? Chances are good that your first thoughts were "nurse" and "cat," respectively. Semantic priming is the finding that reading times and decision speeds are facilitated for words that follow a closely related semantic associate. Although the method has been around for many decades, Kuipers and Thierry (2011) were the first to simultaneously record pupillometry with electroencephalography (EEG) to examine how people perceive and understand meaning. In their study, participants saw pairs of semantically related or unrelated words in sequential order. For example, in some trials, participants would hear the word *duck* and subsequently see a picture of a duck (related) or a bed (unrelated). In other trials, they would see a picture of a duck and subsequently see the word *duck* or *bed*. With no further instructions for processing or memorizing material, this sort of paradigm is well suited to examining semantic processing via EEG recording (e.g., Relander et al., 2009).

Kuipers and Thierry (2011) were interested in both pupillary responses and the N400 event-related potential (ERP) component. ERPs are averaged neural responses occurring after the presentation of a specific stimulus or event; the N400 is a negative amplitude ERP component occurring between 300 and 500 ms poststimulus onset that previous research has demonstrated gets more negative in response to more difficult semantic processing. As predicted, there was a difference in cognitive processing for matched versus mismatched item pairs: The N400 amplitudes during the critical time window were more negative, indicating more effortful processing during the mismatch trials. These trials were also associated with an increase in pupil diameter. Moreover, the N400 and pupillary responses were negatively correlated, such that trials with greater pupil dilation were associated with a less negative N400. The authors interpreted this as reflecting processing efficiency in the LC-NE system.

Ledoux et al. (2016) continued to investigate semantic memory using the N400 and pupil size in their study of receptive vocabulary in neurotypical adults. Participants completed two tasks, both of which involved the presentation of an auditory known or unknown word. In one task, participants were to select a pictorial match for that word from a lineup of four options, and in the other task, they were simply to determine whether a single picture shown on-screen was a good match for that word. As in other paradigms, the amplitude of the N400 was reduced when participants were presented with congruent word/picture pairs, but only for words that they were expected to know. The N400, after all, relies on word knowledge, so this effect is to be expected. Pupils, on the other hand, were larger when participants encountered unknown, relative to known, words. Ledoux et al. (2016) also observed a suggestive, but non-reliable, negative correlation between pupil size and the N400, replicating Kuipers and Thierry (2011).

Categorical Information Another heavily researched area of semantic memory is how associations within semantic space are structured and what impact they might have on information processing and memory. People routinely rely on these sorts of spaces when encountering new instances of familiar objects. For example, your prior knowledge of dogs will impact how you interact with a dog you have never met before. Relationships among concepts are generally organized either taxonomically or thematically. The taxonomic organization relies on featural similarity and category membership (e.g., *whale* and *dog* are both animals; Rogers & Mcclelland, 2004), and thematic organization relies on mutual involvement in events (e.g., *whale* and *binoculars* are both involved in whale watching, or so you hope when you spend a lot of money on tour; Estes et al., 2011). Although debate exists regarding whether these two organization schemes exist as separate semantic memory systems, the debate is outside the focus of this chapter (interested readers are directed to Mirman et al., 2017; Thompson et al., 2017). Our focus is on what can be learned about semantic memory by studying these organization schemes via pupillometry.

Geller et al. (2019) examined whether cognitive control demands differ when people process thematic versus taxonomic relationships, using pupil size to quantify control. Participants judged the relatedness of word pairs that differed in association type (taxonomic vs. thematic) and relatedness (strong vs. weak). They found that the strength of the relationship between taxonomic and thematic similarity across word pairs correlated with pupil dilation, with weaker links inducing *more* need for control, as evidenced by steeper pupil slopes. This trend was most pronounced for weakly related taxonomic word pairs. This result contradicts earlier findings in which thematic relationships were found to require more cognitive control (e.g., Thompson et al., 2017). The authors suggest that these differences arose due to changes in task demands across studies and that thematic relationships are more flexible. For example, whereas detecting thematic relationships may be more challenging than detecting taxonomic relationships (as in Thompson et al., 2017), judging item pairs based on similarity (as in Geller et al., 2019) seems to produce enhanced challenges for the taxonomic pairs. Regardless of whether thematic or taxonomic relationships impose greater demand, Geller et al. (2019) concluded that the primary determinant of cognitive control demands is the strength of the relationship between the two items under consideration.

3.4 Implicit Measures

What is the fate of forgotten memories? Are they ever *truly* forgotten? These questions are nearly impossible to answer, but decades of research on memory cues and implicit memory paradigms offer some clues that forgotten memories may just be consciously inaccessible. Not all memory measures rely on tasks that require conscious retrieval. For example, free word naming, preference judgments, and even viewing behaviors can reveal memories without individuals needing to consciously declare their memory states. These measures can be important when dealing with supposedly forgotten memories.

A striking example of implicit recognition observed via pupillary changes can be found in case study research with amnesic patients with hippocampal lesions and an impaired ability to explicitly learn new facts (Laeng et al., 2007). Amnesic patients and matched controls were read factual statements (e.g., *"Penguins lay blue eggs*") while viewing a screen with four quadrants. During each statement, an image would appear in one of the four quadrants. Patients were later asked to answer questions about the statements (e.g., "*What is the color of a penguin's egg?*"). Eye movements were recorded during learning and recall. Although the amnesic patients could not provide details about facts they recently heard, eye tracking and pupillometry revealed evidence of implicit recognition. During the testing phase, participants spent more time looking at the "correct" quadrant that corresponded to the fact, as compared to the other three "incorrect" quadrants. When the amnesic participants completed an old/new recognition test for old objects that appeared during the study intermixed with new objects, they responded new to a majority of the pictures. However, the researchers reported a pupil old/new effect, with increased pupil sizes for new compared to old items. Careful readers will note that this pupil's old/new effect is in the opposite direction of the typical old/new effect (old > new), but also that old/new effects depend on engaging in memory retrieval. Patients with amnesia are unlikely to be consciously accessing memories and, as such, their pupil response to old items is more likely to reflect processing fluency than declarative retrieval. Despite being in the opposite direction of standard pupil old/new effects, these findings suggest that pupillary measures of memory may reveal intact implicit memories where explicit memories no longer exist.

4 Concluding Remarks

According to the Multiple Memory Systems framework (Squire, 2009), declarative long-term memories, which include both episodic and semantic memories, rely on neural structures primarily located in the medial temporal lobes. Other areas (e.g., frontal structures, the diencephalon) participate in learning and memory through cognitive control and focusing attention. Despite being coordinated through relatively restricted brain structures, declarative memories are nevertheless remarkably diverse. This diversity has inspired countless experiments and theoretical debates – how can one brain region produce an episodic memory rich with contextual detail and another relatively devoid of detail? In this chapter, we reviewed evidence linking the LC-NE system to memory formation and suggest that differential engagement of the LC produces the diversity of declarative memories. Although we reviewed abundant evidence in which pupil size, as an indirect proxy for LC activity, was related to core aspects of learning and memory, we did not provide conclusive evidence that the LC is responsible for variations in memory strength and specificity. This, naturally, will be the job of future research.

References

Aggleton, J. P., Vann, S. D., Denby, C., Dix, S., Mayes, A. R., Roberts, N., & Yonelinas, A. P. (2005). Sparing of the familiarity component of recognition memory in a patient with hippocampal pathology. *Neuropsychologia, 43*, 1810–1823. https://doi.org/10.1016/j.neuropsychologia.2005.01.019

Ariel, R., & Castel, A. D. (2014). Eyes wide open: Enhanced pupil dilation when selectively studying important information. *Experimental Brain Research*. https://doi.org/10.1007/s00221-013-3744-5

Aston-Jones, G., & Cohen, J. D. (2005a). Adaptive gain and the role of the locus coeruleus–norepinephrine system in optimal performance. *Journal of Comparative Neurology, 493*(1), 99–110. https://doi.org/10.1002/CNE.20723

Aston-Jones, G., & Cohen, J. D. (2005b). An integrative theory of locus coeruleus-norepinephrine function: Adaptive gain and optimal performance. *Annual Review of Neuroscience, 28*(1), 403–450. https://doi.org/10.1146/annurev.neuro.28.061604.135709

Bergado, J. A., Frey, S., López, J., Almaguer-Melian, W., & Frey, J. U. (2007). Cholinergic afferents to the locus coeruleus and noradrenergic afferents to the medial septum mediate LTP-reinforcement in the dentate gyrus by stimulation of the amygdala. *Neurobiology of Learning and Memory, 88*(3), 331–341. https://doi.org/10.1016/j.nlm.2007.05.003

Berntson, G. G., Sarter, M., & Cacioppo, J. T. (2003). Ascending visceral regulation of cortical affective information processing. *European Journal of Neuroscience, 18*(8), 2103–2109. https://doi.org/10.1046/J.1460-9568.2003.02967.X

Berridge, C. W., & Waterhouse, B. D. (2003). The locus coeruleus-noradrenergic system: Modulation of behavioral state and state-dependent cognitive processes. *Brain Research Reviews*. Elsevier. https://doi.org/10.1016/S0165-0173(03)00143-7

Bickler, P. E., & Hansen, B. M. (1996). α2-Adrenergic agonists reduce glutamate release and glutamate receptor-mediated calcium changes in hippocampal slices during hypoxia. *Neuropharmacology, 35*(6). https://doi.org/10.1016/0028-3908(96)84639-9

Bliss, T. V. P., & Collingridge, G. L. (1993). A synaptic model of memory: Long-term potentiation in the hippocampus. *Nature*. https://doi.org/10.1038/361031a0

Bouret, S., & Sara, S. J. (2005). Network reset: A simplified overarching theory of locus coeruleus noradrenaline function. *Trends in Neurosciences, 28*(11), 574–582. https://doi.org/10.1016/j.tins.2005.09.002

Brocher, A., & Graf, T. (2017). Decision-related factors in pupil old/new effects: Attention, response execution, and false memory. *Neuropsychologia, 102*, 124–134. https://doi.org/10.1016/j.neuropsychologia.2017.06.011

Castel, A. D. (2007). The adaptive and strategic use of memory by older adults: Evaluative processing and value-directed remembering. *Psychology of Learning and Motivation - Advances in Research and Theory*. https://doi.org/10.1016/S0079-7421(07)48006-9

Cirelli, C., & Tononi, G. (2000). Differential expression of plasticity-related genes in waking and sleep and their regulation by the noradrenergic system. *Journal of Neuroscience, 20*(24). https://doi.org/10.1523/jneurosci.20-24-09187.2000

Cirelli, C., Pompeiano, M., & Tononi, G. (1996). Neuronal gene expression in the waking state: A role for the locus coeruleus. *Science, 274*(5290). https://doi.org/10.1126/science.274.5290.1211

Clewett, D., Sakaki, M., Nielsen, S., Petzinger, G., & Mather, M. (2017). Noradrenergic mechanisms of arousal's bidirectional effects on episodic memory. *Neurobiology of Learning and Memory, 137*, 1–14. https://doi.org/10.1016/j.nlm.2016.10.017

Clewett, D., Huang, R., Velasco, R., Lee, T.-H., & Mather, M. (2018). Locus coeruleus activity strengthens prioritized memories under arousal. *The Journal of Neuroscience, 38*(6), 2097–2017. https://doi.org/10.1523/JNEUROSCI.2097-17.2017

Davis, M., & Whalen, P. J. (2001). The amygdala: Vigilance and emotion. *Molecular Psychiatry, 6*, 13–34. Retrieved from https://www.nature.com/mp

Delaney, A. J., Crane, J. W., & Sah, P. (2007). Noradrenaline modulates transmission at a central synapse by a presynaptic mechanism. *Neuron, 56*(5). https://doi.org/10.1016/j.neuron.2007.10.022

Devauges, V., & Sara, S. J. (1991). Memory retrieval enhancement by locus coeruleus stimulation: Evidence for mediation by β-receptors. *Behavioural Brain Research, 43*(1). https://doi.org/10.1016/S0166-4328(05)80056-7

Dittrich, L. (2017). *Patient HM: A story of memory, madness, and family secrets*. Random House Trade Paperbacks.

Dunn, J. C. (2004). Remember-know: A matter of confidence. *Psychological Review, 111*(2), 524–542. https://doi.org/10.1037/0033-295X.111.2.524

Dunn, J. C. (2008). The dimensionality of the remember-know task: A state-trace analysis. *Psychological Review, 115*(2). https://doi.org/10.1037/0033-295X.115.2.426

Egli, R. E., Kash, T. L., Choo, K., Savchenko, V., Matthews, R. T., Blakely, R. D., & Winder, D. G. (2005). Norepinephrine modulates glutamatergic transmission in the bed nucleus of the stria terminalis. *Neuropsychopharmacology, 30*(4). https://doi.org/10.1038/sj.npp.1300639

El Haj, M., Janssen, S., Gallouj, K., & Lenoble, Q. (2019). Autobiographical memory increases pupil dilation. *Translational Neuroscience, 10*(1), 280–287. https://doi.org/10.1515/TNSCI-2019-0044/ASSET/GRAPHIC/J_TNSCI-2019-0044_FIG_002.JPG

Eldridge, L. L., Engel, S. A., Zeineh, M. M., Bookheimer, S. Y., & Knowlton, B. J. (2005). Behavioral/systems/cognitive a dissociation of encoding and retrieval processes in the human hippocampus. https://doi.org/10.1523/JNEUROSCI.3420-04.2005

Estes, Z., Golonka, S., & Jones, L. L. (2011). Thematic thinking: The apprehension and consequences of thematic relations. *Psychology of Learning and Motivation - Advances in Research and Theory, 54*, 249–294. https://doi.org/10.1016/B978-0-12-385527-5.00008-5

Foote, S. L., Freedman, R., & Oliver, A. P. (1975). Effects of putative neurotransmitters on neuronal activity in monkey auditory cortex. *Brain Research, 86*(2). https://doi.org/10.1016/0006-8993(75)90699-X

Freedman, R., Hoffer, B. J., Woodward, D. J., & Puro, D. (1977). Interaction of norepinephrine with cerebellar activity evoked by mossy and climbing fibers. *Experimental Neurology, 55*(1). https://doi.org/10.1016/0014-4886(77)90175-3

Fries, P., Reynolds, J. H., Rorie, A. E., & Desimone, R. (2001). Modulation of oscillatory neuronal synchronization by selective visual attention. *Science, 291*(5508). https://doi.org/10.1126/science.1055465

Geller, J., Landrigan, J.-F., & Mirman, D. (2019). A pupillometric examination of cognitive control in taxonomic and thematic semantic memory. *Journal of Cognition, 2*(1). https://doi.org/10.5334/joc.56

Gilsbach, R., & Hein, L. (2008). Presynaptic metabotropic receptors for acetylcholine and adrenaline/noradrenaline. *Handbook of Experimental Pharmacology, 184*. https://doi.org/10.1007/978-3-540-74805-2_9

Goldinger, S. D., & Papesh, M. H. (2012). Pupil dilation reflects the creation and retrieval of memories. *Current Directions in Psychological Science, 21*(2), 90–95. https://doi.org/10.1177/0963721412436811

Gross, M. P., & Dobbins, I. G. (2020). Pupil dilation during memory encoding reflects time pressure rather than depth of processing. *Journal of Experimental Psychology: Learning Memory and Cognition*. https://doi.org/10.1037/xlm0000818

Hämmerer, D., Hopkins, A., Betts, M. J., Maaß, A., Dolan, R. J., & Düzel, E. (2017). Emotional arousal and recognition memory are differentially reflected in pupil diameter responses during emotional memory for negative events in younger and older adults. *Neurobiology of Aging, 58*, 129–139. https://doi.org/10.1016/J.NEUROBIOLAGING.2017.06.021

Harley, C. W. (1987). A role for norepinephrine in arousal, emotion and learning?: Limbic modulation by norepinephrine and the kety hypothesis. *Progress in Neuropsychopharmacology and Biological Psychiatry, 11*(4). https://doi.org/10.1016/0278-5846(87)90015-7

Hauser, T. U., Eldar, E., Purg, N., Moutoussis, M., & Dolan, R. J. (2019). Distinct roles of dopamine and noradrenaline in incidental memory. *The Journal of Neuroscience, 39*(39), 7715–7721. https://doi.org/10.1523/JNEUROSCI.0401-19.2019

Heaver, B., & Hutton, S. B. (2011). Keeping an eye on the truth? pupil size changes associated with recognition memory. *Memory, 19*(4), 398–405. https://doi.org/10.1080/09658211.2011.575788

Holdstock, J. S., Mayes, A. R., Gong, Q. Y., Roberts, N., & Kapur, N. (2005). Item recognition is less impaired than recall and associative recognition in a patient with selective hippocampal damage. *Hippocampus, 15*(2), 203–215. https://doi.org/10.1002/HIPO.20046

Joshi, S., Li, Y., Kalwani, R. M., & Gold, J. I. (2016). Relationships between pupil diameter and neuronal activity in the locus coeruleus, colliculi, and cingulate cortex. *Neuron, 89*(1), 221–234. https://doi.org/10.1016/j.neuron.2015.11.028

Kafkas, A., & Montaldi, D. (2011). Recognition memory strength is predicted by pupillary responses at encoding while fixation patterns distinguish recollection from familiarity. *Quarterly Journal of Experimental Psychology, 64*(10), 1971–1989. https://doi.org/10.1080/17470218.2011.588335

Kafkas, A., & Montaldi, D. (2012). Familiarity and recollection produce distinct eye movement, pupil and medial temporal lobe responses when memory strength is matched. *Neuropsychologia, 50*(13), 3080–3093. https://doi.org/10.1016/j.neuropsychologia.2012.08.001

Kahneman, D., & Peavler, W. S. (1969). Incentive effects and pupillary changes in association learning. *Journal of Experimental Psychology, 79*(2 PART 1), 312–318. https://doi.org/10.1037/H0026912

Kilpatrick, L., & Cahill, L. (2003). Amygdala modulation of parahippocampal and frontal regions during emotionally influenced memory storage. *Neuroimage, 20*, 2091–2099. https://doi.org/10.1016/j.neuroimage.2003.08.006

Kishiyama, M. M., Yonelinas, A. P., & Lazzara, M. M. (2004). The von Restorff effect in amnesia: The contribution of the hippocampal system to novelty-related memory enhancements. *Journal of Cognitive Neuroscience, 16*(1), 15–23. Retrieved from http://direct.mit.edu/jocn/article-pdf/16/1/15/1763904/089892904322755511.pdf?casa_token=PjP9pWW7CpIAAAAA:B7Hhj3UdLlxVVtPAhFnvZzmAuMuUE_YPtwQnh9En4mL556ccYbQ-5J8rbFYhOeQD4oHgmqS9

Knight, R. T. (1996). Contribution of human hippocampal region to novelty detection. *Nature, 383*(6597), 256–259. https://doi.org/10.1038/383256a0

Kuipers, J. R., & Thierry, G. (2011). N400 amplitude reduction correlates with an increase in pupil size. *Frontiers in Human Neuroscience, 5*(June), 1–5. https://doi.org/10.3389/fnhum.2011.00061

Laeng, B., Waterloo, K., Johnsen, S. H., Bakke, S. J., Låg, T., Simonsen, S. S., & Høgsaet, J. (2007). The eyes remember it: Oculography and pupillometry during recollection in three amnesic patients. *Journal of Cognitive Neuroscience, 19*(11), 1888–1904. https://doi.org/10.1162/jocn.2007.19.11.1888

Laeng, B., Sirois, S., & Gredebäck, G. (2012). Pupillometry: A window to the preconscious? *Perspectives on Psychological Science, 7*(1), 18–27. https://doi.org/10.1177/1745691611427305

Langer, S. Z. (2008). Presynaptic autoreceptors regulating transmitter release. *Neurochemistry International, 52*(1). https://doi.org/10.1016/j.neuint.2007.04.031

Ledoux, K., Coderre, E., Bosley, L., Buz, E., Gangopadhyay, I., & Gordon, B. (2016). The concurrent use of three implicit measures (eye movements, pupillometry, and event-related potentials) to assess receptive vocabulary knowledge in normal adults. *Behavior Research Methods, 48*(1), 285–305. https://doi.org/10.3758/S13428-015-0571-6/METRICS

Lynch, M. A. (2004). Long-term potentiation and memory. *Physiological Reviews*. https://doi.org/10.1152/physrev.00014.2003

Mandler, G. (1980). Recognizing: The judgment of previous occurrence. *Psychological Review, 87*(3), 252–271.

Mather, M., Clewett, D., Sakaki, M., & Harley, C. W. (2016). Norepinephrine ignites local hotspots of neuronal excitation: How arousal amplifies selectivity in perception and memory. *Behavioral and Brain Sciences, 39*, 1–75. https://doi.org/10.1017/S0140525X15000667

Mickes, L., Hwe, V., Wais, P. E., & Wixted, J. T. (2011). Strong memories are hard to scale. *Journal of Experimental Psychology: General, 140*(2), 239–257. https://doi.org/10.1037/a0023007

Miller, A. L., Gross, M. P., & Unsworth, N. (2019). Individual differences in working memory capacity and long-term memory: The influence of intensity of attention to items at encoding as measured by pupil dilation. *Journal of Memory and Language, 104*, 25–42. https://doi.org/10.1016/j.jml.2018.09.005

Mirman, D., Landrigan, J.-F., & Britt, A. E. (2017). Taxonomic and thematic semantic systems. *Psychological Bulletin T, 143*(5), 499. https://doi.org/10.1037/bul0000092

Moncada, D., & Viola, H. (2007). Induction of long-term memory by exposure to novelty requires protein synthesis: Evidence for a behavioral tagging. *Journal of Neuroscience, 27*(28). https://doi.org/10.1523/JNEUROSCI.1083-07.2007

Moncada, D., Ballarini, F., Martinez, M. C., Frey, J. U., & Viola, H. (2011). Identification of transmitter systems and learning tag molecules involved in behavioral tagging during memory formation. *Proceedings of the National Academy of Sciences of the United States of America, 108*(31). https://doi.org/10.1073/pnas.1104495108

Murchison, C. F., Zhang, X. Y., Zhang, W. P., Ouyang, M., Lee, A., & Thomas, S. A. (2004). A distinct role for norepinephrine in memory retrieval. *Cell, 117*(1). https://doi.org/10.1016/S0092-8674(04)00259-4

Murphy, P. R., Boonstra, E., & Nieuwenhuis, S. (2016). Global gain modulation generates time-dependent urgency during perceptual choice in humans. *Nature Communications, 7*(1), 1–15. https://doi.org/10.1038/ncomms13526

Naber, M., Frassle, S., Rutishauser, U., & Einhauser, W. (2013). Pupil size signals novelty and predicts later retrieval success for declarative memories of natural scenes. *Journal of Vision, 13*(2), 11–11. https://doi.org/10.1167/13.2.11

Oliveira, J., Fernandes, M., Rosa, P. J., & Gamito, P. (2021). Is pupil activity associated with the strength of memory signal for words in a continuous recognition memory paradigm? *Frontiers in Psychology, 12*, 5306. https://doi.org/10.3389/FPSYG.2021.686183/BIBTEX

Otero, S. C., Weekes, B. S., & Hutton, S. B. (2011). Pupil size changes during recognition memory. *Psychophysiology, 48*(10), 1346–1353. https://doi.org/10.1111/j.1469-8986.2011.01217.x

Papesh, M. H., Goldinger, S. D., & Hout, M. C. (2012). Memory strength and specificity revealed by pupillometry. *International Journal of Psychophysiology, 83*(1), 56–64. https://doi.org/10.1016/j.ijpsycho.2011.10.002

Polack, P. O., Friedman, J., & Golshani, P. (2013). Cellular mechanisms of brain state-dependent gain modulation in visual cortex. *Nature Neuroscience, 16*(9). https://doi.org/10.1038/nn.3464

Rajkowski, J., Kubiak, P., & Aston-Jones, G. (1994). Locus coeruleus activity in monkey: Phasic and tonic changes are associated with altered vigilance. *Brain Research Bulletin, 35*(5–6). https://doi.org/10.1016/0361-9230(94)90175-9

Rajkowski, J., Majczynski, H., Clayton, E., & Aston-Jones, G. (2004). Activation of monkey locus coeruleus neurons varies with difficulty and performance in a target detection task. *Journal of Neurophysiology, 92*(1). https://doi.org/10.1152/jn.00673.2003

Ramos, B. P., & Arnsten, A. F. T. (2007). Adrenergic pharmacology and cognition: Focus on the prefrontal cortex. *Pharmacology and Therapeutics*. https://doi.org/10.1016/j.pharmthera.2006.11.006

Relander, K., Rämä, P., & Kujala, T. (2009). Word semantics is processed even without attentional effort. *Journal of Cognitive Neuroscience, 21*(8), 1511–1522. https://doi.org/10.1162/JOCN.2009.21127

Rinne, A., Birk, A., & Búnemann, M. (2013). Voltage regulates adrenergic receptor function. *Proceedings of the National Academy of Sciences of the United States of America, 110*(4). https://doi.org/10.1073/pnas.1212656110

Rogers, T. T., & Mcclelland, J. L. (2004). *Semantic cognition: A parallel distributed processing approach*. MIT Press. https://doi.org/10.7551/mitpress/6161.001.0001

Rotello, C. M., & Zeng, M. (2008). Analysis of RT distributions in the remember–know paradigm CAREN. *Psychonomic Bulletin & Review, 15*(4), 825–832.

Sara, S. J. (2009). The locus coeruleus and noradrenergic modulation of cognition. *Nature Reviews Neuroscience, 10*(3), 211–223. https://doi.org/10.1038/nrn2573

Sara, S. J., & Devauges, V. (1988). Priming stimulation of locus coeruleus facilitates memory retrieval in the rat. *Brain Research, 438*(1–2). https://doi.org/10.1016/0006-8993(88)91351-0

Scoville, W. B., & Milner, B. (1957). Loss of recent memory after bilateral hippocampal lesions. *Journal of Neurology, Neurosurgery, and Psychiatry, 20*, 11.

Segal, M., & Bloom, F. E. (1976). The action of norepinephrine in the rat hippocampus. IV. The effects of locus coeruleus stimulation on evoked hippocampal unit activity. *Brain Research, 107*(3), 513–525. https://doi.org/10.1016/0006-8993(76)90141-4

Seol, G. H., Ziburkus, J., Huang, S. Y., Song, L., Kim, I. T., Takamiya, K., et al. (2007). Neuromodulators control the polarity of spike-timing-dependent synaptic plasticity. *Neuron, 55*(6). https://doi.org/10.1016/j.neuron.2007.08.013

Smith, C. N., & Squire, L. R. (2009). Medial temporal lobe activity during retrieval of semantic memory is related to the age of the memory. *The Journal of Neuroscience, 29*(4), 930–938. https://doi.org/10.1523/JNEUROSCI.4545-08.2009

Squire, L. R. (2009). Memory and brain systems: 1969-2009. *Journal of Neuroscience, 29*(41), 12711–12716. https://doi.org/10.1523/JNEUROSCI.3575-09.2009

Starke, K. (2001). Presynaptic autoreceptors in the third decade: Focus on α2-adrenoceptors. *Journal of Neurochemistry*. https://doi.org/10.1046/j.1471-4159.2001.00484.x

Sterpenich, V., D'Argembeau, A., Desseilles, M., Balteau, E., Albouy, G., Vandewalle, G., et al. (2006). The locus ceruleus is involved in the successful retrieval of emotional memories in humans. *Journal of Neuroscience, 26*(28), 7416–7423. https://doi.org/10.1523/JNEUROSCI.1001-06.2006

Strange, B. A., & Dolan, R. J. (2004). β-Adrenergic modulation of emotional memory-evoked human amygdala and hippocampal responses. *Proceedings of the National Academy of Sciences of the United States of America, 101*(31), 11454–11458. https://doi.org/10.1073/pnas.0404282101

Strange, B. A., Hurlemann, R., & Dolan, R. J. (2003). An emotion-induced retrograde amnesia in humans is amygdala- and β-adrenergic-dependent. *Proceedings of the National Academy of Sciences of the United States of America, 100*(23), 13626–13631. https://doi.org/10.1073/pnas.1635116100

Squire, L. R. (2004). Memory systems of the brain: a brief history and current perspective. *Neurobiology of Learning and Memory, 82*(3), 171–177.

Thompson, H., Davey, J., Hoffman, P., Hallam, G., Kosinski, R., Howkins, S., et al. (2017). Semantic control deficits impair understanding of thematic relationships more than object identity. *Neuropsychologia, 104*, 113–125. https://doi.org/10.1016/J.NEUROPSYCHOLOGIA.2017.08.013

Traynelis, S. F., Wollmuth, L. P., McBain, C. J., Menniti, F. S., Vance, K. M., Ogden, K. K., et al. (2010). Glutamate receptor ion channels: Structure, regulation, and function. *Pharmacological Reviews*. https://doi.org/10.1124/pr.109.002451

Tsukahara, J. S., Harrison, T. L., & Engle, R. W. (2016). The relationship between baseline pupil size and intelligence. *Cognitive Psychology, 91*, 109–123. https://doi.org/10.1016/j.cogpsych.2016.10.001

Tulving, E. (1983). *Elements of episodic memory*. Retrieved from https://philpapers.org/rec/TULEOE

Tulving, E. (1985). Memory and consciousness. *Canadian Psychology/Psychologie Canadienne, 26*(1), 1–12. https://doi.org/10.1037/H0080017

Tulving, E. (2002). Episodic memory: From mind to brain. *Annual Review of Psychology, 53*, 1–25. https://doi.org/10.1146/annurev.psych.53.100901.135114

Uncapher, M. R., & Rugg, M. D. (2005). Behavioral/systems/cognitive encoding and the durability of episodic memory: A functional magnetic resonance imaging study. *Journal of Neuroscience, 25*(31), 7260–7267. https://doi.org/10.1523/JNEUROSCI.1641-05.2005

Urgolites, Z. J., Wixted, J. T., Goldinger, S. D., Papesh, M. H., Treiman, D. M., Squire, L. R., & Steinmetz, P. N. (2022). Two kinds of memory signals in neurons of the human hippocampus. *Proceedings of the National Academy of Sciences of the United States of America, 119*(19), e2115128119. https://doi.org/10.1073/PNAS.2115128119/SUPPL_FILE/PNAS.2115128119.SAPP.PDF

van Rijn, H., Dalenberg, J. R., Borst, J. P., & Sprenger, S. A. (2012). Pupil dilation co-varies with memory strength of individual traces in a delayed response paired-associate task. *PLoS One, 7*(12). https://doi.org/10.1371/journal.pone.0051134

Vankov, A., Hervé-Minvielle, A., & Sara, S. J. (1995). Response to novelty and its rapid habituation in locus coeruleus neurons of the freely exploring rat. *European Journal of Neuroscience, 7*(6). https://doi.org/10.1111/j.1460-9568.1995.tb01108.x

Verfaellie, M., Rajaram, S., Fossum, K., & Williams, L. (2008). Not all repetition is alike: Different benefits of repetition in amnesia and normal memory. *Journal of the International Neuropsychological Society : JINS, 14*(3), 365. https://doi.org/10.1017/S1355617708080612

Võ, M. L. H., Jacobs, A. M., Kuchinke, L., Hofmann, M., Conrad, M., Schacht, A., & Hutzler, F. (2008). The coupling of emotion and cognition in the eye: Introducing the pupil old/new effect. *Psychophysiology, 45*(1), 130–140. https://doi.org/10.1111/j.1469-8986.2007.00606.x

Wais, P. E., Squire, L. R., & Wixted, J. T. (2010). In search of recollection and familiarity signals in the hippocampus. *Journal Cognitive Neuroscience, 22*(1), 109–123. https://doi.org/10.1162/jocn.2009.21190.In

Waterhouse, B. D., & Woodward, D. J. (1980). Interaction of norepinephrine with cerebrocortical activity evoked by stimulation of somatosensory afferent pathways in the rat. *Experimental Neurology, 67*(1), 11–34. https://doi.org/10.1016/0014-4886(80)90159-4

Wixted, J. T. (2009). Remember/know judgments in cognitive neuroscience: An illustration of the underrepresented point of view. *Learning and Memory, 16*, 406–412. https://doi.org/10.1101/lm.1312809

Wixted, J. T., & Mickes, L. (2010). A continuous dual-process model of remember/know judgments. *Psychological Review, 117*(4), 1025–1054. https://doi.org/10.1037/a0020874

Wixted, J. T., Goldinger, S. D., Squire, L. R., Kuhn, J. R., Papesh, M. H., Smith, K. A., et al. (2018). Coding of episodic memory in the human hippocampus. *Proceedings of the National Academy of Sciences of the United States of America, 115*(5). https://doi.org/10.1073/pnas.1716443115

Yonelinas, A. P. (2001). Components of episodic memory: The contribution of recollection and familiarity. *Philosophical Transactions of the Royal Society B: Biological Sciences, 356*(1413), 1363–1374. https://doi.org/10.1098/rstb.2001.0939

Yonelinas, A. P. (2002). The nature of recollection and familiarity: A review of 30 years of research. *Journal of Memory and Language, 46*(3), 441–517. https://doi.org/10.1006/jmla.2002.2864

Yonelinas, A. P., Kroll, N. E. A., Quamme, J. R., Lazzara, M. M., Sauvé, M.-J., Widaman, K. F., & Knight, R. T. (2002). Effects of extensive temporal lobe damage or mild hypoxia on recollection and familiarity. *Nature Neuroscience, 5*(11), 1236–1241. https://doi.org/10.1038/nn961

Zola-Morgan, S., & Squire, L. R. (1982). The neuroanatomy of amnesia. *Science, 218*, 1337–1339. https://doi.org/10.1093/brain/101.3.403

Pupil Dilation as a Dependent Variable in Language Research

Jens Schmidtke and Stephen J. Tobin

When searching the term "pupillometry" and "pupillometrics," on Google's Ngram viewer, one finds the earliest mentions in the corpus around 1882, but it is not until 1960 that one can make out an increase of interest in the method. This increased interest seems to have been stirred by research conducted by Hess and Polt (1960). These researchers discovered that the pupils of subjects dilated in response to pictures that were interesting to the viewer but not in response to uninteresting pictures. However, the effect may also be interpreted as a surprise reaction to the stimuli in question, which were provocative images of naked male and female subjects among otherwise rather innocuous images. This finding sparked "enormous interest" in researchers investigating diverse issues such as "public opinion, consumer research, lie detection, personality research, sexual behavior, and social attitudes" (Hakerem, 1973b, p. 239). However, this interest soon declined due to problems in the interpretation of results (ibid.). For example, the notion that the pupil constricts in response to unpleasant stimuli, propagated by Hess (1965), was soon dismissed (Peavler & McLaughlin, 1967). Four years after their initial paper, Hess and Polt (1964) published another groundbreaking study by demonstrating a correlation between the difficulty of an arithmetic problem and the increase in pupil size of a person while solving the problem. This line of research proved more fruitful and was extended by Kahneman and Beatty (1966) who demonstrated that the pupil dilates with (moment-to-moment) increasing load on short-term memory. In the same publication, these researchers introduced the construct of *processing load*, which turned out to be rather fruitful in cognitive psychology (cf. Kahneman, 1973). Other early studies expanded the field of inquiry even further. One study showed greater dilation when

J. Schmidtke (✉)
Haifa Centre for German and European Studies, University of Haifa, Mount Carmel, Haifa, Israel
e-mail: schmi474@msu.edu

S. J. Tobin
Department of Computer Science, University of Potsdam, Potsdam, Germany

M. H. Papesh, S. D. Goldinger (eds.), *Modern Pupillometry*,
https://doi.org/10.1007/978-3-031-54896-3_7

participants were asked to create mental images of more abstract nouns compared to more concrete nouns (Paivio & Simpson, 1966), thus demonstrating a pupillary response not only to the quantity of stimuli but also the quality (i.e., concreteness). Yet another study suggested a correlation between pupil dilation and the difficulty of a decision. In this study (Kahneman & Beatty, 1967), participants had to decide whether two tones had the same or a different pitch. The closer together the tones were in pitch, the more the pupil dilated during the decision period and the response.

From this brief survey of early studies in cognitive psychology that used pupillometry, it is already obvious that the pupil response as a physiological signal is indicative of diverse underlying cognitive processes. In a recent review of the literature, Einhäuser (2017) identified nine cognitive factors that influence pupil response, three of which are relevant to the present chapter: cognitive load, violated expectations or surprise, and decision-making. We structure this chapter around these three cognitive factors to familiarize researchers in psycholinguistics with pupillometry and to provide a basis for easier decision-making when designing linguistic experiments that involve these factors.

When describing studies, we always use the terms *pupil response* and *pupil dilation* to signify a *task-evoked* pupil response, that is, the amount of dilation that occurs in response to a certain event that is time-locked to the onset of the event. This is usually achieved by subtracting a baseline value, the pre-trial pupil size prior to any event, from all subsequent measurements in the same trial. This procedure is important because the baseline dilation can differ from trial to trial and from participant to participant based on luminance and variation in participants' general alertness. Task-evoked pupil responses have been shown to be independent of baseline dilation (Reilly et al., 2019).

1 Pupil Response as an Indicator of Surprise or Expectancy Violation

At around the same time as the first pupillometry studies started to appear in the 1960s, researchers also made advances in the field of electroencephalography (EEG) to monitor the cortical response to stimuli, specifically, responses to specific events (i.e., event-related potentials (ERPs)). One influential study by Sutton (1965) found that the response of one component of the signal, the P300 or P3, depended on the probability or the uncertainty of the occurrence of an event. Hakerem, who was a colleague of Sutton, used a similar paradigm but with variation in pupil size as the dependent variable (reported in Hakerem, 1973a) and found a corresponding response pattern of the pupil: the less probable an event appeared to participants, the more their pupils would dilate. In Friedman et al. (1973), these researchers specifically compared the P3 and pupil responses in the same auditory oddball paradigm experiment and found that the amplitude of both responses was related to stimulus probability. In a more recent review, Nieuwenhuis et al. (2011) concluded that the

P3 and pupil dilation are *complementary* responses to "mobilization for action following motivationally significant stimuli," a hypothesis that was corroborated in a later study using the oddball paradigm. In this study, the pupil dilation response and the P3, despite showing a similar pattern, were uncorrelated on a trial-by-trial basis (Kamp & Donchin, 2015)[1]. Yet, it may be that both responses are caused by the release of norepinephrine (NE) in the locus-coeruleus (LC). Thus, researchers approach the study of pupil response as an indicator of surprise from two perspectives: on the one hand, there are those who are interested in the function of the LC-NE system and measure pupil response as a proxy for LC activity. On the other hand, there are those who use the pupil response as a dependent variable to investigate whether certain individuals can detect deviations from a regular pattern, for example, in speech perception. Pupillometry has many advantages over EEG, not least of which is the greater simplicity of data collection and analysis. Yet, the number of pupillometric studies is far outnumbered by ERP studies using some variant of the oddball paradigm.

Whereas in the oddball paradigm, listener expectations are both formed and violated by the structure of the experimental task, our next group of investigations depends critically on participants' prior knowledge and experience of linguistic structure and object reference. In an investigation of meaning integration, Kuipers and Thierry (2011) presented adult native English speakers with pictures of highly familiar objects and animals with a spoken word label while capturing EEG and pupillometric data. Stimulus sequence was varied by block, and pictures were preceded or followed by the spoken word: either the name of the object (match) or an unrelated word (mismatch). The authors observed an N400 of greater magnitude in mismatch trials than in match trials, irrespective of the order of presentation. A parallel effect was observed in participants' pupil dilation, but only when the word preceded the picture. Correlational analyses of pupil size with N400 amplitude further revealed a positive relation between these two measures. Given that the N400 is a negative component, this positive relation indicates that a lower amplitude N400 was associated with greater pupil dilation. This association is consistent with both physiological measures being reflexes of the LC-NE system. Renner and Włodarchak (2017) extended this lexical line of pupillometric inquiry, presenting a picture of objects for 6 s accompanied by matching vs. mismatching auditory words 2.5 s after the picture appeared. By applying functional principal component analysis, among other functional data analytic tools, they found that the overall magnitude of pupil dilation (PC1) differed significantly between the match vs. mismatch trials. Although the second principal component (PC2) clearly indexed the relative latency of peak pupil dilation, it was not found to differ between conditions. While Renner and Włodarchak classify the effect as an index of lexical retrieval, they acknowledge that this process has not yet been disentangled from more general processes of semantic integration.

---

[1] One more recent study using the same paradigm found a correlation between baseline pupil dilation and the P3 on a trial-by-trial basis (LoTemplio et al., 2021).

Fritzsche and Höhle (2015) applied a similar match/mismatch design, assessing 30-month-old native German acquiring children's pupil responses to mispronunciations consisting of a change in place of articulation (e.g., Fisch /fɪʃ/ → [zɪʃ]). In this experiment, spoken words were presented 1 s after the pictures appeared, and the pictures remained on screen for the full 7 s of each trial. Both among the children and a group of adult controls, mean pupil size in response to mispronunciations was significantly larger than in response to correct word forms. Tamási et al. (2019) extended this finding using a similar paradigm, showing that mean pupil size in the window of interest (200–3000 ms) mirrors gradient variation in the degree of featural mismatch (1–3 features) of a mispronunciation: the greater the mismatch, the larger the mean pupil size.

Hochmann and Papeo (2014) focused on still finer a level of granularity, testing whether 3- and 6-month-olds could recognize that consonants produced in different vocalic contexts were indeed of the same category. Although the task they used was an oddball task, its design was such that performance on the task depended on prior linguistic experience, consistent with the other investigations we are considering in this second group. Given that formant transitions for /di/ (rising second formant) and /du/ (falling second formant) are heteromorphic with respect to the single alveolar target of the tongue tip (Liberman & Whalen, 2000; Mattingly & Liberman, 1988), insufficient sensorimotor experience might prevent younger infants from recognizing these onsets as instances of the same consonant. The infants heard a series of four monosyllabic words beginning with /b/ or /d/, the last of which was either consistent with the first three or began with the other consonant (deviant). As expected, the 6-month-olds, but not the 3-month-olds, showed significantly greater pupil dilation when the initial consonant was deviant vs. consistent.

Like Hochmann and Papeo (2014), Kinzuka et al. (2020) used an oddball task that drew both on participants' prior experience with English and, presumably, on sensitization within the experimental session. Kinzuka et al.'s stimuli consisted of /r/-/l/ minimal pairs, such as *pray/play*, and their participants were native speakers of Japanese, a language in which this contrast is not present. Participants' ability to perceive the /r/-/l/ contrast was first determined from their accuracy in a two-alternative forced choice task. Subsequently, participants' pupil responses to the /r/-/l/ lexical oddball task were measured. While the finding for the entire sample was that participants demonstrated a greater pupil dilation in response to deviant than typical stimuli, those participants who were more sensitive to the /r/-/l/ contrast showed a greater difference in pupil dilation between the two conditions with an earlier onset than participants who were less sensitive to the contrast, as determined in the forced choice test. One interesting question for a follow-up study would be whether a difference in pupil response would also be detected if participants are unaware of the purpose of the study.

These lexical and phonological findings are fully consistent with more general cognitive findings of pupillary responses to met and unmet experience-based expectations. Pätzold and Liszkowski (2019) report, for example, that 18-month-old pupil size is greater when a toy unexpectedly disappears after being occluded behind a closed door than when the toy is still visible after the door reopens (object permanence). This is in contrast to a group of 10-month-olds who showed no such systematic variation in pupil dilation, consistent with theories of cognitive development (Piaget, 1977).

While pupillometry appears well suited for eliciting distinctive responses to very fine linguistic mismatches within a short time window, Scheepers et al. (2013) investigated violations of expectation built up over the time course of limericks. Poetry is a domain that is by its nature highly dependent on absolute and partial regularities at various levels of linguistic structure and may well prove a valuable testing ground for pupillometric investigation. Scheepers and colleagues manipulated a series of limericks by introducing a violation of semantics, syntax, rhyme, or meter at the very end of the last line, by which point listeners/readers have developed strong expectations about the remaining structure of the limerick. Stimuli were presented in the auditory modality. Interestingly, only the violation of rhyme yielded significantly greater pupillary dilation than the unmanipulated limerick. Given that each one of their manipulations constituted a violation of listener expectation, the authors attribute pupil dilation in the rhyme violation trials to an emotional response rather than to the detection of an anomaly per se. We also note that the multisentence stimuli likely yield substantial variation in tonic and phasic pupil size over the course of a trial. This may, in turn, reduce the capacity to detect effects even if the analysis window is limited to a small portion of the full duration of the stimulus. As such, shorter poetic sequences may yet reveal effects not attested in longer ones.

Differential pupillary responses to met and unmet expectations, whether these expectations are built while acquiring a first language, learning a second language, or performing an experimental task, depend crucially on the perceivers' capacity to distinguish the expected pattern from an unexpected one. Quirins et al. (2018) presented participants with both local and global regularities and violations in a syllable (/ba/ vs. /di/) oddball task (cf. Hochman & Papeo, 2014). At the local level, regularity vs. violation constituted repetition of the same syllable five times (/ba, ba, ba, ba/) vs. a change of syllable in the final, fifth item of the syllable sequence /ba, ba, ba, di/. Whether the constant or final-syllable-change sequences constituted the frequent (typical) vs. infrequent (deviant) sequence was manipulated between groups. At the global level, regularity vs. violation constituted maintenance of the frequent pattern of five syllables vs. a switch to the infrequent pattern. Among their manipulations, the experimenters included versions of the task with and without instructions. Though participants showed no significant effects of local violation, there was a significant effect of global violation. The overall effect was smaller when no instructions were given than when they were given specific instructions since the instructions (e.g., to count violations) highlighted some of the regularities among the stimuli. Critically, among participants in the no-instruction condition who had recognized and were able to explain the regularity in a follow-up interview, a significant difference between pupillary responses to typical and deviant sequences (at the global level) was present. No such difference was present among participants who did not report the regularity. The authors deduce that the differences in pupil dilation are due to the presence vs. absence of conscious awareness and processing of the regularity. This is consistent with those of Kinzuka (2020). While both of

Kinzuka's participant groups showed awareness of the contrast between /r/ and /l/, their pupillary dilation response varied as a function of their sensitivity to the contrast. Quirins et al. (2018) present this group-wise variation at its logical extreme, comparing participants who explicitly identified a regularity with those who were, for all intents and purposes, unaware of it. As indicated by Quirins et al. (2018), such findings hold promise as the basis for a possible means to noninvasively test comatose patients for cognitive activity.

2 Pupil Response as an Indicator of Cognitive Load

Whereas the effect of violated expectation may seem rather simple and crude in terms of attentional demands, the pupil also dilates as a function of cognitive resources required for solving a task. The first observations of this effect were made more than a century ago, possibly longer. Heinrich (1896) measured the pupil of two of his colleagues under different lighting conditions. His experiment included a condition in which the subjects were asked to solve mathematical problems, originally with the intent of distracting their attention away from visual input (p. 363). Between different conditions, Heinrich observed the largest readings on the ophthalmometer when his subjects were engaged in mental arithmetic. Even earlier, in 1849, other psychologists had observed the so-called *cortical pupil reflex*, a constriction of the pupil when subjects imagined a bright object and a dilation when they imagined a dark object (Behr, 1924). These early findings pointed to a link between cognition and pupil response and have proven quite robust despite the lack of advanced technical equipment and extremely small sample sizes. For example, Mathôt et al. (2017b) found that hearing and reading words that are associated with brightness (e.g., *day*) evoked greater dilation, and words associated with darkness (e.g., *night*) evoked less dilation than neutral control stimuli.

As already mentioned in the introduction to this chapter, *processing load* was a term introduced by Kahneman and Beatty (1966), the idea being that tasks differ in terms of the amount of mental effort required to solve them. Effort, then, is what mediates between task performance on the one hand and the characteristics of the task together with the individual's information-processing capacity on the other hand (Shenhav et al., 2017, p. 100).

Most studies reviewed in this section deal with the processing of spoken and written sentences, a domain in which pupillometry is becoming increasingly popular. The processing capacity recruited to understand the message of a sentence can vary as a function of signal quality (i.e., different types of signal distortions in auditory sentence processing), the content and structure of a sentence (e.g., syntactic complexity, sentence length), or listener variables (e.g., unfamiliarity with an accent; hearing loss; limited proficiency in a language) (Zekveld et al., 2018). While listening effort attributable to signal distortions and hearing loss is a construct that has received attention in many recent studies, this topic is covered in Chap. 8 and will therefore not be discussed here.

2.1 Processing Load During Sentence Processing

Not all pupillometric investigations have augured well for its wider application as a tool for linguistic research. Indeed, one early study was rather pessimistic in its outlook on pupillometry as a tool to study processing load during reading and listening. In four experiments, Carver (1971) had participants read or listen to 150-word long passages of varying difficulty but found that the mean pupil dilation during reading was not larger than in the control conditions, which consisted of either looking at numbers dispersed around the screen to emulate eye movements during reading and looking at the same text with all words turned upside down. Text difficulty did not emerge as a significant predictor of pupil dilation during reading or listening either. However, a limitation of this study is that pupil measurements were only taken at three-time points in each condition and the precision of the measurement was only 0.14 mm (differences between conditions can be much smaller than this) and, taken together, this may explain why no significant effects were found in this study. Despite these methodological weaknesses, Carver made an important observation: pupil size varied systematically as a function of the region of the screen that a participant was fixating on, which changes the angle of the pupil to the camera, and so he made corrections for this potentially confounding, systematic variation.

In a more recent study (Schultheis & Jameson, 2004) with a similar setup, participants also read texts of varying difficulties while their pupil dilation was tracked. Although the investigators used more accurate equipment and continuous measures, the investigators came to the same basic conclusion as Carver (1971), namely, that pupil size is not a good indicator of cognitive load outside of tightly controlled experiments. A caveat when studying language understanding at the paragraph level is finding an appropriate baseline since what one is measuring when looking at the pupil response over longer periods are changes in tonic (sustained) and phasic (transitory) pupil dilation, whereas only phasic changes are of interest when looking at cognitive load (see Chap. 2). As such, it is much more common to investigate cognitive load in reading and speech understanding at the sentence level.

In another early study of sentence processing, Wright and Kahneman (1971) used shorter, single-sentence stimuli. Their study demonstrated that pupil dilation during auditory sentence processing is always an interaction between the stimuli used and the task demands. In advance of the presentation of each auditory stimulus sentence, participants were either told to repeat the sentence aloud or to answer a question given before or after listening to the sentence. In the repeat-aloud condition, pupils tended to dilate more toward the end of the sentence and during a retention interval compared to the other two conditions. Pupils dilated least when the comprehension question was asked before the sentence was presented, presumably because participants were then able to listen for just the relevant part of the sentence and only had to retain that part.

2.1.1 Pupillometry and Reading

One dimension that has often been manipulated in pupillometric investigations of sentence processing is sentence complexity (Schluroff, 1982). Just and Carpenter (1993) compared sentences containing subject-relative with object-relative clauses and, in a second experiment, sentences containing filler-gap dependencies with linearly interpretable sentences. In each case, the mean pupil dilation of participants reading these sentences was higher for the more complex compared to the simpler sentence structure. In addition, implausible sentences were associated with greater pupil dilation than plausible sentences. One potential confound in this study is that the authors did not correct for any potential distortions of the pupil while participants were moving their eyes from left to right. However, mitigating this confound is the fact that vertical eye movements are more susceptible to this distortion than horizontal eye movements as they occur when reading a sentence. All the same, the majority of more recent pupillometric investigations of sentence processing have used auditory stimuli. One exception is Frank and Thompson (2012). Applying a rather sophisticated design, they investigated the relationship between word surprisal and pupil dilation, hypothesizing that words that are less predictable from previous context will evoke greater effort during the reading process. The stimuli were 205 sentences taken from novels likely unknown to the participants, and about half of these sentences were followed by a comprehension question to ensure participants were indeed processing these sentences. The authors obtained a surprising value for each word in the sentence based on an analysis of a large corpus on which they trained a recurrent neural network based on co-occurrences. Having applied a correction procedure to their pupil measurements to account for systematic variation in pupil size attributable to the angle between the eye and the camera, pupil size for each sentence was rescaled as a percentage of the average pupil size over the course of that sentence. Then, pupil size was analyzed in the time window beginning 500 ms before and ending 1000 ms after the first fixation of a word, in 2 ms increments (corresponding to the resolution of the eye tracker). In distinction to most other studies, the predictor of interest was a continuous variable and so the authors could not simply compare two conditions as in a factorial design. Instead, they ran a mixed-effects regression model with the surprisal value and a few control variables, such as the location of the word in the sentence, as predictors of pupil size for each 2 ms time bin, plotting the effect of surprisal over the analysis time window. The effect was first significant 250 ms before the first fixation and remained significant for about 400 ms. Frank and Thompson's (2012) surprisal effect is quite similar to Just and Carpenter's (1993) plausibility effect, the difference being that Frank and Thompson measured plausibility on a continuous scale. In addition, Frank and Thompson were able to demonstrate that the surprisal effect occurred before the first fixation of a word.

2.1.2 Pupillometry and Visual Scenes in Auditory Sentence Processing

Auditory sentence processing has been widely studied using eye tracking in combination with pictorial visual information, also referred to as the visual-world paradigm. The idea here is that as a sentence unfolds, eye movements to pictures related to the auditory information are time-locked to what someone is currently processing or anticipating next. For example, hearing "will eat" triggers more looks to an image of something edible than hearing "will move" (Altmann & Kamide, 1999). When using the visual-world paradigm in combination with pupillometry, on the other hand, the purpose is to investigate the processing effort associated with comprehending a sentence. The experimental task may be to integrate auditory and visual information (responding differently, for example, depending on whether there is a match vs. a mismatch between the two) or to identify an object that, or a person who, is both referenced in the sentence and is present in the visual scene.

Scheepers and Crocker (2004) investigated syntactic priming in speakers of German. German syntax allows OVS word order, but while nominative and accusative cases are distinguished by the form of the article (in some cases also the noun) for masculine nouns, there are no such explicit case differences between nominative and accusative for feminine nouns, rendering such OVS sentences temporarily ambiguous. Since the more frequent SVO word order is the default and thus the expected structure, listeners have to reinterpret the first encountered noun phrase as the object of the sentence (i.e., the garden-path effect). The research question was whether participants would be less likely to expect a subject-first sentence when they were primed with an OVS sentence. They found an effect of word order on pupil dilation, with greater dilation in the OVS relative to the SVO condition, presumably because the reanalysis of OVS sentences required more effort. The effect of prime type was substantial in the gaze data but quite subtle in the pupil data.

Garden-path sentences combined with visual information were also the object of Engelhardt et al.'s (2010) study. In sentences such as "while the woman cleaned the dog that was big and brown stood in the yard," the listener could interpret *the dog* as the object of the verb, which then requires a reanalysis upon encountering the second verb. However, inserting a brief pause between "cleaned" and "the dog" immediately disambiguates the sentence. Engelhardt et al. asked whether the absence of such prosodic information would lead to greater processing effort (Exp. 1) and whether a visual depiction of a sentence would ameliorate the effect of mismatching prosody on processing effort (Exp. 2). In Experiment 1, conflicting prosody indeed seemed to be associated with greater processing effort as the slope of the pupil response was steeper in this condition compared to the matching prosody condition. This effect disappeared in Experiment 2 when the visual information was consistent with the sentence but remained when it was inconsistent (i.e., a woman cleaning a dog).

Wendt et al. (2016) also used visual information but as a way to test comprehension of their sentences. They showed a picture depicting two characters, one being the agent and the other the patient of an action. The picture was only shown prior to hearing the sentence, and after the sentence presentation, participants were asked whether the content of the picture and the sentence matched. Processing effort was manipulated by sentence complexity (canonical SVO vs. less typical OVS word order) and noise level (a signal-to-noise ratio [SNR] of +12 dB vs. –6 dB). As in German, OVS sentences are noncanonical in Danish, the language of the experiment, and, as in Scheepers and Crocker (2004), the sentence was ambiguous as to the role assignment of agent and patient until the verb, which provided enough information for participants to decide whether there was a match between visual and auditory information. There was an effect of noise level in the first part of the sentence (up to the verb) and an effect of sentence complexity in the second part extending over a 3 s retention interval (pupil dilation was averaged for each epoch).

Wagner et al. (2016) employed the classic visual-world paradigm with four distinct pictures to study the effects of degraded speech on lexical access. They manipulated durational cues to word boundaries that normally help listeners to differentiate *bok* from *bokser* (Dutch for goat and boxer) by cross-splicing *bok* onto the second syllable in *bokser* (mismatch condition). Signal quality was manipulated through sinusoid vocoding with eight channels, which reduces spectral information while largely maintaining the speech envelope. Eye-movement patterns suggested increased competition in the mismatch condition compared to the match condition. The pupil data showed increased pupil dilation in the mismatch compared to the match condition for normal speech. Interestingly, in the vocoded speech condition, there was no effect of mismatch condition and the percentage of pupil dilation relative to the baseline was less than in the normal speech condition. This last finding is counterintuitive and may have to do with the fact that this was a between-participants design and so the two degradation conditions may not be directly comparable. Furthermore, the authors analyzed the percent of pupil dilation, which is dependent on the baseline pupil size. With a larger baseline, a larger absolute value can become smaller in terms of percent increase compared to a smaller baseline and a smaller absolute value.

2.1.2.1 Analyzing Pupil Data in Sentence Processing

What stands out when comparing these four studies is that they all used a different method to analyze the pupil data. Scheepers and Crocker (2004) used a repeated-measures ANOVA with a time bin as the repeated measure to test for differences in dilation across time bins and conditions across participants and items. Engelhardt et al. (2010), on the other hand, fitted a regression model to the pupil data of each participant per condition and then used the resulting slope estimates for the *time bin* as the dependent variable in a second analysis to test for differences in slopes between conditions. Wendt et al. (2016) divided each trial into three epochs and analyzed each epoch separately with repeated-measures ANOVAs. Lastly, Wagner et al. (2016) used linear mixed-effects models with polynomial terms to model (nonlinear) differences in pupil dilation between conditions over time. In all four studies, the authors were interested in changes over time and did not use one of the more traditional analyses, mean pupil dilation or amplitude and latency of the peak dilation (i.e., the baseline-corrected maximum dilation in a trial). For a time-series analysis, the method used by Wagner et al. is probably the most appropriate one of the four, however, a big disadvantage is the interpretation of effects, especially when dealing with second- and third-order polynomials necessary to capture the curvature of pupil dilation over time. Vogelzang et al. (2016) introduced yet another method for analyzing pupil data, generalized additive models (GAMs), that has several advantages over the previously discussed methods (van Rij et al., 2019). Their research question was whether the resolution of pronouns during sentence processing is more effortful compared to full noun phrases (in subject position) or reflexive objects (in object position). Participants heard short stories consisting of three clauses that always contained two characters, which were also depicted on the screen. After the sentence, they had to answer a comprehension question by clicking on one of the characters. The window of analysis was always aligned with the subject in the third clause, which was the region of interest. One advantage of GAMs is that the model results can be plotted in such a way that regions of significant differences between conditions can be easily identified. When comparing full noun phrases to a pronoun in the same position they found a divergence in pupil size between the two conditions from about 1 s after the onset of the subject, suggesting that pronouns were more effortful to process. The same was true when comparing object pronouns to reflexive pronouns (which referred to the immediately preceding subject).

2.2 *Cognitive Effort Due to a Suboptimal Signal*

As mentioned in the introduction, cognitive effort during listening can also arise due to a suboptimal signal. Whereas listening under optimal conditions appears effortless, encoding and processing a distorted signal seem to require additional effort. For example, memory for words presented in noise is worse compared to words presented in quiet, even when the words are repeated correctly (Rabbitt, 1966). Also, inducing a cognitive load through a secondary task changes speech perception, presumably because of a reduced capacity for a detailed analysis on a perceptual level (Mattys & Wiget, 2011). Thus, the effects of listening effort, as revealed in pupillometric investigations, are not only interesting for audiologists working with people with hearing impairments but can also inform models of speech perception or language understanding.

One model of language understanding that specifically incorporates listening effort is the Ease of Language Understanding (ELU) model (Rönnberg et al., 2013, 2019). It assumes that mismatches between the acoustic signal and representations of words in long-term memory lead to increased processing effort. This is because additional attentional resources in working memory are required to recreate the intended message, for example, when acoustic information is missing due to a competing signal. Another factor leading to a mismatch can be an accent that the listener is unfamiliar with. Porretta and Tucker (2019) presented Chinese-accented English words of varying intelligibility to listeners with varying familiarity with Chinese accents. The participant's task was to repeat the word. The authors found that better intelligibility and greater familiarity were associated with reduced pupil dilation (McLaughlin & Van Engen, 2020). Borghini and Hazan (2020) investigated the influence of speaking style (a clear speaking style vs. plain speech) and semantic plausibility (implausible sentences were semantically anomalous but grammatically well formed) on the pupil response when sentences were presented in noise. Each sentence contained four keywords, and by means of a staircase procedure, the SNR was adapted for each participant to obtain a mean accuracy rate of 50%. While both a clear speaking style and plausible sentences resulted in greater tolerance for noise (i.e., a lower SNR), only speaking style had an effect on peak and mean pupil dilation. This is contrary to earlier findings on plausibility (see the Sect. 2.1.1) and may have to do with the fact that in Borghini and Hazan, the SNR for plausible sentences was lower than for implausible sentences (Borghini and Hazan also tested nonnative listeners, but these results will be discussed in a separate section in this chapter).

2.3 Cognitive Effort and Speech Production

When it comes to pupillometry and speech production, there are surprisingly few studies. One of the earliest studies comes from Bernick and Oberlander (1968) who found that verbalizing thoughts was associated with greater pupil dilation than simply thinking about something, which suggests that cognitive load during speaking should always be tested against another speaking condition with lower cognitive load. Hyönä et al. (1995) tested interpreters in training who were native speakers of Finnish. Participants heard a paragraph (Exp. 1) or single words (Exp. 2) from either language. In Experiment 2, the task was to (1) simply listen to the words, (2) to shadow the words, or (3) to translate them into the other language. Words were categorized into *easy to translate* or *difficult to translate* depending on whether a direct translation of the word existed or the word had to be translated with a paraphrase. The pupil response was measured from the onset of the word until a response was given. Results showed that listening elicited the smallest pupil response, followed by shadowing, and translating showed the largest response. Also, the pupil response was greater for words that were difficult to translate compared to the easy ones. Lastly, English words were associated with larger pupil responses compared to Finnish words.

Seeber and Kerzel (2012) also investigated simultaneous interpretation, but instead of comparing interpreting to shadowing, they compared the pupil response to two different syntactic structures during simultaneous interpretation. The source language was German, and the target language was English. One challenge for interpreters is that in German subordinate clauses, the verb moves to the end of the clause and so the syntax of the German sentence and the equivalent English translation is asymmetrical. Indeed, the authors found increased dilation for asymmetrical sentences compared to symmetrical ones, and the effect was most pronounced during a 2000 ms pause immediately following each sentence.

The Hyönä et al. (1995) study showed that cognitive effort was evident in the pupil response when participants were planning an utterance: greater dilation was present when the to-be-translated word was more difficult. Papesh and Goldinger (2012) showed that this was also true when participants were preparing to produce a low-frequency word compared to a high-frequency word. The frequency effect was also present during the response phase and even past the response phase but not when participants were silently reading the to-be-uttered word on the screen or during a waiting period between word presentation and response cue. This suggests that the pupil response was indicative of post-lexical-access processes, that is, after lexical access was complete.

The basic results of the Papesh and Goldinger (2012) study were replicated in Lõo et al. (2016) who also asked participants to read words of varying frequency; however, there was no delay in naming. Instead, participants were asked to immediately say the word. Lõo et al. found substantial individual differences in the pupil response pattern, with some participants showing two peaks and others just one. For their analysis, they put participants' data into three different groups based on their individual average response patterns. In two groups, the frequency effect was present from about the time the articulation started although it was less pronounced in one group than the other. In the third group, there was no frequency effect. The authors speculate that participants engaged with the task differently, with some adopting a shallower lexical processing strategy.

Sevilla et al. (2014) investigated speech planning processes in sentence production in Spanish. Participants were asked to describe a picture involving two characters, the agent and the patient of some action. On each trial, they were cued to use one of three syntactic structures: canonical SVO word order, a passive construction, or a so-called clitic left dislocated sentence where the object moves to the front. Differences between the SVO condition and either of the two other conditions emerged even before vocalization and emerged earlier when the cue was given before participants saw the picture, suggesting that they started to plan their utterance even in the absence of lexical information.

3 Pupil Response as an Indicator of Decision Difficulty

Most pupillometry studies in the language sciences are framed within a cognitive effort paradigm. Yet, sometimes it may be more beneficial to view an experiment in terms of a decision-making process, as pupil dilation has been linked to decision uncertainty (Urai et al., 2017). In an early study, Kahneman and Beatty (1967) showed that pupil dilation was related to decision difficulty in a simple pitch discrimination task. They presented participants with a 250 ms standard tone, which was always of the same frequency, and a second tone, which could be the same, higher, or lower in frequency. The closer the two tones were to one another, the more difficult the task was, as indicated by error rates. Pupil dilation also varied as a function of the difference between the tone frequencies, the time at which participants heard the comparison tone and the time at which they were prompted to give their response. Because frequencies higher and lower than the standard tone elicited the same pupil response depending on their distance to the standard tone, it was not the physical properties of the tones that caused the pupil to dilate but rather the difficulty of the decision. But even quite simple decisions can prompt a pupil's response. For example, Simpson and Hale (1969) simply asked participants to move a lever. Half of the participants were instructed to move the lever in a particular direction, and the other half were free to decide between two directions. During a 7-s period preceding the motor event, the pupils of those in the decision group dilated more than those of the instructed group, presumably induced by the decision process. Interestingly, the actual hand movement was accompanied by a significant increase in pupil dilation in both groups, suggesting that executing the motor action induced pupil dilation.

Coming back to language processing, Kuchinke et al. (2007) asked whether lexical frequency and emotional valence effects would be visible in the pupil response during a lexical decision task. Participants saw a string of letters and had to decide whether it was a real German word or a nonword. In the analysis of reaction times (RTs), they were able to replicate previous findings: words with higher emotional valence (whether positive or negative) were processed faster than words of neutral valence, and high-frequency words were processed faster than low-frequency words. In the pupil response, on the other hand, only the lexical frequency effect was present. The authors also conducted a principal component analysis of the pupil data (with data averaged by participant and condition) and extracted the first three principal components. Lexical frequency was associated with variation in the third, latest component that emerged in the time window from 450 to 1500 ms after word onset. This time window corresponds roughly to the average RT, which was around 760 ms, and the investigators therefore speculated that lexical frequency effects were associated with response selection and execution processes (Zellin et al., 2011). This raises the question of whether a distinct pupil response to high- and low-frequency words could still be observed if the word recognition/decision phase and the response phase are separated.

To test the effect of response execution, Haro et al. (2016) compared the regular lexical decision task to a delayed-response lexical decision task. In the delayed-response condition, participants had to wait for 2000 ms following the onset of the word or nonword until they could submit their response. The effect of the delay on RTs was that the frequency effect disappeared, suggesting that participants had decided what response they were going to make before the signal to respond. In the pupil response, the peak latency was longer when the response was delayed and there was a more pronounced difference between high- and low-frequency words. In the analysis of peak amplitude, the frequency effect was only present in the delayed-response condition. Thus, this study confirms that the lexical frequency effect in the pupil response is also present when no immediate response is necessary. In this case, the effect was even more pronounced compared to the immediate-response condition.

The frequency effect in the lexical decision task was again replicated by Geller et al. (2016). They found that peak pupil dilation as well as mean pupil dilation were larger in response to low-frequency compared to high-frequency words. In addition, they investigated the effect of lexical inhibition. To this end, every letter string was preceded by a prime presented for 40 ms. The prime was either an orthographic neighbor of the target word or an unrelated word. When the prime was a neighbor, it was a high-frequency word for low-frequency targets and a low-frequency word for high-frequency targets. The analysis of RTs showed an interaction between prime type and target type. Recognition of low-frequency targets was delayed when primed with a high-frequency prime (i.e., inhibition), but there was no effect of prime type for high-frequency targets. The pupil data, on the other hand, demonstrated a main effect of prime type and no interaction, that is, priming with an orthographic neighbor was always associated with a larger pupil response, regardless of target type.

Based on these studies, it seems fair to conclude that the pupil response during lexical decision is indicative of word recognition processes. However, it is not clear why there is sometimes a divergence between RTs and the pupil response. One possibility is that the pupil response is indicative of later processes, whereas RTs reflect the sum of early and late processes in word recognition. By analyzing effects over the whole course of a trial, as opposed to a peak analysis (van Rij et al., 2019), researchers may be better able to determine when effects of lexical variables such as frequency, length, or neighborhood density become significant, which, in turn, may inform theories of word recognition.

4 Comparison of Paradigms: Overt Response Versus No Response

The Haro et al. (2016) study, described above, showed that the pupil response differed depending on whether participants responded as soon as they made a decision or had to withhold their response until they were cued. The question then is what happens when no response at all is required.

Word recognition without an overt response was investigated by Chapman and Hallowell (2015). These authors simply presented participants with pictures along with their auditory labels. They also inserted foils where there was a mismatch between the picture and its label, but these trials were only used as attention catchers and were not analyzed. The words they presented were categorized as "easy" or "hard," depending on their lexical frequency, familiarity, age of acquisition, and naming latency, based on previously established norms. In doing so, the authors were not interested in the effect of any one of these variables per se, rather, they were interested in group differences between healthy controls and people with aphasia. The analysis of mean pupil dilation during the 3 s trial presentation showed a significant effect of word difficulty and no interaction with the group, suggesting that the effect was present in both groups. Given these results, it seems that a response is not necessary to study language processing with pupillometry. The authors suggest that this passive viewing paradigm is therefore especially suitable to study language processes in populations that would have difficulties giving an overt response (such as people with aphasia). However, one disadvantage of this paradigm is that the researcher cannot confirm whether someone was really engaged in the task. Also, the stimuli differed on several dimensions that are known to influence word recognition, and lexical effects with small effect sizes may not show up in a passive viewing paradigm.

Chapman (2018) specifically compared a no-response paradigm with a response paradigm. Participants heard sentences with canonical and noncanonical word order. In the first block, they just listened to the sentences, and in the second block, they had to answer a yes/no comprehension question after some sentences. The results suggested that processing differences between sentence types (canonical vs. noncanonical word order) only emerged in the second block, which included occasional comprehension questions. This indicates that a task of some kind may be necessary so that participants are more engaged and process the stimuli more than superficially.

5 Pupillometry Across the Lifespan and in Special Populations

While much research on language processing and cognitive effort is aimed at testing a particular theory of language processing or of how the mind works, other research looks at differences between typical and various special, including clinical, populations to understand their particular challenges. As seen in the Chapman and Hallowell (2015) study cited in the previous section, one advantage of pupillometry is that no overt response may be necessary. Another advantage is that pupillometry can provide an objective measure of a construct, such as *listening effort*, a variable that is otherwise hard to quantify. Piquado et al. (2010) tested younger and older adults with sentences of differing complexity (Exp. 2). Similar to Just and Carpenter (1993), the sentences contained a subject relative clause or a more complex object relative clause and consisted of either 9 or 12 words. The participant's task was to recall each sentence after a 2-s retention interval. To account for the fact that older adults' pupils are generally smaller and dilate less, Piquado et al. first normalized each participant's dilation pattern based on minimum and maximum values observed while participants looked at a bright and a dark computer screen, respectively. Then, they fitted linear regression lines to each of the four conditions per participant to calculate the slope of the pupil response over the course of the sentence, similar to Engelhardt et al. (2010). For the younger participants, there was an effect of sentence length with a steeper slope for longer sentences but no effect of complexity. For the older participants, neither effect reached significance. This suggests that linguistic complexity per se may not be the reason why older adults often perceive listening under adverse conditions as more effortful.

Older adults (with mild to moderate hearing loss) were also the participants in Kuchinsky et al.'s (2013) study. Participants heard words presented in noise, and the task was to select the correct response from among four words (a modified version of the visual world paradigm), while SNR (easier and harder, individually adjusted for each participant) and response difficulty (presence or absence of a phonological competitor) were manipulated. In general, the pupil response was sensitive to both task manipulations. The harder level of SNR generally elicited a greater and more prolonged pupil response, while the harder level of SNR in the presence of a competitor seemed to have been the most challenging condition.

Lum et al. (2017) investigated the effect of sentence complexity in typically developing children and children with specific language impairments (mean age ~ 6 years). Easy sentences consisted of SV(−complement) constructions, the verb being intransitive or copular, such as "The ball is yellow," and the hard sentences consisted of SVO constructions, with a ditransitive verb or a prepositional object, such as "The clown sits on the box." The foil pictures depicted some variant of the correct picture, for example, a red ball (instead of a yellow ball). The children heard these sentences, and their task was to decide which of the four pictures matched the content of the sentence. The pupil response in the group of typically developing children was similar to previous studies with adults: longer, more complex sentences elicited a larger pupil response during the listening as well as the response phase. In the group of children with specific language impairment, on the other hand, there was no difference in the pupil response between conditions. Although the results of this study were unexpected in that the processing of more complex sentences did not appear to incur a greater processing cost relative to less complex sentences in children with specific language impairment compared to typically developing children, the study nevertheless demonstrates that pupillometry can be used to assess sentence processing in children.

6 Pupillometry and Bilingual/Second Language Speakers

Several studies have used pupillometry to investigate differences between processing one's first language (L1) compared to a second language (L2). Borghini and Hazan (2018) presented native English speakers and L2 speakers of English with sentences in a low- and a high-noise condition, with individual intelligibility levels adjusted so that participants would understand 40% and 80%, respectively, correctly. They found larger pupil dilation for L2 speakers compared to the native speakers, which was replicated in Borghini and Hazan (2020; see section on sentence processing) with a different task. In addition, in the 2020 study, they found a significant effect of noise. The authors concluded that listening at a lower SNR was more effortful, which is in line with previous research. They also concluded that listening was more effortful for the L2 speakers, but in the absence of a significant interaction, this has to be taken with caution. Group comparisons are in general not straightforward with pupillometry because, although baseline correction provides a degree of normalization, large differences may still be present in the pupil response between individuals and this variation may be unevenly distributed between groups.

A way around the problem of comparing groups is to test the same participants in their L1 and L2, which is what Francis et al. (2018) did. The results are somewhat complex but overall, listening in their L2 did not seem to be more effortful for participants, as demonstrated by pupil dilation. Rather, listening appeared to be most effortful when the masker language and the target language were the same, and the largest pupil dilations were observed when the target language was unpredictable from one trial to the next (i.e., English and Dutch trials were intermixed).

Duñabeitia and Costa (2015) showed that speaking in a foreign language was associated with a greater increase in pupil dilation compared to speaking in one's native language. In addition, they investigated the effect of lying. In their particular task, participants had to describe one of three animals by saying its color and the number of legs. Each animal was repeated 20 times, but in half of the trials, participants were to give the attributes of one of the other two animals instead, thus making a false statement. These false statements were associated with greater pupil dilation compared to the true statements, suggesting that giving a false response was somewhat more effortful. However, there was no interaction between language and statement veracity, suggesting that the effect was the same across languages. This study demonstrates how pupillometry can be used to investigate mental processes during second language speech production, which would not be possible with non-concurrent/offline methods.

Lastly, word recognition in second language and bilingual speakers has been studied with pupillometry. Guash et al. (2016) used a standard lexical decision task to investigate the effect of cognate status on orthographic word recognition. Catalan-Spanish bilinguals responded faster to identical cognates than nonidentical cognates and to nonidentical cognates than to noncognates. The same trend was observable in peak pupil dilations, but only the difference between identical cognates and noncognates reached statistical significance.

Toivo and Scheepers (2019) explored the effect of emotional arousal (examples for low and high arousal are *swamp* and *riot*, respectively) on word recognition in L1 and L2 (in different blocks). On each trial, participants saw a mask of Xs for 500 ms followed by the target word (between 110 ms and 250 ms depending on word length) and another mask of Xs for 1700 ms. Participants then had to decide whether they had recognized the word. This design is different from a standard lexical decision task in that it was untimed. Thus, there was no effect of response execution on the pupil response, as was the case in Haro et al.'s (2016) delayed lexical decision task. Toivo and Scheepers found that when participants were tested in their L1 (English, Finnish, or German), high-arousal words were associated with a higher pupil response than the low-arousal words. However, when tested in their L2 (English), neither the Finnish nor the German native speakers showed an effect of arousal on the pupil response. An interesting second observation in this study was that in the bilingual groups, there was no main effect of language, suggesting that word recognition in L2 was not more effortful than in L1, with the caveat that stimuli were not exactly matched across languages on factors known to influence word recognition such as lexical frequency.

7 Outstanding Issues in Task-Evoked Pupil Responses in Language Research

As this review has shown, there is now a rich literature in language research in which the pupil response is a dependent variable. While the studies reviewed are not an exhaustive list of the extant literature, we have endeavored to provide a broad overview of the kinds of research problems that psycholinguists have tackled and are currently tackling with pupillometry. Most of these studies have been carried out within the past 10 years, yet it is astonishing that much of the groundwork was done more than 50 years ago, at a time when the technological possibilities were a small fraction of what they are today. In the following section, we highlight some issues that should be addressed in future research (for important considerations when setting up an experiment, we refer readers to Mathôt and Vilotijević (2022), Steinhauer et al. (2022), Winn et al. (2018) and Chaps. 14 and 15 in this volume).

7.1 Differences in Task Demands and Participant Characteristics

When comparing the different studies reviewed here, the diversity of tasks that researchers have used is particularly striking. This may not be surprising, given that the selection of studies represents a cross-section of psycholinguistic research in a way. However, it is important to bear in mind that identical stimuli can potentially evoke quite different pupil responses depending on the tasks in which they are embedded. It is therefore important for the researcher to consider the specific task demands when designing an experiment. Let us say we have two sets of sentences and we want to determine whether two groups of participants process the sets of sentences differently. The task could focus on speed (identify a picture that matches a word in the sentence that is predictable to varying degrees), memory (repeat the sentence), comprehension (answer a question about the content of the sentence), or the task could be simply to listen to or to read the sentence, either silently or aloud. These tasks would have quite different attentional demands, and the presence or absence of group differences may well depend on the task.

Of course, besides researcher-intended task demands, individuals also differ in how they approach a task. For example, some participants may be disengaged and process the sentences in a shallow manner, whereas others may be much more engaged. But even if all participants are demonstrably engaged in a task, there appear to be large individual differences in pupil response between participants even with baseline-corrected values, something that is not usually discussed in pupillometric investigations (Lõo et al., 2016). In an attempt to normalize the pupil response, some researchers have addressed the problem of between-participant differences by analyzing pupil dilation in proportion (i.e., percent increase) to a baseline, typically the pupil size prior to stimulus onset. A second practice with regard to normalization is to set the pupil size as a portion of the range of pupil size attested in individual participants. As mentioned earlier, Piquado et al. (2010) opted for the latter procedure in order to account for the fact that the range of variation in pupil size gradually decreases with age. However, using proportional dilation rather than absolute dilation as the dependent variable is not without complications (Mathôt et al., 2017a; Reilly et al., 2019). The best solution, therefore, may be to expose all participants to all experimental conditions, that is, to use a within-participant design in pupillometric investigations as a general plan of attack. This way, the effect of interest can be statistically disentangled on the basis of within-participant variance. Furthermore, when group differences themselves are the focus of a study, the more interesting question may be whether the size of an effect is different between groups rather than whether there are absolute differences in pupil size between groups.

Given the presence of between-subject differences, one question is how much of this variance is *noise*, that is, pupil dilation in response to some variable unrelated to the experimental manipulation, and how much of this variance is systematic. In his influential book *Attention and Effort*, Kahneman (1973) proposed that a perfect physiological measure of mental effort would not only reflect between-task and within-task variation in difficulty but also "differences in the amount of effort that different people invest in a given task" (p. 19). With the third condition, Kahneman seems to suggest that if the pupil response was a perfect measure of effort, the more effort an individual exerts, the more the pupils would dilate. At a group level, this has been shown to be true. In a working memory experiment, participants who received monetary compensation based on performance, something that would presumably increase participants' effort, showed better performance and a larger pupil response (Heitz et al., 2008). In the same study, the investigators also divided the sample into low- and high-working memory span groups and found that in the group that did not receive a monetary incentive, the mean (task-evoked) pupil response of the low-span group was larger than that of the high span group. This suggests that individuals with greater ability, say in working memory, exert less effort on a task requiring this ability compared to those with lower ability (Ahern & Beatty, 1979). These findings seem to contradict a study conducted by van der Meer et al. (2010) who found that individuals with high intelligence had a larger pupil response during a geometric analogies task compared to individuals with average intelligence. In addition to these apparently contradictory findings, a further recent study found no effect of cognitive abilities on the task-evoked pupil response (Aminihajibashi et al., 2020).

Although there is currently no agreement about the relationship between individual differences in cognitive ability and pupil response, individual differences that are not controlled for can potentially introduce confounds into an experiment, given the results of the studies cited above. Therefore, when comparing groups, researchers should at least ensure that groups are comparable on background variables such as level of education and that all participants are engaged in the task to a high level. A possible problem in language research with different populations is that a certain task can be too easy for a typical, native-speaking control group or too hard for an L2 group or a group with a clinical condition. Both of these scenarios could lead to task disengagement.

In research with second language speakers, some studies sought to relate language proficiency to processing effort as measured by the pupil response, but results are mixed. Schmidtke (2014) found that in an English word recognition task, a higher English vocabulary score was associated with a shorter peak dilation latency in native Spanish speakers, but there was no effect of vocabulary on peak amplitude. Francis et al. (2018) found that higher English proficiency was associated with a higher peak amplitude in Dutch speakers when listening to English sentences, but only when the target and the masker language were English and the target language was fixed rather than mixed with Dutch target language trials. Lastly, Borghini and Hazan (2020) found no effect of native Italian listeners' English listening proficiency on mean or peak dilation amplitude during a sentence comprehension task in English.[2] One distinctive characteristic of Francis et al.'s and Borghini and Hazan's investigations is that they both used an adaptive procedure to adjust the SNR to each participant so that performance, as measured by comprehension accuracy, was equated across participants. In both studies, English proficiency predicted SNR, indicating that participants with better English proficiency were able to tolerate more background noise. In future research, it may be advisable to test all participants under the same conditions as is usually done in research on individual differences. This means, for example, that all participants see the same sequence of trials, and, in the case of speech perception in noise, that the SNR is presented at the same level or range of levels for all participants.

Another potential shortcoming in the range of pupillometric studies we have considered is the tendency of researchers to simultaneously test for effects while also testing the method itself. For example, it is not rare to read conclusions such as *We found evidence for hypothesis X and demonstrated that pupillometry is a suitable method to investigate this hypothesis*. However, if there is no evidence for the hypothesis, it could be that the hypothesis is wrong or, potentially, that there were confounds in the design. It would therefore be desirable to have more methodologically oriented studies. For example, Reilly et al. (2019) investigated a certain effect at different screen illumination levels and demonstrated that the effect size was independent of luminance-induced differences in baseline pupil dilation. Another

[2] It should be noted that these three studies are not directly comparable due to differences in the experimental paradigms, the language contexts, and the proficiency tests that were used.

important study would be to expand the Reilly study by investigating changes in luminance during a trial. Although best practice is to have no changes in luminance, this is not always possible when visual information is given. This notwithstanding, it is not currently clear whether or how changes in luminance interact with language-related effects. Thus, it is up to the subjective judgment of researchers and reviewers to decide what is acceptable.

8 Conclusion

In this review, we have summarized and discussed a wide range of studies to give readers a broad overview of the various applications of pupillometry in language research. By grouping these studies according to three main types of cognitive effects associated with pupil response, we hope that this chapter will both inspire new research in each of these paradigms and encourage researchers to consider the results of pupillometric investigations from different angles. For example, the relationship between the pupil response and lexical frequency in a lexical decision task could be interpreted in three different ways: surprise (infrequently encountered words are less expected), cognitive effort (processing low-frequency words is more effortful), or decision difficulty (low-frequency words have a weaker memory trace and therefore participants have less certainty as to whether they constitute a real word). Pupillometric research since 1960, and especially within the past 15 years, has revealed that the complexity of cognitive processes underlying the pupil response goes far beyond the initial somewhat simplistic claims. Yet many unanswered questions remain about the nature and specificity vs. generality of these effects. We look forward to further research on linguistic effects on pupil size and to further investigations of the methodology of pupillometry.

References

Ahern, S., & Beatty, J. (1979). Pupillary responses during information processing vary with scholastic aptitude test scores. *Science, 205*(4412), 1289–1292. https://doi.org/10.1126/science.472746

Altmann, G. T. M., & Kamide, Y. (1999). Incremental interpretation at verbs: Restricting the domain of subsequent reference. *Cognition, 73*(3), 247–264. https://www.ncbi.nlm.nih.gov/pubmed/10585516

Aminihajibashi, S., Hagen, T., Andreassen, O., Laeng, B., & Espeseth, T. (2020). The effects of cognitive abilities and task demands on tonic and phasic pupil sizes. *Biological Psychology*. https://doi.org/10.1016/j.biopsycho.2020.107945

Behr, C. (1924). Die Lehre von den Pupillenbewegungen. In *Die Untersuchungsmethoden*. Springer. https://doi.org/10.1016/s0002-9394(24)90764-1

Bernick, N., & Oberlander, M. (1968). Effect of verbalization and two different modes of experiencing on pupil size. *Perception & Psychophysics, 3*(5A), 327–330.

Borghini, G., & Hazan, V. (2018). Listening effort during sentence processing is increased for non-native listeners: A pupillometry study. *Frontiers in Neuroscience, 12*(152), 1–13. https://doi.org/10.3389/fnins.2018.00152

Borghini, G., & Hazan, V. (2020). Effects of acoustic and semantic cues on listening effort during native and non-native speech perception. *The Journal of the Acoustical Society of America, 147*(6), 3783–3794. https://doi.org/10.1121/10.0001126

Carver, R. P. (1971). Pupil dilation and its relationship to information processing during reading and listening. *Journal of Applied Psychology, 55*(2), 126–134. https://doi.org/10.1037/h0030664

Chapman, L. (2018). *Using pupillometry to index cognitive effort in sentence processing in people with and without aphasia*. Ohio University.

Chapman, L., & Hallowell, B. (2015). A novel pupillometric method for indexing word difficulty in individuals with and without aphasia. *Journal of Speech, Language, and Hearing Research, 58*, 1508–1520. https://doi.org/10.1044/2015

Duñabeitia, J. A., & Costa, A. (2015). Lying in a native and foreign language. *Psychonomic Bulletin & Review, 22*(4), 1124–1129. https://doi.org/10.3758/s13423-014-0781-4

Einhäuser, W. (2017). The pupil as marker of cognitive processes. In Q. Zhao (Ed.), *Computational and cognitive neuroscience of vision* (pp. 141–169). Springer Science. https://doi.org/10.1007/978-981-10-0213-7_7

Engelhardt, P. E., Ferreira, F., & Patsenko, E. G. (2010). Pupillometry reveals processing load during spoken language comprehension. *Quarterly Journal of Experimental Psychology, 63*(4), 639–645. https://doi.org/10.1080/17470210903469864

Francis, A. L., Tigchelaar, L. J., Zhang, R., & Zekveld, A. A. (2018). Effects of second language proficiency and linguistic uncertainty on recognition of speech in native and nonnative competing speech. *Journal of Speech, Language, and Hearing Research, 61*(7), 1815–1830. https://doi.org/10.1044/2018_JSLHR-H-17-0254

Frank, S. L., & Thompson, R. L. (2012). Early effects of word surprisal on pupil size during reading. *Proceedings of the 34th Annual Conference of the Cognitive Science Society* (pp. 1554–1559).

Friedman, D., Hakerem, G., Sutton, S., & Fleiss, J. L. (1973). Effect of stimulus uncertainty on the pupillary dilation response and the vertex evoked potential. *Electroencephalography and Clinical Neurophysiology, 34*(5), 475–484. https://doi.org/10.1016/0013-4694(73)90065-5

Fritzsche, T., & Höhle, B. (2015). 30-Month-Olds and adults measured by pupillometry. *Proceedings of the 18th International Congress of Phonetic Sciences*.

Geller, J., Still, M. L., & Morris, A. L. (2016). Eyes wide open: Pupil size as a proxy for inhibition in the masked-priming paradigm. *Memory & Cognition, 44*(4), 554–564. https://doi.org/10.3758/s13421-015-0577-4

Guasch, M., Ferré, P., & Haro, J. (2016). Pupil dilation is sensitive to the cognate status of words: Further evidence for non-selectivity in bilingual lexical access. *Bilingualism: Language and Cognition, 20*(1), 49–54. https://doi.org/10.1017/S1366728916000651

Hakerem, G. (1973a). Conceptual stimuli, pupillary dilation and evoked cortical potentials: A review of recent advances. In M. Janisse (Ed.), *Pupillary dynamics and behavior* (pp. 135–158). Springer Science. https://doi.org/10.1017/CBO9781107415324.004

Hakerem, G. (1973b). The effect of cognitive manipulation on pupillary diameter and evoked vertex potentials. In E. Dodt & K. E. Schrader (Eds.), *Die normale und die gestörte Pupillenbewegung/Normal and disturbed pupillary movements* (pp. 239–245). J.F. Bergmann-Verlag. https://doi.org/10.1007/978-3-642-80488-5_30

Haro, J., Guasch, M., Vallès, B., & Ferré, P. (2016). Is pupillary response a reliable index of word recognition? Evidence from a delayed lexical decision task. *Behavior Research Methods*. https://doi.org/10.3758/s13428-016-0835-9

Heinrich, W. (1896). Die Aufmerksamkeit und die Funktion der Sinnesorgane. *Zeitschrift Für Psychologie Und Physiologie Der Sinnesorgane, 9*, 342–388.

Heitz, R. P., Schrock, J. C., Payne, T. W., & Engle, R. W. (2008). Effects of incentive on working memory capacity: Behavioral and pupillometric data. *Psychophysiology, 45*(1), 119–129. https://doi.org/10.1111/j.1469-8986.2007.00605.x

Hess, E. H. (1965). Attitude and pupil size. *Scientific American, 212*(4), 46–55. https://doi.org/10.1038/scientificamerican1004-76

Hess, E. H., & Polt, J. M. (1960). Pupil size as related to interest value of visual stimuli. *Science, 132*, 349–350.

Hess, E. H., & Polt, J. M. (1964). Pupil size in relation to mental activity during simple problem-solving. *Science, 143*(3611), 1190–1192.

Hochmann, J.-R., & Papeo, L. (2014). The invariance problem in infancy: A pupillometry study. *Psychological Science, 25*(11), 2038–2046. https://doi.org/10.1177/0956797614547918

Hyönä, J., Tommola, J., & Alaja, A. (1995). Pupil dilation as a measure of processing load in simultaneous interpretation and other language tasks. *The Quarterly Journal of Experimental Psychology Section A: Human Experimental Psychology, 43*(3), 598–612. https://www.tandfonline.com/doi/full/10.1080/14640749508401407

Just, M. A., & Carpenter, P. A. (1993). The intensity dimension of thought: Pupillometric indices of sentence processing. *Canadian Journal of Experimental Psychology, 47*(2), 310–339. https://doi.org/10.1037/h0078820

Kahneman, D. (1973). Attention and effort. *The American Journal of Psychology, 88*(2). Prentice-Hall. https://doi.org/10.2307/1421603

Kahneman, D., & Beatty, J. (1966). Pupil diameter and load on memory. *Science, 154*(3756), 1583–1585. https://doi.org/10.1126/science.154.3756.1583

Kahneman, D., & Beatty, J. (1967). Pupillary responses in a pitch-discrimination task. *Perception & Psychophysics, 2*(3), 101–105. https://doi.org/10.3758/BF03210302

Kamp, S. M., & Donchin, E. (2015). ERP and pupil responses to deviance in an oddball paradigm. *Psychophysiology, 52*(4), 460–471. https://doi.org/10.1111/psyp.12378

Kinzuka, Y., Minami, T., & Nakauchi, S. (2020). Pupil dilation reflects English /l//r/ discrimination ability for Japanese learners of English: A pilot study. *Scientific Reports, 10*(1), 1–9. https://doi.org/10.1038/s41598-020-65020-1

Kuchinke, L., Võ, M. L.-H., Hofmann, M., & Jacobs, A. M. (2007). Pupillary responses during lexical decisions vary with word frequency but not emotional valence. *International Journal of Psychophysiology, 65*(2), 132–140. https://doi.org/10.1016/j.ijpsycho.2007.04.004

Kuchinsky, S. E., Ahlstrom, J. B., Vaden, K. I., Cute, S. L., Humes, L. E., Dubno, J. R., & Eckert, M. A. (2013). Pupil size varies with word listening and response selection difficulty in older adults with hearing loss. *Psychophysiology, 50*(1), 23–34. https://doi.org/10.1111/j.1469-8986.2012.01477.x

Kuipers, J. R., & Thierry, G. (2011). N400 amplitude reduction correlates with an increase in pupil size. *Frontiers in Human Neuroscience, 5*(61). https://doi.org/10.3389/fnhum.2011.00061

Liberman, A. M., & Whalen, D. H. (2000). On the relation of speech to language. *Trends in Cognitive Sciences, 4*(5), 187–196. https://doi.org/10.1016/S1364-6613(00)01471-6

Lõo, K., Hall, A., Tg, A. B., & Baayen, H. (2016). Individual differences in pupil dilation during naming task. *Proceedings of the 38th Annual Conference of the Cognitive Science Society* (pp. 550–555).

LoTemplio, S., Silcox, J., Federmeier, K. D., & Payne, B. R. (2021). Inter- and intra-individual coupling between pupillary, electrophysiological, and behavioral responses in a visual oddball task. *Psychophysiology, 58*(4), 1–14. https://doi.org/10.1111/psyp.13758

Lum, J. A. G., Youssef, G. J., & Clarka, G. M. (2017). Using pupillometry to investigate sentence comprehension in children with and without specific language impairment. *Journal of Speech, Language, and Hearing Research, 60*(6), 1648–1660. https://doi.org/10.1044/2017_JSLHR-L-16-0158

Mathôt, S., & Vilotijević, A. (2022). Methods in cognitive pupillometry: Design, preprocessing, and statistical analysis. *Behavior Research Methods*. https://doi.org/10.3758/s13428-022-01957-7

Mathôt, S., Fabius, J., van Heusden, E., & Van der Stigchel, S. (2017a). *Safe and sensible baseline correction of pupil-size data*. 94–106. https://doi.org/10.7287/peerj.preprints.2725.

Mathôt, S., Grainger, J., & Strijkers, K. (2017b). Pupillary responses to words that convey a sense of brightness or darkness. *Psychological Science, 28*(8), 1116–1124. https://doi.org/10.1177/0956797617702699

Mattingly, I., & Liberman, A. M. (1988). Specialized perceiving systems for speech and other biologically significant sounds. In G. Edelman (Ed.), *Auditory function* (pp. 775–793). Wiley.

Mattys, S. L., & Wiget, L. (2011). Effects of cognitive load on speech recognition. *Journal of Memory and Language, 65*(2), 145–160. https://doi.org/10.1016/j.jml.2011.04.004

McLaughlin, D. J., & Van Engen, K. J. (2020). Task-evoked pupil response for accurately recognized accented speech. *Journal of the Acoustical Society of America, 147*(2), EL151–EL156. https://doi.org/10.1121/10.0000718

Nieuwenhuis, S., de Geus, E., & Aston-Jones, G. (2011). The anatomical and functional relationship between the P3 and autonomic components of the orienting response. *Psychophysiology, 48*(2), 162–175. https://doi.org/10.1016/j.physbeh.2017.03.040

Paivio, A., & Simpson, H. M. (1966). The effect of word abstractness and pleasantness on pupil size during an imagery task. *Psychonomic Science, 5*(2), 55–56. https://doi.org/10.3758/BF03328277

Papesh, M. H., & Goldinger, S. D. (2012). Pupil-BLAH-metry: Cognitive effort in speech planning reflected by pupil dilation. *Attention, Perception & Psychophysics, 74*(4), 754–765. https://doi.org/10.3758/s13414-011-0263-y

Pätzold, W., & Liszkowski, U. (2019). Pupillometry reveals communication-induced object expectations in 12- but not 8-month-old infants. *Developmental Science, 22*(6). https://doi.org/10.1111/desc.12832

Peavler, W. S., & McLaughlin, J. P. (1967). The question of stimulus content and pupil size. *Psychonomic Science, 8*(12), 505–506. https://doi.org/10.3758/BF03331723

Piaget, J. (1977). *The development of thought: Equilibration of cognitive structures* (A. Rosin, Trans.). Viking.

Piquado, T., Isaacowitz, D., & Wingfield, A. (2010). Pupillometry as a measure of cognitive effort in younger and older adults. *Psychophysiology, 47*(3), 560–569. https://doi.org/10.1111/j.1469-8986.2009.00947.x

Porretta, V., & Tucker, B. V. (2019). Eyes wide open: Pupillary response to a foreign accent varying in intelligibility. *Frontiers in Communication, 4*(February), 1–12. https://doi.org/10.3389/fcomm.2019.00008

Quirins, M., Marois, C., Valente, M., Seassau, M., Weiss, N., El Karoui, I., Hochmann, J.-R., & Naccache, L. (2018). Conscious processing of auditory regularities induces a pupil dilation. *Scientific Reports, 8*(1), 14819. https://doi.org/10.1038/s41598-018-33202-7

Rabbitt, P. (1966). Recognition: Memory for words correctly heard in noise. *Psychonomic Science, 6*(8), 383–384. http://link.springer.com/article/10.3758/BF03330948

Reilly, J., Kelly, A., Kim, S. H., Jett, S., & Zuckerman, B. (2019). The human task-evoked pupillary response function is linear: Implications for baseline response scaling in pupillometry. *Behavior Research Methods, 51*(2), 865–878. https://doi.org/10.3758/s13428-018-1134-4

Renner, L. F., & Włodarczak, M. (2017). When a dog is a cat and how it changes your pupil size: Pupil dilation in response to information mismatch. *Proceedings of the Annual Conference of the International Speech Communication Association, INTERSPEECH, 2017-Augus* (pp. 674–678). https://doi.org/10.21437/Interspeech.2017-353

Rönnberg, J., Lunner, T., Zekveld, A. A., Sörqvist, P., Danielsson, H., Lyxell, B., Dahlström, O., Signoret, C., Stenfelt, S., Pichora-Fuller, M. K., & Rudner, M. (2013). The ease of language understanding (ELU) model: Theoretical, empirical, and clinical advances. *Frontiers in Systems Neuroscience, 7*, 1–31. https://doi.org/10.3389/fnsys.2013.00031

Rönnberg, J., Holmer, E., & Rudner, M. (2019). Cognitive hearing science and ease of language understanding. *International Journal of Audiology, 58*(5), 247–261. https://doi.org/10.1080/14992027.2018.1551631

Scheepers, C., & Crocker, M. W. (2004). Constituent order priming from reading to listening: A visual-world study. *The On-Line Study of Sentence Comprehension: Eyetracking, ERPs and Beyond, 44*, 167–184. https://doi.org/10.4324/9780203509050

Scheepers, C., Mohr, S., Fischer, M. H., & Roberts, A. M. (2013). Listening to limericks: A pupillometry investigation of perceivers' expectancy. *PLoS One, 8*(9), e74986. https://doi.org/10.1371/journal.pone.0074986

Schluroff, M. (1982). Pupil responses to grammatical complexity of sentences. *Brain and Language, 17*(1), 133–145. https://doi.org/10.1016/0093-934X(82)90010-4

Schmidtke, J. (2014). Second language experience modulates word retrieval effort in bilinguals: Evidence from pupillometry. *Frontiers in Psychology, 5*(137). https://doi.org/10.3389/fpsyg.2014.00137

Schultheis, H., & Jameson, A. (2004). Assessing cognitive load in adaptive hypermedia systems: Physiological and behavioral methods. In W. Nejdl & P. De Bra (Eds.), *Lecture notes in computer science* (Vol. 3137, pp. 225–234). Springer. https://doi.org/10.1007/978-3-540-27780-4_26

Seeber, K. G., & Kerzel, D. (2012). Cognitive load in simultaneous interpreting: Model meets data. *International Journal of Bilingualism, 16*(2), 228–242. https://doi.org/10.1177/1367006911402982

Sevilla, Y., Maldonado, M., & Shalóm, D. E. (2014). Pupillary dynamics reveal computational cost in sentence planning. *Quarterly Journal of Experimental Psychology, 67*(6), 1041–1052. https://doi.org/10.1080/17470218.2014.911925

Shenhav, A., Musslick, S., Lieder, F., Kool, W., Griffiths, T. L., Cohen, J. D., & Botvinick, M. M. (2017). Toward a rational and mechanistic account of mental effort. *Annual Review of Neuroscience, 40*, 99–124. https://doi.org/10.1146/annurev-neuro-072116

Simpson, H. M., & Hale, S. M. (1969). Pupillary changes during. *Perceptual and Motor Skills, 29*(2), 495–498.

Steinhauer, S. R., Bradley, M. M., Siegle, G. J., Roecklein, K. A., & Dix, A. (2022). Publication guidelines and recommendations for pupillary measurement in psychophysiological studies. *Psychophysiology, 59*(4), 1–36. https://doi.org/10.1111/psyp.14035

Sutton, S., Braren, M., Zubin, J., & John, E. R. (1965). Evoked-potential correlates of stimulus uncertainty. *Science, 150*(3700), 1187–1188. https://doi.org/10.1126/science.150.3700.1187

Tamási, K., McKean, C., Gafos, A., & Höhle, B. (2019). Children's gradient sensitivity to phonological mismatch: Considering the dynamics of looking behavior and pupil dilation. *Journal of Child Language, 46*(1), 1–23. https://doi.org/10.1017/S0305000918000259

Toivo, W., & Scheepers, C. (2019). Pupillary responses to affective words in bilinguals' first versus second language. *PLoS One, 14*(4), 1–20. https://doi.org/10.1371/journal.pone.0210450

Urai, A. E., Braun, A., & Donner, T. H. (2017). Pupil-linked arousal is driven by decision uncertainty and alters serial choice bias. *Nature Communications, 8*(14637). https://doi.org/10.1038/ncomms14637

van der Meer, E., Beyer, R., Horn, J., Foth, M., Bornemann, B., Ries, J., Kramer, J., Warmuth, E., Heekeren, H. R., & Wartenburger, I. (2010). Resource allocation and fluid intelligence: Insights from pupillometry. *Psychophysiology, 47*(1), 158–169. https://doi.org/10.1111/j.1469-8986.2009.00884.x

van Rij, J., Hendriks, P., van Rijn, H., Baayen, R. H., & Wood, S. N. (2019). Analyzing the time course of pupillometric data. *Trends in Hearing, 23*, 1–22. https://doi.org/10.1177/2331216519832483

Vogelzang, M., Hendriks, P., & van Rijn, H. (2016). Pupillary responses reflect ambiguity resolution in pronoun processing. *Language, Cognition and Neuroscience, 31*(7), 876–885. https://doi.org/10.1080/23273798.2016.1155718

Wagner, A. E., Toffanin, P., & Baskent, D. (2016). The timing and effort of lexical access in natural and degraded speech. *Frontiers in Psychology, 7*(398), 1–14. https://doi.org/10.3389/fpsyg.2016.00398

Wendt, D., Dau, T., & Hjortkjaer, J. (2016). Impact of background noise and sentence complexity on processing demands during sentence comprehension. *Frontiers in Psychology, 7*(345), 1–12. https://doi.org/10.3389/fpsyg.2016.00345

Winn, M. B., Wendt, D., Koelewijn, T., & Kuchinsky, S. E. (2018). Best practices and advice for using pupillometry to measure listening effort: An introduction for those who want to get started. *Trends in Hearing, 22*, 1–32. https://doi.org/10.1177/2331216518800869

Wright, P., & Kahneman, D. (1971). Evidence for alternative strategies of sentence retention. *The Quarterly Journal of Experimental Psychology, 23*(2), 197–213. https://doi.org/10.1080/14640747108400240

Zekveld, A. A., Koelewijn, T., & Kramer, S. E. (2018). The pupil dilation response to auditory stimuli: Current state of knowledge. *Trends in Hearing, 22*, 233121651877717. https://doi.org/10.1177/2331216518777174

Zellin, M., Pannekamp, A., Toepel, U., & van der Meer, E. (2011). In the eye of the listener: Pupil dilation elucidates discourse processing. *International Journal of Psychophysiology, 81*(3), 133–141. https://doi.org/10.1016/j.ijpsycho.2011.05.009

Pupillometry Studies of Listening Effort: Implications for Clinical Audiology

Stefanie E. Kuchinsky and Kristina DeRoy Milvae

1 What Brings You to the Audiology Clinic Today?

Patient-centered care in audiology considers the experiences and goals of the patient to be critical for diagnosing and treating a hearing condition. The first step of this process can begin simply by asking the patient: "What brings you in today?" (Grenness et al., 2015). One of the most common answers is that patients have difficulty understanding a spouse, relative, or friend; they mishear words in a conversation or feel exhausted after chatting in a noisy restaurant. Such answers reveal that hearing matters to patients because they value social interactions (Hughes et al., 2018). These responses also suggest that patient problems extend beyond how many words are understood and include the experience of increased listening-related effort and fatigue (Hétu et al., 1988). For example, the consequences of untreated hearing loss include not only poorer speech recognition accuracy (Dubno et al., 1984) but also an increased risk of falls (Viljanen et al., 2009) and declines in cognitive functioning (McCoy et al., 2005).

One possible cause for observed associations between hearing loss and nonauditory-related deficits is that recognizing speech with hearing loss imposes greater demands on the limited pool of mental resources that must be divided among tasks (Kahneman, 1973). Thus, the more resources allocated to understanding speech (i.e., listening effort), the fewer resources that can be devoted to other

S. E. Kuchinsky (✉)
National Military Audiology and Speech Pathology Center, Walter Reed National Military Medical Center, Bethesda, MD, USA

Department of Hearing and Speech Sciences, University of Maryland, College Park, MD, USA
e-mail: stefanie.e.kuchinsky.civ@health.mil

K. DeRoy Milvae
Department of Communicative Disorders and Sciences, University at Buffalo, Buffalo, NY, USA

M. H. Papesh, S. D. Goldinger (eds.), *Modern Pupillometry*,
https://doi.org/10.1007/978-3-031-54896-3_8

functions important for daily life (Lin & Albert, 2014). Particularly problematic for individuals with hearing loss is that communicating in real-world settings commonly involves complex environments that heavily tax listeners' auditory, linguistic, and cognitive functions.

As a visual analogy, Fig. 1 depicts that the ability to understand the final word of the sentence ("bloom") is more challenging both when the signal quality is poor (such as in the presence of noise) and particularly when contextual support is limited. To compensate for a poorer quality signal, greater engagement of linguistic or cognitive processes may be needed to help fill in missing information or revise misinterpretations. As this chapter highlights, the sum total of these interactive demands can give rise to error-prone and/or effortful listening in even normal-hearing, healthy individuals, but particularly for patient populations who may be unable to engage the compensatory mechanisms needed to understand degraded speech (Wingfield & Grossman, 2006).

Audiologists have an array of diagnostic tools to assess the health of the peripheral and central auditory systems, but have fewer options for examining interactions with central linguistic and cognitive systems that are likely to be engaged in real-world listening environments and contribute to effortful listening. There has been increasing recognition that cognitive screenings are important to consider in patient counseling (Shen et al., 2016) and device settings (Souza & Sirow, 2014). Clinical assessments will likely also benefit from regularly including standardized tests of speech-in-noise intelligibility, especially those with more ecologically valid communicative demands (Beechey et al., 2019; Lunner et al., 2020). However, these measures do not quantify the effort that listeners expend to achieve a given level of speech intelligibility.

The goal of the current chapter is to highlight pupillometry research that has aimed to support clinicians' ability to assess and address issues related to listening effort. Rather than presenting an in-depth review of pupillometry studies of listening effort (Van Engen & McLaughlin, 2018; Zekveld et al., 2018), this chapter focuses on how research may inform evidence-based practice and/or be integrated into the clinic with the goal of more comprehensively addressing the problems that patients face in their daily lives. Although our focus is on listening, it should be noted that pupillometry has also been used to examine effort in conditions that would be relevant to vestibular audiologists. For example, even among healthy

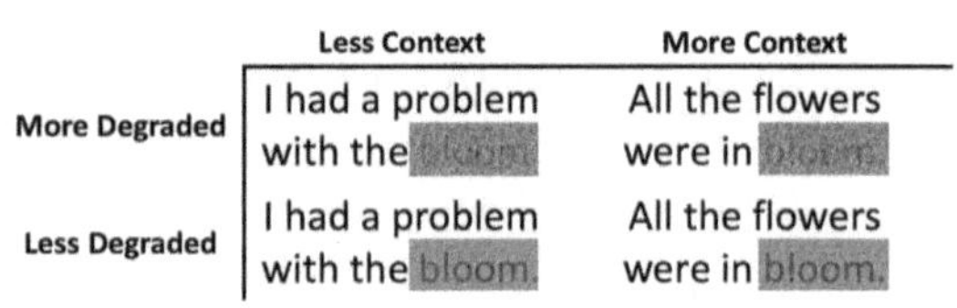

Fig. 1 Demands on auditory, linguistic, and cognitive processes interactively contribute to effortful listening. In this visual analogy, the ease of recognizing the final word of the sentence as "bloom" varies with both the level of signal degradation (such as when listening in noise or with a hearing loss) and the availability of sentence context information

younger adults, increasing postural demands have been observed to lead to a larger pupil response, suggesting that patients with balance disorders may experience even greater cognitive effort (Kahya et al., 2018).

With regards to listening effort, critical challenges for audiologists are answering (1) what is the best way to assess listening effort in a way that is practical and appropriate for the clinic and (2) if effort can be assessed, how should test results inform counseling or intervention plans. This chapter first outlines the foundational work that motivated the use of pupillometry to study listening effort before delving into how this method can be used to begin to address these two challenges, assessments and interventions of effort, in greater detail.

2 Defining and Measuring Listening Effort

2.1 Pupil Dilation as a Measure of Listening Effort

A commonly cited definition of listening effort is given by the Framework for Understanding Effortful Listening (FUEL; Pichora-Fuller et al., 2016). It defines mental effort as "the deliberate allocation of mental resources to overcome obstacles in goal pursuit when carrying out a task" (p. 10S), with listening being one such task. This framework, based on capacity-limited models of attention (Kahneman, 1973), provides a foundation for understanding how mental resources are allocated to speech understanding in a goal-directed fashion. Factors that influence the allocation of effort include the demands of the task, the listener's individual resource capacity, and his/her level of motivation.

Other chapters of this book provide a comprehensive overview of the measure of pupillometry and how it can be used to track moment-by-moment changes in mental effort. Furthermore, previous reviews have highlighted design considerations and major findings from pupillometry studies of speech and language processing (Winn et al., 2018; Schmidtke, 2018). This section provides a brief review of the literature that has motivated the use of pupillometry to study listening effort, particularly as it applies to the field of clinical audiology.

In short, pupillometry has been associated with the functioning of the locus coeruleus-norepinephrine (LC-NE) system. LC-NE activity modulates states of attention (Aston-Jones & Cohen, 2005; Joshi et al., 2016), in part, via its connections within regions of the cingulo-opercular cortical network, which supports performance monitoring and adjustment, including in challenging listening tasks for individuals with and without hearing loss (Vaden et al., 2016). Critically, pupillometry tracks changes in both modes of LC activity: baseline arousal-related activity (tonic firing pattern) has been observed to correlate with average dilation (Aston-Jones & Cohen, 2005), while task-driven bursts of activity (phasic firing pattern) have been associated with the task-evoked pupil response (Joshi et al., 2016). Successful task performance is associated with a balance between these two states:

focused attention and optimal task performance occur at moderate levels of arousal but high levels of task engagement. As will be discussed, the ability to track both modes of activity may be critical for understanding the mechanisms that underlie the deployment of effort during listening (Gilzenrat et al., 2010).

Though pupillometry has been used to measure mental processing for over half a century (e.g., Hess & Polt, 1964), Kramer et al. (1997) were the first to apply pupillometry to study changes in effort associated with listening to speech in noise, particularly in individuals with hearing loss. Participants with and without hearing loss repeated sentences in fluctuating noise at varying signal-to-noise ratios (SNRs). These SNRs were either +10, +5, or the same SNR relative to each participant's speech-reception threshold (SRT; i.e., the SNR associated with 50% intelligibility, determined in a separate task). They observed that, in general, pupil size decreased in better SNRs relative to one's SRT, indicative of reduced listening effort. However, individuals with hearing loss benefited less (i.e., exhibited less reduction in pupil size) from better SNRs compared to normal-hearing individuals, despite accounting for individuals' own SRTs. Furthermore, they found a significant association between self-reported ratings of perceived hearing handicap and larger pupil size when listening to speech. These results highlight the interactive role of acoustic demands and hearing loss on listening effort, highlighting what would be later formalized in FUEL (Pichora-Fuller et al., 2016), that both stimulus-related and listener-related factors impact effortful listening.

2.2 *Considerations for Using Pupillometry Measures of Listening Effort to Inform Clinical Audiology*

Numerous subjective, behavioral, neural, and physiological measures have been proposed to index listening effort. Investigations that have employed these measures and their relative advantages and disadvantages have been discussed in previous reviews (e.g., McGarrigle et al., 2014; Francis & Oliver, 2018; Strand et al., 2018). This chapter focuses on the relative strengths and weaknesses of pupillometry compared to the most commonly used measures of listening effort in studies that inform the practice of clinical audiology.

2.2.1 Pupillometry Versus Subjective Measures

Surveys have been developed with the goal of systematically probing effortful listening in patient populations (e.g., Gatehouse & Noble, 2004; Alhanbali et al., 2017; Krueger et al., 2017; Hughes et al., 2018). Although clinicians are motivated to address the effort-related complaints of their patients, these measures are rarely implemented in clinics due to time constraints as well as limited evidence of how survey results should guide counseling and treatment plans. Instead, clinicians may simply ask patients how effortful a test was or observe that a patient was slow to respond or otherwise struggled to repeat back words.

Studies that have compared subjective and objective measures of effort, including pupillometry, have shown little correspondence, despite high within-test reliability (Strand et al., 2018; Alhanbali et al., 2019). Some models of listening effort explain this disconnect in part by distinguishing between listening engagement and effort. While the former describes the recruitment of mental resources to support a listener's ability to perform a valued task, effort is indicative of one's subjective experience, including enjoyment, frustration, and boredom, which results from engaging in a listening task (Herrmann & Johnsrude, 2020).

The lack of strong correlations may also not be surprising given that subjective reports are limited by an individual's ability to accurately remember and reflect on their effort (Scheiter et al., 2020) and appear to be more tightly related to an individual's own perceived performance, rather than effort (Moore & Picou, 2018). Stronger associations, however, have been observed between subjective effort and later epochs of the pupil response, as participants were holding responses in memory before responding (Russo et al., 2020). This suggests that the time course information provided by pupillometry may be important for elucidating the complex relationship between subjective and objective measures of effort.

To avoid subjective biases, laboratory-based researchers may simply opt to focus on objective measures of effort. However, clinicians do not have this luxury. How should audiologists proceed if a patient reports that listening is draining, but no diagnostic test reveals a deficit? Problems with speech recognition, particularly in noise (Pienkowski, 2017), in the absence of deficits on standard audiograms or tests of speech recognition in quiet are common in a variety of hearing-related disorders (Tremblay et al., 2015), such as central auditory processing disorder (CAPD) and auditory neuropathy spectrum disorder (ANSD). The ability to link subjective and pupillometry indicators of effort could be validating for these patients. However, to the extent that these measures diverge, pupillometry may help to quantify deficits that are otherwise unreported. For example, new cochlear implant (CI) users often experience such large improvements in speech recognition following implantation that they may be less likely to report high listening effort.

Another question for audiologists is what should be done if, following an intervention such as hearing aids, patients do not "feel" that listening is any easier. Could pupillometry provide evidence that would convincingly demonstrate that they are, in fact, benefiting? Conversely, might pupillometry help validate patients' reports that they like an intervention, even if standard measures show little or no benefit? For example, in individuals with CAPD or other auditory processing deficits, there has been increasing use of hearing aids that provide only a low amount of acoustic gain compared to traditional hearing aids (Gallun et al., 2012), with these patients reporting benefit from their use (Roup et al., 2018). It is currently unknown to what extent patient satisfaction with low-gain hearing aids is related to a reduction in listening effort.

2.2.2 Pupillometry Versus Other Objective Measures of Listening Effort

2.2.2.1 Behavioral Measures

Behavioral measures of effort include the observation of poorer performance or slower reaction times under high listening demands (McGarrigle et al., 2014). In particular, the dual-task paradigm has been commonly used as a behavioral measure of listening effort. This method operationalizes effort in terms of decrements in the performance on a secondary task (e.g., visual probe identification) as a concurrent, primary listening task becomes increasingly challenging. Though easy to implement, a number of design considerations are thought to impact dual-task measures' sensitivity to listening effort, potentially leading to the inconsistent results observed in the hearing literature (Gagné et al., 2017) as well as a lack of correspondence to pupillary measures (Strand et al., 2018).

Pupillometry offers several advantages over behavioral measures of effort, including that a single pupil trace can provide finer-grained temporal information about how effort and arousal change within and across experimental trials. For example, pupillometry can reveal how effort unfolds across an entire listening event: throughout pre-stimulus baseline, listening, retention, response, and recovery epochs. Additionally, the same recordings can be analyzed to examine changes in arousal/alertness, task engagement/effort, and fatigue across an entire experiment.

Pupillometry design considerations that are more complex than those of behavioral measures include issues related to luminance and reductions in pupillary dynamic range with age (for a review see Winn et al., 2018). The need for specialized eye tracking equipment and analytical expertise has largely limited pupillometry to laboratory-based studies. However, given that vestibular audiologists use eye tracking as part of videonystagmography and rotary chair assessments of balance disorders, the gap between lab and clinical-based measures may be smaller than one might initially think.

2.2.2.2 Pupillometry Versus Neural Measures

Pupillometry is generally more practical to implement (e.g., less costly, fewer contraindications) than most neural measures of effort. However, its relatively lower temporal and spatial resolution results in pupillometry providing more limited mechanistic insight into the systems that underlie effortful listening. To close this gap, a number of studies have sought to link pupillometry to neural measures of listening effort.

Electroencephalography (EEG) studies have found a correlation between pupil size and the P3 event-related potential reflecting focused attention in an auditory oddball task (Murphy et al., 2011). Associations between spectral EEG components, such as alpha power, and pupillary measures during listening have been less consistent and may depend on the specifics of the task demands (Miles et al., 2017). A study that directly compared alpha-power EEG, pupillary, skin conductance, and self-reported listening effort observed variable relationships among these measures (Alhanbali et al., 2019). Similarly, Kramer et al. (2016) examined three physiological indicators of cognitive load (peak pupil dilation, salivary cortisol, and chromogranin A), which were all differentially sensitive to speech recognition demands or hearing loss.

In a simultaneous functional magnetic resonance imaging (fMRI) and pupillometry study, Zekveld et al. (2014a) observed that increasing pupil dilation while listening to speech in noise was correlated with activity in regions related to auditory processing (i.e., bilateral superior temporal cortex) as well as a region within the cingulo-opercular performance-monitoring network (i.e., the left anterior insula). However, this relationship was not sensitive to varying listening demands. Thus, additional research is needed to understand the conditions in which pupillometry reflects the neural systems that underlie effortful listening (Peelle, 2018). Greater insight into the mechanisms that drive patient complaints related to speech recognition in adverse conditions could facilitate the validation of clinical assessments of effort as well as more targeted remediations.

2.3 *Defining and Measuring Listening Effort: Interim Summary*

The application of pupillometry to the study of listening effort, particularly in conditions and patients relevant for audiologists, encompasses a relatively short time in the over 50-year history of using pupil dilation to index mental effort. This section discusses a number of methodological considerations relevant to the design of studies that aim to inform or be incorporated into the practice of clinical audiology. In sum, pupillometry provides bias-free information about how effort unfolds over time (both within or across trials) to a greater degree than subjective or behavioral measures. Though less costly than neural measures, pupillometry alone may not provide as clear insight into the underlying systems that contribute to effortful listening. Laboratory studies that collect and aim to link multiple measures simultaneously may thus provide key insights into the extent to which different purported measures of listening effort explain unique or overlapping variance in the study of listening effort.

However, given the noted disconnect among measures of effort in previous studies, the remainder of this chapter asks what then is the role for pupillometry in informing clinical practice. To foreshadow, the current literature suggests that pupillometry can help (1) quantify the challenges associated with speech understanding, particularly those that extend beyond currently available measures of hearing or speech intelligibility and (2) guide the development of interventions that not only improve audibility and intelligibility but also optimize listening effort.

3 Applications to Assessment

Studies of listening effort have examined how varying listening demands impact the pupillary response, particularly across different groups or individuals, such as those with varying degrees of hearing loss, controlling for other variables, like age (Zekveld et al., 2018). However, the reliability and validity of pupillary measures of listening effort at the individual level, which is critical for diagnostics and assessing longitudinal changes in clinical settings, warrants further investigation (for discussions, see Winn et al., 2018; Brungart et al., 2022). Initial studies addressing this issue have been promising. Good within-individual reliability has been observed in at least some portions of the pupillary response in normal and hearing-impaired individuals (e.g., rise and fall, mean dilation; Neagu et al., 2019). Giuliani et al. (2020) reported that only pupillometry measures of effort in varying listening conditions were consistent across different test sessions, unlike subjective ratings, dual-task reaction times, or skin conductance measures. Furthermore, Wagner et al. (2019) found that pupillometry is sensitive to the variability in speech processing that would be expected of clinical patients.

Researchers have begun to explore the benefits of integrating pupillometry into clinical assessments. Bala et al. (2020) have explored whether tone-evoked pupillary responses could be used in lieu of a manual response in hearing assessments, a core test in the audiology clinic. They examined an early portion of the pupil response to quantify pupil dilation associated with early, automatic orienting to sound, rather than a later, effortful decision-making response. Hearing thresholds obtained from the pupil measure were highly comparable to those of a standard hearing test. Thus, pupillometry could directly benefit clinical practice by providing an index that can stand in place of the audiogram for patients who have difficulty making manual responses. Future studies may benefit from examining the full time-course of the pupil response during a standard audiogram. Within a single trial, pupillometry may allow for the examination of the difficulty associated with both detecting and responding to challenging auditory stimuli (Kuchinsky et al., 2013), particularly as the intensity levels of the tones approach the listener's sensation level threshold.

3.1 Beyond Clinical Measures of Intelligibility

For objective measures of listening effort to be of value to audiologists, they need to provide information about communication beyond the performance measures already at their disposal. Effort has indeed been shown to diverge from speech recognition accuracy in ways that could prove useful for clinical assessments. Pupillometry studies have shown that individuals engage their maximum level of effort when a listening task is moderately difficult – not too hard or too easy. While performance tends to decrease along a psychometric function with more challenging listening demands, a nonlinear function characterizes the relationship between listening demands and effort. Effort, as measured via the pupil response, tends to asymptote (Zekveld & Kramer, 2014) or decrease as task demands become overwhelmingly difficult and listeners begin to give up (Ohlenforst et al., 2017b; Ayasse & Wingfield, 2018). The dissociation between effort and speech intelligibility is schematized in Fig. 2.

This nonlinear relationship can help explain why two patients might achieve similar scores on a speech recognition test yet exert substantially different levels of effort to do so. Studies have reported that differences in listening effort can be observed even when individuals have normal hearing (Zekveld et al., 2010) or when speech recognition performance is near ceiling or matched across different conditions or individuals (Zekveld et al., 2010; Kuchinsky et al., 2013). As discussed in Sect. 4, understanding the complex relationship between task demands, intelligibility, and effort will be critical for making a priori predictions about the benefits one would expect with a given intervention.

The results of Winn and Teece (2021b) suggest that dissociations between speech intelligibility scores and pupillary measures of effort depend on the nature of the recognition errors made. Among other factors, they observed that the pupil response was more sensitive to how sensical a sentence recognition response was, regardless of whether it was objectively correct. That is, errors are not necessarily associated with effortful listening if they make sense in the context of the rest of the sentence, and thus, the listener has likely not perceived that an error was made. Their results align with the described neuroimaging results that indicate the involvement of domain-general performance monitoring processes in the generation of a pupillary response, namely the cingulo-opercular network (Zekveld et al. 2014a).

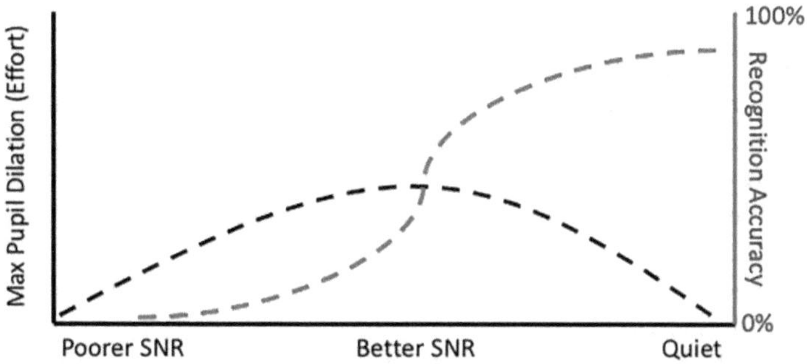

Fig. 2 Schematic showing the divergence between the functions for recognition accuracy and listening effort across varied listening conditions. While task demands and performance are generally related along a psychometric function (gray dashed line), pupil size tends to be maximal at moderate levels of task difficulty (e.g., near 50% intelligibility; black dashed line). Thus, relatively lower effort may be observed in very easy and very difficult conditions, indicative of the need to interpret pupillometry measures of effort in the context of overall task difficulty

3.2 Factors That Contribute to Effortful Listening

FUEL (Pichora-Fuller et al., 2016) highlighted that effort is driven by an interaction between the demands imposed by the listening task, the resource capacity available to the listener, and their motivation or alertness. Patients reporting effortful speech recognition may be experiencing difficulties along one or more of these dimensions. Comprehensive clinical assessments may need to consider all of these factors in identifying the source of a patient's problems to select the most appropriate intervention. For example, someone experiencing high levels of listening effort due to low cognitive capacities would likely need a different treatment plan than a patient whose effort stems from low motivation.

Previous reviews have examined the factors driving effort laid out by FUEL in adults (Van Engen & McLaughlin, 2018; Zekveld et al., 2018) and in children (Gómez-Merino et al., 2020). This chapter cites representative studies reviewed in those papers as well as more recent publications.

3.2.1 Stimulus-Related Factors: Stimulus Quality and Task Demands

Speech comprehension involves multiple hierarchical and interactive processes, including auditory, linguistic, and cognitive processing (Mattys et al., 2012). Challenges can arise at each level. Pupillometry studies have provided evidence that increased demands on these functions contribute to more effortful listening, until demands become exceedingly challenging and people begin to give up (Ohlenforst et al., 2017b; Ayasse & Wingfield, 2018). In many cases, these effects have been dissociable from intelligibility or other task performance metrics.

Pupil dilation has been observed to be generally larger, slower to peak, and/or more sustained with increasing acoustic demands, such as greater stimulus degradation (Winn et al., 2015; Miles et al., 2017) or background noise (Zekveld et al., 2010; Kuchinsky et al., 2013), especially when noise contains intelligible speech (Zekveld & Kramer, 2014; Koelewijn et al. 2014b; Ohlenforst et al., 2017b) or is a same-gender competing talker (Zekveld et al. 2014b). While speech rate has not yet been found to influence the pupil response to stimulus presentation (Koch & Janse, 2016; Winn & Teece, 2021a), there is evidence that faster speech rates are associated with longer recovery times of the response to baseline, indicative of more sustained effortful processing (Winn & Teece, 2021a).

Studies have also generally found larger, slower, and/or more sustained pupil responses with increasing linguistic demands, including listening to foreign-accented speech (Schmidtke, 2014), syntactically complex sentences (Ayasse & Wingfield, 2018), or sentences with low semantic context (Winn, 2016). Larger, slower, and/or more sustained responses have also been observed with increasing cognitive demands, such as memory load (Piquado et al., 2010) and attentional demands (Koelewijn et al. 2014a; McCloy et al., 2018).

Auditory stimuli with an emotional valence (Partala & Surakka, 2003), particularly aversive ones (Babiker et al., 2015), have been shown to yield larger pupil responses. This may be relevant to patient assessments, as clinicians report that some individuals exert substantial effort even for simple pure tone detection during an audiogram, especially when there is an emotional component to their visit to the clinic, such as with the realization that their performance is poorer than they expected or desired.

Interactions among these stimulus-related factors may be important to consider when trying to understand patient functioning in daily life, which often involves communicating in acoustically, visually, and cognitively complex environments. Studies examining the impact of multiple task-demands on listening effort have reported even larger pupil responses than would be predicted for each demand alone. Interactions have been reported between acoustic and linguistic demands at the level of lexical (Kuchinsky et al., 2013; Wagner et al., 2016), syntactic (Wendt et al., 2016), and semantic processing (Winn, 2016). Especially large, slow, and/or sustained pupil responses have also been found with high levels of acoustic and memory demands (Piquado et al., 2010; Otero et al., 2011; Zekveld et al., 2019). However, such interactions have a greater potential to overload individuals' processing capacities. Indeed, Zekveld et al. (2019) observed that when memory load demands of a task began to exceed listeners' memory capacities, typical effects of SNR on the pupil response were obscured.

3.2.2 Listener-Related Factors: Individual and Group Differences

Patients vary greatly in their auditory, linguistic, and cognitive abilities, as well as in their personalities, all of which can influence their ability to effectively allocate effort to speech recognition. It can be complicated, or even impossible, to account for these factors in diagnostic testing, such as when a patient is a nonnative English speaker or exhibits signs of mild cognitive impairment (Chermak et al., 2017). Pupillometry could provide a means to quantify how individual differences impact the extent to which listeners experience effortful listening in a variety of conditions.

Group and individual differences in baseline pupil dilation and task-evoked pupillary responses to challenging auditory stimuli have been investigated in populations relevant to audiology with somewhat mixed results. Some studies have observed an association between smaller task-evoked pupil dilation and poorer auditory processing, indicating that these individuals were less able to exert the effort required in adverse listening conditions. These investigations have included individuals with hearing loss (Zekveld et al., 2011; Kuchinsky et al., 2014; Kramer et al., 2016; Ohlenforst et al., 2017b) or tinnitus (Juul Jensen et al., 2018). However, other studies have noted the opposite relationship, with hearing loss linked to a larger task-evoked pupil response, at least in some conditions (Winn, 2016; Ayasse et al., 2017), or even no relationship (Koelewijn et al., 2017). Larger pre-stimulus baseline pupil diameter has been recorded in individuals with poorer hearing thresholds, suggesting they experience greater arousal as they anticipate the upcoming difficult listening task (Ayasse & Wingfield, 2020).

Pupil dilation has also been investigated with variation in linguistic experience. Schmidtke (2014) found that bilingual English-Spanish speakers, particularly those with lower English proficiency, exhibited a slower pupil response when listening to English words compared to monolingual speakers. Compared to native English speakers, nonnative speakers listening to English sentences in background noise have been reported to exhibit a larger pupil response indicative of greater listening effort (Borghini & Hazan, 2018).

The impact of variation in cognitive abilities on pupillary measures of listening effort has been examined in a number of domains, including short-term memory (Wendt et al., 2016), working memory (Zekveld et al., 2019), and vocabulary knowledge (Kuchinsky et al., 2014). Group-difference analyses have also been conducted in populations that may have, on average, reduced cognitive capacities compared to younger, normal-hearing, healthy adults. Again, the results have been somewhat mixed. Some studies of challenging listening have observed larger task-evoked pupil responses with aging (Piquado et al., 2010), others a smaller response (Zekveld et al., 2011), and others no difference (Ayasse et al., 2017). In a study of listening effort in children, larger responses occurred in older versus younger children (Steel et al., 2015). Although there has been interest in evaluating listening effort in individuals with injuries or disorders that impact central systems and communication, limited work has been done in these populations. One study has been conducted in individuals with traumatic brain injury who had deficits recognizing speech in noise despite normal audiograms. The study showed no difference in pupil dilation in these individuals compared to controls overall, but participants with both a TBI and poor SRTs exhibited relatively smaller task-evoked pupil dilations (Koelewijn et al. 2018a).

The influence of personality factors on listening effort has been examined to a lesser extent. Antikainen and Niemi (1983) reported a stronger relationship between stimulus loudness and larger pupil size for more neurotic individuals. However, no relationship has been observed between listening effort and extraversion or sensitivity personality measures (Ohlenforst et al., 2017b; Strand et al., 2018).

As noted, the evidence for the sensitivity of pupil dilation to individual and group differences has been mixed (Zekveld et al. 2018). Inconsistencies may arise in part because individuals cannot be randomly assigned to these factors of interest. Additionally, some special populations can be hard to recruit into research studies, limiting sample sizes. In those cases, it may be difficult to balance or control sources of variation that differ across individuals (e.g., outcomes following cochlear implantation; Pisoni et al., 2017) or within individuals (e.g., fluctuations in tinnitus symptom severity; Dauman et al., 2015).

A larger issue when studying group or individual differences with pupillometry is that predicting whether individuals with better auditory or working memory skills, for example, will exert more or less effort during listening is not straightforward. One might predict individuals with larger capacities should be able to more efficiently allocate cognitive resources to listening, thereby experiencing less effort and reduced pupil responses (Ohlenforst et al., 2017b; Strand et al., 2018). Alternatively, individuals with larger capacities may have the ability to apply greater cognitive resources to listening, especially in more challenging conditions. These individuals thus should exert relatively more effort than those with smaller capacities and exhibit increased pupil responses (Zekveld & Kramer, 2014; Kuchinsky et al., 2014). A takeaway from this mixed evidence is that the relationship between individual/group differences and listening effort may depend on the nature of the capacity of interest (e.g., working memory, vocabulary knowledge), the details of the stimuli or task design (e.g., speech in noise versus a competing speaker), and how challenging the task is (Zekveld et al., 2011; for further discussion see Strand et al., 2018).

3.2.3 Within-Individual Factors: Fluctuations in Fatigue and Motivation

Having sufficient processing capacity is only part of the equation in predicting whether an individual may engage in effortful listening (Pichora-Fuller et al., 2016; Peelle, 2018). High fatigue or low motivation may limit one's ability or willingness to allocate mental resources to communication. Importantly, these factors may vary systematically in the groups of patients that seek care in audiology clinics.

Fatigue has been defined in terms of weariness, tiredness, lack of energy, or decreased motivation to perform a task (reviewed in Hornsby et al., 2016). Objective measures of listening-related fatigue have included declines in performance with time on task, and because of its correlation with arousal-related tonic LC activity, reductions in baseline pupil dilation (Alhanbali et al., 2020). Smaller task-evoked pupillary responses have also been used to index a decrease in listening task engagement (Kuchinsky et al., 2016; Wang et al., 2018), suggestive of increasing levels of fatigue throughout a challenging listening task (McGarrigle et al., 2017).

Fatigue is a common complaint among individuals with hearing loss and may lead to social isolation, among other negative consequences (Hétu et al., 1988; Davis et al., 2020). High levels of listening effort have been thought to contribute to the end-of-the-day fatigue that individuals report. Though a causal relationship between task-related effort and general fatigue seems intuitive, there is little evidence to support a direct link between spending more energy on listening and having less available for the rest of daily life activities (Alhanbali et al., 2017). However, Wang et al. (2018) observed that self-reported fatigue was associated with a smaller pupil response when listening to speech in noise, beyond the variance explained by hearing acuity.

There is some evidence to support a link between listening-related effort and fatigue on a shorter time scale, namely fatigue throughout the duration of an experiment. McGarrigle et al. (2021) collected pupillometry during a listening task that varied in its acoustic and attentional demands. They found a significant relationship between the task-evoked pupil response and periodically collected ratings of "tiredness from listening" (i.e., subjective listening fatigue), but no such relationship between dilation and ratings of effort. The association between pupillary measures of effort and fatigue was specific to the effort experienced during listening, as they did not observe a relationship between baseline pupil dilation and fatigue or effort.

There may be many reasons for the inconsistent correlations between listening effort and fatigue. Hornsby et al. (2016) noted that there are multiple potential sources of fatigue (e.g., physical, mental, emotional) that emerge across multiple time scales (within trial, across experiment, end of day, chronic fatigue). Furthermore, motivation appears to be a key modulator of the relationship between effort and fatigue (Inzlicht & Schmeichel, 2012). Like McGarrigle et al. (2017), decrements in task-evoked pupil responses, but not baseline dilation, have been observed after performing a two-hour cognitively demanding task (Hopstaken et al., 2015). However, when presented with a sufficiently motivating reward at the end of the task, levels of effort, as indexed by pupil dilation, were restored. Of course, there are individuals for whom no amount of motivation would be sufficient to overcome their fatigue (e.g., individuals with chronic fatigue syndrome).

Previous reviews and empirical papers have described motivation, or a willingness to recruit mental resources, as playing a critical role in the allocation of effort to listening (Richter, 2016; Strauss & Francis, 2017; Peelle, 2018). Neuroeconomic models of mental effort, in particular, note that individuals may not engage in an effortful task if the cost of allocating effort outweighs foreseeable benefits (Botvinick & Braver, 2015; Shenhav et al., 2017), including for speech recognition tasks (Eckert et al., 2016). The Model of Listening Engagement (MoLE; Herrmann & Johnsrude, 2020) further stresses that the decision to engage in effortful listening is determined by both the perceived intelligibility of the speech and the listener's experience (e.g., boredom, enjoyment, frustration).

Researchers have begun to explore factors that might help individuals with low motivation engage in listening in adverse listening conditions. Koelewijin et al. (2018b) used pupillometry to investigate the extent to which an extrinsic reward could facilitate engagement in a speech-in-noise task. While the amount of monetary reward was unrelated to recognition performance, peak pupil dilation was significantly larger when individuals were given higher rewards for listening to speech in noise.

Importantly, motivating factors can vary across different groups of individuals seeking audiology care. For example, older and younger adults appear to be differentially motivated by monetary reward to participate in effortful listening, independent of self-reported socioeconomic status. McLaughlin et al. (2021) observed that, compared to younger adults, normal-hearing older adults chose to engage in less demanding SNRs, forgoing a larger monetary reward. A greater tendency to exhibit this pattern of discounting behavior was found among older adults with poorer hearing thresholds and smaller working memory capacities.

3.3 Applications to Assessment: Interim Summary

Listening-related effort and fatigue are common complaints in the audiology clinic, and thus developing reliable and valid measures of listening effort in individuals experiencing hearing difficulties is of high concern to audiologists. Current clinical measures of speech recognition difficulties focus on the intelligibility of speech in quiet or in noise. However, as detailed in this section, effort and intelligibility diverge in important ways that can be tracked with pupillometry. As a result, two patients with similar intelligibility performance may have to work differentially hard to achieve those scores and exhibit very different pupil responses.

Pupillometry studies provide evidence for capacity-limited models of effort (e.g., FUEL) that specify that external factors, such as stimulus degradation, as well as internal factors, such as individual or group differences in cognitive capacity, motivation, and fatigue, can interactively impact how effortful processing unfolds when listening to speech in adverse conditions. The work reviewed in this section highlights how basic research studies using pupillometry have helped to validate patients' hearing complaints in the clinic and support the development of clinically feasible measures of effort.

4 Applications to Treatment

Aural rehabilitation aims to improve hearing-related functions and ultimately patient quality of life through interventions that include the management of sensory symptoms (e.g., hearing aids, CIs), speech-perception or other types of training, and patient instruction and counseling (Boothroyd, 2007). Behavioral performance measures have been the gold standard for assessing outcomes of treatments for hearing loss. Their importance remains, but as interest in cognitive aspects of audiology has grown, so have questions about how hearing loss and hearing assistive devices impact listening effort. Because hearing loss leads to a degraded auditory signal, it would seem intuitive that individuals with hearing loss experience increased listening effort compared to individuals with normal hearing. This is consistent with the common patient complaint that listening in challenging acoustic environments is effortful. Additionally, it might be assumed that given that hearing aid and CI devices are beneficial to quality of life (e.g., Cohen et al., 2004; Kochkin, 2011), their use would lead to a reduction in listening effort with treatment.

However, a systematic review showed that there is not conclusive evidence that hearing loss increases listening effort or that hearing assistive technology decreases listening effort (Ohlenforst et al., 2017a). As noted in Sect. 3, pupillometry studies have resulted in a positive, a negative, or no relationship between hearing loss and listening effort (Zekveld et al. 2018). This lack of conclusive evidence points toward a framework for thinking about hearing loss and listening effort in which the goal should be to *optimize* performance and listening effort with treatment. A patient with hearing loss may exert little effort prior to treatment by staying home and limiting interactions to one-on-one conversations, since they know that communication will not be successful in a crowded restaurant with a group of friends. Hearing technology may empower this patient to face the challenge of a restaurant, where listening effort will be high, but with the benefit of their devices, they do not have to give up.

To most effectively counsel and treat audiology patients, it would be critical to know whether an intervention is likely to result in an increase or decrease in listening effort. The nonlinear relationship between task demands and pupillary measures of effort (Ohlenforst et al., 2017b) leads to the prediction that interventions that improve speech intelligibility could, in fact, yield an increase in listening effort in highly challenging conditions. Figure 3 depicts hypothesized intervention-related changes plotted for speech intelligibility (gray) and effort (black) from pre- (dashed lines) to post-intervention (solid lines) along a continuum of listening demands (from poorer to better SNRs to quiet). If, prior to intervention, patients are overwhelmed by the demands of a challenging listening environment, such as in poorer SNRs, they tend to exhibit both low effort and low performance (Ohlenforst et al., 2017b; Ayasse & Wingfield, 2018). Following an intervention, when listening in poorer SNRs, improvements in speech recognition accuracy may co-occur with increases in listening effort (arrow 1): individuals can now engage in listening conditions which they may have previously chosen to avoid (e.g., a busy restaurant). However, in relatively easier SNRs, improvements in speech recognition accuracy following intervention may be associated with reductions in listening effort as listening becomes more efficient and begins to require fewer mental resources (arrow 2).

As noted in Sect. 3.2.2, pupillometry has become a useful tool to assess listening effort in populations that are relevant to the audiology clinic, particularly individuals experiencing greater challenges with speech recognition than would be predicted by their hearing loss. The availability of reliable and valid measures of listening effort is likely to also lead to technological developments and treatment strategies to help these patients optimize listening effort and performance and to experience the best hearing possible. Pupillometry measures of effort could support the development of interventions in at least two ways: (1) at the group level, as an outcome measure evaluating the impact of the intervention on listening effort and (2) at the individual level, supporting the identification of patients who are most likely to benefit from a given intervention. The following section outlines how pupillometry measures of effort have been used to evaluate audiologic interventions, including device usage, configurations, technological developments, and training programs.

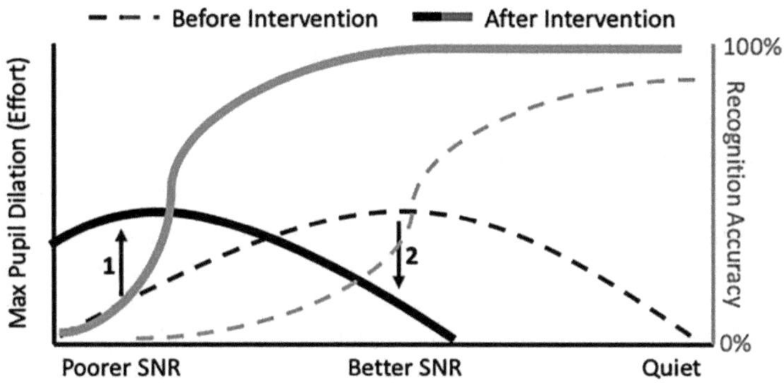

Fig. 3 Schematic expanding Fig. 2 showing that predicted benefits of audiological interventions (e.g., hearing aids, speech-perception training) may differ for recognition accuracy and listening effort across varied listening conditions. Dashed lines represent pre-intervention functions relating listening demands, such as SNR, to recognition accuracy (gray lines) and to maximum pupil size (black lines). Diverging patterns for effort and performance indicate that improvements in recognition accuracy with intervention (gray solid line > gray dashed line) may be associated with increased listening effort at poorer SNRs in which individuals had been giving up (arrow 1: black solid line > black dashed line), but reduced listening effort at better SNRs (arrow 2: black solid line < black dashed line)

4.1 Device Usage

Hearing aids amplify sound for the listener and are commonly used to provide audibility of sounds to listeners with hearing loss. However, pupillometry has seldom been used to investigate the benefits of listening effort with these devices; studies have been primarily limited to examining the effects of specific algorithm use (discussed in Sect. 4.3) rather than overall device use. Pupillometry could be a useful tool to better understand the changes in listening effort with hearing aids, and this is an area needing further research.

CIs are auditory prosthetic devices that deliver sound by directly stimulating the auditory nerve with electrical pulses. This technology is used in individuals who do not experience functional benefits from hearing aids. CIs improve access to sound and often result in improved speech understanding (e.g., Holden et al., 2013). However, cochlear implantation does not restore normal hearing, with CIs delivering a highly degraded signal, the degree of which can vary across users. This puts users at risk for experiencing high levels of listening effort, especially in challenging acoustic environments.

The level of degradation has been parametrically adjusted with CI simulations to explore how variability associated with CI devices might affect listening effort. Young listeners with normal hearing have shown that more degradation (poorer spectral resolution) is associated with higher pupil dilation and growth of pupil dilation over time, interpreted as higher listening effort (Winn et al., 2015), even with high performance. It is possible that listeners with CIs and poorer spectral resolution, perhaps those with longer durations of deafness or poorer electro-neural interfaces, require greater listening effort in everyday listening to maintain accurate speech understanding. In addition, peak pupil dilation on a speech-in-noise task was higher in noise than in quiet (Russo et al., 2020) and remained high with decreases in signal-to-noise ratio in listeners with CIs, supporting overall effortful listening with CIs (Dingemanse & Goedegebure, 2022). The effect of signal quality on listening effort has also been explored with bone-anchored hearing systems, another device used to treat hearing loss. A higher maximum force output in these devices is associated with less distortion of the signal by the output-limiting compressor. Listeners with conductive or mixed hearing loss showed reduced pupil dilation and improved performance when using a device with a higher maximum force output on a speech intelligibility task (Bianchi et al., 2019). This suggests that higher maximum output devices may have benefits for both performance and listening effort, regardless of the degree of hearing loss.

The greater signal degradation that these device users experience may result in a greater reliance on top-down (e.g., linguistic, cognitive) processes to help fill in missing information or repair errors. For example, pupillometry has been used to demonstrate that higher sentence context is associated with a reduction in effort in listeners with normal hearing and listeners with CIs (Winn, 2016). In addition, the timing of the release from effort differed across listener groups; release from effort with higher context was observed to occur more slowly in the CI group, even with perfect intelligibility. Further, this ability to depend on context can be disrupted by a subsequently attended sentence (Winn & Moore, 2018), which suggests that a context benefit may not occur in conversation when there is a steady stream of information to attend. Conversely, a slower speaking rate improves the ability of CI listeners to benefit from context effects on listening effort (Winn & Teece, 2021a). In addition, the types of errors made or mental repairs necessary to understand speech can impact effort in a way that dissociates from performance in CI listening (Winn & Teece, 2021b, 2022).

4.2 Device Configuration

Bilateral CIs are often recommended for individuals meeting CI candidacy criteria in both ears, as auditory input to two ears gives access to binaural benefits. These benefits include an improved ability to locate sound sources (Grantham et al., 2007) and improved speech-in-noise listening (Loizou et al., 2009). However, the binaural benefits that bilateral CIs impart are modest (Litovsky et al., 2009; Dorman et al., 2016) and many bilateral candidates choose a single CI (Holder et al., 2018). If the benefits of bilateral CIs include reduced listening effort, this knowledge would be useful for candidacy discussions and could lead to improved outcomes in listeners who otherwise would have chosen a single device. Similarly, pupillometry could be informative for audiologic recommendations with other device configurations, such as with bilateral hearing aids, but the study of binaural benefits to listening effort measured with pupillometry has been primarily focused on CI device configurations.

Binaural fusion, the ability to perceive a single sound rather than two separate sounds, has also been explored to investigate listening effort with bilateral CIs (Steel et al., 2015). Children with bilateral CIs and with normal hearing were asked how many sounds they heard as level, timing, and pitch cues were varied across the ears. The children with bilateral CIs did not perceive a fused image as often as their normal-hearing peers and had higher peak pupil dilation during the task, providing evidence that listening with CIs is effortful, and that a reduced ability to perceive a fused auditory image may contribute to effortful listening. Although they did not test fusion directly, a study in adult CI listeners found increased listening effort on a speech recognition task in a bilateral condition compared to the better-performing ear only, suggesting increased listening effort with integration across two asymmetric ears (Burg et al., 2023).

Another two-ear listening task was used to further explore questions about bilateral listening effort (DeRoy Milvae et al., 2021). A dichotic digit recall task, with target and interferer digits presented to separate ears, was used to examine how spectral resolution of the target and interfering digits impacts listening effort. Improved spectral resolution in both the target and interferer, implemented with increased channels of a vocoder, led to an increase in listening effort without a change in digit recall performance. This result suggests that the increased effort involved in ignoring a clearer interferer outweighs the benefit from a clearer target.

Device configuration benefits have not yet been explored in situations more similar to everyday listening; it is not yet known if spatial separation of target and interfering speech leads to a reduction in listening effort as compared to a co-located condition for listeners with bilateral CIs. Pupillometric examination of this in listeners with normal hearing has suggested no benefit with spatial separation (Zekveld et al. 2014b), but this has not yet been explored in listeners with CIs who are largely unable to rely on pitch cues for stream segregation, in contrast to listeners with normal hearing.

4.3 Technological Developments

Pupillometry has also been used to help validate the efficacy of algorithms implemented in digital hearing aids, which is especially important when these algorithms do not result in a performance benefit. For example, digital noise reduction algorithms have been shown to reduce task-evoked pupil dilation, interpreted as a reduction in listening effort, even when the algorithm does not change performance (Wendt et al., 2017; Bönitz et al., 2021; Fiedler et al., 2021), but the opposite has also been observed (Dingemanse & Goedegebure, 2022). An increase in baseline pupil dilation, interpreted as greater resources allocated for memory, has also occurred with noise reduction algorithms when recall accuracy also improved (Micula et al. 2021). Juul Jensen et al. (2018) also observed that individuals with tinnitus exhibited smaller pupil responses when using hearing aids with activated noise reduction, suggesting that hearing aids were associated with improvements in listening effort. However, no such change in pupil dilation was found for an algorithm that aims to enhance the perception of soft speech in CI users by reducing speech processor system noise (Stronks et al., 2021). The use of pupillometry to investigate listening effort in these examples permits more ecological assessments of the benefit from such algorithms beyond intelligibility, especially since noisy acoustic environments are typically at positive signal-to-noise ratios, where performance is expected to be high (Smeds et al., 2015).

4.4 Training-Based Interventions

The improvements in audibility provided by devices unfortunately do not address all of the challenges that individuals experience in real-world listening scenarios. As a result, training programs, used in aural rehabilitation plans, have been developed as relatively low-cost, self-directed supplements to other interventions. Auditory training generally involves challenging phoneme, word, or sentence discrimination or recognition tasks that provide the users with performance feedback. Auditory-cognitive training programs furthermore incorporate tasks that place greater demands on cognitive functioning and are thought to yield improved outcomes compared to auditory training alone (Ferguson & Henshaw, 2015).

A limited set of studies have used pupillometry to evaluate training-based interventions aimed at improving speech recognition. Kuchinsky et al. (2014) observed that, compared to controls, speech-perception trainees exhibited a faster, but overall larger pupillary response, even for correctly identified words in noise. Though trainees' speech-in-noise recognition improved, the difficulty of the listening task was quite challenging, with performance near 50% accuracy. Following arrow 1 in Fig. 2, the authors interpreted these results in the context of greater engagement in the listening task with training. Indeed, a follow-up study with the same participants found that individuals with a larger pupil response exhibited more vigilant, task-engaged patterns of behavior and neural activity (Kuchinsky et al., 2016).

Although not a training study, Brown et al. (2020) observed changes in pupil dilation with adaptation to unfamiliar speech sounds. In particular, the extra effort associated with processing nonnative-relative to native-accented speech was found to decrease as participants gained more experience with the speech materials over the course of the experiment.

The results of Brown et al. (2020) generally align with a study that evaluated changes in listening effort associated with training native English speakers to learn Mandarin lexical tone contrasts. Lexical tones are notoriously challenging for native

English speakers. Pandža et al. (2020) examined the extent to which noninvasive peripheral nerve stimulation paired with a learning paradigm could boost outcomes for these sounds. Transcutaneous vagal nerve stimulation (tVNS) is thought to modulate activity within the LC-NE system and, depending on the way in which stimulation is delivered, may enhance learning outcomes by optimizing the learner's state of attention or their general arousal level. Pupillometry was recorded as participants listened to unfamiliar Mandarin lexical tone words paired with visually presented English translations. Compared to a sham control, bursts of stimulation time-locked to the onset of the auditory stimulus (i.e., peristimulus stimulation) led to a faster peaking, less sustained pupillary response from day 1 to day 2 of learning, indicative of reductions in listening effort. Weaker evidence was observed for 10 minutes of pre-task stimulation (i.e., priming) compared to sham, resulting in overall larger pupil dilation from day 1 to day 2, reminiscent of the increasing task engagement noted in Kuchinsky et al. (2014). These results suggest that peristimulus and priming tVNS may impact effortful listening by altering phasic and tonic modes of LC activity, respectively.

Together, these studies suggest that the likelihood of an intervention resulting in reductions versus increases in listening effort depends on the overall level of difficulty of the listening conditions being evaluated as well as the specific auditory and cognitive mechanisms that are thought to be targeted by the intervention.

4.5 Applications to Treatment: Interim Summary

For patients experiencing hearing difficulties, intervention outcomes that only include intelligibility scores may not fully characterize treatment benefits or costs. As speech recognition performance improves – or even in cases where there are no changes in performance – with the use of hearing aids, CIs, or auditory training, pupillometry studies have revealed intervention-related decreases in effort. However, treatment-related performance benefits have also been accompanied by increases in listening effort as individuals become increasingly able to engage in previously overwhelmingly difficult listening situations.

The pupillometry studies described in this section highlight the potential challenge of predicting the direction of benefits and costs expected from an intervention and emphasize the consideration of the overall level of listening difficulty that a group or individual experiences pre-intervention. The ability to make predictions regarding the nature of treatment outcomes is not only critical from a hypothesis testing perspective but also from a clinical counseling perspective. Patients must be appropriately informed about what they can expect as they invest their precious time and money in a treatment plan. As described in this section, research studies that have used pupillometry to measure listening effort in the laboratory provide a foundation for understanding the potential impact of different remediations in the clinic.

5 Conclusions

Pupillometry, as an objective measure of listening effort, has potential clinical value in providing a more holistic picture of a patient's experience with hearing loss. Changes in pupil dilation appear to track changes in effort associated with listening in acoustically, linguistically, and cognitively demanding situations that are common in the real-world settings that most concern patients. Pupillary measures also can be sensitive to individual and group differences that influence the likelihood of experiencing effortful listening, though additional research is needed to clarify the conditions under which patient populations are most likely to be impacted. Additionally, a major advantage of pupillary measures of effort is that they provide insight into the deployment of effort across different timescales, with a single measure allowing for the examination of how hearing issues impact arousal, listening task engagement, and fatigue.

The research reviewed in this chapter suggests a role for pupillometry in the audiology clinic for both assessment and treatment. Currently, pupillometry studies of listening effort inform evidence-based practice by identifying speech recognition challenges that are not well characterized with current clinical diagnostics. In addition, these studies identify groups of individuals who are most likely to be at risk for experiencing these challenges. This literature can inform counseling of patients by helping to validate their reports of speech recognition difficulties in the absence of hearing loss.

Furthermore, pupillometry research on listening effort has the potential to guide audiologic recommendations and treatment plans. For example, evidence of reduced listening effort in adverse conditions with bilateral cochlear implantation could affect the standard of care. This research could also enable more personalized care, for example, in the selection of specific hearing aid settings (e.g., compression settings or noise reduction algorithms) for patients who are particularly affected by listening effort in complex settings.

Longer-term, there is potential for pupillometry measures to be integrated into audiological batteries, perhaps by leveraging eye-tracking tools already in use by vestibular audiologists. However, a better understanding of pupillometry's sensitivity and specificity for measuring effort at the individual level would be needed.

Clinical practice that is either informed by or includes pupillometry measures of effort is likely to provide a better view of the challenges that individuals with hearing loss experience and lead to more effective treatment. With the goal of addressing one of the most common clinical complaints, quantifying listening effort may be critical for effective patient-centered care and, ultimately, improved patient satisfaction and quality of life.

Acknowledgments The authors are grateful for helpful comments on a previous version of this chapter from Miranda Cleary and Jordan Abramowitz. This work was supported in part by grants from the National Institutes of Health National Institute on Deafness and Other Communication Disorders R03 DC15059 (SEK) and K01 DC018064 (KDM).

For Dr. Kuchinsky: The views expressed in this chapter are those of the authors and do not reflect the official policy of the Department of Army/Navy/Air Force, Department of Defense, or U.S. Government.

References

Alhanbali, S., Dawes, P., Lloyd, S., & Munro, K. J. (2017). Self-reported listening-related effort and fatigue in hearing-impaired adults. *Ear and Hearing, 38*, e39–e48. https://doi.org/10.1097/AUD.0000000000000361

Alhanbali, S., Dawes, P., Millman, R. E., & Munro, K. J. (2019). Measures of listening effort are multidimensional. *Ear and Hearing, 40*, 1084–1097. https://doi.org/10.1097/AUD.0000000000000697

Alhanbali, S., Munro, K. J., Dawes, P., et al. (2020). Dimensions of self-reported listening effort and fatigue on a digits-in-noise task, and association with baseline pupil size and performance accuracy. *International Journal of Audiology, 0*, 1–11. https://doi.org/10.1080/14992027.2020.1853262

Antikainen, J., & Niemi, P. (1983). Neuroticism and the pupillary response to a brief exposure to noise. *Biological Psychology, 17*(2–3), 131–135. https://doi.org/10.1016/0301-0511(83)90013-3

Aston-Jones, G., & Cohen, J. D. (2005). An integrative theory of locus coeruleus-norepinephrine function: Adaptive gain and optimal performance. *Annual Review of Neuroscience, 28*, 403–450. https://doi.org/10.1146/annurev.neuro.28.061604.135709

Ayasse, N. D., & Wingfield, A. (2018). A tipping point in listening effort: Effects of linguistic complexity and age-related hearing loss on sentence comprehension. *Trends in Hearing, 22*, 233121651879090. https://doi.org/10.1177/2331216518790907

Ayasse, N. D., & Wingfield, A. (2020). Anticipatory baseline pupil diameter is sensitive to differences in hearing thresholds. *Frontiers in Psychology, 10*, 1–7. https://doi.org/10.3389/fpsyg.2019.02947

Ayasse, N. D., Lash, A., & Wingfield, A. (2017). Effort not speed characterizes comprehension of spoken sentences by older adults with mild hearing impairment. *Frontiers in Aging Neuroscience, 8*, 329. https://doi.org/10.3389/fnagi.2016.00329

Babiker, A., Faye, I., Prehn, K., & Malik, A. (2015). Machine learning to differentiate between positive and negative emotions using pupil diameter. *Frontiers in Psychology, 6*, 1–10. https://doi.org/10.3389/fpsyg.2015.01921

Bala, A. D. S., Whitchurch, E. A., & Takahashi, T. T. (2020). Human auditory detection and discrimination measured with the pupil dilation response. *Journal of the Association for Research in Otolaryngology, 21*, 43–59. https://doi.org/10.1007/s10162-019-00739-x

Beechey, T., Buchholz, J. M., & Keidser, G. (2019). Eliciting naturalistic conversations: A method for assessing communication ability, subjective experience, and the impacts of noise and hearing impairment. *Journal of Speech, Language, and Hearing, 62*, 470–484. https://doi.org/10.1044/2018_JSLHR-H-18-0107

Bianchi, F., Wendt, D., Wassard, C., et al. (2019). Benefit of higher maximum force output on listening effort in bone-anchored hearing system users: A pupillometry study. *Ear and Hearing, 40*, 1220–1232. https://doi.org/10.1097/AUD.0000000000000699

Bönitz, H., Lunner, T., Finke, M., Fiedler, L., Lyxell, B., Riis, S. K., Ng, E., Lopez Valdes, A., Büchner, A., & Wendt, D. (2021). How do we allocate our resources when listening and memorizing speech in noise? A pupillometry study. *Ear and Hearing, 42*(4), 846–859.

Boothroyd, A. (2007). Adult aural rehabilitation: What is it and does it work? *Trends in Amplification, 11*, 63–71. https://doi.org/10.1177/1084713807301073

Borghini, G., & Hazan, V. (2018). Listening effort during sentence processing is increased for non-native listeners: A pupillometry study. *Frontiers in Neuroscience, 12*, 1–13. https://doi.org/10.3389/fnins.2018.00152

Botvinick, M., & Braver, T. (2015). Motivation and cognitive control: From behavior to neural mechanism. *Annual Review of Psychology, 66*, 83–113. https://doi.org/10.1146/annurev-psych-010814-015044

Brown, V. A., McLaughlin, D. J., Strand, J. F., & Van Engen, K. J. (2020). Rapid adaptation to fully intelligible nonnative-accented speech reduces listening effort. *The Quarterly Journal of Experimental Psychology, 73*, 1431–1443. https://doi.org/10.1177/1747021820916726

Brungart, D. S., Sherlock, L. P., Kuchinsky, S. E., et al. (2022). Assessment methods for determining small changes in hearing performance over time. *The Journal of the Acoustical Society of America, 151*, 3866–3885. https://doi.org/10.1121/10.0011509

Burg, E. A., Thakkar, T. D., & Litovsky, R. Y. (2023). Interaural speech asymmetry predicts bilateral speech intelligibility but not listening effort in adults with bilateral cochlear implants. *Frontiers in Neuroscience, 16*, 1038856. https://doi.org/10.3389/fnins.2022.1038856

Chermak, G. D., Bamiou, D.-E., Iliadou, V., & Musiek, F. E. (2017). Practical guidelines to minimise language and cognitive confounds in the diagnosis of capd: A brief tutorial. *International Journal of Audiology, 56*, 499–506. https://doi.org/10.1080/14992027.2017.1284351

Cohen, S. M., Labadie, R. F., Dietrich, M. S., & Haynes, D. S. (2004). Quality of life in hearing-impaired adults: The role of cochlear implants and hearing aids. *Otolaryngology – Head, Neck and Surgery, 131*, 413–422. https://doi.org/10.1016/j.otohns.2004.03.026

Dauman, N., Erlandsson, S., Lundlin, L., & Dauman, R. (2015). Intra-individual variability in tinnitus patients. *HNO, 63*, 302–306. https://doi.org/10.1007/s00106-014-2978-2

Davis, H., Schlundt, D., Bonnet, K., et al. (2020). Understanding listening-related fatigue: Perspectives of adults with hearing loss. *International Journal of Audiology, 0*, 1–11. https://doi.org/10.1080/14992027.2020.1834631

DeRoy Milvae, K., Kuchinsky, S. E., Stakhovskaya, O. A., & Goupell, M. J. (2021). Dichotic listening performance and effort as a function of spectral resolution and interaural symmetry. *The Journal of the Acoustical Society of America, 150*(2), 920–935. https://doi.org/10.1121/10.0005653

Dingemanse, G., & Goedegebure, A. (2022). Listening effort in cochlear implant users: The effect of speech intelligibility, noise reduction processing, and working memory capacity on the pupil dilation response. *Journal of Speech, Language, and Hearing Research, 65*(1), 392–404. https://doi.org/10.1044/2021_JSLHR-21-00230

Dorman, M. F., Loiselle, L. H., Cook, S. J., et al. (2016). Sound source localization by normal-hearing listeners, hearing-impaired listeners and cochlear implant listeners. *Audiology and Neurotology, 21*, 127–131. https://doi.org/10.1159/000444740

Dubno, J. R., Dirks, D. D., & Morgan, D. E. (1984). Effects of age and mild hearing loss on speech recognition in noise. *The Journal of the Acoustical Society of America, 76*, 87–96. https://doi.org/10.1121/1.391011

Eckert, M. A., Teubner-Rhodes, S., & Vaden, K. I. (2016). Is listening in noise worth it? The neurobiology of speech recognition in challenging listening conditions. *Ear and Hearing, 37*, 101S–110S. https://doi.org/10.1097/AUD.0000000000000300

Ferguson, M. A., & Henshaw, H. (2015). How does auditory training work? Joined-up thinking and listening. *Seminars in Hearing, 36*, 237–249. https://doi.org/10.1055/s-0035-1564456

Fiedler, L., Seifi Ala, T., Graversen, C., Alickovic, E., Lunner, T., & Wendt, D. (2021). Hearing aid noise reduction lowers the sustained listening effort during continuous speech in noise—A combined pupillometry and EEG study. *Ear and Hearing, 42*(6), 1590. https://doi.org/10.1097/AUD.0000000000001050

Francis, A. L., & Oliver, J. (2018). Psychophysiological measurement of affective responses during speech perception. *Hearing Research, 369*, 103–119. https://doi.org/10.1016/j.heares.2018.07.007

Gagné, J.-P., Besser, J., & Lemke, U. (2017). Behavioral assessment of listening effort using a dual-task paradigm. *Trends in Hearing, 21*, 1–25. https://doi.org/10.1177/2331216516687287

Gallun, F. J., Diedesch, A. C., Kubli, L. R., et al. (2012). Performance on tests of central auditory processing by individuals exposed to high-intensity blasts. *Journal of Rehabilitation Research and Development, 49*, 1005. https://doi.org/10.1682/JRRD.2012.03.0038

Gatehouse, S., & Noble, W. (2004). The speech, spatial and qualities of hearing scale (ssq). *International Journal of Audiology, 43*, 85–99. https://doi.org/10.1080/14992020400050014

Gilzenrat, M. S., Nieuwenhuis, S., Jepma, M., & Cohen, J. D. (2010). Pupil diameter tracks changes in control state predicted by the adaptive gain theory of locus coeruleus function. *Cognitive, Affective, & Behavioral Neuroscience, 10*, 252–269. https://doi.org/10.3758/CABN.10.2.252

Giuliani, N. P., Brown, C. J., & Wu, Y.-H. (2020). Comparisons of the sensitivity and reliability of multiple measures of listening effort. *Ear and Hearing*. Publish Ah:1–10. https://doi.org/10.1097/AUD.0000000000000950

Gómez-Merino, N., Gheller, F., Spicciarelli, G., & Trevisi, P. (2020). Pupillometry as a measure for listening effort in children: A review. *Hearing, Balancing, and Communication, 18*, 152–158. https://doi.org/10.1080/21695717.2020.1807256

Grantham, D. W., Ashmead, D. H., Ricketts, T. A., et al. (2007). Horizontal-plane localization of noise and speech signals by postlingually deafened adults fitted with bilateral cochlear implants. *Ear and Hearing, 28*, 524–541. https://doi.org/10.1097/AUD.0b013e31806dc21a

Grenness, C., Hickson, L., Laplante-Lévesque, A., et al. (2015). Communication patterns in audiologic rehabilitation history-taking: Audiologists, patients, and their companions. *Ear and Hearing, 36*, 191–204. https://doi.org/10.1097/AUD.0000000000000100

Herrmann, B., & Johnsrude, I. S. (2020). A model of listening engagement (mole). *Hearing Research, 397*, 108016. https://doi.org/10.1016/j.heares.2020.108016

Hess, E. H., & Polt, J. M. (1964). Pupil size in relation to mental activity during simple problem-solving. *Science, 143*, 1190–1192. https://doi.org/10.1126/science.143.3611.1190

Hétu, R., Riverin, L., Lalande, N., et al. (1988). Qualitative analysis of the handicap associated with occupational hearing loss. *British Journal of Audiology, 22*, 251–264. https://doi.org/10.3109/03005368809076462

Holden, L. K., Finley, C. C., Firszt, J. B., et al. (2013). Factors affecting open-set word recognition in adults with cochlear implants. *Ear and Hearing, 34*, 342–360.

Holder, J. T., Reynolds, S. M., Sunderhaus, L. W., & Gifford, R. H. (2018). Current profile of adults presenting for preoperative cochlear implant evaluation. *Trends in Hearing, 22*, 1–16.

Hopstaken, J. F., van der Linden, D., Bakker, A. B., & Kompier, M. A. J. (2015). The window of my eyes: Task disengagement and mental fatigue covary with pupil dynamics. *Biological Psychology, 110*, 100–106. https://doi.org/10.1016/j.biopsycho.2015.06.013

Hornsby, B. W. Y., Naylor, G., & Bess, F. H. (2016). A taxonomy of fatigue concepts and their relation to hearing loss. *Ear and Hearing, 37*, 136S–144S. https://doi.org/10.1097/AUD.0000000000000289

Hughes, S. E., Hutchings, H. A., Rapport, F. L., et al. (2018). Social connectedness and perceived listening effort in adult cochlear implant users: A grounded theory to establish content validity for a new patient-reported outcome measure. *Ear and Hearing, 39*, 922–934. https://doi.org/10.1097/AUD.0000000000000553

Inzlicht, M., & Schmeichel, B. J. (2012). What is ego depletion? Toward a mechanistic revision of the resource model of self-control. *Perspectives on Psychological Science, 7*, 450–463. https://doi.org/10.1177/1745691612454134

Joshi, S., Li, Y., Kalwani, R. M., & Gold, J. I. (2016). Relationships between pupil diameter and neuronal activity in the locus coeruleus, colliculi, and cingulate cortex. *Neuron, 89*, 221–234. https://doi.org/10.1016/j.neuron.2015.11.028

Juul Jensen, J., Callaway, S. L., Lunner, T., & Wendt, D. (2018). Measuring the impact of tinnitus on aided listening effort using pupillary response. *Trends in Hearing, 22*, 233121651879534. https://doi.org/10.1177/2331216518795340

Kahneman, D. (1973). *Attention and effort*. Prentice-Hall.

Kahya, M., Wood, T. A., Sosnoff, J. J., & Devos, H. (2018). Increased postural demand is associated with greater cognitive workload in healthy young adults: A pupillometry study. *Frontiers in Human Neuroscience, 12*, 1–8. https://doi.org/10.3389/fnhum.2018.00288

Koch, X., & Janse, E. (2016). Speech rate effects on the processing of conversational speech across the adult life span. *The Journal of the Acoustical Society of America, 139*, 1618–1636. https://doi.org/10.1121/1.4944032

Kochkin, S. (2011). MarkeTrak viii patients report improved quality of life with hearing aid usage. *The Hearing Journal, 64*, 25–26. https://doi.org/10.1097/01.HJ.0000399150.30374.45

Koelewijn, T., Shinn-Cunningham, B. G., Zekveld, A. A., & Kramer, S. E. (2014a). The pupil response is sensitive to divided attention during speech processing. *Hearing Research, 312*, 114–120. https://doi.org/10.1016/j.heares.2014.03.010

Koelewijn, T., Zekveld, A. A., Festen, J. M., & Kramer, S. E. (2014b). The influence of informational masking on speech perception and pupil response in adults with hearing impairment. *The Journal of the Acoustical Society of America, 135*, 1596–1606. https://doi.org/10.1121/1.4863198

Koelewijn, T., Versfeld, N. J., & Kramer, S. E. (2017). Effects of attention on the speech reception threshold and pupil response of people with impaired and normal hearing. *Hearing Research, 354*, 56–63. https://doi.org/10.1016/j.heares.2017.08.006

Koelewijn, T., van Haastrecht, J. A. P., & Kramer, S. E. (2018a). Pupil responses of adults with traumatic brain injury during processing of speech in noise. *Trends in Hearing, 22*, 233121651881144. https://doi.org/10.1177/2331216518811444

Koelewijn, T., Zekveld, A. A., Lunner, T., & Kramer, S. E. (2018b). The effect of reward on listening effort as reflected by the pupil dilation response. *Hearing Research, 367*, 106–112. https://doi.org/10.1016/j.heares.2018.07.011

Kramer, S. E., Kapteyn, T. S., Festen, J. M., & Kuik, D. J. (1997). Assessing aspects of auditory handicap by means of pupil dilatation. *International Journal of Audiology, 36*, 155–164. https://doi.org/10.3109/00206099709071969

Kramer, S. E., Teunissen, C. E., & Zekveld, A. A. (2016). Cortisol, chromogranin a, and pupillary responses evoked by speech recognition tasks in normally hearing and hard-of-hearing listeners. *Ear and Hearing, 37*, 126S–135S. https://doi.org/10.1097/AUD.0000000000000311

Krueger, M., Schulte, M., Zokoll, M. A., et al. (2017). Relationship between listening effort and speech intelligibility in noise. *American Journal of Audiology, 26*, 378–393.

Kuchinsky, S. E., Ahlstrom, J. B., Vaden, K. I., et al. (2013). Pupil size varies with word listening and response selection difficulty in older adults with hearing loss. *Psychophysiology, 50*, 23–34. https://doi.org/10.1111/j.1469-8986.2012.01477.x

Kuchinsky, S. E., Ahlstrom, J. B., Cute, S. L., et al. (2014). Speech-perception training for older adults with hearing loss impacts word recognition and effort. *Psychophysiology, 51*, 1046–1057. https://doi.org/10.1111/psyp.12242

Kuchinsky, S. E., Vaden, K. I., Ahlstrom, J. B., et al. (2016). Task-related vigilance during word recognition in noise for older adults with hearing loss. *Experimental Aging Research, 42*, 50–66. https://doi.org/10.1080/0361073X.2016.1108712

Lin, F. R., & Albert, M. (2014). Hearing loss and dementia – Who is listening? *Aging Ment Heal, 18*, 671–673. https://doi.org/10.1080/13607863.2014.915924

Litovsky, R. Y., Parkinson, A., & Arcaroli, J. (2009). Spatial hearing and speech intelligibility in bilateral cochlear implant users. *Ear and Hearing, 30*, 419–431. https://doi.org/10.1097/AUD.0b013e3181a165be

Loizou, P. C., Hu, Y., Litovsky, R., et al. (2009). Speech recognition by bilateral cochlear implant users in a cocktail-party setting. *The Journal of the Acoustical Society of America, 125*, 372–383. https://doi.org/10.1121/1.3036175

Lunner, T., Alickovic, E., Graversen, C., et al. (2020). Three new outcome measures that tap into cognitive processes required for real-life communication. *Ear and Hearing, 41*, 39S–47S. https://doi.org/10.1097/AUD.0000000000000941

Mattys, S. L., Davis, M. H., Bradlow, A. R., & Scott, S. K. (2012). Speech recognition in adverse conditions: A review. *Language & Cognitive Processes, 27*, 953–978. https://doi.org/10.1080/01690965.2012.705006

McCloy, D. R., Larson, E., & Lee, A. K. C. (2018). Auditory attention switching with listening difficulty: Behavioral and pupillometric measures. *The Journal of the Acoustical Society of America, 144*, 2764–2771. https://doi.org/10.1121/1.5078618

McCoy, S. L., Tun, P. A., Cox, L. C., et al. (2005). Hearing loss and perceptual effort: Downstream effects on older adults' memory for speech. *Quarterly Journal of Experimental Psychology, 58*, 22–33. https://doi.org/10.1080/02724980443000151

McGarrigle, R., Munro, K. J., Dawes, P., et al. (2014). Listening effort and fatigue: What exactly are we measuring? *International Journal of Audiology, 53*, 433–440. https://doi.org/10.3109/14992027.2014.890296

McGarrigle, R., Dawes, P., Stewart, A. J., et al. (2017). Pupillometry reveals changes in physiological arousal during a sustained listening task. *Psychophysiology, 54*, 193–203. https://doi.org/10.1111/psyp.12772

McGarrigle, R., Rakusen, L., & Mattys, S. (2021). Effortful listening under the microscope: Examining relations between pupillometric and subjective markers of effort and tiredness from listening. *Psychophysiology, 58*, 1–22. https://doi.org/10.1111/psyp.13703

McLaughlin, D. J., Braver, T. S., & Peelle, J. E. (2021). Measuring the subjective cost of listening effort using a discounting task. *Journal of Speech, Language, and Hearing Research, 64*, 337–347. https://doi.org/10.1044/2020_JSLHR-20-00086

Miles, K., McMahon, C., Boisvert, I., et al. (2017). Objective assessment of listening effort: Coregistration of pupillometry and eeg. *Trends in Hearing, 21*, 233121651770639. https://doi.org/10.1177/2331216517706396

Moore, T. M., & Picou, E. M. (2018). A potential bias in subjective ratings of mental effort. *Journal of Speech, Language, and Hearing Research, 61*, 2405–2421. https://doi.org/10.1044/2018_JSLHR-H-17-0451

Micula, A., Rönnberg, J., Fiedler, L., Wendt, D., Jørgensen, M. C., Larsen, D. K., & Ng, E. H. N. (2021). The Effects of task difficulty predictability and noise reduction on recall performance and pupil dilation responses. *Ear and Hearing, 42*(6), 1668–1679. https://doi.org/10.1097/AUD.0000000000001053

Murphy, P. R., Robertson, I. H., Balsters, J. H., & O'Connell, R. G. (2011). Pupillometry and p3 index the locus coeruleus-noradrenergic arousal function in humans. *Psychophysiology, 48*, 1532–1543. https://doi.org/10.1111/j.1469-8986.2011.01226.x

Neagu, M. B., Dau, T., Hyvärinen, P., et al. (2019). Reliability of pupillometry as a measure of listening effort. In A. Kressner, J. Regev, J. Christensen-Dalsgaard, et al. (Eds.), *Proceedings of the international symposium on auditory and audiological research*. Danavox Jubilee Foundation.

Ohlenforst, B., Zekveld, A. A., Jansma, E. P., et al. (2017a). Effects of hearing impairment and hearing aid amplification on listening effort: A systematic review. *Ear and Hearing, 38*, 267–281.

Ohlenforst, B., Zekveld, A. A., Lunner, T., et al. (2017b). Impact of stimulus-related factors and hearing impairment on listening effort as indicated by pupil dilation. *Hearing Research, 351*, 68–79.

Otero, S. C., Weekes, B. S., & Hutton, S. B. (2011). Pupil size changes during recognition memory. *Psychophysiology, 48*, 1346–1353. https://doi.org/10.1111/j.1469-8986.2011.01217.x

Pandža, N. B., Phillips, I., Karuzis, V. P., et al. (2020). Neurostimulation and pupillometry: New directions for learning and research in applied linguistics. *Annual Review of Applied Linguistics, 40*, 56–77. https://doi.org/10.1017/S0267190520000069

Partala, T., & Surakka, V. (2003). Pupil size variation as an indication of affective processing. *International Journal of Human Computer Studies, 59*, 185–198. https://doi.org/10.1016/S1071-5819(03)00017-X

Peelle, J. E. (2018). Listening effort. *Ear and Hearing, 39*, 204–214. https://doi.org/10.1097/AUD.0000000000000494

Pichora-Fuller, M. K., Kramer, S. E., Eckert, M. A., et al. (2016). Hearing impairment and cognitive energy: The framework for understanding effortful listening (fuel). *Ear and Hearing, 37*, 5S–27S. https://doi.org/10.1097/AUD.0000000000000312

Pienkowski, M. (2017). On the etiology of listening difficulties in noise despite clinically normal audiograms. *Ear and Hearing, 38*, 135–148. https://doi.org/10.1097/AUD.0000000000000388

Piquado, T., Isaacowitz, D., & Wingfield, A. (2010). Pupillometry as a measure of cognitive effort in younger and older adults. *Psychophysiology, 47*, 560–569. https://doi.org/10.1111/j.1469-8986.2009.00947.x

Pisoni, D. B., Kronenberger, W. G., Harris, M. S., & Moberly, A. C. (2017). Three challenges for future research on cochlear implants. *World Journal of Otorhinolaryngology-Head and Neck Surgery, 3*, 240–254. https://doi.org/10.1016/j.wjorl.2017.12.010

Richter, M. (2016). The moderating effect of success importance on the relationship between listening demand and listening effort. *Ear and Hearing, 37*, 111S–117S. https://doi.org/10.1097/AUD.0000000000000295

Roup, C. M., Post, E., & Lewis, J. (2018). Mild-gain hearing aids as a treatment for adults with self-reported hearing difficulties. *Journal of the American Academy of Audiology, 29*, 477–494. https://doi.org/10.3766/jaaa.16111

Russo, F. Y., Hoen, M., Karoui, C., et al. (2020). Pupillometry assessment of speech recognition and listening experience in adult cochlear implant patients. *Frontiers in Neuroscience, 14*, 1–13. https://doi.org/10.3389/fnins.2020.556675

Scheiter, K., Ackerman, R., & Hoogerheide, V. (2020). Looking at mental effort appraisals through a metacognitive lens: Are they biased? *Educational Psychology Review, 32*, 1003–1027. https://doi.org/10.1007/s10648-020-09555-9

Schmidtke, J. (2014). Second language experience modulates word retrieval effort in bilinguals: Evidence from pupillometry. *Frontiers in Psychology, 5*, 1–16. https://doi.org/10.3389/fpsyg.2014.00137

Schmidtke, J. (2018). Pupillometry in linguistic research. *Studies in Second Language Acquisition, 40*, 529–549. https://doi.org/10.1017/S0272263117000195

Shen, J., Anderson, M. C., Arehart, K. H., & Souza, P. E. (2016). Using cognitive screening tests in audiology. *American Journal of Audiology, 25*(4), 319–331. https://doi.org/10.1044/2016_AJA-16-0032

Shenhav, A., Musslick, S., Lieder, F., et al. (2017). Toward a rational and mechanistic account of mental effort. *Annual Review of Neuroscience, 40*, 99–124. https://doi.org/10.1146/annurev-neuro-072116-031526

Smeds, K., Wolters, F., & Rung, M. (2015). Estimation of signal-to-noise ratios in realistic sound scenarios. *Journal of the American Academy of Audiology, 26*, 183–196.

Souza, P. E., & Sirow, L. (2014). Relating working memory to compression parameters in clinically fit hearing aids. *American Journal of Audiology, 23*(4), 394–401. https://doi.org/10.1044/2014_AJA-14-0006

Steel, M. M., Papsin, B. C., & Gordon, K. A. (2015). Binaural fusion and listening effort in children who use bilateral cochlear implants: A psychoacoustic and pupillometric study. *PLoS One, 10*, 1–29.

Strand, J. F., Brown, V. A., Merchant, M. B., et al. (2018). Measuring listening effort: Convergent validity, sensitivity, and links with cognitive and personality measures. *Journal of Speech, Language, and Hearing Research, 61*, 1463–1486. https://doi.org/10.1044/2018_JSLHR-H-17-0257

Strauss, D. J., & Francis, A. L. (2017). Toward a taxonomic model of attention in effortful listening. *Cognitive, Affective, & Behavioral Neuroscience, 17*, 809–825. https://doi.org/10.3758/s13415-017-0513-0

Stronks, H. C., Apperloo, E., Koning, R., et al. (2021). SoftVoice improves speech recognition and reduces listening effort in cochlear implant users. *Ear and Hearing, 42*(2), 381–392. https://doi.org/10.1097/AUD.0000000000000928

Tremblay, K. L., Pinto, A., Fischer, M. E., et al. (2015). Self-reported hearing difficulties among adults with normal audiograms: The beaver dam offspring study. *Ear and Hearing, 36*, e290–e299. https://doi.org/10.1097/AUD.0000000000000195

Vaden, K. I., Kuchinsky, S. E., Ahlstrom, J. B., et al. (2016). Cingulo-opercular function during word recognition in noise for older adults with hearing loss. *Experimental Aging Research, 42*, 67–82. https://doi.org/10.1080/0361073X.2016.1108784

Van Engen, K. J., & McLaughlin, D. J. (2018). Eyes and ears: Using eye tracking and pupillometry to understand challenges to speech recognition. *Hearing Research, 369*, 56–66. https://doi.org/10.1016/j.heares.2018.04.013

Viljanen, A., Kaprio, J., Pyykkö, I., et al. (2009). Hearing as a predictor of falls and postural balance in older female twins. *The Journals of Gerontology. Series A, Biological Sciences and Medical Sciences, 64*, 312–317. https://doi.org/10.1093/gerona/gln015

Wagner, A. E., Toffanin, P., & Baskent, D. (2016). The timing and effort of lexical access in natural and degraded speech. *Frontiers in Psychology, 7*, 1–14. https://doi.org/10.3389/fpsyg.2016.00398

Wagner, A. E., Nagels, L., Toffanin, P., et al. (2019). Individual variations in effort: Assessing pupillometry for the hearing impaired. *Trends in Hearing, 23*, 233121651984559. https://doi.org/10.1177/2331216519845596

Wang, Y., Naylor, G., Kramer, S. E., et al. (2018). Relations between self-reported daily-life fatigue, hearing status, and pupil dilation during a speech perception in noise task. *Ear and Hearing, 39*, 573–582. https://doi.org/10.1097/AUD.0000000000000512

Wendt, D., Dau, T., & Hjortkjær, J. (2016). Impact of background noise and sentence complexity on processing demands during sentence comprehension. *Frontiers in Psychology, 7*, 345. https://doi.org/10.3389/fpsyg.2016.00345

Wendt, D., Hietkamp, R. K., & Lunner, T. (2017). Impact of noise and noise reduction on processing effort. *Ear and Hearing, 38*, 690–700. https://doi.org/10.1097/AUD.0000000000000454

Wingfield, A., & Grossman, M. (2006). Language and the aging brain: Patterns of neural compensation revealed by functional brain imaging. *Journal of Neurophysiology, 96*, 2830–2839. https://doi.org/10.1152/jn.00628.2006

Winn, M. B. (2016). Rapid release from listening effort resulting from semantic context, and effects of spectral degradation and cochlear implants. *Trends in Hearing, 20*, 233121651666972. https://doi.org/10.1177/2331216516669723

Winn, M. B., & Moore, A. N. (2018). Pupillometry reveals that context benefit in speech perception can be disrupted by later-occurring sounds, especially in listeners with cochlear implants. *Trends in Hearing, 22*, 1–22. https://doi.org/10.1177/2331216518808962

Winn, M. B., & Teece, K. H. (2021a). Slower speaking rate reduces listening effort among listeners with cochlear implants. *Ear and Hearing, 42*(3), 584–595. https://doi.org/10.1097/AUD.0000000000000958

Winn, M. B., & Teece, K. H. (2021b). Listening effort is not the same as speech intelligibility score. *Trends in Hearing, 25*, 233121652110276. https://doi.org/10.1177/23312165211027688

Winn, M. B., & Teece, K. H. (2022). Effortful listening despite correct responses: The cost of mental repair in sentence recognition by listeners with cochlear implants. *Journal of Speech, Language, and Hearing Research, 65*, 3966–3980. https://doi.org/10.1044/2022_JSLHR-21-00631

Winn, M. B., Edwards, J. R., & Litovsky, R. Y. (2015). The impact of auditory spectral resolution on listening effort revealed by pupil dilation. *Ear and Hearing, 36*, e153–e165.

Winn, M. B., Wendt, D., Koelewijn, T., & Kuchinsky, S. E. (2018). Best practices and advice for using pupillometry to measure listening effort: An introduction for those who want to get started. *Trends in Hearing, 22*, 233121651880086. https://doi.org/10.1177/2331216518800869

Zekveld, A. A., & Kramer, S. E. (2014). Cognitive processing load across a wide range of listening conditions: Insights from pupillometry. *Psychophysiology, 51*, 277–284. https://doi.org/10.1111/psyp.12151

Zekveld, A. A., Kramer, S. E., & Festen, J. M. (2010). Pupil response as an indication of effortful listening: The influence of sentence intelligibility. *Ear and Hearing, 31*, 480–490. https://doi.org/10.1097/AUD.0b013e3181d4f251

Zekveld, A. A., Kramer, S. E., & Festen, J. M. (2011). Cognitive load during speech perception in noise: The influence of age, hearing loss, and cognition on the pupil response. *Ear and Hearing, 32*, 498–510. https://doi.org/10.1097/AUD.0b013e31820512bb

Zekveld, A. A., Heslenfeld, D. J., Johnsrude, I. S., et al. (2014a). The eye as a window to the listening brain: Neural correlates of pupil size as a measure of cognitive listening load. *NeuroImage, 101*, 76–86. https://doi.org/10.1016/j.neuroimage.2014.06.069

Zekveld, A. A., Rudner, M., Kramer, S. E., et al. (2014b). Cognitive processing load during listening is reduced more by decreasing voice similarity than by increasing spatial separation between target and masker speech. *Frontiers in Neuroscience, 8*, 1–11. https://doi.org/10.3389/fnins.2014.00088

Zekveld, A. A., Koelewijn, T., & Kramer, S. E. (2018). The pupil dilation response to auditory stimuli: Current state of knowledge. *Trends in Hearing, 22*, 233121651877717. https://doi.org/10.1177/2331216518777174

Zekveld, A. A., Kramer, S. E., Rönnberg, J., & Rudner, M. (2019). In a concurrent memory and auditory perception task, the pupil dilation response is more sensitive to memory load than to auditory stimulus characteristics. *Ear and Hearing, 40*, 272–286. https://doi.org/10.1097/AUD.0000000000000612

What's a Pupil Worth? The Promise and Challenges of Cognitive Pupillometry in the Wild

Russell A. Cohen Hoffing and Steven M. Thurman

1 The Promise

Opportunistic sensing harbors the promise of changing our relationship with technology use from mere tools to members of a productive team. Opportunistic sensing refers to the passive collection of data from sensors that can be leveraged for other purposes (Campbell et al., 2006; Lance et al., 2020). Here, we discuss opportunistic sensing in relation to physiology, specifically pupil size fluctuations. In the context of pupil size dynamics, eye-tracking technology can be used to passively record and report pupil size. An opportunistic sensing system would then leverage this data to infer a mental state of the user and subsequently enable an intervention to modify that mental state or otherwise aim to optimize behavior and performance. Physiological data, and pupil size in particular, may hold the key to enabling the seamless integration of technology into human-agent teaming contexts because (i) it allows for the ability to measure changes in the autonomic nervous system, which has been shown to allow inference of human mental states (Mathôt, 2018; Sirois & Brisson, 2014) and (ii) it can be collected at low cost with minimally invasive sensors that can function in low-power and ambulatory environments. These features are particularly important for user adoption so as not to interfere with other wearable technologies, behaviors, and ongoing task performance.

Another important advantage of using physiological data to infer human mental states in an opportunistic sensing context is that it can provide objective information about the user. Objective reporting of internal mental states may be helpful in situations where reporting an internal state would interrupt natural behavior, where the user is unable to report at all (i.e., due to injury), or does not wish to report their internal state (i.e., a football player may have motivation not to report a concussion).

R. A. Cohen Hoffing (✉) · S. M. Thurman
U.S. DEVCOM Army Research Laboratory, Adelphi, MD, USA
e-mail: russell.a.cohenhoffing.civ@army.mil

M. H. Papesh, S. D. Goldinger (eds.), *Modern Pupillometry*,
https://doi.org/10.1007/978-3-031-54896-3_9

In this review, we focus on the promise and challenges of using pupil size data for opportunistic sensing in the wild.

Pupils are of particular interest here because eyes are one of only two organs (the other being the skin) that are directly visible to the outside world and are influenced by both the autonomic and central nervous systems (see Mathôt, 2018; Sirois & Brisson, 2014; and Laeng & Alnaes, 2019 for reviews). Carefully controlled experiments have shown that pupil size changes reliably and robustly reflect a panoply of meaningful internal state changes. For example, paradigms that investigate cognitive load, attention, surprise, decision making, memory processes, target recognition, prediction errors, stress, fatigue, vigilance, creativity, and the exploration/exploitation trade-off have all shown that these cognitive processes can measurably influence pupil size.

As pupil size measurements relate to the development of opportunistic sensing systems, it is important to consider whether the given signal can track a particular state of interest at an actionable temporal scale, and with a suitable level of both sensitivity and specificity. For example, sub-components of the decision-making process like information accumulation, decision selection, and decision output, given their rapid temporal profile, might be insufficiently distinguishable using pupil size due to the slower temporal dynamics of pupil size that varies on the order of seconds rather than milliseconds. There is, therefore, a limit to the ability of pupil size to accurately track state changes that occur in rapid succession or that last only fractions of a second. In this review, we use a temporal framework to discuss the time scales at which pupil size may be able to track internal state changes, specific examples of how this information may be exploited in opportunistic sensing systems, and what significant gaps and challenges remain that prevent the development and deployment of such systems.

2 Temporal Dynamics of Pupil Size

To first understand which state changes have the potential to be extracted from the pupillary response, one must understand the constraints of the anatomical factors and neural circuitry that control pupil dynamics. Below, we discuss pupillary dynamics in terms of transient, or phasic, responses, followed by longer-term changes or tonic responses.

2.1 Phasic Pupillary Dynamics

Pupil size changes, at their fastest, are constrained by the speed of musculature activation, as well as the input-response delay of the pupil circuit. Studies have quantified the delay by looking at the time to the initial constriction of the pupil in response to a brighter stimulus, which typically ranges from 180 to 350 ms (Bergamin & Kardon, 2003; Bos, 1991; Ellis, 1981; Lowenstein & Loewenfeld, 1964). Depending on the stimuli and experimental conditions, reports have shown that the speed of the constriction response in control subjects can range from 2.79 to 9.27 mm/s, whereas the dilation response is much slower ranging from 0.76 to 3.84 mm/s (Bremner, 2012; Tekin et al., 2018; Thiagarajan & Ciuffreda, 2015; Truong & Ciuffreda, 2016). This speed asymmetry is why the pupil response is better fit by asymmetric functions like the gamma probability density function, compared to symmetric distributions like the normal Gaussian distribution, due to the shallower slope of the dilatory response (Hoeks & Levelt, 1993). Studies that have investigated the component influence of rod, cone, and melanopsin ganglion cells on pupil size using repetitive sinusoidal light stimuli have found that while pupil size can synchronize with flickering stimuli up to 9 Hz, peak signal synchronization occurs in the range of 0.5–1 Hz (Adhikari et al., 2019). Similarly, studies that have used frequency analysis of pupil size to derive cognitive features typically focus on the 0.01–5 Hz frequency range for an optimal balance of time resolution and signal strength of cognitive-related modulations (Duchowski et al., 2020; Onorati et al., 2013). These results suggest a temporal limit for detecting robust phasic pupillary responses to one, or at most two, cognitively-induced dilatory events per second.

Research has found that despite the pupil being a sluggish signal, the pupil can respond to a single, very briefly presented stimulus (see Fig. 1). For example, Lee et al. (2017) reported robust pupillary light responses to flashes of light that lasted only 1 ms. Paradigms investigating cognitive effects such as surprise or target detection on pupil size (while controlling for luminance) find that a single, very brief stimulus (i.e., 133 ms) such as an auditory oddball can induce a cognitive-related transient dilation of the pupil (Cohen Hoffing & Seitz, 2015; Gilzenrat et al., 2010; Murphy et al., 2011; Nassar et al., 2012). Yet, given the input-response delay and sluggish musculature, it generally takes about 0.5–2 s to reach a maximal pupil dilation response, even for very brief events (Mathôt, 2018; Sirois & Brisson, 2014). Thus, while transient cognitive processes can be inferred from pupil size in isolation in controlled laboratory studies, the influence on pupil size may be confounded by

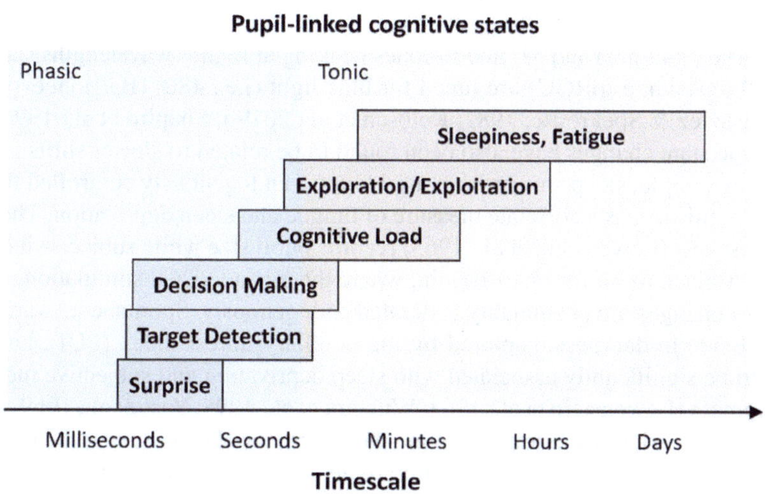

Fig. 1 Overview of cognitive states and their timescale

subsequent internal and/or external factors that immediately follow. As we will discuss later (Fig. 3), these influences can either depress, shift, or combine in a linear or non-linear fashion with previous and ongoing pupil signal modulations. These features of the pupil circuit exemplify why it is important in controlled laboratory settings to allow for long inter-stimulus intervals, thereby allowing researchers to capture the maximal response to a stimulus without interference from preceding or forthcoming trials. It also illustrates one of the challenges of employing pupillometry for state estimation in uncontrolled or real-world contexts where stimuli and inter-stimulus intervals cannot be predetermined by the experimenter.

2.2 *Tonic Pupillary Dynamics*

Tonic pupillary dynamics refer to slower shifts in pupil size that reflect longer-term changes in luminance and/or slower fluctuations in state. Understanding of both tonic and phasic dynamics comes from the study of the photoreceptor contribution to pupil size. Using the "silent substitution" technique, researchers have studied how rods, cones, and ipRGC signals influence pupil size by making use of light adaptation periods and briefly presented monochromatic light stimuli (Barrionuevo et al., 2014; Estévez & Spekreijse, 1982; Kelbsch et al., 2019a; Shapiro et al., 1996). For example, "dark adaptation" consists of subjects being exposed to scotopic conditions for 5–15 min to deactivate cones and activate rod photoreceptors. These adapting light protocols not only cause changes in absolute pupil size due to the amount of ambient light but also lead to differences in the phasic response to briefly presented color stimuli. These relative phasic changes are then used to infer the photoreceptor contribution to pupil size dynamics. Findings from this work have elucidated the conditions in which photoreceptors drive pupil size and have found that rods influence pupil size in scotopic conditions, whereas cones and ipRGCs influence pupil size under photopic conditions. Additionally, cones and ipRGCs are tuned to respond to particular wavelengths with S-cones peaking at lower wavelengths (i.e., 442 nm) and M- and L-cones peaking at higher wavelengths (i.e., 543 and 570 nm) while ipRGCs are tuned for blue light (i.e., 480) (Barrionuevo et al., 2014; Estévez & Spekreijse, 1982; Kelbsch et al., 2019b; Shapiro et al., 1996).

Unique state changes have also been found to be related to slower shifts in pupil size. For example, sleep studies have used hippus, endogenously controlled fluctuations in pupil size, as a correlate measure of fatigue and sleep deprivation. The pupil sleepiness test (Lowenstein et al., 1963) records pupil size while subjects sit quietly in a pitch-black room for up to 10 min, where there is no visual stimulation, and all pupillary changes are presumably generated endogenously. Spontaneous variability of pupil size in darkness, captured by the pupillary unrest index (PUI), has been found to be significantly associated with sleep deprivation and subjective measures of sleepiness (Lowenstein et al., 1963; Wilhelm et al., 1998; Yoss et al., 1969, 1970).

Similarly, variations in tonic baseline pupil size have been found to be related to time-on-task effects reflecting shifts in fatigue, engagement, and performance over

Table 1 Overview of cognitive states, their temporal scale and example application space

Cognitive state	Temporal scale	Application
Decision making	ms-seconds	"Just-in-time" intervention
Surprise, target detection, arousal	ms-seconds	Automatically annotate world-events
Cognitive load	seconds-minutes	Thresholds guide offloading of tasks
Exploration/Exploitation	seconds-minutes	Imitation learning of adaptive human traits
Fatigue, sleepiness, Circadian cycles	minutes-hours-days	Predictive allocation of tasks; return to duty

the course of several minutes of the task (van den Brink et al., 2016; Hopstaken et al., 2015). Sleep-related studies have also shown that baseline pupil size fluctuates gradually and predictably over the course of several hours of wakefulness to reflect both homeostatic sleep pressure (linear decrease in baseline pupil size) and cyclical circadian effects depending on the time of day (Daguet et al., 2019). In a longitudinal study, baseline pupil size was shown to vary with response times within subjects across multiple sessions that spanned 16 weeks (Cohen Hoffing et al., 2020).

Other studies investigating exploration-exploitation behavior and the neural structures (i.e., locus coeruleus–norepinephrine system [LC-NE]) driving these behaviors use the pupil as an indirect measure of LC-NE activity. In these studies, baseline pupil size varied with exploration and exploitation, where leading up to a shift from exploitation to exploration, there was a large increase in pupil size followed by a reduction in pupil size after exploration was initiated (Algermissen et al., 2019; Gilzenrat et al., 2010). Together, these studies suggest that pupil size is also influenced by longer-term environmental and endogenous influences ranging from minutes to hours.

Taken together, the pupil, while sluggish, is a signal that can respond to temporally disparate stimuli and be used to infer internal states that unfold over a range of time scales. Next, we provide a more comprehensive review of literature associating pupil dynamics with state changes, starting with the shortest time scales and continuing to longer time scales, with commentary to discuss potential applications for opportunistic sensing systems (Table 1).

3 Phasic State Changes

3.1 Decision Making

Pupil dynamics reflect components of the decision-making process on the order of milliseconds to seconds. Recently, research has cast a different light on understanding cognitive influences on pupil size using a theoretical model of decision-making and operationalized using the drift diffusion model (Ratcliff et al., 2016). A key component of this model posits that preceding a decision (i.e., between two alternative choices), the brain accumulates information until a threshold is met and a choice (operationalized by reaction time) is made (for review see Ratcliff et al., 2016). Pupil size has been found to track multiple facets of the decision-making process. For example, in a mental arithmetic task, the latency of peak dilation after stimulus onset was strongly correlated with response times on a trial-by-trial basis (Cohen Hoffing et al., 2020). This result was interpreted as pupil size dilating until a response was made and then returning to the pre-stimulus baseline level. Similarly, a number of studies have found that pupil size tracks motor responses (Hupe et al., 2009), perceptual decisions (Einhauser et al., 2008), and can explain individual differences in decision biases (de Gee et al., 2014). Interestingly, other work has found that even in the absence of an overt motor response, phasic pupillary responses are associated with the mental decision-making process (Einhauser et al., 2010). These results suggest that pupil size can be used as an indicator of decision-making and may reflect evidence accumulation. In line with this interpretation of pupil size, Zénon (2019) suggests a framework where phasic dilations in pupil size can be thought of as tracking "information gain" when internal models are updated after new information is acquired (Zénon, 2019).

In an opportunistic sensing context where the decision-making state can be inferred, a system can engage in just-in-time information management. For example, information presentation can be delayed until after a user is done making a decision. This type of system would be able to manage overload of information and distractions so as to protect ongoing decisions. In times when an immediate task must be completed, task allocation may be distributed based on the cognitive availability of a user and could help to distribute tasks among members of a team based on decision-making capacity.

3.2 Surprise, Target Detection, and Phasic Arousal Changes

Pupil dynamics reflect surprise, target detection, and phasic arousal changes on the order of milliseconds to seconds. Pupillometry research has found that target detection (Cohen Hoffing & Seitz, 2015; Liao et al., 2016; Privitera et al., 2010; Strauch et al., 2020) and surprise (Kamp & Donchin, 2015; Lavin et al., 2014; Preuschoff et al., 2011) can cause a transient dilation of the pupil. Paradigms studying target detection and surprise focus on inducing deviant stimuli. For example, in the oddball paradigm, participants are told to identify a target among distractors. What makes the target deviant in these paradigms is that there are rare targets and frequent distractors. In paradigms targeting surprise mechanisms, deviant stimuli are unexpected, such as when a low probability gamble pays off. Typical findings in these paradigms indicate pupil dilations in response to deviant stimuli. These pupillary responses have been interpreted as reflecting increased transient attention allocation and activation of the underlying norepinephrine system when processing a "surprising" deviant stimulus.

An opportunistic system can exploit many of these state-linked pupillary responses via visualization by automatically annotating events (Epps & Chen, 2018; Solon et al., 2019) that correspond to the inferred state. Developing this opportunistic visualization has been the motivation behind our group's Tactical Awareness via Collective Knowledge project (*TACK*) that seeks to develop a system capable of automatically annotating objects seen in the environment as being particularly salient or of human interest. This concept proposes using human cognition and behavior as a filter to segment world-data into important objects and events. An exemplar system that may make use of pupillometry would be an Augmented Reality heads-up display that incorporates eye tracking. This system could implement computer vision that can be applied to the world camera feed to segment objects. Next, a combination of fixation data and pupillary responses could then down-select detected objects and automatically label the world object as an object of human interest.

4 Phasic to Tonic State Changes

4.1 Mental Effort and Cognitive Load

Pupil dynamics reflect mental effort and cognitive load on the order of seconds to minutes. Pupillometry for cognitive-related research originally found its legs in studies that focused on understanding how mental effort and task difficulty influenced pupil size dynamics, including several early and influential studies (Beatty, 1982; Hess & Polt, 1964; Kahneman et al., 1967). In a variety of tasks manipulating stimulus difficulty and the amount of information in working memory, studies have reliably shown that the pupil dilates in response to increases in mental effort and working memory load. These findings have been interpreted as a compensatory response to task demands, where the pupil is tracking onto the increased allocation of attention or "energetic" resources (Laeng & Alnaes, 2019).

A major focus and interest in this compensatory response is due to the behavioral errors that arise when attentional and resource demands are high. For example, target detection (Helton & Russell, 2011), air traffic control (Shorrock, 2005), and driving behavior (Ross et al., 2014) are often impaired under conditions of high load. Because of the real-world life and death consequences of cognitive load errors in accident-prone contexts, opportunistic sensing of excessive demand is desirable to intervene before catastrophic errors are made. In a closed-loop system, pupillometric features of mental load could be used to identify and respond to these contexts. For example, in a helicopter pilot scenario requiring a difficult maneuver in low vision conditions, a system could offload navigational task demands by activating auto-pilot if it detects that the pilot is overloaded. Pupil size metrics are a candidate feature because they can track cognitive demand as it develops over the course of seconds.

Another potential method of exploiting the task-demand compensatory response is in an adaptive learning system. Research has found intraindividual differences in the compensatory response by expertise, where experts show reduced responses to increased task demands (Allen et al., 2004; Camp et al., 2001; Sweller, 1988; see Box 1). As such, reductions in the compensatory response to a loading stimulus may be used to infer learning and the mastery of a task. For example, software may leverage opportunistic sensing of this signal to ascertain mastery not just based on task performance (i.e., the standard way) but by objective physiological indicators of mastery before moving on to the next lesson.

Box 1

After Action Reviews: Eye movement tracking research has identified differences between experts and novices (Gegenfurtner et al., 2011). For example, expert radiologists are able to identify cancer in an image faster than non-expert radiologists and show accompanying shorter fixation durations and search times. These eye movement differences have been related to performance enhancement and are interpreted to signify subject-matter expertise and efficient strategy selection (Gegenfurtner et al., 2011). Subsequent studies have used these differences to inform and train novices to adopt expert-like eye movements to improve performance (Tien et al., 2014). These types of training tools are reminiscent of the After Action Review (AAR) used in military settings where performance on an operation is discussed to identify what happened, why it happened, and how performance can be improved (Mastiglio et al., 2011; Morrison & Meliza, 1999). Pupil-linked states can be used to supplement an AAR as a way to identify expert-like unfolding of cognition in a task or operation. For example, pupil-linked state inference could be used to “tag” temporally specific portions of a task to guide discussion about the unfolding cognitive processes that were present or absent.

4.2 Exploration/Exploitation

Pupil dynamics can reflect exploration and exploitation behavioral states on the order of seconds to minutes. Exploration/exploitation behavior describes an adaptive control state that organisms have evolved to navigate the trade-off between either staying engaged with the current task and behaviors or, at a cost, exploring alternative and possibly more rewarding tasks and behaviors (Cohen et al., 2007; Hills et al., 2015; Mehlhorn et al., 2015). Recent theories have implicated the locus coeruleus-norepinephrine system, thought to be a major source of cognitive influences on the pupillary system, in supporting adaptive behavior on the exploration/exploitation continuum (Aston-Jones & Cohen, 2005). In support of this connection, studies have found that pupil size can track transitions along the continuum of these two states. For example, in a study by Jepma and Nieuwenhuis (2011), participants completed a four-armed bandit task, where a reward was given after choosing one of four slot machines. Participants were instructed to earn the greatest number of points while the reward value of slot machines varied independently over time. In this task, participants had to decide between choosing a previously known slot machine of high value (exploitation) or switching to find another potentially high-value slot machine (exploration), particularly as value changed dynamically over time (Jepma & Nieuwenhuis, 2011). Pupil size was found to vary with the exploration/exploitation state, with exploration showing a higher average pupil size in comparison to pupil size during exploitation. Additionally, as participants transitioned between the two states, results indicated a gradual increase in average pupil size from exploitation to exploration and a gradual decrease in pupil size from exploration to exploitation. Similarly, other studies have found similar changes; in one notable study, when participants freely chose when to switch to another task, average pupil size was found to increase during this transition stage (Algermissen et al., 2019; Gilzenrat et al., 2010).

The ability to track an adaptive human state can be useful for developing intelligent, adaptive systems. For example, reinforcement learning algorithms have been used to enable the imitation of human behaviors (Cèsar-Tondreau et al., 2021). These types of reinforcement learning algorithms could be guided by pupil-linked exploration and exploitation states to learn different sets of behaviors and the contexts in which to appropriately switch between them. A system with knowledge of the context and environment could also adaptively encourage a user to explore other options if they have spent too much time in exploitation mode, or vice versa, as a means of pushing users toward a more optimal balance of exploration/exploitation behaviors. This could be especially useful for novices that are learning a new task domain or in extremely stressful situations in which a user's exploration/exploitation strategy may be thrown off-kilter by intense cognitive demands, time pressure, or stress.

5 Tonic and Beyond State Changes

5.1 Fatigue

Pupil dynamics can reflect fatigue on the order of minutes to hours to days. Sleep research has made use of pupillometry as a measure of autonomic arousal to investigate how performance and arousal are affected by sleep deprivation (McClelland et al., 2010; Wilhelm et al., 1998). Sleep studies have shown that steady-state (baseline) pupil size tracks very closely with changes in subjective sleepiness and arousal associated with circadian phase (Van Egroo et al., 2019). A recent study reported that baseline pupil size reduced linearly over time with sustained wakefulness, reflecting homeostatic sleep pressure, but also showed a sinusoidal component reflecting the influence of circadian processes (Daguet et al., 2019). Though this study reported a phase difference in the circadian modulation between baseline pupil size and subjective measures of sleepiness, indicating that the pupil may not reliably reflect sleepiness per se at all times of the day, the results do confirm that baseline pupil size varies throughout the day consistent with homeostatic and circadian influences on arousal. There is also evidence that pupil size variability, quantified by the pupillary unrest index (PUI), varies systematically with EEG power and subjective sleepiness during sustained wakefulness (Regen et al., 2013).

Fatigue effects have been viewed similarly to mental effort and cognitive load, where increases in fatigue are related to performance decrements in demanding tasks such as driving or piloting a vehicle. Systems that focus on cognitive load and mental effort could intervene to reduce the demands causing load, but because fatigue is a state that manifests on a longer time scale, systems that focus on preventative measures may be more valuable. For example, a system that can control task allocation may benefit from fatigue detection where a pilot would be re-tasked or disallowed from performing critical operations. This may be necessary in instances where a user may be unmotivated or unable to subjectively assess levels of fatigue accurately. Another less reactive and more preventative measure would be in defining the boundaries of alertness levels. If individualized variations in fatigue can be predicted (see Box 2), then predictive tasking can be used to reallocate personnel to operations at the time of predicted increases or decreases in fatigue.

Box 2

Intra- and Inter-individual differences in the PLR: Much of the understanding gained about underlying processes and factors that drive the pupil response has been conducted in the field of ophthalmology. Research has focused on clinical applications to diagnose abnormalities in the brain as indicated by deviations from a normal PLR (see Hall & Chilcott, 2018 for review). For example, one can infer neural complications if one or both pupils do not respond to a flash of light. Intraindividual differences in the PLR have been used to infer drug use, concussions (Hall & Chilcott, 2018), and as a signal for brain–computer interfaces (Mathôt et al., 2016). Variation in the PLR has also been used to infer spatial attention. For example, Mathot (2016) paired letter stimuli with black and white patches for users to fixate on covertly (i.e., while keeping fixation on the center of the screen) while these patches flashed. The resulting pupillary response to changes in luminance was modulated by covert attention and could be used to select the corresponding letter, thereby demonstrating its applicability to patients with movement paralysis. In addition to intraindividual variation in the PLR, interindividual PLR variation has been used to classify a number of conditions such as aging, autism spectrum, Parkinson's disease, Alzheimer's disease, and schizophrenia among others (Hall & Chilcott, 2018).

6 The Challenge

6.1 *The Inverse Problem and a Dilation by Any Other Name*

The cognitive pupil pathway is thought to be primarily mediated via activation of the sympathetic system, and/or inhibition of the parasympathetic pathway (Hall & Chilcott, 2018; Mathôt, 2018; Sirois & Brisson, 2014). Because the dilator muscles of the pupil are activated via sympathetic activation, researchers have typically categorized pupillary dilations as reflecting cognitive underpinnings (however, see Naber et al., 2013, which look at reductions in the PLR). This logic is underpinned by carefully controlled studies that account for confounds such as luminance stimulation (Naber & Nakayama, 2013; Sulutvedt et al., 2021; Zavagno et al., 2017). However, in the wild, this control is not possible, such that there are many confounding factors when attempting to infer mental states. The scientific challenge, then, for opportunistic sensing systems to exploit pupil size is in dealing with the inverse problem.

A "dilation by any other name" refers to the inverse problem; the notion that it is difficult, if not impossible, to assign causality to a component influence of a signal because the true relationship between the component and measured signal is unknown. With regards to pupil size, measured pupil size is a unitary signal that is ultimately controlled by the same parasympathetic and sympathetic muscles regardless of cognitive or non-cognitive influence (e.g., changes in brightness). Thus, the grand pupillometric challenge, given the laundry list of cognitive and non-cognitive processes that have access to the circuit that modulates pupil size, is in attributing a given pupillary change to a specific cognitive process or external event.

Traditionally, computerized pupillometry studies have assigned cognitive influence on pupil size by controlling for confounds using experimental controls such as eye movements (Brisson et al., 2013; Hayes & Petrov, 2016; Mathôt et al., 2015), motor movements (Einhauser et al., 2010), visual stimulation (Cohen Hoffing & Seitz, 2015; Nassar et al., 2012) and luminance (Naber & Nakayama, 2013; Sulutvedt et al., 2021; Willenbockel et al., 2010; Zavagno et al., 2017). Eye movements, including variation in saccades and fixations, as well as the change in visual stimulation that accompanies the eye movements, are controlled for via gaze-contingent stimulus presentation, whereby participants are required to maintain fixation at a specified area such as the center of the screen (Mathôt et al., 2015). To control for motor movements, like a button press used in typical computerized cognitive studies, researchers have decoupled a button press and pupillary response by asking participants to make a response after a trial is completed, thereby delaying when a motor response is made (Einhauser et al., 2010). Researchers can control for visual stimulation on the pupil by using auditory instead of visual stimuli to deliver cognitive tasks (Cohen Hoffing & Seitz, 2015; Nassar et al., 2012). For luminance controls, simple stimuli (e.g., a math equation, color patches) can be equated by matching stimulus luminance between conditions within a participant. For more complex stimuli-like images, gray scaling and matching by luminance using tools such as the SHINE toolbox are possible (Willenbockel et al., 2010).

When experimental control of non-cognitive pupillary confounds is not possible, statistical methods have been used. However, the applications of statistical methods to estimate pupil-linked states vary by the temporal scale of the state. Below, we further discuss how the estimation of shorter and longer time scale pupil-linked state changes face different challenges.

6.2 Phasic to Tonic State Changes

Using phasic changes in pupil size to estimate transient cognitive processes is difficult due to the sluggish nature of the pupil. Moreover, the difficulty of assigning a given dilation or constriction to a cognitive influence is compounded within complex environments due to the fact that during naturalistic behavior stimuli can rapidly change. For example, visual stimulation can change dramatically every time an eye movement is made. Research focusing on naturalistic visual search behavior indicates that participants, on average, make 2–4 fixations per second (Cronin et al., 2020; Hwang et al., 2011). In addition to local luminance variation due to gaze changes, evidence suggests that eye movements like saccades and blinks also influence pupil size (Knapen et al., 2016; Yoo et al., 2021). In combination with the input-to-pupil response lag, there are at least two properties of pupil size that make the statistical accounting of non-cognitive influences and subsequent assignment of pupil dilations to cognition difficult: (1) cognitive influences have a smaller signal-to-noise ratio than non-cognitive influences and (2) cognitive-related dilations are relatively slower.

6.2.1 Cognitive Influences Have Smaller Signal-to-Noise Ratio

A classical finding from the pupillometry literature is that mental effort and decision processes during mental arithmetic tasks can be reliably related to pupil dilations (Beatty, 1982; Cohen Hoffing et al., 2020; Kahneman et al., 1967). However, consider Fig. 2, illustrating a single trial each of a bright stimulus presentation on a dark background and an easy mental arithmetic task of evaluating the truth of a modular math statement (i.e., 36 = 4 [mod 2]). The discrepancy in one trial between the non-cognitive mediated pupillary light response and the cognitive-mediated mental processing is large. This discrepancy is likely because pupil circuitry evolved to protect and focus light onto the retina, as well as coordinate both eyes, and thus reflexively responds to light (Mathôt, 2018), whereas, in relative comparison, the cognitive influence on the pupil may not have had such direct evolutionary pressures (however, see Prochazkova et al., 2018). The reduced signal-to-noise ratio of cognitive influences on the pupil adds to the difficulty of accounting for concomitant non-cognitive influences.

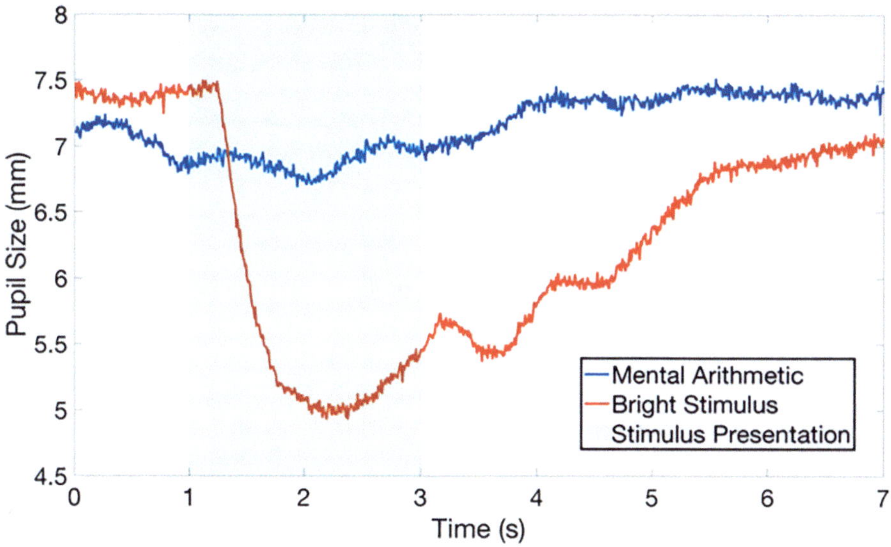

Fig. 2 Pupil dynamics from one subject illustrate the difference in magnitude of the effect from one trial of solving an easy mental arithmetic problem (blue) to one trial of viewing a bright stimulus (red) over two seconds. The pupil response induced by light is both faster and larger in magnitude than the change induced by mental arithmetic

6.2.2 Dilations Are Slow

It has been well documented that pupillary dilations and constrictions present a velocity asymmetry, with dilations being slower than constrictions (see temporal dynamics section above). To illustrate this point, compare a pupillary constriction (Fig. 3a) in response to a bright stimulus change with a pupillary dilation in response to a dark stimulus change (Fig. 3b). These figures illustrate a much slower dilation rate when compared with the faster constriction rate. This is likely due to the differences in characteristics of the sympathetic and parasympathetic systems that, respectively, control the dilation and constriction muscles (Hall & Chilcott, 2018). These differences create a challenge when attempting to statistically model the cognitive component. For example, typical analytic methods consist of peak scoring (e.g., max amplitude of post-stimulus response) and averaging windows (e.g., average pupil size pre-stimulus) of pupil size data. In controlled settings, peaks and stimulus times are clear and consistent. However, in unconstrained settings, the cognitive components will likely coincide with both constrictions and dilations due to the multitude of concurrent influences on the pupil. To specify the max/min or an average window using peak scoring, a time window must be specified, and different time windows will need to be specified to match the shape of the dilation or constriction. The designation of a time window will be a challenge for generalization to other task structures, especially when windows do not match, and stimulus event timing is not under experimenter control.

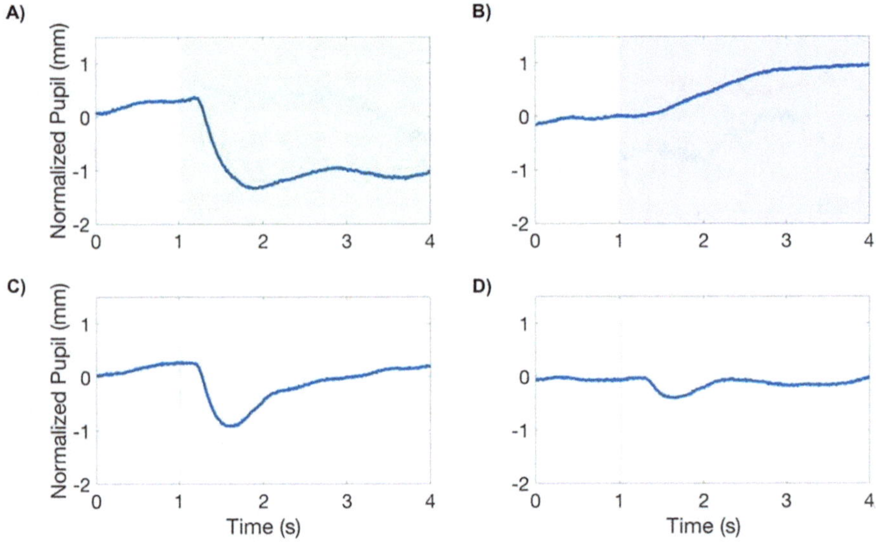

Fig. 3 Pupil responses to grayscale stimuli for 3 s (**a, b**) and 0.03 s (**c, d**). (**a, c**) A lighter stimulus (compared with the previous stimulus) leads to a rapid constriction of the pupil. (**b**) A darker stimulus leads to a slower dilation of the pupil. (**d**) A briefly presented dark stimulus, and a subsequent brighter stimulus at the dark stimulus offset, leads to a reduced constriction. (Reproduced based on Korn and Bach (2016))

Additionally, the peak scoring technique does not make headway against the inverse problem as it does not dissociate any concomitant non-cognitive events from cognitive events. To further illustrate how the temporal dynamics of stimuli impact the resulting pupillary dilations and constrictions, consider when stimuli are rapidly presented (e.g., every 0.03 s). In Fig. 3c, the pupil exhibits a robust constriction in response to the briefly presented bright stimulus. By comparison, the pupil exhibits no observable dilation in response to the briefly presented dark stimulus (Fig. 3d). However, upon offset of the dark stimulus (thereby leading to a bright stimulus change back to the baseline background color), a constriction can be observed. This constriction is of a lesser magnitude than the constriction to the single pulse of light (Fig. 3c), even though the magnitude of brightness onset is the same as in Fig. 3c. One explanation for the reduced constriction is that the interaction between the delayed activation of the dilation muscle and faster activation of the constriction muscle leads to an overall reduced constriction (Korn & Bach, 2016).

General linear modeling (GLM) is a promising method that has recently been borrowed from signal processing methods used in EEG and fMRI (Hoeks & Levelt, 1993; Wierda et al., 2012) to model the pupil. This methodology conceives the pupil response as a combination of various impulse signals over time, which are convolved with an impulse response function to predict the measured output signal. Much of the pupil response modeling work has been influenced by early work by Hoeks and Levelt (1993). In this work, an underlying transformation from internal events activating relevant pupil-linked circuitry is abstracted out and marked as an "event." This event is then convolved with a canonical pupil response function (Ehrlang Gamma) and then used to fit the resulting pupillary response. This approach is advantageous because it makes use of all data points collected and is flexible enough to model different kinds of events and variations in the pupil impulse response functions. For example, in Knapen et al. (2016) deconvolution methods were used to model the influence of eye movements on the pupil response including saccades and blinks (Knapen et al., 2016). In Denison et al. (2020), sequential pupil-linked mental components of a temporal attention task were estimated using a general linear model that assumed the canonical shape of the pupil response as reported by Hoeks and Levelt, namely a latency of 930 ms and a shape parameter of 10.1 (Denison et al., 2020). A key finding from this work was that model fits were greatly improved by allowing event latency to vary in the parameter fitting procedure, revealing anticipatory event-related pupil dilations that occurred prior to the onset of temporally predictable pre-cue and stimulus events at the start of each trial. In Korn and Bach (2016), luminance changes are modeled as events using two canonical pupil response forms to fit the different constriction and dilation shapes due to an increase or decrease in brightness, respectively.

While this methodology allows for modeling of many different sub-component influences, it still remains to be seen whether sub-processes can be dissociated using this method. For example, can a dilation caused by a darker stimulus be disambiguated from a dilation induced by a cognitive process? One open research question is whether the underlying assumptions of GLM modeling methodology are reflective of the processes that create the resulting pupillary response. GLM modeling assumes that sub-component processes are additive, in that the summation of each component signal will give rise to the final signal. At face value this cannot always be true, as the pupil cannot infinitely dilate or constrict. But it remains to be seen whether this methodology is adequate for predicting pupil responses given an "event." An additional minor disadvantage of GLM methodology is the complexity and specification issues associated with parametric statistical models, whereas in contrast, this matter is avoided by using simple peak scoring methods. Overall, GLM modeling and similar methodologies are promising statistical modeling approaches to dealing with the obstacles that exist for estimating short-time scale pupil-linked states, but their full potential has not yet been realized.

6.3 Tonic Changes and Beyond

While using short-term phasic changes in pupil size for state estimation faces a signal-to-noise ratio issue, state estimation using tonic changes in pupil size may not be similarly plagued. For example, if 30-second windowed averages of pupil size are used to estimate cognitive load, a single "event" (i.e., a rapid change in stimulus luminance) will not likely substantially influence the cognitive load estimation over such a long time window. In contrast, if a large enough change in stimulus luminance coincides with a target detection cognitive event, the resulting cognitive signal will be washed out. In other words, the effect of moment-to-moment non-cognitive influences on pupil size will be averaged out given a long enough time window average. However, this will remain true only if the non-cognitive events do not incur a substantial, consistent, and/or frequent unidirectional dilation or constriction. For example, a change in ambient lighting when going from indoor lighting to unlit outdoors during the night will cause a shift in baseline pupil size on the order of several millimeters (Atchison et al., 2011). The brightness of a prolonged change in ambient light will influence not only the baseline pupil size but also the types of photoreceptors that are activated (see Sect. 2). This change in active photoreceptors will also influence the response profile of pupil size, given that spectral properties differentially impact each photoreceptor, a characteristic that would mostly impact short-term phasic changes in pupil size.

While changes in ambient light will influence baseline pupil size, perceived brightness changes will also similarly influence pupil size (Castellotti et al., 2020; Laeng & Endestad, 2012; Naber & Nakayama, 2013; Sulutvedt et al., 2021; Suzuki et al., 2019; Thurman et al., 2021; Zavagno et al., 2017). For example, Naber and Nakayama (2013) presented cartoon scenes, depicting a sunny day, to participants and found that there were larger constrictions when the scene was upright in comparison to upside-down versions of the same scenes, despite brightness changes being equal. A number of studies have found consistent findings of perceptual brightness effects on pupil constrictions. Explanations of the perceptual effect on pupil size have relied on the ecological hypothesis, whereby evolutionary processes have led to the development of expectations that light from the sun comes from above. Thus, humans have learned to associate light with upper spatial locations and constrict their pupils in preparation of engaging with the brighter environment.

While these studies primarily examined phasic responses to perceived luminance in controlled settings, recently these findings have been found to generalize to naturalistic contexts. In Thurman et al. (2021), participants navigated a naturalistic virtual outdoor environment to identify targets (e.g., motorcycles) while pupil size was recorded. To investigate the relationship between luminance and pupil size, pixel-wise correlations of the viewed scenes and pupil size were conducted and visualized as correlation maps. While a foveal bias hypothesis was predicted, where luminance changes near fixation would correlate the strongest with pupil size, the results instead indicated that luminance variation above fixation led to the largest constrictions and accounted for the most variance in the pupil signal over time. This result was interpreted to be caused by the pupil's inherent sensitivity to perceptual brightness due to the presence of the blue sky during the outdoor visual search task.

In the study by Thurman et al. (2021), the blue sky hypothesis was confirmed in a controlled follow-up experiment with luminance-matched color stimuli (i.e., red, blue, and green) paired with grayscale patches in which each color was presented above and below a central fixation point. Results indicated that when blue light was presented above fixation, it induced larger pupillary constrictions relative to when blue light was presented below fixation. In contrast, this relative difference was not found in the other color conditions or even when blue light was presented to the left or right of fixation. This result supported the interpretation that the visual system is sensitive to the global contextual structure of the environment and that the appearance of blue overhead can drive pupil size even more strongly than luminance changes near the fovea.

Another route through which non-cognitive influences can shift tonic pupil size is through consistent eye movements related to task goals. Classic work by Yarbus (1967) has shown that eye movements differ as a function of the task (Yarbus, 1967). If the task goal-induced behavior consistently interacts with environmental stimuli, it will affect baseline pupil size. For example, exploratory search behavior consists of larger and spatially divergent eye movements in comparison to exploitative search behavior. These consistencies in where, when, and what eye movements are being made have the potential to confound baseline shifts in pupil size, which might otherwise be inferred to be directly influenced by cognitive processes.

With respect to the influence of eye movement and shifts in ambient light, methods to tackle the inverse problem with the goal of disentangling non-cognitive influences on tonic changes in pupil size are needed. Using models of sustained changes in pupil size may be applied (Spitschan, 2019; Watson & Yellott, 2012), but these models were developed using data that ranged on the seconds to minutes time scale and might not generalize to longer-term shifts in active photoreceptor populations. However, even if a non-cognitive influence effect size can be estimated, there is very little research on how this influence should be accounted for statistically (Wong et al., 2020). Some methods include ignoring the influence (via assumptions of randomness and averaging out non-cognitive effects), excluding data with known non-cognitive events, linearly subtracting out the effect, including non-cognitive influences as linear covariates, or using non-linear methods like machine learning. Another strategy to estimate pupil-linked states is to use non-cognitive influences as features themselves rather than excluding or accounting for them (Naber et al., 2013). In this method, variation in the PLR due to concomitant non-cognitive and cognitive pupil size would need to be sufficient to infer a latent pupil-linked state. As a proof of concept of this method, Bhatt (2018) found that using pupillary light response features improved the classification of cognitive load (Bhatt, 2018). Overall, more research is needed to understand the shape and temporal profile of how various co-occurring pupil-influencing events combine to create the unitary pupillary response.

6.3.1 Future Directions

Cognitive pupillometry has great promise and faces challenges before this technique can be used in real-world contexts and opportunistic sensing systems (see Open Questions Box). Pupillary fluctuations betray a range of cognitive states that unfold over a few seconds to days. However, the most challenging aspect of using cognitive pupillometry in opportunistic sensing contexts is mapping the temporal dynamics of non-cognitive and cognitive influences to the pupil. For example, the relative weighting of non-cognitive (e.g., blinks, saccades, and luminance) and cognitive factors (e.g., cognitive load, surprise, and fatigue) on the unitary pupillary signal is unclear. While progress has been made in predicting steady-state pupil size (Spitschan, 2019; Watson & Yellott, 2012) and modeling the temporal unfolding of cognitive processes on pupil size (Denison et al., 2020), more work is needed to generalize to naturalistic contexts and different temporal scales. In addition to understanding the weighting of exogenous and endogenous events on the pupil, a further understanding of how simultaneous signals are combined is needed. For example, the GLM modeling technique assumes an additive model where upstream signals are linearly summed to generate the predicted influence on pupil size. While this may predict pupil size in a simplified laboratory experimental context where luminance levels are accurately known, it may be difficult to generalize this technique to complex visual scenes where luminance varies across the whole scene or where fine-grained luminance measurements are infeasible. Specifically, investigations onto the weighting of simultaneous light wavelengths from the periphery and fovea onto pupil size fluctuations are needed. Furthermore, because work has demonstrated that perceived luminance also influences pupil size, the context of the visual scene must also be integrated into predictive modeling of the pupil.

While much of the above discussion has focused on knowledge gaps in pupillary dynamics and computational approaches to filling in these gaps, in part these gaps are due to a lack of naturalistic datasets. In turn, these datasets are unavailable in part due to technological limitations. For example, many eye-tracking systems rely on infrared technology, making it difficult to collect data under sunlight due to the infrared signal being washed out. Additionally, the form factor of existing eye-tracking systems does not lend itself to many realistic contexts or for extended use. For example, mobile eye-tracking systems, typically in the form of wearable glasses, may not be appropriate for highly mobile contexts where large movements are made, such as a pilot undergoing turbulence or an aerial maneuver. Large movements can potentially misalign the camera angle relative to the face and diminish the quality of the pupillary signal. It is worth noting that very few longitudinal pupillary datasets exist that examine continuous variation over the course of a day or repeated measurements across days or across human development. These types of datasets will be crucial for understanding variation in tonic pupillary changes and their respective cognitive and non-cognitive influences. Overall, new eye-tracking technologies that are comfortable, unobtrusive, and robust to movement and noise will enable the ability to collect longitudinal and naturalistic datasets.

Open Questions Box

1. What are the weightings and temporal dynamics of cognitive and non-cognitive events on the pupillary system?
2. Do concomitant influences on the pupil combine linearly or non-linearly?
3. What luminance (i.e., various wavelengths) from a naturalistic scene (i.e., peripheral vs. foveal) predicts pupillary dynamics?
4. How does perceived luminance in naturalistic scenes influence pupil size?
5. What kind of eye-tracking systems need to be developed to track pupil size longitudinally in rugged and ambulatory environments?

References

Adhikari, P., Feigl, B., & Zele, A. J. (2019). The flicker Pupil Light Response (fPLR). *Translational Vision Science & Technology, 8*(5), 29. https://doi.org/10.1167/tvst.8.5.29

Algermissen, J., Bijleveld, E., Jostmann, N. B., & Holland, R. W. (2019). Explore or reset? Pupil diameter transiently increases in self-chosen switches between cognitive labor and leisure in either direction. *Cognitive, Affective, & Behavioral Neuroscience, 19*(5), 1113–1128. https://doi.org/10.3758/s13415-019-00727-x

Allen, R., Mcgeorge, P., Pearson, D., & Milne, A. B. (2004). Attention and expertise in multiple target tracking. *Applied Cognitive Psychology, 18*(3), 337–347. https://doi.org/10.1002/acp.975

Aston-Jones, G., & Cohen, J. D. (2005). An integrative theory of locus coeruleus-norepinephrine function: Adaptive gain and optimal performance. *Annual Review of Neuroscience, 28*(1), 403–450. https://doi.org/10.1146/annurev.neuro.28.061604.135709

Atchison, D. A., Girgenti, C. C., Campbell, G. M., Dodds, J. P., Byrnes, T. M., & Zele, A. J. (2011). Influence of field size on pupil diameter under photopic and mesopic light levels. *Clinical and Experimental Optometry, 94*(6), 545–548. https://doi.org/10.1111/j.1444-0938.2011.00636.x

Barrionuevo, P. A., Nicandro, N., McAnany, J. J., Zele, A. J., Gamlin, P., & Cao, D. (2014). Assessing rod, cone, and melanopsin contributions to human pupil flicker responses. *Investigative Ophthalmology & Visual Science, 55*(2), 719–727. https://doi.org/10.1167/iovs.13-13252

Beatty, J. (1982). Task-evoked pupillary responses, processing load, and the structure of processing resources. *Psychological Bulletin, 91*(2), 276.

Bergamin, O., & Kardon, R. H. (2003). Latency of the pupil light reflex: Sample rate, stimulus intensity, and variation in normal subjects. *Investigative Ophthalmology & Visual Science, 44*(4), 1546–1554. https://doi.org/10.1167/iovs.02-0468

Bhatt, K. (2018). *Inferring task load from the pupil light reflex*. University of New South Wales – Australia.

Bos, J. E. (1991). Detection of the pupil constriction latency. *Medical and Biological Engineering and Computing, 29*(5), 529–534. https://doi.org/10.1007/BF02442326

Bremner, F. D. (2012). Pupillometric evaluation of the dynamics of the pupillary response to a brief light stimulus in healthy subjects. *Investigative Ophthalmology & Visual Science, 53*(11), 7343–7347. https://doi.org/10.1167/iovs.12-10881

van den Brink, R. L., Murphy, P. R., & Nieuwenhuis, S. (2016). Pupil diameter tracks lapses of attention. *PLoS One, 11*(10), e0165274. https://doi.org/10.1371/journal.pone.0165274

Brisson, J., Mainville, M., Mailloux, D., Beaulieu, C., Serres, J., & Sirois, S. (2013). Pupil diameter measurement errors as a function of gaze direction in corneal reflection eyetrackers. *Behavior Research Methods, 45*(4), 1322–1331.

Camp, G., Paas, F., Rikers, R., & van Merrienboer, J. (2001). Dynamic problem selection in air traffic control training: A comparison between performance, mental effort and mental efficiency. *Computers in Human Behavior, 17*(5), 575–595. https://doi.org/10.1016/S0747-5632(01)00028-0

Campbell, A. T., Eisenman, S. B., Lane, N. D., Miluzzo, E., & Peterson, R. A. (2006). People-centric urban sensing. *Proceedings of the 2nd Annual International Workshop on Wireless Internet*, 18-es. https://doi.org/10.1145/1234161.1234179

Castellotti, S., Conti, M., Feitosa-Santana, C., & Del Viva, M. M. (2020). Pupillary response to representations of light in paintings. Journal of Vision, 20(10), 14–14.

Cèsar-Tondreau, B., Warnell, G., Stump, E., Kochersberger, K., & Waytowich, N. R. (2021). Improving autonomous robotic navigation using imitation learning. *Frontiers in Robotics and AI, 8*, 627730. https://doi.org/10.3389/frobt.2021.627730

Cohen Hoffing, R., & Seitz, A. R. (2015). Pupillometry as a glimpse into the neurochemical basis of human memory encoding. *Journal of Cognitive Neuroscience, 27*(4), 765–774. https://doi.org/10.1162/jocn_a_00749

Cohen Hoffing, R. A., Lauharatanahirun, N., Forster, D. E., Garcia, J. O., Vettel, J. M., & Thurman, S. M. (2020). Dissociable mappings of tonic and phasic pupillary features onto cognitive processes involved in mental arithmetic. *PLoS One, 15*(3), e0230517. https://doi.org/10.1371/journal.pone.0230517

Cohen, J. D., McClure, S. M., & Yu, A. J. (2007). Should I stay or should I go? How the human brain manages the trade-off between exploitation and exploration. *Philosophical Transactions of the Royal Society B: Biological Sciences, 362*(1481), 933–942.

Cronin, D. A., Hall, E. H., Goold, J. E., Hayes, T. R., & Henderson, J. M. (2020). Eye movements in real-world scene photographs: General characteristics and effects of viewing task. Frontiers in Psychology, 10, 2915.

Daguet, I., Bouhassira, D., & Gronfier, C. (2019). Baseline pupil diameter is not a reliable biomarker of subjective sleepiness. *Frontiers in Neurology, 10*, 108. https://doi.org/10.3389/fneur.2019.00108

Denison, R. N., Parker, J. A. & Carrasco, M. (2020). Modeling pupil responses to rapid sequential events. *Behav Res* 52, 1991–2007. https://doi.org/10.3758/s13428-020-01368-6

Duchowski, A. T., Krejtz, K., Gehrer, N. A., Bafna, T., & Bækgaard, P. (2020). The Low/High Index of Pupillary Activity. *Proceedings of the 2020 CHI Conference on Human Factors in Computing Systems* (CHI '20). Association for Computing Machinery, 1–12. https://doi.org/10.1145/3313831.3376394

Einhauser, W., Stout, J., Koch, C., & Carter, O. (2008). Pupil dilation reflects perceptual selection and predicts subsequent stability in perceptual rivalry. *Proceedings of the National Academy of Sciences of the United States of America, 105*(5), 1704–1709. https://doi.org/10.1073/pnas.0707727105

Einhauser, W., Koch, C., & Carter, O. L. (2010). Pupil dilation betrays the timing of decisions. *Frontiers in Human Neuroscience, 4*, 18. https://doi.org/10.3389/fnhum.2010.00018

Ellis, C. J. (1981). The pupillary light reflex in normal subjects. *British Journal of Ophthalmology, 65*(11), 754–759.

Epps, J., & Chen, S. (2018). Automatic task analysis: Toward wearable behaviometrics. *IEEE Systems, Man, and Cybernetics Magazine, 4*(4), 15–20. https://doi.org/10.1109/MSMC.2018.2822846

Estévez, O., & Spekreijse, H. (1982). The "silent substitution" method in visual research. *Vision Research, 22*(6), 681–691. https://doi.org/10.1016/0042-6989(82)90104-3

de Gee, J. W., Knapen, T., & Donner, T. H. (2014). Decision-related pupil dilation reflects upcoming choice and individual bias. *Proceedings of the National Academy of Sciences of the United States of America, 111*(5), E618–E625. https://doi.org/10.1073/pnas.1317557111

Gegenfurtner, A., Lehtinen, E., & Säljö, R. (2011). Expertise differences in the comprehension of visualizations: A meta-analysis of eye-tracking research in professional domains. *Educational Psychology Review, 23*(4), 523–552. https://doi.org/10.1007/s10648-011-9174-7

Gilzenrat, M. S., Nieuwenhuis, S., Jepma, M., & Cohen, J. D. (2010). Pupil diameter tracks changes in control state predicted by the adaptive gain theory of locus coeruleus function. *Cognitive, Affective, & Behavioral Neuroscience, 10*(2), 252–269. https://doi.org/10.3758/CABN.10.2.252

Hall, C. A., & Chilcott, R. P. (2018). Eyeing up the future of the pupillary light reflex in neurodiagnostics. *Diagnostics, 8*(1), 19.

Hayes, T. R., & Petrov, A. A. (2016). Mapping and correcting the influence of gaze position on pupil size measurements. *Behavior Research Methods, 48*(2), 510–527. https://doi.org/10.3758/s13428-015-0588-x

Helton, W. S., & Russell, P. N. (2011). Working memory load and the vigilance decrement. *Experimental Brain Research, 212*(3), 429–437. https://doi.org/10.1007/s00221-011-2749-1

Hess, E. H., & Polt, J. M. (1964). Pupil size in relation to mental activity during simple problem-solving. *Science, 143*(3611), 1190–1192.

Hills, T. T., Todd, P. M., Lazer, D., Redish, A. D., & Couzin, I. D. (2015). Exploration versus exploitation in space, mind, and society. *Trends in Cognitive Sciences, 19*(1), 46–54. https://doi.org/10.1016/j.tics.2014.10.004

Hoeks, B., & Levelt, W. J. (1993). Pupillary dilation as a measure of attention: A quantitative system analysis. *Behavior Research Methods, Instruments, & Computers, 25*(1), 16–26.

Hopstaken, J. F., van der Linden, D., Bakker, A. B., & Kompier, M. A. J. (2015). A multifaceted investigation of the link between mental fatigue and task disengagement. *Psychophysiology, 52*(3), 305–315. https://doi.org/10.1111/psyp.12339

Hupe, J. M., Lamirel, C., & Lorenceau, J. (2009). Pupil dynamics during bistable motion perception. *Journal of Vision, 9*(7), 10. https://doi.org/10.1167/9.7.10

Hwang, A. D., Wang, H.-C., & Pomplun, M. (2011). Semantic guidance of eye movements in real-world scenes. *Vision Research, 51*(10), 1192–1205.

Jepma, M., & Nieuwenhuis, S. (2011). Pupil diameter predicts changes in the exploration–exploitation trade-off: Evidence for the adaptive gain theory. *Journal of Cognitive Neuroscience, 23*(7), 1587–1596. https://doi.org/10.1162/jocn.2010.21548

Kahneman, D., Beatty, J., & Pollack, I. (1967). Perceptual deficit during a mental task. *Science, 157*(3785), 218–219. https://doi.org/10.1126/science.157.3785.218

Kamp, S.-M., & Donchin, E. (2015). ERP and pupil responses to deviance in an oddball paradigm. *Psychophysiology, 52*(4), 460–471. https://doi.org/10.1111/psyp.12378

Kelbsch, C., Stingl, K., Kempf, M., Strasser, T., Jung, R., Kuehlewein, L., Wilhelm, H., Peters, T., Wilhelm, B., & Stingl, K. (2019a). Objective measurement of local rod and cone function using gaze-controlled chromatic pupil campimetry in healthy subjects. *Translational Vision Science & Technology, 8*(6), 19–19.

Kelbsch, C., Strasser, T., Chen, Y., Feigl, B., Gamlin, P. D., Kardon, R., Peters, T., Roecklein, K. A., Steinhauer, S. R., & Szabadi, E. (2019b). Standards in pupillography. *Frontiers in Neurology, 10*, 129.

Knapen, T., de Gee, J. W., Brascamp, J., Nuiten, S., Hoppenbrouwers, S., & Theeuwes, J. (2016). Cognitive and ocular factors jointly determine pupil responses under equiluminance. *PLoS One, 11*(5), e0155574.

Korn, C. W., & Bach, D. R. (2016). A solid frame for the window on cognition: Modeling event-related pupil responses. *Journal of Vision, 16*(3), 28–28.

Laeng, B., & Alnaes, D. (2019). Pupillometry. In C. Klein & U. Ettinger (Eds.), *Eye movement research: An introduction to its scientific foundations and applications* (pp. 449–502). Springer International Publishing. https://doi.org/10.1007/978-3-030-20085-5_11

Laeng, B., & Endestad, T. (2012). Bright illusions reduce the eye's pupil. *Proceedings of the National Academy of Sciences, 109*(6), 2162. https://doi.org/10.1073/pnas.1118298109

Lance, B. J., Larkin, G. B., Touryan, J. O., Rexwinkle, J. T., Gutstein, S. M., Gordon, S. M., Toulson, O., Choi, J., Mahdi, A., Hung, C. P., & Lawhern, V. J. (2020). Minimizing data requirements for soldier-interactive AI/ML applications through opportunistic sensing. *Artificial Intelligence and Machine Learning for Multi-Domain Operations Applications II, 11413*, 19–27. https://doi.org/10.1117/12.2564515

Lavin, C., San Martín, R., & Rosales Jubal, E. (2014). Pupil dilation signals uncertainty and surprise in a learning gambling task. *Frontiers in Behavioral Neuroscience, 7*, 218. https://doi.org/10.3389/fnbeh.2013.00218

Liao, H.-I., Yoneya, M., Kidani, S., Kashino, M., & Furukawa, S. (2016). Human pupillary dilation response to deviant auditory stimuli: Effects of stimulus properties and voluntary attention. *Frontiers in Neuroscience, 10*, 43. https://doi.org/10.3389/fnins.2016.00043

Lowenstein, O., & Loewenfeld, I. E. (1964). The sleep-waking cycle and pupillary activity. *Annals of the New York Academy of Sciences, 117*(1), 142–156. https://doi.org/10.1111/j.1749-6632.1964.tb48169.x

Lowenstein, O., Feinberg, R., & Loewenfeld, I. E. (1963). *Pupillary movements during acute and chronic fatigue: A new test for the objective evaluation of tiredness*. Federal Aviation Agency, Office of Aviation Medicine.

Mastiglio, T., Wilkinson, J., Jones, P. N., Bliss, J. P., & Barnett, J. S. (2011). *Current practice and theoretical foundations of the after action review*. MYMIC LLC.

Mathôt S, Melmi JB, van der Linden L, Van der Stigchel S (2016) The Mind-Writing Pupil: A Human-Computer Interface Based on Decoding of Covert Attention through Pupillometry. *PLOS ONE 11*(2): e0148805. https://doi.org/10.1371/journal.pone.0148805

Mathôt, S. (2018). Pupillometry: Psychology, physiology, and function. *Journal of Cognition, 1*(1), 16.

Mathôt, S., van der Linden, L., Grainger, J., & Vitu, F. (2015). The pupillary light response reflects eye-movement preparation. *Journal of Experimental Psychology: Human Perception and Performance, 41*(1), 28.

Mathôt, S., Melmi, J.-B., van der Linden, L., & Van der Stigchel, S. (2016). The mind-writing pupil: A human-computer Interface based on decoding of covert attention through pupillometry. *PLoS One, 11*(2), e0148805. https://doi.org/10.1371/journal.pone.0148805

McClelland, L. E., Pilcher, J. J., & Moore, D. D. (2010). Oculomotor measures as predictors of performance during sleep deprivation. *Aviation, Space, and Environmental Medicine, 81*(9), 833–842.

Mehlhorn, K., Newell, B. R., Todd, P. M., Lee, M. D., Morgan, K., Braithwaite, V. A., Hausmann, D., Fiedler, K., & Gonzalez, C. (2015). Unpacking the exploration–exploitation tradeoff: A synthesis of human and animal literatures. *Decision, 2*(3), 191–215. https://doi.org/10.1037/dec0000033

Morrison, J. E., & Meliza, L. L. (1999). *Foundations of the after action review process*. Institute for Defense Analyses.

Murphy, P. R., Robertson, I. H., Balsters, J. H., & O'connell, R. G. (2011). Pupillometry and P3 index the locus coeruleus–noradrenergic arousal function in humans. *Psychophysiology, 48*(11), 1532–1543.

Naber, M., & Nakayama, K. (2013). Pupil responses to high-level image content. *Journal of Vision, 13*(6), 7.

Naber, M., Frassle, S., Rutishauser, U., & Einhauser, W. (2013). Pupil size signals novelty and predicts later retrieval success for declarative memories of natural scenes. *Journal of Vision, 13*(2), 11. https://doi.org/10.1167/13.2.11

Nassar, M. R., Rumsey, K. M., Wilson, R. C., Parikh, K., Heasly, B., & Gold, J. I. (2012). Rational regulation of learning dynamics by pupil-linked arousal systems. *Nature Neuroscience, 15*(7), 1040–1046. https://doi.org/10.1038/nn.3130

Lee, S., Muto, N., Shimomura, Y. et al. Human pupillary light reflex during successive irradiation with 1-ms blueand green-pulsed light. *J Physiol Anthropol 36, 37* (2017). https://doi.org/10.1186/s40101-017-0153-7

Onorati, F., Barbieri, R., Mauri, M., Russo, V., & Mainardi, L. (2013). Characterization of affective states by pupillary dynamics and autonomic correlates. *Frontiers in Neuroengineering, 6*, 9. https://doi.org/10.3389/fneng.2013.00009

Preuschoff, K., 't Hart, B. M., & Einhauser, W. (2011). Pupil dilation signals surprise: Evidence for noradrenaline's role in decision making. *Frontiers in Neuroscience, 5*, 115.

Privitera, C. M., Renninger, L. W., Carney, T., Klein, S., & Aguilar, M. (2010). Pupil dilation during visual target detection. *Journal of Vision, 10*(10), 3. https://doi.org/10.1167/10.10.3

Prochazkova, E., Prochazkova, L., Giffin, M. R., Scholte, H. S., De Dreu, C. K. W., & Kret, M. E. (2018). Pupil mimicry promotes trust through the theory-of-mind network. *Proceedings of the National Academy of Sciences, 115*(31), E7265–E7274. https://doi.org/10.1073/pnas.1803916115

Ratcliff, R., Smith, P. L., Brown, S. D., & McKoon, G. (2016). Diffusion decision model: Current issues and history. *Trends in Cognitive Sciences, 20*(4), 260–281. https://doi.org/10.1016/j.tics.2016.01.007

Regen, F., Dorn, H., & Danker-Hopfe, H. (2013). Association between pupillary unrest index and waking electroencephalogram activity in sleep-deprived healthy adults. *Sleep Medicine, 14*(9), 902–912. https://doi.org/10.1016/j.sleep.2013.02.003

Ross, V., Jongen, E. M. M., Wang, W., Brijs, T., Brijs, K., Ruiter, R. A. C., & Wets, G. (2014). Investigating the influence of working memory capacity when driving behavior is combined with cognitive load: An LCT study of young novice drivers. *Accident Analysis & Prevention, 62*, 377–387. https://doi.org/10.1016/j.aap.2013.06.032

Shapiro, A. G., Pokorny, J., & Smith, V. C. (1996). Cone-rod receptor spaces with illustrations that use CRT phosphor and light-emitting-diode spectra. *Journal of the Optical Society of America. A, Optics, Image Science, and Vision, 13*(12), 2319–2328. https://doi.org/10.1364/josaa.13.002319

Shorrock, S. T. (2005). Errors of memory in air traffic control. *Safety Science, 43*(8), 571–588. https://doi.org/10.1016/j.ssci.2005.04.001

Sirois, S., & Brisson, J. (2014). Pupillometry. *Wiley Interdisciplinary Reviews: Cognitive Science, 5*(6), 679–692.

Solon, A. J., Lawhern, V. J., Touryan, J., McDaniel, J. R., Ries, A. J., & Gordon, S. M. (2019). Decoding P300 variability using convolutional neural networks. *Frontiers in Human Neuroscience, 13*, 201. https://doi.org/10.3389/fnhum.2019.00201

Spitschan, M. (2019). Photoreceptor inputs to pupil control. *Journal of Vision, 19*(9), 5. https://doi.org/10.1167/19.9.5

Strauch, C., Koniakowsky, I., & Huckauf, A. (2020). Decision making and oddball effects on pupil size: Evidence for a sequential process. *Journal of Cognition, 3*(1), 7. https://doi.org/10.5334/joc.96

Sulutvedt, U., Zavagno, D., Lubell, J., Leknes, S., de Rodez Benavent, S. A., & Laeng, B. (2021). Brightness perception changes related to pupil size. *Vision Research, 178*, 41–47. https://doi.org/10.1016/j.visres.2020.09.004

Suzuki, Y., Minami, T., Laeng, B., & Nakauchi, S. (2019). Colorful glares: Effects of colors on brightness illusions measured with pupillometry. *Acta Psychologica, 198*, 102882. https://doi.org/10.1016/j.actpsy.2019.102882

Sweller, J. (1988). Cognitive load during problem solving: Effects on learning. *Cognitive Science, 12*(2), 257–285. https://doi.org/10.1016/0364-0213(88)90023-7

TACK. (n.d.). *ARL*. Retrieved January 4, 2022, from https://www.arl.army.mil/cast/CaNCTA_Transition_TACK

Tekin, K., Sekeroglu, M. A., Kiziltoprak, H., Doguizi, S., Inanc, M., & Yilmazbas, P. (2018). Static and dynamic pupillometry data of healthy individuals. *Clinical and Experimental Optometry, 101*(5), 659–665. https://doi.org/10.1111/cxo.12659

Thiagarajan, P., & Ciuffreda, K. J. (2015). Pupillary responses to light in chronic non-blast-induced mTBI. *Brain Injury*. https://www.tandfonline.com/doi/full/10.3109/02699052.2015.1045029

Thurman, S., Cohen, R. H., Madison, A., Ries, A., Gordon, S., & Touryan, J. (2021). "Blue sky effect": Contextual influences on pupil size during naturalistic visual search. *Frontiers in Psychology, 12*, 748539.

Tien, T., Pucher, P. H., Sodergren, M. H., Sriskandarajah, K., Yang, G.-Z., & Darzi, A. (2014). Eye tracking for skills assessment and training: A systematic review. *Journal of Surgical Research, 191*(1), 169–178. https://doi.org/10.1016/j.jss.2014.04.032

Truong, J. Q., & Ciuffreda, K. J. (2016). Comparison of pupillary dynamics to light in the mild traumatic brain injury (mTBI) and normal populations. *Brain Injury*. https://www.tandfonline.com/doi/full/10.1080/02699052.2016.1195922

Van Egroo, M., Gaggioni, G., Cespedes-Ortiz, C., Ly, J. Q. M., & Vandewalle, G. (2019). Steady-state pupil size varies with circadian phase and sleep homeostasis in healthy young men. *Clocks & Sleep, 1*(2), 240–258. https://doi.org/10.3390/clockssleep1020021

Watson, A. B., & Yellott, J. I. (2012). A unified formula for light-adapted pupil size. *Journal of Vision, 12*(10), 12.

Wierda, S. M., van Rijn, H., Taatgen, N. A., & Martens, S. (2012). Pupil dilation deconvolution reveals the dynamics of attention at high temporal resolution. *Proceedings of the National Academy of Sciences, 109*(22), 8456–8460.

Wilhelm, B., Wilhelm, H., Lüdtke, H., Streicher, P., & Adler, M. (1998). Pupillographic assessment of sleepiness in sleep-deprived healthy subjects. *Sleep, 21*(3), 258–265. https://doi.org/10.1093/sleep/21.3.258

Willenbockel, V., Sadr, J., Fiset, D., Horne, G. O., Gosselin, F., & Tanaka, J. W. (2010). Controlling low-level image properties: The SHINE toolbox. *Behavior Research Methods, 42*(3), 671–684.

Wong, H. K., Epps, J., & Chen, S. (2020). A comparison of methods for mitigating within-task luminance change for eyewear-based cognitive load measurement. *IEEE Transactions on Cognitive and Developmental Systems, 12*(4), 681–694. https://doi.org/10.1109/TCDS.2018.2876348

Yarbus, A. L. (1967). Eye movements during perception of complex objects. In *Eye movements and vision* (pp. 171–211). Springer.

Yoo, K., Ahn, J., & Lee, S.-H. (2021). The confounding effects of eye blinking on pupillometry, and their remedy. *PLoS One, 16*(12), e0261463. https://doi.org/10.1371/journal.pone.0261463

Yoss, R. E., Moyer, N. J., & Ogle, K. N. (1969). The pupillogram and narcolepsy: A method to measure decreased levels of wakefulness. *Neurology, 19*(10), 921–928. https://doi.org/10.1212/WNL.19.10.921

Yoss, R. E., Moyer, N. J., & Hollenhorst, R. W. (1970). Pupil size and spontaneous pupillary waves associated with alertness, drowsiness, and sleep. *Neurology, 20*(6), 545–554. https://doi.org/10.1212/WNL.20.6.545

Zavagno, D., Tommasi, L., & Laeng, B. (2017). The eye pupil's response to static and dynamic illusions of luminosity and darkness. *I-Perception, 8*(4), 2041669517717754. https://doi.org/10.1177/2041669517717754

Zénon, A. (2019). Eye pupil signals information gain. *Proceedings of the Royal Society B: Biological Sciences, 286*(1911), 20191593. https://doi.org/10.1098/rspb.2019.1593

Pupillometry in Schizophrenia: What We Know and Where We Could Go

Jessica E. McGovern

1 Introduction

Schizophrenia is a severe and chronic mental illness with a lifetime prevalence rate of about 0.5–1% worldwide and a global prevalence of over 20 million people (Charlson et al., 2018). It has a typical age of onset in late adolescence or early adulthood and tends to be associated with lifelong functional disability, diminished quality of life, and premature mortality (American Psychiatric Association, 2013; Olfson et al., 2015). Despite its relatively low prevalence, schizophrenia is ranked among the top 15 most disabling disorders in the world (Vos et al., 2015) as well as one of the most costly in terms of health-related, social, and economic impact to individuals, families, and societies (Chong et al., 2016). In the United States alone, economic costs of the illness range from $94 million to $102 billion annually (Chong et al., 2016) and account for 75% of all mental health spending, as well as around 40% of all publicly funded disability payments (Martin & Miller, 1998). Despite advances in psychopharmacological and psychotherapy interventions, much of the functional impairment remains. In short, schizophrenia is a major health problem by just about any standard, which underscores the pressing need to better understand this disorder.

Schizophrenia is diagnosed based on the presence of positive, negative, and/or disorganized symptoms. Positive symptoms reflect the presence of unusual types of thinking and experiences not usually seen in healthy individuals (i.e., delusions and hallucinations). Negative symptoms reflect a reduction or loss of normal functions, such as deficits in motivation (i.e., avolition), enjoyment/interest (i.e., anhedonia), social interest or behavior (i.e., asociality), emotional expression (i.e., blunted

J. E. McGovern (✉)
Veterans Affairs San Diego Healthcare System, San Diego, CA, USA

Department of Psychiatry, University of California at San Diego, La Jolla, CA, USA
e-mail: Jessica.Mcgovern@va.gov

M. H. Papesh, S. D. Goldinger (eds.), *Modern Pupillometry*,
https://doi.org/10.1007/978-3-031-54896-3_10

affect), and speech (i.e., alogia). Disorganized symptoms reflect odd or unusual speech (e.g., loose associations, tangentiality, thought disorder) or behavior (e.g., catatonia). While the positive and disorganized symptoms can sometimes be more episodic, the negative symptoms tend to be relatively more enduring and largely unresponsive to currently available treatments (Kirkpatrick et al., 2006). In addition, these symptom classes (i.e., positive, negative, disorganized) and even subclasses of symptoms (e.g., avolition versus blunted affect) are thought to have different neurophysiological underpinnings and different associations with functioning (Kirkpatrick et al., 2006). Lastly, based on the current diagnostic criteria, it is possible for people with no overlap in symptoms to receive the same diagnosis; for instance, a person with paranoid delusions and avolition and another with auditory hallucinations and disorganized behavior would both receive a schizophrenia diagnosis. Such a heterogeneous disorder requires approaches that account for group-level and within-group differences in order to characterize this variability.

Despite not being a formal diagnostic criterion of schizophrenia, cognitive impairments occur in approximately 80% of individuals who carry the diagnosis, often predating the onset of positive symptoms, and tend to be relatively enduring across the lifespan and across phases of illness (McCleery & Nuechterlein, 2019; Reichenberg et al., 2009). Cognitive deficits in schizophrenia occur in just about every cognitive domain, with relatively larger deficits (Hedge's g effect sizes ranging from 0.89 to 1.83) in the domains of processing speed, attention/vigilance, working memory, learning/memory, and executive functioning, and relatively smaller effect-size deficits (*g's* between 0.43 and 0.79) for language, vocabulary, and spatial reasoning (Dickinson et al., 2007; Glahn et al., 2005; McCleery & Nuechterlein, 2019; Reichenberg et al., 2009; Schaefer et al., 2013). These deficits emerge at some of the earliest phases of information processing (e.g., visual and auditory perceptual deficits) and may contribute to cascade effects on "higher-level" or downstream processes (e.g., motivation and social cognition) (Green et al., 2012, 2019). Importantly, these cognitive deficits are not explained by antipsychotic medications or their side effects, occur independently from the psychiatric symptoms, and they are detectable at attenuated levels in unaffected first-degree relatives and in prodromal samples (i.e., samples at putatively higher risk for developing a psychotic illness) (for literature reviews, see Green et al., 2019; McCleery & Nuechterlein, 2019).

Pupillometry is uniquely well-suited to shedding light on the mechanisms underlying this complex and deleterious disorder for at least several reasons. First, because this disorder is unusually heterogeneous, pupillometry may help parse how the allocation of available resources (measured by pupillary dilation) may contribute to different symptoms or deficits. Second, pupillometry is most sensitive to the specific cognitive domains where this population tends to show the largest deficits (e.g., attention, executive functioning, processing speed, and working memory). Third, more recent theories have highlighted the role of the locus coeruleus-norepinephrine system in the pathophysiology of schizophrenia (especially with regard to the cognitive and motivational deficits) (Mäki-Marttunen et al., 2020; Yamamoto & Hornykiewicz, 2004), and these pathways play a prominent role in the

Table 1 Pupillometry tasks used in schizophrenia patients sorted by relevant cognitive domains, task names, and task descriptions

Domain/subdomain	Task name	Task description
Sensory perception		
Early auditory processing	Modified auditory oddball task[a]	In auditory oddball tasks, sequences of auditory stimuli (usually tones) are presented one at a time. The "rare" (i.e., deviant) tone occurs far less frequently than the "frequent" (i.e., "standard") tone. In traditional oddball tasks, the S does not make a motor response (which makes it a purer measure of early auditory processing), and responses are measured via a psychophysiological measure (e.g., Mismatch Negativity [MMN] or P300 waveform in event-related potential (ERP) tasks). In this modified version, there were two conditions that both required the S to make motor responses: (1) In the *""Counting" condition*, Ss were played a series of brief tones, and they were asked to count the number of rare tones. (2) In the "C*hoice Reaction Time (RT)" condition*, Ss responded to each stimulus tone with one button for "rare" and another button for the "frequent" tone. Because a response is required for each stimulus, every stimulus is relevant; thus, whereas traditional oddball tasks may conflate a stimulus's "value" or "importance" with its "rareness" or "unexpectedness," this modified version isolates the pupillary responses due to the latter while better controlling for the former. Of note, both the counting and choice RT conditions have a larger working memory load than more traditional oddball tasks because they require more than passive viewing. *Primary pupillary DV*: Average pupillary dilation amplitude measured from stimulus onset.
Early visual information processing	Span of apprehension task[b]	Ss are shown very briefly (e.g., 50–100 ms) a screen with varying numbers of letters (e.g., 1, 3, 6, 8, 10, or 12) on a screen and must report which of two target letters (e.g., T or F) is present among the letters displayed very briefly. In other variations of this procedure, Ss must report the number of target stimuli present, whether a target stimulus is present vs absent, or which quadrant contained the target stimulus. This task requires serial scanning to rapidly reject irrelevant letters and identify target one(s), since there is not an obvious strategy to guide top-down control of the visual search. The task examines the amount of information that can be processed when presented simultaneously. Accuracy of detecting the target stimulus decreases as the array size increases. People with schizophrenia tend to show accuracy and pupillary response decrements at larger (e.g., 8 or more) but not smaller (e.g., 1–3) array sizes. *Primary pupillary DV*: Average total pupil response for the first 3 seconds after trial onset (from fixation through array presentation).

(continued)

Table 1 (continued)

Domain/subdomain	Task name	Task description
Early visual informationprocessing	Visual backward masking (VBM)[c]	Ss were presented with a *target* stimulus (e.g., a specific letter or judge which of two vertical lines is longer) that is rapidly followed by a *mask* (e.g., patterned or random lines) to determine the effect of the mask on perception of the target. Typically, Ss are asked to either identify the target or its location. The amount of time between onset of the target and onset of the mask (stimulus onset asynchrony or SOA) is varied (e.g., 50, 67, 100, 134, and 217 milliseconds apart and no-mask condition). When the mask and target are presented at certain SOAs (~100–200 ms in HC), it is thought to interfere with target accuracy because attention must be shifted and shared between the target and mask. Thus, one must selectively attend to the target and ignore the mask to correctly respond to the target. *Primary pupillary DV*: With primary component analyses, one can isolate pupillary responses to targets and to masks. For instance, in a 3-second trial, the early factor occurs 0–0.7 s, the middle factor occurs 0.7–1.55 s, and the late factor occurs 1.55–3.0 s. The middle factor indexes allocation of resources to target processing, while the late factor indexes allocation to masks (since the mask occurs after the target).
Complex visual processing	Simple vs complexvisual processing[d]	Ss were instructed to look at photos of either a simple line drawing (from the abbreviated version of the Boston Naming Test) or a complex/abstract visual stimulus (Rorschach ink blot image). Images were displayed on a computer screen for 30 seconds, while the S looked at the image. After 30 seconds, the image was removed, and then Ss verbalized their responses. *Primary pupillary DV*: Average pupillary dilation was calculated from the first 10 seconds of stimulus presentation.
Attention		
Attention span	Digit span- digits forward[e]	Single-digit numbers are presented in spans of various lengths (e.g., 3–12 digits long) via audio recording, and then Ss must report the numbers back aloud in the same order. Ss must actively maintain and then retrieve relevant digits to perform accurately. *Primary pupillary DV*: Average pupillary response during the presentation of the last digit in each trial prior to the S's verbal response.

Divided attention/ dual-tasking	Divided attention test[f]	This test consisted of an auditory digits-forward *digit span (DS) task* and a *simple visual response time (RT)* task. This test had four conditions: participants performed *neither task (no task), DS-only, RT-only, or both tasks simultaneously (dual task).* During the DS task, Ss heard the word "ready," followed by 4-, 6-, or 8-digit sequences with a digit presented every 2 s. During the RT task, Ss were presented with the picture of a small rock that appeared randomly for 200 ms in one of the quadrants of the screen. In the RT-only and dual task conditions, Ss pressed a button with the index finger of their dominant hand as soon as they saw the rock. The rock occurred at pseudorandom intervals that, during the dual task, would occur between and never during auditory DS stimuli. The word go was presented 2 s after the onset of the last digit. The only difference between conditions was the instruction on what stimuli to pay attention to. During the no-task condition, Ss were told to look at the center of the screen and listen to the digits without trying to remember them and to ignore the visual stimuli. During the DS- or RT-only conditions, they were told to only pay attention to the relevant stimulus type (e.g., auditory digits or visual "rocks," respectively) and ignore the stimuli of the other type. During the dual-task condition, Ss performed both tasks at the same time and were encouraged to emphasize both accuracy and speed. *Primary pupillary DV*: Task-evoked pupillary dilation was defined as the magnitude of the adjusted maximum dilation during the 1500 ms from onset of each auditory stimulus. Adjusted pupillary dilation was calculated for each auditory stimulus and then averaged across trials for each sequence of the same length in each condition. Dilation in the single- and dual-task conditions was defined as the proportional increase in dilation compared with the no-task condition. Because the no-task condition was used to standardize the data, it was not used in the primary analyses.
Selective attention	Auditory digit span distractibility task (ADSDT)[g]	Ss were presented with single-digit strings of numbers under two different conditions and asked to report aloud the target numbers in the same order presented. The distraction and non-distraction (or neutral) conditions were equated for discriminating power by manipulating both task difficulty (assessed via task accuracy of the healthy controls) and processing resources required (assessed via pupillary dilation). In the *distraction condition*, participants had to repeat five digits in a target string of numbers spoken by a female voice while ignoring numbers spoken by a male distractor voice saying a non-target digit between each target digit. In the *neutral condition*, there was no distractor voice, but participants had to repeat six digits. Thus, the neutral condition required retaining more digits in working memory than the distraction condition, but the distraction condition required both working memory and selective attention skills. *Primary pupillary DV*: Pupillary responses were compared during presentation of the final target digit or sixth epoch, to capture maximum input load in both conditions.

(continued)

Table 1 (continued)

Domain/subdomain	Task name	Task description
Verbal fluency		
Phonemic fluency	Letter fluency[h]	Ss are instructed to say as many words aloud as they can within a specific period of time (typically 60 seconds) that start with a specific letter (e.g., F, A, or S). Such tasks require mental flexibility, efficient storage, manipulation, retrieval of words, cognitive self-control, reaction initiation, and reaction inhibition of already stated or inaccurate words. *Primary pupillary DV*: Pupillary responses were measured throughout the 60 s and binned into 15-second intervals.
Semantic fluency	Category fluency[d]	Ss are instructed to say as many words aloud as they can within a specific period of time (typically 60 seconds) that fit into a particular category (e.g., animals, vegetables). Such tasks require mental flexibility, efficient storage, manipulation, retrieval of words, cognitive self-control, reaction initiation, and reaction inhibition of already stated or inaccurate words. *Primary pupillary DV*: Pupillary responses were measured throughout the 60 s and binned into 15-second intervals.
Executive functioning		

Planning/preparatory action	Double-step task[i]	This task is comprised of randomly interleaved *no-step* (60%) and *step* (40%) trials. Ss were instructed to make a saccade as quickly as possible from a central fixation point (variable fixation display period between 500–1000 ms) to a visual target (T1; grey square) displayed in one of eight positions on the screen (*no-step trials*). During *step trials*, a first target (T1) was quickly followed (94 ms ± 47 ms) by a second target (T2) that flashed at a new location, and Ss were instructed to ignore T1 (i.e., rapidly inhibit the planned saccade to T1) and instead direct their gaze towards T2. Trials where Ss successfully went from fixation to T2 (without detouring to T1) were referred to as *compensated*, and trials where Ss erroneously looked first towards T1 before redirecting to T2 were referred to *as non-compensated*. The probability of correctly compensating becomes more difficult with longer delay times between T1 and T2 (i.e., longer target step delays or TSDs). TSDs were manipulated using a staircase procedure to ensure ~50% accuracy on step trials. This is considered a reactive cognitive control task because Ss were not informed in advance whether it would be a step or no-step trial, so they needed to reactively inhibit an initial saccade to T1 if T2 appeared. Ss performed a practice block of 60 trials and 4 experimental blocks of 120 trials each. *Primary pupillary DV*: Because pupil size is confounded with eye position when using the video-based eye tracker, the epoch of interest was the change in pupil size (i.e., slope) during the 500–1000 ms from trial fixation onset, which preceded the appearance of any target (T1 or T2). Thus, Ss had no indication at this stage as to whether a given trial would require cognitive control to alter a prepared eye movement. Steeper slope reflects greater effort exertion during motor action preparation and can be averaged across trial types.
Response inhibition	Anti-saccade[j]	In this version of the task (for Ss 8–20 years old), a visual stimulus appeared for 1500 milliseconds after a fixation cross either to the right or left of center. In the *pro-saccade condition*, Ss were instructed to look at the stimulus (an ant) as soon as it appeared. In the *anti-saccade condition*, Ss were instructed to look in the opposite direction (right or left) of where the stimulus (a different bug) appears. Because it is more automatic to follow the direction pointed toward (as in the pro-saccade condition), one must inhibit this prepotent response to make the correct eye movement in the anti-saccade condition. Ss were instructed to emphasize both speed and accuracy. *Primary pupillary DV*: Average pupillary responses were measured during the 800 ms after the onset of the stimulus.

(continued)

Table 1 (continued)

Domain/subdomain	Task name	Task description
Set-shifting/switching	Deck choice task[k]	Ss complete blocks ("decks") of forced-choice responses to single-digit stimuli (trials or "cards"). During *Easy decks*, participants applied a single rule throughout the deck based on the number's color (i.e., odd/even if blue; or </>5 if yellow). During *Hard decks*, color switches on 90% of trials, requiring frequent rule-shifting. Of note, both the easy and hard decks have a significant working memory component to maintain the appropriate rule(s) in mind. However, only the hard decks require actively holding and shifting between rules within a single deck. *Primary pupillary DV*: The median difference in pupillary dilation between the average Hard and average Easy deck timecourses.
Updating	Spatial N-back[l]	Ss completed a spatial 0- and 1-back task (similar to the Continuous Performance Test-Identical Pairs task), in which Ss are presented a stream of visual stimuli where each stimulus occurs in one of eight different locations. In the *0-back (control) condition*, participants pressed the right or left button depending on whether the stimulus was in one pre-designated location (upper right quadrant) (yes or no, respectively). In the *1-back condition*, Ss are asked to press the right or left button depending on whether the stimulus was in the same or different location, respectively, as on the previous trial. The 1-back task requires one to hold the location of a stimulus in working memory for 2 s and updating its representation on every trial. To adapt the procedure for children, the instructions were framed within a story about chipmunks playing "catch me if you can." A small picture of a chipmunk appeared for 200 ms in 1 of 8 locations (distributed in an imaginary circle around the center) 300 ms after the onset of the fixation cross with 2 s inter-stimulus intervals. There were 20 practice trials and 200 experimental trials in each condition; targets appeared on 25% of trials in each in condition. There were 50 targets and 150 distractors in each condition. *Primary pupillary DV*: Magnitude of the maximum pupillary dilation during the 1500 ms after the onset of the stimulus.

Probabilistic learning	Probabilistic decision-making task[m]	For each trial, Ss were shown a vertically striped Gabor patch for 2 s. Then, there was a 2 s delay where Ss had to predict via button press whether an upcoming Gabor patch would be tilted to the left or to the right from the center (i.e., predict which direction is the current majority stimulus), and finally the outcome was presented and remained on screen for another 2 s. The "correct" stimulus would change (i.e., reverse) approximately once per 20 ± 4 trials. In the *High Noise* (60:40; 40:60) *condition*, 40% of stimuli were tilted one direction and 60% were tilted the other, making the ability to guess correctly closer to chance. In the *Low Noise* (85:15; 15:18) *condition*, 85% of stimuli would be tilted one direction, and 15% the other; thus, the rewarded stimulus should be more easily learned. Ss were instructed to track the probabilities and the changes as best they could and to minimize errors. The task also had cued and uncued blocks. In the *uncued* (i.e., *High Volatility) block*, the contingencies would change without warning. For the *cued* (i.e., *Low Volatility) block*, Ss were advised to "reset" their beliefs about the distribution of stimuli at every announced change point (via a screen with the word "change!" on it) and to relearn the new underlying distribution through choice-outcome observations. Volatility here refers to the subjectively perceived volatility or instability of the environment (i.e., of the contingency/reward structure). *Primary pupillary DV*: Pupillary responses were measured as the trial-wise maximum pupil dilation during outcome presentation (square root transformed to normalize model residuals). The pupillary response signal was Z-scored per task block and participant and was baseline-corrected per trial through subtraction of the average signal of the 500 ms preceding outcome onset.
Social cognition		
Emotion perception	Facial affect recognition task[n]	In this task, stimuli included 16 happy, 16 afraid, and 16 neutral pictures of Caucasian and Asian people, with an equal number of males and females per condition. As a non-face control condition, 16 "scrambled" images were created from the original files. Each trial began with the word "Ready?" presented on screen until the S verbally indicated their readiness. Next, a 3 s fixation cross was presented followed by a 5 s target stimulus. Last, Ss indicated which of the 4 stimulus types (happy, afraid, neutral, scrambled) they saw. A total of 64 stimuli were presented in a randomized order, where each of the 4 conditions contained 8 stimuli and was presented twice. *Primary pupillary DV*: Average change in pupil size between 1 s and 3 s of the 5 s stimulus presentation. This time window was chosen to exclude the initial pupil contraction due to the light reflex from the onset of the stimulus.

(continued)

Table 1 (continued)

Domain/subdomain	Task name	Task description
Emotion regulation		
Directed attention	Emotion regulation task[o]	Ss passively viewed a total of 20 neutral (e.g., household objects, social interactions) and 20 unpleasant (e.g., threat, mutilation) images (from the International Affective Picture System [IAPS] library) for 3 s each. Attention was directed to one highlighted 250x250 pixel subregion or *Area of Interest (AOI)* within each image where that window was clear and the rest of the image was faded out (with 30% opacity). Each unpleasant image was presented twice: once with an *arousing AOI (A-AOI)* and once with a *non-arousing AOI (N-AOI)*. For instance, in an image of a building on fire, an A-AOI might include fire coming out of a building window and a N-AOI might include a fireman nearby without fire in the frame. This procedure allowed the AOI to direct attention without obstructing the view of any part of the image. Ss were instructed to focus their attention only within the target window. *Primary pupillary DV*: Mean pupil size at fixation (in pixels) within the AOI.

Reappraisal and distraction	Emotion regulation task[p]	Ss viewed neutral and unpleasant IAPS images during which they were instructed to passively view or apply one of two emotion regulation techniques (distraction or cognitive reappraisal) to decrease their negative emotional responses. After being taught how to use reappraisal and distraction to reduce negative affect, Ss were guided through a series of training trials where the experimenter provided verbal feedback. Next, Ss completed a series of practice trials (6 reappraisals, 6 distractions, 4 unpleasant passive viewings, and 4 neutral viewings), where they had to verbalize the exact manner in which they used reappraisal or distraction to decrease negative emotion while viewing unpleasant photographs. The experimental trials included 112 images (28 neutral, 84 negative) that were not used during the training trials. They were administered in 2 blocks of 56 trials each. There were 4 trial types: 14 Watch (unpleasant passive viewing), 14 View (neutral passive viewing), 28 Distract, and 28 Reappraise. Reappraise and Distract conditions were presented in separate blocks to reduce task demands and confusion. During *View* and *Watch* trials, participants were instructed to look at the pictures and allow their thoughts and emotions to unfold naturally. On *Reappraise* trials, Ss were instructed to feel neutral by reinterpreting the meaning of the unpleasant image to something more neutral (e.g., people in the image crying tears of joy rather than tears of sadness). On *Distract* trials, Ss were instructed to feel neutral by thinking of something unrelated to the image on the screen (e.g., think about a complex geometric shape). Each trial began with a central fixation target that lasted 1000 ms followed by an instruction cue (View, Watch, Distract, or Reappraise) presented for 2000ms. Next, the IAPS image was presented for 5 seconds while eye movements and pupil dilation were recorded. After the image disappeared, Ss had unlimited time to self-report how negatively that trial made them feel from 1 (not at all) to 5 (extremely negative). The neutral and unpleasant passive viewing conditions control for pupil dilation due to sympathetic nervous system changes from emotional reactivity. Cognitive reappraisal is thought to be highly cognitively demanding in that it requires initially directing attention to relevant stimulus features, holding reappraisal goals and the initial stimulus content in working memory, inhibiting the prepotent appraisal, and generating a goal-relevant reappraisal by drawing on semantic memory. Distraction involves trying to feel more neutral by generating a mental representation of something unrelated to the presented stimulus. Distraction potentially requires less effort than reappraisal because the realm of possible unrelated stimuli is greater and does not require one to pay much attention to the unpleasant stimuli. *Primary pupillary DV*: Average pupil dilation during the 5 seconds while the IAPS image was displayed. They also examined pupil dilation at the initial fixation within the emotional area of interest (E-AOI) for that image and segmented pupillary dilation into 5 one-second epochs to evaluate differences over the time course of viewing the image.

(continued)

Table 1 (continued)

Note. DV Dependent Variable, *Ss* Research subjects or participants

[a]Steinhauer and Zubin (1982) and Zahn et al. (1991)
[b]Granholm et al. (2009) and Granholm and Verney (2004)
[c]Granholm et al. (2009) and Granholm and Verney (2004)
[d]Minassian et al. (2004)
[e]Granholm et al. (1997), Granholm et al., 2016, Kreis et al. (2020), Morris et al. (1997), and Straube (1982)
[f]Karatekin et al. (2008)
[g]Fish and Granholm (2008)
[h]Granholm et al. (1998)
[i]Thakkar et al. (2018)
[j]Karatekin et al. (2010)
[k]McGovern et al. (2020) and Reddy et al. (2018)
[l]Karatekin et al. (2009)
[m]Kreis et al. (2021)
[n]Wynn et al. (2019)
[o]Strauss et al. (2015)
[p]Bartolomeo et al. (2020) and Strauss et al. (2016)

Table 2 Relationship between pupillary dilation and symptomatology

Study	Task	SZn	Age mean (SD)[a]	Positive symptoms	Negative symptoms	Disorganized symptoms
Sensory perception						
Minassian et al. (2004)	Complex visual processing (Rorschach inkblots)	24	39.8 (10.8)	NS	(–)	(–)
Granholm et al. (2009)	Visual backward masking	51	54.9 (8.2)	(–)[b]	NS	
Granholm and Verney (2004)	Visual backward masking	16	50.0 (8.4)	NS	NS	
Attention						
Fish and Granholm (2008)	Auditory digit span distractibility task	35	57.09 (9.36)	(+)	NS	
Granholm et al. (1997)	Digit span forward (auditory)	24	45.2 (11.8)			
Granholm et al. (2016)	Digit span forward (auditory)	149	44.4 (10.5)		(–)[c]	
Morris et al. (1997)	Digit span forward (auditory)	29[d]	47.0 (12.7)	NS	NS	NS
Kreis et al. (2020)	Digit span forward (visual)	29	47.55 (11.66)		NS[e]	
Karatekin et al. (2008)	Divided attention	34	170 (43) months			
Granholm et al. (2000)	Span of apprehension task	31	56.9 (9.1)	NS	NS	
Granholm et al. (2007)	Span of apprehension task	58	54.0 (7.9)	NS	(–)	NS
Working memory & executive functioning						
Karatekin et al. (2010)	Anti-saccade task	29	178 (41) months			
McGovern et al. (2020)	Deck choice effort task	66	54.85 (8.80)	./NS[f]	(–)	
Reddy et al. (2018)	Deck choice effort task	33	54.6 (7.9)	.	(–)[g]	
Thakkar et al. (2018)	Double-step (actionpreparation) task	16	39.9 (9.4)	NS	(–)	.
Karatekin et al. (2009)	N-back (0- and 1-back)	29	174 (43) months	.	.	.
Kreis et al. (2021)	Probabilistic decision-making task	31	47.13(11.43)	.	.	.

(continued)

Table 2 (continued)

Study	Task	SZn	Age mean (SD)[a]	Positive symptoms	Negative symptoms	Disorganized symptoms
Verbal fluency						
Granholm et al. (1998)	Verbal fluency	15	49.1 (11.0)	NS	NS/(−)[h]	NS
Emotion regulation						
Strauss et al. (2015)	Directed attention	28	45.4 (12.0)	NS	NS	NS
Bartolomeo et al. (2020)	Distraction	25	37.2 (12.9)	NS	NS	NS
Bartolomeo et al. (2020)	Reappraisal	25	37.2 (12.9)	(+)	NS	NS

Notes. A study was included if it provided at least one analysis on the relationship between pupillary dilation and symptoms or medication effects. (+) = Statistically significant positive association between higher symptoms and greater pupillary dilation. (−) = Statistically significant negative or inverse association (i.e., higher symptoms, lower dilation). "." = relationship not reported. NS = relationship was examined but statistically not significant. SZn = sample size of the schizophrenia group used in the pupillometry analyses

[a]Mean age in the schizophrenia sample, in years, unless otherwise stated

[b]Analyses were performed on subgroups that were formed by a median split of the pupil dilation amplitude. In the high-dilation schizophrenia subgroup, more severe positive symptoms were associated with less dilation to masks. There were no significant correlations with symptoms in the low-dilation schizophrenia subgroup. Both the high-dilation and low-dilation schizophrenia subgroups showed abnormally low dilation to targets relative to healthy controls, and only the high-dilation schizophrenia subgroup showed higher dilation to masks relative to targets

[c]The direct correlation between pupillary dilation and negative symptoms was not statistically significant; however, greater defeatist beliefs, which are thought to play a causal role in negative symptoms, were inversely correlated with negative symptoms. In this study, where subgroups were formed based on a median split of the defeatist beliefs score, the high defeatist beliefs subgroup had higher negative symptoms than the low defeatist beliefs subgroup, which also provides indirect evidence of a relationship between negative symptoms and reduced pupillary dilation

[d]Although 29 Ss were included in the overall pupillary analyses, only 12 had symptom data, rendering it difficult to draw firm conclusions from these data

[e]The authors noted the low average symptom severity in their sample, which may have restricted the range in detecting relationships

[f]This analysis was not reported in the published manuscript, but unpublished data available from the authors was provided; positive symptoms were not correlated with pupillary dilation, $r = -0.04$, $p = 0.89$

[g]The correlation was statistically significant after controlling for overall cognitive ability

[h]Correlation was within the medium effect-size range ($r > 0.35$) but did not survive alpha value adjustment for multiple comparisons ($p < 0.01$)

pupillary dilation response to arousal and mental effort (Mathôt, 2018; Steinhauer et al., 2004). This chapter reviews the pupillometry findings within schizophrenia populations, including the pupillary light response and other early work, next task-evoked pupillary responses in relation to cognition and resource allocation more broadly, then relationships between pupillary findings and symptomatology, and concludes with special considerations and future directions.

2 The Pupillary Light Response and Other Early Work in Schizophrenia

The earliest published studies examining the pupillary response in individuals with schizophrenia date back to the early twentieth century with observations of the abnormal pupillary light response (PLR) by the German scientists Bach (1908), Bumke (1911), and Westphal (1907). Westphal (1907) used the term "catatonic pupil" to describe the reduced reactivity to light (i.e., reduced PLR amplitude) of individuals with psychosis in a "catatonic" (unreactive) state; however, it has since been recognized that this PLR abnormality extends beyond catatonic states to the broader schizophrenia population, including inpatient and outpatient, acutely ill and symptomatically in-remission, first episode and chronically ill, as well as medicated and unmedicated samples (Hakerem et al., 1964; Ikushima & Matsunga, 1975; May, 1947; Rubin, 1962, 1970; Steinhauer et al., 1979). Subsequent studies have replicated findings of reduced PLR amplitude as well as evidence of a faster constriction rate and shorter latency to the end of the constriction process to light (Hakerem et al., 1964; Ikushima & Matsunga, 1975; Lidsky et al., 1971; May, 1947; Okada et al., 1978; Steinhauer et al., 1979; Steinhauer & Zubin, 1982).

Some of this earlier work also examined pupillary responses to acute transient stressors, such as mild pain (e.g., a pinch on the neck; May, 1947), immersion of the participant's hand in ice-cold water (Rubin & Barry, 1972), and application of physical strain (e.g., squeezing the experimenter's fingers, flexing the arm against resistance by the experimenter; May, 1947[1]). These early studies generally found reduced pupillary dilation responses to acute stressors in individuals with schizophrenia relative to non-psychiatric healthy comparison samples (referred to herein as "healthy controls") and contributed to early hypotheses of an autonomic imbalance in individuals with psychosis (Zahn et al., 1991). These findings are consistent with theories on the role of blunted stress responses and impairments in the hypothalamic-pituitary-adrenal (HPA) axis in the onset and maintenance of schizophrenia symptoms (Corcoran et al., 2003; Dawson et al., 1994; Gispen-de Wied, 2000; Jansen et al., 1998).

[1]The May (1947) study is also notable for being one of the earliest to measure reduced pupillary responses to physical effort exertion and to observe reduced dilation during preparatory action by individuals with schizophrenia, which will be expanded upon later in this chapter.

3 Pupillometry as a Window into How Resource Allocation Relates to Cognition in Schizophrenia

As in the broader pupillometry literature, more recent decades have emphasized the measurement of pupillary responses within the context of cognitive tasks. As might be expected based on the pattern of significant impairments across a wide variety of cognitive domains among individuals with schizophrenia, this population tends to show worse behavioral task performance and a reduced amplitude of pupillary dilation responses relative to healthy controls across a large variety of cognitive domains (Fish & Granholm, 2008; Granholm et al., 1997, 2007, 2016; Karatekin et al., 2010; Kreis et al., 2020; Minassian et al., 2004; Morris et al., 1997; Thakkar et al., 2018), although there are exceptions (Granholm et al., 1998; Straube, 1982). See Table 2 for descriptions of cognitive constructs, tasks, and relevant references that have included participants with schizophrenia. When tasks included conditions with varying levels of difficulty or complexity (e.g., the number of digits to be recalled or stimuli to be processed), the schizophrenia group tends to show reduced pupillary dilation (and impaired task performance) only at higher processing loads. The reduced pupillary dilation amplitudes in schizophrenia have generally been interpreted to mean that individuals with schizophrenia either (1) have inadequate control over the allocation of available resources (resulting in misallocation or inefficient utilization of available resources that leaves fewer resources available for task-relevant processing; i.e., misallocation account) (Fish & Granholm, 2008; Granholm et al., 2009) and/or (2) do not have sufficient processing resources available to perform tasks with higher processing demands, regardless of cognitive abilities tapped by tasks (i.e., insufficient capacity account) (Granholm et al., 2000, 2007; Morris et al., 1997).

The misallocation account would be supported if individuals with schizophrenia showed *more* pupillary dilation relative to healthy controls on at least some aspects of the task, such as to distractor or mask stimuli relative to target stimuli, despite similar or worse behavioral performance. In such cases, excessive pupillary dilation may indicate that the necessary "horsepower" exists and impaired performance cannot be attributed to insufficient effort. Thus, it is more likely that available resources are being utilized inefficiently, diverted to other processes, and/or require additional recruitment of compensatory processes not typically involved in the task at hand. This account is consistent with the theory that a central problem in schizophrenia is an attentional or executive control deficit that interferes with the appropriate deployment of available resources (Luck & Gold, 2008).

The insufficient capacity account (also known as the resource-limitations hypothesis) would be supported if individuals with schizophrenia show *less* pupillary dilation and impaired behavioral performance relative to healthy controls, especially as processing demands increase and exceed available resources. According to this account, cognitive impairments reduce the overall pool of available resources, so there are fewer resources available to allocate to processing complex stimuli, regardless of the particular abilities tapped by the tasks (Granholm et al., 1996, 1997;

Nuechterlein et al., 1998; Nuechterlein & Dawson, 1984). This account is often linked with diathesis-stress-related models of schizophrenia, which suggest that a biological predisposition to cognitive impairments increases vulnerability to developing and maintaining psychosis; for instance, cognitive deficits heighten the impact of early stressors, which then more easily overwhelm the system (Nuechterlein et al., 1998; Nuechterlein & Dawson, 1984). That said, the misallocation and inadequate capacity accounts are not mutually exclusive; both may be true at the same time (fewer resources and impaired deployment of available resources), or each may be accurate under certain conditions.

In a series of pupillometry studies intended to pit the insufficient capacity and misallocation accounts against each other, the researchers found some support for each, even within the context of a single task. For instance, on a visual backward masking (VBM) task that required participants to identify a target (e.g., specific letter) that was followed closely in time by a mask (e.g., patterned or random lines) (see Table 1), although the healthy control group had greater pupillary dilation amplitude than the schizophrenia group overall (suggestive of insufficient capacity), individuals with schizophrenia also tended to allocate relatively more resources (i.e., greater pupillary dilation) to mask or distractor stimuli than to target stimuli (suggestive of misallocation of resources as well) (Granholm & Verney, 2004). The misallocation of attentional resources to irrelevant stimuli (i.e., excessive dilation to masks) likely contributed to their impaired behavioral performance (i.e., worse target detection accuracy) in that it left fewer resources available to correctly identify target stimuli. A follow-up study with the same VBM paradigm in a larger sample found two schizophrenia subgroups with different pupillometry patterns: one with normal overall pupillary response amplitudes relative to healthy controls but greater dilation to masks relative to targets (i.e., normal capacity with misallocation of resources) and a second group with abnormally small overall pupillary response amplitudes with reduced dilation to both target and mask stimuli (i.e., reduced overall resources) (Granholm et al., 2009). In other words, one subgroup of schizophrenia participants appeared to better fit the misallocation account while the other better fit the insufficient resources account.

In a different paradigm pitting these accounts against each other, an auditory digit span distractibility task (ADSDT) was used to compare behavioral performance and pupillary responses on a 5-digits-long distraction condition (which had a high selective attention demand) versus a 6-digits-long non-distraction condition (which had no selective attention demand but a higher memory span demand) (see task details in Table 1). On this task, the healthy control group had better behavioral performance (more digits recalled) and larger pupillary dilation amplitudes in the distraction condition (which had fewer target digits overall) relative to the non-distraction condition; this pattern suggested that the distraction condition was "easier" (based on accuracy) despite requiring more processing resources (based on pupillary dilation) (Fish & Granholm, 2008). In contrast, the schizophrenia group had lower pupillary dilation overall (i.e., across both conditions relative to the healthy control group), performed equally poorly in both the distraction and non-distraction conditions, and was disproportionately impacted by the distractor stimuli (Fish & Granholm, 2008). The disproportionate impact of the distractor stimuli in the schizophrenia group but not the healthy control group suggested there was a role for misallocation of resources in the schizophrenia group. However, there was also evidence consistent with the inadequate capacity account. That is, the schizophrenia group did not differ in pupillary dilation from healthy controls at lower processing loads (i.e., during the presentation of the initial relative to later digits within a full digit string), but failed to further increase dilation after the fourth target digit (of the 5- or 6-digit strings), suggesting an earlier depletion of resources in the schizophrenia group. Thus, schizophrenia participants can show evidence of both insufficient resources and misallocation of resources within the same task.

In summary, tasks that systematically increase processing load are likely to reveal patterns of overall insufficient resources (i.e., reduced dilation), while tasks that require filtering out irrelevant information may reveal inadequate capacity and/or misallocation-related deficits. Subgroup analyses, as in Granholm et al. (2009), suggest that certain individuals may fall more into one category or the other, which raises the question of what factors determine subgroup type (e.g., symptomatology, cognitive impairment profile).

4 Relating Pupillometry to Symptomatology

4.1 Theory Underlying the Pathophysiology of Symptoms in Schizophrenia

To inform how we might hypothesize pupillometry findings to relate to symptoms, it is helpful to step back and discuss the leading theories on the pathophysiology of schizophrenia. In brief, the positive symptoms of psychosis have been associated with erroneous spontaneous increases in neural activity (i.e., dopamine and/or norepinephrine) in response to neutral or irrelevant stimuli during laboratory tasks (i.e., increased noise), while negative (and to some extent disorganized) symptoms tend to be associated with insufficient neural activation to relevant stimuli (for reviews, see Maia & Frank, 2017; Mäki-Marttunen et al., 2020). Given the positive relationship between resource allocation and pupillary dilation responses, this leads to the prediction that positive symptoms may be associated with wasteful or inefficient allocation (i.e., excessive pupillary dilation) to distractor or task-irrelevant stimuli, as in the misallocation account above, whereas the negative and disorganized symptoms may be associated with reduced pupillary dilation, as in the inadequate capacity account above.

While the theories about the pathophysiology of schizophrenia with the greatest research support differ in their emphases on the roles of specific neurotransmitters (e.g., dopamine, norepinephrine), all tend to implicate dysregulation in the neural systems that govern communication between the frontal cortex (especially prefrontal cortex) and subcortex (e.g., striatum, ventral tegmental area, thalamus, locus coeruleus) and in the neural systems involved in fine-tuning signal-to-noise ratios to parse relevant from irrelevant information (for theories that emphasize the role of dopamine, see Grace, 1991; Heinz & Schlagenhauf, 2010; Howes & Kapur, 2009; Kapur, 2003; Maia & Frank, 2017; Millard et al., 2022; for those that emphasize norepinephrine, see Mäki-Marttunen et al., 2020; Yamamoto & Hornykiewicz, 2004).

Although the task-evoked pupillary response is more directly related to norepinephrine, dopamine and norepinephrine interact in several key ways, including with regard to coding salient events, uncertainty processing, selective attention, memory consolidation, feedback-based learning, motivation, and stress responses, which all tend to be impaired in individuals with schizophrenia (Aston-Jones & Cohen, 2005; Chandler et al., 2014; Joshi et al., 2016; Nieoullon, 2002; Sara & Bouret, 2012; Valentino & Van Bockstaele, 2008). While both dopamine and norepinephrine are activated by salient stimuli relevant to predicting future behavior, dopamine tends to be more related to reward valuation and error prediction, whereas norepinephrine is more related to arousal, vigilance, and fine-tuning the signal-to-noise ratio of processes necessary for optimal behavior (e.g., selective attention) (Chandler et al., 2014; Mäki-Marttunen et al., 2020; Robbins & Arnsten, 2009). In general, we expect a positive relationship between pupillary dilation and dopamine or norepinephrine activity.

4.1.1 Positive Symptoms Reflect Neural Misfiring Under Stress

Positive symptoms are thought to stem from a combination of high tonic locus coeruleus norepinephrine activation and phasic spontaneous dopamine and norepinephrine misfiring (Kapur, 2003; Mäki-Marttunen et al., 2020). High tonic activation in response to stress increases general arousal and hypervigilance. When this tonic increase in activation combines with transient dopamine and norepinephrine misfiring to neutral or irrelevant stimuli (aberrantly increasing their salience), selective attention and other processes are altered such that the mind is more likely to conclude (falsely) that the stimuli associated with that firing are personally relevant or meaningful (Kapur, 2003; Mäki-Marttunen et al., 2020). Thus, if you are stressed and the misfiring occurs while someone looks your way, you might jump to the conclusion that the person is watching you and wants to harm you (i.e., paranoid and referential delusions). If the misfiring occurs during internally-generated speech (or thoughts), you might falsely conclude that the speech was externally-generated (i.e., auditory hallucinations). Thus, increased pupillary dilation during neutral or irrelevant stimuli relative to target stimuli is hypothesized to be associated with positive symptoms.

In addition to being primed to jump to conclusions quickly and (sometimes) falsely based on limited information, these false conclusions are thought to be maintained by a deficit in the mechanisms that update mental value representations to integrate new information (Schlagenhauf et al., 2009); as a result, these false beliefs are exceptionally resistant to changes or updates despite exposure to contradictory or disconfirming evidence. If these failures reflect insufficient updating capacity, this particular deficit may be associated with reduced pupillary dilation and increased positive symptoms.

In sum, positive symptoms may be linked to increased pupillary dilation during the presentation of task-irrelevant or distracting stimuli (consistent with the misallocation account described previously) and may be linked to decreased pupillary dilation on tasks that require updating value representations (as with the insufficient capacity account).

4.1.2 Negative and Disorganized Symptoms Reflect Neural Underactivity

If the neural firing (especially striatal dopamine) is under-reactive to stimuli associated with (social or nonsocial) goals, this insufficient neural "push" decreases the likelihood that these stimuli will be subjectively experienced as "worth the effort" to pursue; such frontal-striatal underactivity has been associated with the negative symptoms of schizophrenia, especially social (e.g., asociality) and nonsocial (e.g., avolition) motivational deficits (Culbreth et al., 2016a, b, 2018; Moran et al., 2019). To further compound these effects, if inadequate resources are available to allocate to such complex behavioral goals, those actions will be subjectively appraised as even more difficult and even less rewarding, which further decreases the drive to pursue meaningful goals. According to the cognitive model of schizophrenia, repeated negative experiences or failures can result in the formation and maintenance of dysfunctional attitudes (e.g., defeatist performance beliefs such as, "why bother trying, I'm just going to fail anyway," "failing partly is as bad as being a complete failure") that undermine goal-directed behavior (Beck et al., 2009; Grant & Beck, 2009). Lastly, although not emphasized in these models, disorganized symptoms are thought to reflect a breakdown in the normal organizational processes of the brain and are associated with limited cognitive resources (Beck et al., 2009; Perry & Braff, 1994). Therefore, if negative and disorganized symptoms are associated with insufficient availability of resources, more severe symptoms should be associated with decreased pupillary dilation, especially as processing demands increase. See Table 2 for studies reporting associations between pupillometry and symptoms.

4.2 Positive Symptoms Are Associated with Wasteful Allocation

As follows from the theories outlined above, positive symptoms are hypothesized to be associated with higher pupillary dilation relative to healthy controls for stimuli that are irrelevant or distracting with respect to the primary task goals (indicating wasteful allocation of available resources). This would be most apparent on tasks that can isolate (in space and time) resources allocated to irrelevant or distractor stimuli versus target stimuli (e.g., auditory digit span distractibility task [ADSDT] or visual backward masking [VBM] task; for detailed task descriptions, see Table 1).

Consistent with this hypothesis, in the ADSDT described previously, greater dilation in the distraction condition relative to the non-distraction condition was associated with greater positive symptoms in the schizophrenia group (Fish & Granholm, 2008). Recall that in that study, participants had to ignore the digits stated by a distractor voice and repeat aloud only the digits stated by the target voice; thus, excessive dilation relative to healthy controls for the distractor voice reflected wasteful allocation of resources.

However, on the VBM task, associations with positive symptoms were in the opposite direction expected. That is, when the authors examined resources allocated to mask versus target stimuli, they found the expected reduced overall pupillary dilation in the schizophrenia group as a whole relative to the healthy control group (Granholm et al., 2009; Granholm & Verney, 2004). However, when Granholm et al. (2009) examined schizophrenia subgroups in their larger study, the subgroup with excessive dilation to masks relative to targets (suggestive of wasteful resource allocation) had a correlation between more severe positive symptoms and *less* (not more) resource allocation to masks, which is difficult to reconcile with either account. In addition, the subgroup that showed reduced dilation to both mask and target stimuli showed no relationship to symptoms, although we might have hypothesized this subgroup to have greater negative symptoms. Negative symptoms were not associated with pupillary dilation for either the ADSDT or VBM task.

Although both the ASDT and VBM task revealed impairments in attentional filtering based on behavioral performance deficits in the schizophrenia group relative to the healthy control group, the expected relationship between excessive pupillary dilation and positive symptoms was evident on the ASDT but not the VBM task. These differences may reflect differences in task or sample characteristics. Of note, the schizophrenia samples in the VBM studies had minimal positive symptoms overall, which may have restricted the symptom severity range and obscured the relationship between pupillary dilation and symptoms. In addition, the authors of the VBM study note that about half of the participants in their study had to be excluded due to either near-chance performance on the no-mask (i.e., easiest) condition or excessive artifacts in the pupillary data, which may limit the generalizability of these findings (Granholm et al., 2009). Future studies may also help clarify whether other task characteristics (e.g., auditory versus visual processing, switching versus sequential presentation of distracting or irrelevant stimuli) may account for these different relationships with symptoms.

While misallocation of resources likely plays one role in the formation and maintenance of positive symptoms, inadequate allocation of resources to updating beliefs based on new information may also contribute to positive symptom maintenance. For instance, Kreis et al. (2021) modified a probabilistic reversal-learning task to examine the relationship between resource allocation (measured via pupillometry) and belief updating in response to uncertainty in individuals with schizophrenia. In probabilistic reversal-learning tasks, individuals must choose which of two stimuli (A or B) is more likely to lead to a correct outcome based on hidden choice-outcome contingencies that the participant must ascertain through trial and error with feedback. The contingencies change (i.e., reverse) periodically; for instance, a previously "incorrect" stimulus becomes "correct," or the probabilities of how frequently a correct stimulus (A:B) is rewarded change (e.g., 80:20 to 20:80 or 60:40), so the participant must utilize feedback to adapt to the changing contingencies and figure out the new correct answer (which requires implicit reinforcement learning). In most cases, healthy controls will repeat the same choice if it was previously correct (i.e., win-stay) and try a different subsequent choice if the previous one was incorrect (i.e., lose-shift). On such tasks, individuals with schizophrenia tend to show deficits in detecting rule-changes (a learning deficit that requires updating value representations based on the change in feedback) and show less efficient or less stable decision-making (e.g., increased win-shift and lose-shift behavior) (Culbreth et al., 2016a, b; Schlagenhauf et al., 2014; Waltz et al., 2013). Such behavioral deficits on fMRI-based reversal learning tasks have been associated with *reduced* activity in brain regions associated with cognitive control (e.g., dorsolateral prefrontal cortex, anterior cingulate cortex, and dorsal parietal cortex) and reward (e.g., striatum) (Culbreth et al., 2016a; Schlagenhauf et al., 2009); thus, deficits in learning these contingencies may be reflected in reduced pupillary dilation.

In a pupillometry-based version of this probabilistic reversal learning task (see Table 1), there was an overall effect such that larger pupillary dilation was associated with increased uncertainty and Bayesian surprise (an index of belief updating) in both healthy controls and schizophrenia participants, suggesting that both groups were sensitive to changes in the task conditions and that stimuli with greater informational value resulted in greater pupillary dilation (Kreis et al., 2021). However, the schizophrenia group showed a weaker relationship between uncertainty and pupillary dilation indicating that they showed less pupillary change under more uncertain conditions (i.e., when rule changes were not cued), although these should be the conditions where feedback should have a larger impact (greater informational value) and thus reflect larger pupillary dilation (Kreis et al., 2021). As expected, more severe positive symptoms were associated with a moderate-to-large effect-size decrease in sensitivity to positive feedback (i.e., a greater likelihood of changing their answer after a correct prediction [win-shift]) in the condition with greater certainty (i.e., the condition where rule-changes were cued rather than uncued/less predictable); however, this relationship with positive symptoms did not survive Bonferroni correction for multiple comparisons, and pupil dilation was not associated with psychiatric symptoms in any condition. The group-level finding of reduced pupillary modulation in relation to varying uncertainty is consistent with earlier studies that utilized a probability-based guessing task in schizophrenia participants (Steinhauer et al., 1979; Steinhauer & Zubin, 1982). Together, these results offer modest support that reduced pupillary dilation is associated with impaired belief updating. These results are also consistent with the theory that aberrant norepinephrine signaling impedes learning the discriminative value of a stimulus or outcome, thereby adding "noise" to the signal-to-noise ratio for the valuation of relevant stimuli in individuals with schizophrenia (Mäki-Marttunen et al., 2020). Evidence of a specific association between belief updating and positive symptom severity was more tenuous.

In summary, leading theories of positive symptoms in schizophrenia suggest that while misallocation of resources to irrelevant external stimuli may aid in the formation and maintenance of positive symptoms, failure to update relevant beliefs after receipt of contradictory or inconsistent information may serve to maintain positive symptoms. Available pupillometry data offer modest support for the hypotheses that positive symptoms may be related to (1) increased pupillary dilation to irrelevant stimuli and (2) decreased pupillary dilation during belief updating. Future studies will need to replicate, extend, and refine these hypotheses.

4.3 Negative and Disorganized Symptoms Are Associated with Insufficient Allocation

As mentioned in Sect. 4.1, if negative symptoms are reflective of insufficient availability or utilization of attentional resources, greater negative symptom severity may be associated with reduced pupillary dilation on tasks where there is no competition for relevant versus irrelevant stimuli. This inverse relationship should be most apparent at higher processing loads, where task demands are more likely to be at or beyond individuals' processing capacity, such as on tasks with higher working memory loads (e.g., longer digit span lengths, n-back task with 1- or more back), higher executive functioning demands (e.g., Deck Choice or Anti-Saccade tasks, which require set-shifting and inhibiting a dominant response, respectively), or more attentional processing (e.g., larger spans on the span of apprehension task). If disorganized symptoms reflect a reduction in resources available to organize complex information, these symptoms are more likely to be associated with reduced pupillary dilation during tasks that require complex processing and high organizational demands, such as during the interpretation of ambiguous and abstract stimuli (e.g., the Rorschach inkblot test).

Several studies have found at least some relationship between negative symptoms and reduced pupillary dilation, independent of positive symptom severity and cognitive ability (Granholm et al., 2007, 2016; McGovern et al., 2020; Reddy et al., 2018; Thakkar et al., 2018). While some pupillometry studies have not found a relationship with negative symptoms (Granholm et al., 1998, 2000; Morris et al., 1997), these tended to have smaller samples and/or limited negative symptom range. Negative symptoms (especially motivation deficits) have been associated with reduced willingness to exert effort on behavioral tasks in a large number of studies, independent of cognitive deficits. Because reduced pupillary dilation during cognitive tasks can be due to inadequate resources (i.e., limited capacity or pool of available resources) and/or due to reduced motivation (i.e., reduced willingness or ability to allocate available resources), and individuals with schizophrenia show impairments in both domains, this raises the question of whether these domains make independent or overlapping contributions to resource allocation (pupillary dilation). Using a set-shifting pupillometry task and separate indices of ability and motivation, McGovern et al. (2020) found evidence of strong independent contributions for each domain, even after accounting for the other domain. Thus, there is moderately strong support of an association between reduced pupillary dilation and more severe negative symptoms (especially motivation deficits), independent of other factors.

The motivational factors relevant to negative symptoms are also thought to be influenced by specific types of de-motivating beliefs. According to the cognitive model of schizophrenia (Beck et al., 2009; Grant & Beck, 2009), early negative experiences or failures (which may or may not co-occur with objective cognitive deficits) contribute to the formation and maintenance of self-defeating beliefs (e.g., "why bother trying if I always fail?") that reinforce the behavioral withdrawal characteristic of negative symptoms. Granholm et al. (2016) found that individuals with schizophrenia who are also high in self-defeating beliefs showed reduced pupillary dilation as processing demands increased on a digit span task relative to healthy controls and individuals with schizophrenia who were low in self-defeating beliefs; the high defeatist belief group also had higher negative symptoms.[2] Therefore, negative expectations or beliefs may also play a role in attention allocation in individuals with schizophrenia.

Although low motivation might appear to suggest a voluntary ("giving up") response, it is presently unclear whether these pupillary responses (and negative symptoms more broadly) reflect an intentional or unintentional effort withdrawal response and/or an inability to deploy the necessary resources at the appropriate time. Drawing from Jones, Siegle, and colleagues' work on depression-related rumination and anxiety-related worry (Jones et al., 2010, 2015; Siegle et al., 2001), it may also be that increased negative affect (which is also present in schizophrenia; for review, see Horan et al., 2008) inhibits the recruitment of necessary cognitive control resources. By this logic, heightened negative affect (whether due to self-defeating beliefs, anxiety, or depression) likely usurps or drains available resources that could otherwise be more efficiently allocated to the task at hand.

Regarding disorganized symptoms, only one pupillometry study to date has made explicit symptom-related hypotheses; more specifically, Minassian et al. (2004) found support for their hypothesis that reduced pupillary dilation responses to complex ambiguous visual stimuli (Rorschach inkblots versus simple line drawings) would be associated with disorganized symptoms. The Rorschach test has subscale scores that have been used to quantify the severity of thought disorder in clinical settings based on the principle that complex ambiguous stimuli tax one's capacity to generate, organize, and synthesize ideas (Holzman et al., 2005). Consistent with a resource limitations view, although the healthy control and schizophrenia groups did not differ in pupillary dilation to simple line drawings (Cohen's $d = 0.26$, small but non-statistically significant effects), the schizophrenia group showed less pupillary dilation to the more visually complex Rorschach images

---

[2] That said, see McGovern et al. (2020) for a lack of a direct relationship between self-defeatist beliefs, both before and after accounting for cognitive ability and negative symptoms. It is unclear whether this relationship depends on task type or other factors.

Table 3 Medication effects and baseline group differences in pupil size

Study	Task	Antipsychotic medication effects	Anticholinergic medication effects	Baseline pupil size[a]
Sensory perception				
Minassian et al. (2004)	Complex visual processing (Rorschach inkblots)	NS	(−)	.[b]
Granholm et al. (2009)	Visual backward masking	NS	NS	NS
Granholm and Verney (2004)	Visual backward masking	NS/(+)[c]	NS	NS
Attention				
Fish and Granholm (2008)	Auditory digit span Distractibility task	NS	NS	NS
Granholm et al. (1997)	Digit span forward (auditory)	NS/(+)[d]	NS/(+)[e]	NS
Granholm et al. (2016)	Digit span forward (auditory)	NS	(+)[f]	NS
Morris et al. (1997)	Digit span forward (auditory)			NS
Study	Task	Antipsychotic medication effects	Anticholinergic medication effects	Baseline pupil size[a]
Karatekin et al. (2008)	Divided attention		NS	NS
Granholm et al. (2000)	Span of apprehension task	NS/(−)[g]	NS/(−)[b]	NS
Granholm et al. (2007)	Span of apprehension task	NS	NS	NS
Working memory & executive functioning				
Karatekin et al. (2010)	Anti-saccade task	NS	NS	NS
McGovern et al. (2020)	Deck choice effort task	NS		NA
Reddy et al. (2018)	Deck choice effort task	NS		NA
Thakkar et al. (2018)	Double-step (action preparation) task	NS		.
Karatekin et al. (2009)	N-back (0- and 1-back)	NS	NS	NS/(−)[h]
Kreis et al. (2021)	Probabilistic decision-making task		NS	NS

(continued)

Table 3 (continued)

Study	Task	Antipsychotic medication effects	Anticholinergic medication effects	Baseline pupil size[a]
Verbal fluency				
Granholm et al. (1998)	Verbal fluency	NS	NS	NS
Emotion regulation				
Strauss et al. (2015)	Directed attention			
Bartolomeo et al. (2020)	Distraction			
Bartolomeo et al. (2020)	Reappraisal			

Notes. A study was included if it provided at least one analysis on the relationship between pupillary dilation and symptoms or medication effects. (+) = Statistically significant positive association between higher dosage and greater pupillary dilation. (−) = Statistically significant negative or inverse association (i.e., higher dosage, lower dilation). "." = relationship not reported. NA = not applicable because the study did not include healthy controls. NS = relationship was examined but statistically non-significant

[a]Baseline differences in pupil size between the schizophrenia group and the healthy control group

[b]This study did not report direct baseline comparison effects; however, results were unchanged with versus without including baseline pupil size as a covariate

[c]Greater dosage was associated with a smaller early factor score (i.e., a more normal pupillary response relative to the healthy control group), $r = -0.50$, $p < 0.05$; however, the early factor in this analysis was likely indicative of pupillary constriction to light. The factors relevant to the study hypotheses (i.e., task-evoked responses to targets and masks) were not associated with medications

[d]The authors ran a large number of (>16) linear and quadratic effect analyses to explore the effect of antipsychotic medications on pupillary dilation. Only one quadratic relationship with antipsychotic (CPZE) dose was statistically significant, which offers very weak evidence of an effect on task-related or baseline pupil size

[e]As above, a large number of (>16 additional) linear and quadratic analyses were run to examine any effects of anticholinergic dosage on pupillary dilation, and no effects with digit span were statistically significant. That said, almost all analyses run with baseline pupil area were significant (*r*s ranged from 0.60 to 0.73 in linear correlations, and change-in-R-sq (ΔR^2) ranged from 0.32 to 0.37 for quadratic relationships, offering stronger evidence of an impact of anticholinergic dosage on baseline pupil size

[f]Those prescribed anticholinergic medications had smaller pupils than those not prescribed anticholinergic medications for the 9-digit, but not the 3- or 6-digit, digits-forward conditions (per the study's supplementary data)

[g]Antipsychotic and anticholinergic dosages were not associated with average pupil size in the task-related effects of interest; however, higher dosages of each medication type were associated with reduced peak constriction amplitude in some conditions, suggesting some impact on the pupillary light response

[h]There was not a main effect of the diagnostic group on baseline pupil size, but there was an interaction such that baseline diameter decreased with age in the healthy control group but not in the schizophrenia group

($d = 1.22$, large effect size) (Minassian et al., 2004). In addition, within the schizophrenia group, less pupillary dilation to Rorschach images was associated with more severe disorganized symptoms, reduced complexity of verbal responses to stimuli based on Rorschach subscale scores, and higher negative symptoms.[3] Although the results from this single study are promising, additional research is needed in this area.

5 Special Considerations for Schizophrenia Populations

5.1 *Individual Differences and Implications for Study Design*

As already mentioned, schizophrenia is an unusually heterogeneous psychiatric disorder in terms of phenomenologically-based distinctions (i.e., symptom profiles), type and severity of cognitive deficits, polygenetic contributions, and functional outcomes (Corvin et al., 2013) among other factors, which undoubtedly impact neural pathophysiology and, thereby, the replicability of pupillometry effects across samples. As reviewed above, different symptoms are likely to have different associations with pupillometry findings. While this heterogeneity problem is not unique to pupillometry research and has been highlighted by other schizophrenia researchers (Riehle et al., 2020), it underscores the importance of recruiting larger and more diverse samples based on the construct of interest to ensure an adequate range in severity, especially when studies are interested in individual differences. Within-sample heterogeneity influences not only sensitivity to detecting anticipated effects within a single study but also reduces generalizability and replicability across studies. Thus, to the extent possible, researchers may benefit from oversampling participants at the extremes on the construct of interest (e.g., negative symptoms; cognitive ability) to avoid both ceiling and floor effects in their sample. At the same time, increasing within-sample heterogeneity also decreases power to detect between-group differences (e.g., schizophrenia versus healthy control samples). Thus, researchers are encouraged to weigh these pros and cons with respect to their particular research questions.

[3] Of the other studies that reported disorganized symptoms as an exploratory analysis within the context of other tasks (e.g., digit span or emotion regulation tasks; see Table 2), none have found significant results (Bartolomeo et al., 2020; Morris et al., 1997; Strauss et al., 2015); however, these tasks may not have been particularly well-suited to examining pupillary relationships with disorganized symptoms.

5.2 Medication Effects

Medications are a common treatment component to the management of schizophrenia symptoms, so the large majority of outpatient or inpatient samples are likely to be on at least one medication that acts on neurotransmitters relevant to pupillometry research. Indeed, antipsychotic dosage—whose primary mechanism of action is to block dopamine—has been associated with reduced pupil size in both healthy controls and individuals with schizophrenia (Koller et al., 2018; Loga et al., 1981; Sakalis et al., 1972; Smolen et al., 1975; Steinhauer & Hakerem, 1992). Some antipsychotics (especially first-generation ones) also have anticholinergic effects. In addition, medications commonly used to prevent or reduce antipsychotic side effects (e.g., anticholinergic medications, such as benztropine) can have pupil-dilating effects (Kelbsch et al., 2019).[4] Thus, medication effects may be examined as a potential confound.

Of the pupillometry studies reporting antipsychotic (e.g., chlorpromazine equivalent [CPZE]) or anticholinergic (e.g., benztropine mesylate equivalent [AChE]) dosages, the large majority have found null effects of medications on pupillary responses (see Table 3; e.g., Fish & Granholm, 2008; Granholm et al., 2007, 2009; Karatekin et al., 2010; McGovern et al., 2020). Of the studies that showed a possible medication effect, the strongest effect was for a relationship between higher anticholinergic dosage and smaller baseline pupil dilation (Granholm et al., 1997, 2016). Two studies showed a possible impact of antipsychotic or anticholinergic dosage on the pupillary light response (Granholm et al., 2000; Granholm & Verney, 2004). Importantly, medications do not appear to have a significant effect on task-evoked pupillary responses.

At present, the best research practice is to ensure that study participants are on stable doses of their medication regimen (e.g., no changes in dosage within the last 4 weeks) to reduce this source of variability and report some calculation of antipsychotic and anticholinergic equivalent dosages (Andreasen et al., 2010; Ereshefsky & Richards, 1990; Leucht et al., 2014; Sadock & Sadock, 2000). Authors are encouraged to explore the influence of medication dosage on their pupillometry results, at least as covariates. Although medication equivalency calculations have several limitations, these are currently the best available methods. Some of these limitations include that (1) dosage is often not prescribed independent of symptom severity (rather, higher dosages tend to be prescribed when symptoms do not respond adequately to lower dosages), (2) other non-specific factors such as weight and drug metabolism influence dosage, (3) participants' medication adherence may be variable and this is not readily accounted for by equivalency calculations, and (4) antipsychotics tend to vary in their effects on other neurotransmitters, including acetylcholine (for expanded discussions, see Granholm et al., 1997; Spohn &

[4] The interested reader is referred to Table 2 from Kelbsch et al. (2019) for a list of medications with pupil constricting (i.e., miosis; e.g., antipsychotics) and dilating (i.e., mydriasis; e.g., anticholinergic) effects.

Strauss, 1989; Zahn et al., 1991). Pupillometry studies that include unaffected family members or medication-naïve participants would help rule out medication effects. At a minimum, medication effects do not appear to fully account for any of the task-evoked effects reported to date.

5.3 *Nicotine Effects*

Nicotine enhances cholinergic neurotransmission throughout the brain, with widespread influences on cognition, both transiently and longer-term, and—most relevantly—in the domains of attention, working memory, and sensory gating (Featherstone & Siegel, 2015; Quigley & MacCabe, 2019). Individuals with schizophrenia are 2–3x more likely than the general population to smoke cigarettes (de Leon & Diaz, 2005). By some estimates, as many as 50–88% of people with schizophrenia smoke regularly (American Psychiatric Association, 2013; Hughes et al., 1986). In addition, as reviewed in Sect. 5.2, many individuals with schizophrenia are prescribed medications with anticholinergic properties, which may interact with nicotinic effects. Thirdly, some studies suggest that smokers with schizophrenia may show greater performance improvements than smokers without a mental illness (e.g., Jacobsen et al., 2004), implying that nicotine may affect these groups differently. Lastly, nicotine has been associated with reduced pupil size, at least in non-psychiatric, non-smokers given single doses of nicotine (Wardhani et al., 2020). Despite the potential relevance of nicotine to each cognition, psychosis, and pupillary dilation, no studies known to this author have systematically examined such effects in a schizophrenia sample. Therefore, more research in this area would be informative.

5.4 *Differences in Baseline Pupil Diameter*

Because the majority of the more recent pupillometry articles have emphasized task-evoked changes in pupil dilation and because resting-state pupil diameter is thought to have different neural mechanisms than task-evoked pupillary responses (e.g., reticular activation vs frontal and other cortical activity, respectively) (Granholm & Steinhauer, 2004; Steinhauer et al., 2004, 2015), group-level and individual differences in baseline pupil diameter have largely been treated as a nuisance variable in the schizophrenia literature. However, there may be theoretical reasons for examining baseline or resting-state pupil diameter in this population. For instance, some schizophrenia theories have posited a role for tonic psychophysiological arousal as a state marker of symptom severity (Dawson et al., 1994). In addition, fluctuations in pupillary dilation while at rest have been associated with the salience network (e.g., anterior cingulate cortex and anterior insula) in fMRI resting-state studies (Schneider et al., 2016). The salience network is thought to be involved in recruiting brain regions relevant to processing sensory information, allocation of attention, and initiation of an action (Palaniyappan & Liddle, 2012), and this network is posited to be involved in the production of positive symptoms in schizophrenia (Palaniyappan et al., 2011; Palaniyappan & Liddle, 2012; White et al., 2010). Despite putative theoretical relevance, almost all of the studies that have reported analyses on baseline pupil diameter have not found statistically significant differences between individuals with schizophrenia and healthy controls (see Table 3; e.g., Fish & Granholm, 2008; Granholm et al., 2016; Granholm & Verney, 2004; Karatekin et al., 2010), although there are exceptions (e.g., Lidsky et al., 1971). Thus, the relevance of resting-state pupil size to schizophrenia requires further research.

6 Future Directions

Pupillometry has many potentially fruitful applications in schizophrenia research that have yet to be fully explored. This section offers expanded discussions in a few particularly promising areas, including treatment applications such as measuring pupillometry changes as a function of treatment progress or phase of illness and understudied domains such as social cognition and emotion regulation. Other areas for which there are no current studies include tasks relevant to reward processing and ecological momentary assessment.

6.1 Treatment Applications

Currently, only one peer-reviewed study has examined whether pupillary indices on one or more cognitive tasks change in response to a psychotherapy intervention in individuals with schizophrenia or other psychotic disorders (Reddy et al., 2023). In this study, individuals with moderate or greater negative symptoms were randomized to 12 weeks of either a motivational interviewing with cognitive behavior therapy (MI-CBT) condition or a mindfulness-based control condition. Relative to the pre-treatment baseline, the MI-CBT group showed greater improvement than the control group in motivational negative symptoms at post-treatment and follow-up periods. On a set-shifting task that incorporated pupillometry (see "Deck Choice Task" in Table 1), relative to the pre-treatment baseline, the MI-CBT group showed an increase in average pupil dilation post-treatment in the easy task condition (which required no set-shifting; small effect size; $f^2 = 0.077$, $p = 0.007$), whereas the control group did not ($f^2 = 0.024$, $p = 0.63$). In the hard task condition (which required frequent set-shifting), there was a trend toward a greater increase in pupil dilation in the MI-CBT group relative to the control from pre-treatment to post-treatment ($f^2 = 0.033$, $p = 0.07$). Pupillary changes were not maintained at the 12-week follow-up period. These findings suggest that the intervention, which successfully decreased negative symptoms, also resulted in increased effort exertion (i.e., increased pupillary dilation) on at least some aspects of distinct cognitive tasks. This study offers preliminary support that pupillary responses may have potential as a biomarker of clinical change in the treatment of motivational negative symptoms.

Apart from utilizing pupillometry as a treatment outcome measure, one group has utilized pupillometry-based feedback to tailor treatment progression during cognitive remediation training designed to improve individuals' processing speed (Choi et al., 2017). In that study, individuals at clinical high risk for psychosis (i.e., individuals in the putatively prodromal phase of psychotic illness who may be showing early subthreshold symptoms and/or declines in functioning) were randomized to either processing speed training (PST) that used targeted games and modulated task difficulty based on pupillary responses or to a control condition without pupillometry where participants played commercially-available games (Choi et al., 2017). Traditional approaches to increasing task difficulty tend to be based on behavioral performance, such as the number or percentage of correct responses at a particular level. In this study, in the PST group, pupillometry was used to titrate task difficulty in an effort to keep individuals within their individualized "sweet spot" range of engagement. This "sweet spot" was recalibrated for each individual prior to the start of each training session. The computer adjusted task difficulty when pupillary values fell below an individually-specified range (of constriction) indicative of task disengagement or increased above an individually-specified range (of dilation) indicative of frustration or overextension of available cognitive resources. Choi et al. (2017) found that those in the PST group (which used pupillometry-based feedback) showed faster motoric and non-motoric processing speed than those in the control group on independent tasks at the end of treatment and at two-month follow-up. The authors acknowledge that a more rigorous test of the pupillometry-based feedback would have been to compare PST with and without the pupillometry-based feedback component. Nevertheless, these data suggest that such an approach may be a feasible and useful way to tailor cognitive remediation and similar computer-based treatments and add to the body of work investigating methods of individually tailoring treatment to address the heterogeneous needs of this population (Medalia et al., 2018; Silverstein et al., 2013)

Outside of the psychosis literature, Siegle and colleagues have started to examine treatment-related effects on pupillary indices in individuals with depression or anxiety, such as with a directed-attention cognitive training task to modify attentional biases in these groups (Graur & Siegle, 2013; Siegle et al., 2011, 2014; Woody et al., 2020). Others are starting to utilize pupillometry-based biofeedback to modify cognitive-affective processing, such as while imagining positive or negative scenarios (Ehlers et al., 2016). Existing cognitive remediation (Bowie et al., 2020) and attention shaping (Silverstein et al., 2009) approaches to schizophrenia treatment may be especially well-suited to integrating pupillometry-based feedback or pupillary indices as biomarkers or predictors of treatment response because such treatments tend to emphasize attention, processing speed, and other cognitive tasks that have been associated with pupillary responses in other studies, and they tend to be computer-based trainings, which should make it relatively easier to integrate pupillometry.

6.2 Phases of Illness

Most of the pupillometry studies to date have been conducted in chronic schizophrenia samples. A few studies have examined adolescents with early psychosis (Karatekin et al., 2008, 2009, 2010). One study compared inpatients and outpatients with schizophrenia and found that inpatients were slightly more impaired than outpatients relative to healthy controls (Steinhauer et al., 1993). There are no peer-reviewed studies to date that have examined pupillometry results across phases of illness or changes over time. One study (Steinhauer et al., 1979) examined individuals with schizophrenia and their unaffected siblings and found that siblings tended to have pupillary results intermediate between their schizophrenia probands and the healthy control group on an auditory guessing task. Understanding whether the alterations in resource allocation assessed via pupillometry extend to unaffected family members and those in the prodromal or earliest phases of illness would help ascertain whether any of these deficits reflect genetic vulnerability markers, traits, or disease states.

6.3 Social Cognition

Social cognition refers to the psychological processes involved in perceiving, encoding, storing, retrieving, and regulating information related to oneself or others (Green et al., 2019). In an NIMH consensus statement, experts distinguished five subdomains relevant to (and often impaired in) schizophrenia populations: emotion processing, social perception, social knowledge, attributional bias/style, and theory of mind (Green et al., 2008). Many schizophrenia researchers are interested in whether certain drugs (such as oxytocin) or social cognition training psychotherapies improve social functioning in this population (Kurtz et al., 2016). Pupillometry offers one way to determine whether attention allocation deficits contribute to the behavioral deficits shown on laboratory tasks and whether these change as a function of treatment. In the only peer-reviewed pupillometry study examining an aspect of social cognition in schizophrenia, Wynn et al. (2019) examined whether different doses of intranasal oxytocin impacted facial affect recognition. Individuals with schizophrenia display large effect-size deficits ($d = 0.89$) on facial affect identification tasks relative to healthy controls (Green et al., 2019). Pupillometry studies of healthy controls and other clinical samples (e.g., depression) showed greater dilation when processing facial emotional content compared with non-face content (Prehn et al., 2011; Siegle et al., 2001). In addition, pupillary dilation to facial stimuli increased in healthy controls when they were administered 24 IU of oxytocin relative to placebo, suggesting oxytocin increased the recruitment of attentional resources to social stimuli (Prehn et al., 2013). Wynn et al. (2019) found that schizophrenia participants showed less pupillary dilation when viewing scrambled faces relative to fearful and neutral, although not happy faces, suggesting at least some increase in allocation for socially-relevant stimuli; however, there was no impact of oxytocin dosage on pupillary dilation in this pilot study, at least not after a single dose or on this particular task. Future studies should examine whether pupillometry can shed light on deficits in facial affect recognition, social attention, and other socially-relevant processes.

6.4 Emotion Regulation

Despite the clear ties between pupillometry and emotional arousal and the known abnormalities in emotional regulation in schizophrenia, almost no research has examined this intersection. Emotion regulation refers to the processes by which we modify our positive and negative emotions in terms of their intensity, when and how long they occur, and how they are expressed (Gross, 2002). A growing body of research across multiple modalities (e.g., behavioral tasks, ERP, fMRI, skin conductance) suggests that individuals with schizophrenia have difficulty decreasing their neurophysiological responses to unpleasant stimuli, including when attempting to apply reappraisal, distraction, or directed-attention emotion regulation strategies (Horan et al., 2013; Lincoln et al., 2015; Strauss et al., 2013, 2015; Van Der Meer et al., 2014; Zhang et al., 2020). Consistent with this prior work, when attention was directed to emotionally arousing or non-arousing subregions of unpleasant images (via area of interest windows), individuals with schizophrenia showed reduced pupillary dilation relative to healthy controls (see more detailed task description in Table 1); the eye-tracking data from the same task revealed that the schizophrenia group tended to spend more time attending to emotionally arousing aspects of the image when these fell outside of the non-arousing target region they had been directed toward (Strauss et al., 2015). Thus, individuals with schizophrenia may have difficulty decreasing their neurophysiological responses to unpleasant stimuli at least in part because of a failure to engage effortful cognitive control processes needed to inhibit prepotent responses to look toward unpleasant stimuli and instead direct attention away from unpleasant, emotionally arousing stimuli.

Notably, directed attention tasks cue the participant on exactly where to look, which conflates visual attention and cognitive effort; participant-generated distraction and cognitive reappraisal permit independent examination of visual attention and cognitive effort. In a pupillometry study of healthy controls that examined effort exertion across different emotion regulation strategies, cognitive reappraisal resulted in larger pupil dilation compared to either passively viewing unpleasant images or distraction-based emotion regulation, suggesting that cognitive reappraisal requires more cognitive effort than distraction (Strauss et al., 2016). One plausible explanation is that cognitive reappraisal required more effortful engagement with the unpleasant stimulus in order to generate an alternative response that reinterprets the same stimulus in a new, less threatening way (e.g., reinterpret an image of women crying out of sadness as tears of joy instead), whereas distraction did not require as much engagement with the stimulus in order to generate an alternative mental representation of anything unrelated to the unpleasant stimulus (e.g., a mental image of a house rather than of women crying).

In a follow-up study that included a schizophrenia group, the schizophrenia group did not differ from healthy controls in pupillary dilation during the distraction condition (the putatively "easier" condition) but showed reduced pupil dilation during the cognitive reappraisal condition (Bartolomeo et al., 2020) (see Table 1 for a more detailed description of the Reappraisal and Distraction Emotion Regulation Task). As in other cognitive tasks that vary levels of difficulty, individuals with schizophrenia were unable to allocate the necessary cognitive resources during the more challenging condition (in this case, cognitive reappraisal). When Bartolomeo et al. (2020) integrated their pupillary findings with their eye-tracking data, they learned that the schizophrenia group appeared to deploy an ineffective visual attention strategy. More specifically, the pattern of visual attention from healthy controls revealed that effective cognitive reappraisal involved directing more attention toward arousing content, presumably, to help generate an alternative interpretation of the image. During the distraction condition, healthy controls spent less time looking at the emotionally arousing content. However, individuals with schizophrenia showed the opposite pattern. That is, they spent too little time (i.e., had fewer fixations) looking at emotionally arousing content during reappraisal and spent too much time looking at the emotionally arousing content during the distraction condition. Taken together, individuals with schizophrenia appeared to deploy ineffective strategies during emotion regulation tasks, which may contribute to their impaired ability to down-regulate negative emotions under stress. These and similar paradigms may be useful to examine the efficacy of emotion-regulation based interventions (e.g., Cohen & Ochsner, 2018).

7 Conclusions

In summary, a relatively large number of studies have demonstrated relationships between impaired cognitive functioning and abnormal pupillary responses in individuals with schizophrenia. In the large majority of cases, impairment in the cognitive ability examined was associated with reduced pupillary dilation amplitudes, suggestive of either insufficient attentional resources or impairments in the capacity to deploy the necessary attentional resources. At present, there is limited data to support a relationship between more severe positive symptoms and greater misallocation of available resources (i.e., excessive pupillary dilation) to task-irrelevant or distractor stimuli, modest support for a relationship between more severe disorganized symptoms and reduced dilation to complex stimuli, and growing support for a relationship between more severe negative symptoms and reduced dilation as processing demands increase. In addition, the reduced pupillary dilation on at least some tasks (e.g., Deck Choice) has independent associations with both impaired cognitive ability and increased negative symptom severity, suggesting that these are not merely redundant measures of the same processes. Studying schizophrenia populations requires special attention to several potentially confounding factors (e.g., medications, nicotine, baseline pupil size, and sample heterogeneity). In addition, to ascertain more definitively the relationships between pupillary dilation and psychiatric symptoms, future studies need relatively larger sample sizes and may benefit from oversampling individuals at the more extreme range of symptoms to ensure an adequate range on the symptom(s) of interest.

Pupillary abnormalities have been shown on a variety of cognitive tasks, from "lower level" perceptual discrimination tasks (e.g., oddball, visual backward masking) to "higher level" decision-making tasks (e.g., deck choice, double-step, n-back). However, almost no studies have examined pupillary responses on tasks examining aspects of social cognition, reward processing, emotion regulation, nor in monitoring responses to therapeutic interventions or ecologically-based assessments (e.g., via smartphones), and these areas are especially ripe for future studies.

References

American Psychiatric Association. (2013). *Diagnostic and statistical manual of mental disorders – 5th edition (DSM-5)*. American Psychiatric Publishing. https://doi.org/10.1176/appi.books.9780890425596.744053

Andreasen, N. C., Pressler, M., Nopoulos, P., Miller, D., & Ho, B. C. (2010). Antipsychotic dose equivalents and dose-years: A standardized method for comparing exposure to different drugs. *Biological Psychiatry, 67*(3), 255–262. https://doi.org/10.1016/j.biopsych.2009.08.040

Aston-Jones, G., & Cohen, J. D. (2005). An integrative theory of locus coeruleus-norepinephrine function: Adaptive gain and optimal performance. *Annual Review of Neuroscience, 28*(1), 403–450. https://doi.org/10.1146/annurev.neuro.28.061604.135709

Bach, L. (1908). *Pupillenlehre, anatomie, physiologie und pathologie; Methodik der untersuching*. Karger.

Bartolomeo, L. A., Culbreth, A. J., Ossenfort, K. L., & Strauss, G. P. (2020). Neurophysiological evidence for emotion regulation impairment in schizophrenia: The role of visual attention and cognitive effort. *Journal of Abnormal Psychology, 129*(6), 670–676. https://doi.org/10.1037/abn0000580

Beck, A. T., Rector, N. A., Stolar, N., & Grant, P. (2009). *Schizophrenia: Cognitive theory, research, and therapy*. Guilford Press. https://doi.org/10.1177/002221947600900607

Bowie, C. R., Bell, M. D., Fiszdon, J. M., Johannesen, J. K., Lindenmayer, J. P., McGurk, S. R., Medalia, A. A., Penadés, R., Saperstein, A. M., Twamley, E. W., Ueland, T., & Wykes, T. (2020). Cognitive remediation for schizophrenia: An expert working group white paper on core techniques. *Schizophrenia Research, 215*, 49–53. https://doi.org/10.1016/j.schres.2019.10.047

Bumke, O. (1911). *Die pupillenstörungen bei Geistes- und Nervenkrankheiten (Physiologie und Pathologie der Irisbewegungen)*. Verlag von Gustab Fischer.

Chandler, D. J., Waterhouse, B. D., & Gao, W. (2014). New perspectives on catecholaminergic regulation of executive circuits : evidence for independent modulation of prefrontal functions by midbrain dopaminergic and noradrenergic neurons. *Frontiers in Neural Circuits, 8*, 1–10. https://doi.org/10.3389/fncir.2014.00053

Charlson, F. J., Ferrari, A. J., Santomauro, D. F., Diminic, S., Stockings, E., Scott, J. G., McGrath, J. J., & Whiteford, H. A. (2018). Global epidemiology and burden of schizophrenia: Findings from the global burden of disease study 2016. *Schizophrenia Bulletin, 44*(6), 1195–1203. https://doi.org/10.1093/schbul/sby058

Choi, J., Corcoran, C. M., Fiszdon, J. M., Stevens, M., Javitt, D. C., Deasy, M., Haber, L. C., Dewberry, M. J., & Pearlson, G. D. (2017). Pupillometer-based neurofeedback cognitive training to improve processing speed and social functioning in individuals at clinical high risk for psychosis. *Psychiatric Rehabilitation Journal, 40*(1), 33–42. https://doi.org/10.1037/prj0000217

Chong, H. Y., Teoh, S. L., Wu, D. B. C., Kotirum, S., Chiou, C. F., & Chaiyakunapruk, N. (2016). Global economic burden of schizophrenia: A systematic review. *Neuropsychiatric Disease and Treatment, 12*, 357–373. https://doi.org/10.2147/NDT.S96649

Cohen, N., & Ochsner, K. N. (2018). From surviving to thriving in the face of threats: The emerging science of emotion regulation training. *Current Opinion in Behavioral Sciences, 24*, 143–155. https://doi.org/10.1016/j.cobeha.2018.08.007

Corcoran, C., Walker, E., Huot, R., Mittal, V., Tessner, K., Kestler, L., & Malaspina, D. (2003). The stress cascade and schizophrenia: Etiology and onset. *Schizophrenia Bulletin, 29*(4), 671–692. https://doi.org/10.1093/oxfordjournals.schbul.a007038

Corvin, A., Buchanan, R. W., Carpenter, W. T., Kennedy, J. L., Keshavan, M. S., MacDonald, A. W., Sass, L., & Wessa, M. (2013). Which aspects of heterogeneity are useful to translational success? In S. M. Silverstein, B. Moghaddam, & T. Wykes (Eds.), *Schizophrenia: Evolution and synthesis*. MIT Press. http://europepmc.org/abstract/MED/33886221

Culbreth, A. J., Gold, J. M., Cools, R., & Barch, D. M. (2016a). Impaired activation in cognitive control regions predicts reversal learning in schizophrenia. *Schizophrenia Bulletin, 42*(2), 484–493. https://doi.org/10.1093/schbul/sbv075

Culbreth, A. J., Moran, E. K., & Barch, D. M. (2018). Effort-based decision-making in schizophrenia. *Current Opinion in Behavioral Sciences, 22*, 1–6. https://doi.org/10.1016/j.cobeha.2017.12.003

Culbreth, A. J., Westbrook, A., & Barch, D. (2016b). Negative symptoms are associated with an increased subjective cost of cognitive effort. *Journal of Abnormal Psychology, 125*(4), 528–536. https://doi.org/10.1037/abn0000153

Dawson, M. E., Nuechterlein, K. H., Schell, A. M., Gitlin, M., & Ventura, J. (1994). Autonomic abnormalities in schizophrenia: State or trait indicators? *Archives of General Psychiatry, 51*(10), 813–824. https://doi.org/10.1001/archpsyc.1994.03950100061006

de Leon, J., & Diaz, F. J. (2005). A meta-analysis of worldwide studies demonstrates an association between schizophrenia and tobacco smoking behaviors. *Schizophrenia Research, 76*(2–3), 135–157. https://doi.org/10.1016/j.schres.2005.02.010

Dickinson, D., Ramsey, M. E., & Gold, J. M. (2007). Overlooking the obvious: A meta-analytic comparison of digit symbol coding tasks and other cognitive measures in schizophrenia. *Achives of General Psychiatry, 64*, 532–542. https://doi.org/10.1001/archpsyc.64.5.532

Ehlers, J., Strauch, C., Georgi, J., & Huckauf, A. (2016). Pupil size changes as an active information channel for biofeedback applications. *Applied Psychophysiology Biofeedback, 41*(3), 331–339. https://doi.org/10.1007/s10484-016-9335-z

Ereshefsky, L., & Richards, A. (1990). Psychosis. In L. Y. Young & M. A. Koda-Kimble (Eds.), *Applied therapeutics: The clinical use of drugs* (pp. 1189–1230). Applied Therapeutics, Inc..

Featherstone, R. E., & Siegel, S. J. (2015). The role of nicotine in schizophrenia. In *International review of neurobiology* (Vol. 124, 1st ed., pp. 23–78). Elsevier. https://doi.org/10.1016/bs.irn.2015.07.002

Fish, S. C., & Granholm, E. (2008). Easier tasks can have higher processing loads: Task difficulty and cognitive resource limitations in schizophrenia. *Journal of Abnormal Psychology, 117*(2), 355–363. https://doi.org/10.1037/0021-843X.117.2.355

Gispen-de Wied, C. C. (2000). Stress in schizophrenia: An integrative view. *European Journal of Pharmacology, 405*(1–3), 375–384. https://doi.org/10.1016/S0014-2999(00)00567-7

Glahn, D. C., Ragland, J. D., Abramoff, A., Barrett, J., Laird, A. R., Bearden, C. E., & Velligan, D. I. (2005). Beyond hypofrontality: A quantitative meta-analysis of functional neuroimaging studies of working memory in schizophrenia. *Human Brain Mapping, 25*(1), 60–69. https://doi.org/10.1002/hbm.20138

Grace, A. A. (1991). Phasic versus tonic dopamine release and the modulation of dopamine system responsivity: A hypothesis for the etiology of schizophrenia. *Neuroscience, 41*(1), 1–24. https://doi.org/10.1016/0306-4522(91)90196-U
Granholm, E., Asarnow, R. F., Sarkin, A. J., & Dykes, K. L. (1996). Pupillary responses index cognitive resource limitations. *Psychophysiology, 33*, 457–461. https://doi.org/10.1111/j.1469-8986.1996.tb01071.x
Granholm, E., Chock, D., & Morris, S. (1998). Pupillary responses evoked during verbal fluency tasks indicate semantic network dysfunction in schizophrenia. *Journal of Clinical and Experimental Neuropsychology, 20*(6), 856–872. https://doi.org/10.1076/jcen.20.6.856.1107
Granholm, E., Fish, S. C., & Verney, S. P. (2009). Pupillometric measures of attentional allocation to target and mask processing on the backward masking task in schizophrenia. *Psychophysiology, 46*(3), 510–520. https://doi.org/10.1111/j.1469-8986.2009.00805.x
Granholm, E., Morris, S., Asarnow, R. F., Chock, D., & Jeste, D. V. (2000). Accelerated age-related decline in processing resources in schizophrenia: Evidence from pupillary responses recorded during the span of apprehension task. *Journal of the International Neuropsychological Society, 6*(1), 30–43. https://doi.org/10.1017/S1355617700611049
Granholm, E., Morris, S. K., Sarkin, A. J., Asarnow, R. F., & Jeste, D. V. (1997). Pupillary responses index overload of working memory resources in schizophrenia. *Journal of Abnormal Psychology, 106*(3), 458–467. https://doi.org/10.1037/0021-843X.106.3.458
Granholm, E., Ruiz, I., Gallegos-Rodriguez, Y., Holden, J., & Link, P. C. (2016). Pupillary responses as a biomarker of diminished effort associated with defeatist attitudes and negative symptoms in schizophrenia. *Biological Psychiatry, 80*(8), 581–588. https://doi.org/10.1016/j.biopsych.2015.08.037
Granholm, E., & Steinhauer, S. R. (2004). Pupillometric measures of cognitive and emotional processes. *International Journal of Psychophysiology, 52*, 1–6. https://doi.org/10.1016/j.ijpsycho.2003.12.001
Granholm, E., & Verney, S. P. (2004). Pupillary responses and attentional allocation problems on the backward masking task in schizophrenia. *International Journal of Psychophysiology, 52*(1), 37–51. https://doi.org/10.1016/j.ijpsycho.2003.12.004
Granholm, E., Verney, S. P., Perivoliotis, D., & Miura, T. (2007). Effortful cognitive resource allocation and negative symptom severity in chronic schizophrenia. *Schizophrenia Bulletin, 33*(3), 831–842. https://doi.org/10.1093/schbul/sbl040
Grant, P. M., & Beck, A. T. (2009). Defeatist beliefs as a mediator of cognitive impairment, negative symptoms, and functioning in schizophrenia. *Schizophrenia Bulletin, 35*(4), 798–806. https://doi.org/10.1093/schbul/sbn008
Graur, S., & Siegle, G. (2013). Pupillary motility: Bringing neuroscience to the psychiatry clinic of the future. *Current Neurology and Neuroscience Reports, 13*(8), 365. https://doi.org/10.1007/s11910-013-0365-0
Green, M. F., Hellemann, G., Horan, W. P., Lee, J., & Wynn, J. K. (2012). From perception to functional outcome in schizophrenia: Modeling the role of ability and motivation. *Archives of General Psychiatry, 69*(12), 1216–1224. https://doi.org/10.1001/archgenpsychiatry.2012.652
Green, M. F., Horan, W. P., & Lee, J. (2019). Nonsocial and social cognition in schizophrenia: Current evidence and future directions. *World Psychiatry, 18*(2), 146–161. https://doi.org/10.1002/wps.20624
Green, M. F., Penn, D. L., Bentall, R., Carpenter, W. T., Gaebel, W., Gur, R. C., Kring, A. M., Park, S., Silverstein, S. M., & Heinssen, R. (2008). Social cognition in schizophrenia: An NIMH workshop on definitions, assessment, and research opportunities. *Schizophrenia Bulletin, 34*(6), 1211–1220. https://doi.org/10.1093/schbul/sbm145
Gross, J. J. (2002). Emotion regulation: Affective, cognitive, and social consequences. *Psychophysiology, 39*(3), 281–291. https://doi.org/10.1017/S0048577201393198
Hakerem, G., Sutton, S., & Zubin, J. (1964). Pupillary reactions to light in schizophrenic patients and normals. *Annals of the New York Academy of Sciences, 105*, 820–831. https://doi.org/10.1111/j.1749-6632.1964.tb42965.x

Heinz, A., & Schlagenhauf, F. (2010). Dopaminergic dysfunction in schizophrenia: Salience attribution revisited. *Schizophrenia Bulletin, 36*(3), 472–485. https://doi.org/10.1093/schbul/sbq031
Holzman, P. S., Levy, D. L., & Johnston, M. H. (2005). The use of the Rorschach technique for assessing formal thought disorder. In R. F. Bornstein & J. M. Masling (Eds.), *Scoring the Rorschach: Seven validated systems* (pp. 55–95). Lawrence Erlbaum Associates Publishers.
Horan, W. P., Blanchard, J. J., Clark, L. A., & Green, M. F. (2008). Affective traits in schizophrenia and schizotypy. *Schizophrenia Bulletin, 34*(5), 856–874. https://doi.org/10.1093/schbul/sbn083
Horan, W. P., Hajcak, G., Wynn, J. K., & Green, M. F. (2013). Impaired emotion regulation in schizophrenia: Evidence from event-related potentials. *Psychological Medicine, 43*, 2377–2391. https://doi.org/10.1017/S0033291713000019
Howes, O. D., & Kapur, S. (2009). The dopamine hypothesis of schizophrenia: Version III — The final common pathway. *Schizophrenia Bulletin, 35*(3), 549–562. https://doi.org/10.1093/schbul/sbp006
Hughes, J. R., Hatsukami, K., Mitchell, J. E., & Dahlgren, L. A. (1986). Prevalence of smoking among psychiatric disorders. *American Journal of Psychiatry, 143*(*8*), 993–997.
Ikushima, M., & Matsunga, K. (1975). Pupillary light reactions of medicated and non-medicated schizophrenics. *Kyushu Neuro-Psychiatry, 21*(2), 119–124.
Jacobsen, L. K., D'Souza, D. C., Mencl, W. E., Pugh, K. R., Skudlarski, P., & Krystal, J. H. (2004). Nicotine effects on brain function and functional connectivity in schizophrenia. *Biological Psychiatry, 55*(8), 850–858. https://doi.org/10.1016/j.biopsych.2003.12.023
Jansen, L. M. C., Gispen-de Wied, C. C., Gademan, P. J., De Jonge, R. C. J., Van Der Linden, J. A., & Kahn, R. S. (1998). Blunted cortisol response to a psychosocial stressor in schizophrenia. *Schizophrenia Research, 33*(1–2), 87–94. https://doi.org/10.1016/S0920-9964(98)00066-8
Jones, N. P., Siegle, G. J., & Mandell, D. (2015). Motivational and emotional influences on cognitive control in depression: A pupillometry study. *Cognitive, Affective, & Behavioral Neuroscience, 15*(2), 263–275. https://doi.org/10.3758/s13415-014-0323-6
Jones, N. P., Siegle, G. J., Muelly, E. R., Haggerty, A., & Ghinassi, F. (2010). Poor performance on cognitive tasks in depression: Doing too much or not enough? *Cognitive, Affective, & Behavioral Neuroscience, 10*(1), 129–140. https://doi.org/10.3758/CABN.10.1.129
Joshi, S., Li, Y., Kalwani, R. M., & Gold, J. I. (2016). Relationships between pupil diameter and neuronal activity in the locus coeruleus, colliculi, and cingulate cortex. *Neuron, 89*(1). https://doi.org/10.1016/j.neuron.2015.11.028
Kapur, S. (2003). Psychosis as a state of aberrant salience: A framework linking biology, phenomenology, and pharmacology in schizophrenia. *American Journal of Psychiatry, 160*, 13–23.
Karatekin, C., Bingham, C., & White, T. (2009). Regulation of cognitive resources during an n-back task in youth-onset psychosis and attention-deficit/hyperactivity disorder (ADHD). *International Journal of Psychophysiology, 73*(3), 294–307. https://doi.org/10.1016/j.ijpsycho.2009.05.001
Karatekin, C., Bingham, C., & White, T. (2010). Oculomotor and pupillometric indices of pro- and antisaccade performance in youth-onset psychosis and attention deficit/hyperactivity disorder. *Schizophrenia Bulletin, 36*(6), 1167–1186. https://doi.org/10.1093/schbul/sbp035
Karatekin, C., White, T., & Bingham, C. (2008). Divided attention in youth-onset psychosis and attention deficit/hyperactivity disorder. *Journal of Abnormal Psychology, 117*(4), 881–895. https://doi.org/10.1037/a0013446
Kelbsch, C., Strasser, T., Chen, Y., Feigl, B., Gamlin, P. D., Kardon, R., Peters, T., Roecklein, K. A., Steinhauer, S. R., Szabadi, E., Zele, A. J., Wilhelm, H., & Wilhelm, B. J. (2019). Standards in pupillography. *Frontiers in Neurology, 10*, 129. https://doi.org/10.3389/fneur.2019.00129
Kirkpatrick, B., Fenton, W. S., Carpenter, W. T., & Marder, S. R. (2006). The NIMH-MATRICS consensus statement on negative symptoms. *Schizophrenia Bulletin, 32*(2), 214–219. https://doi.org/10.1093/schbul/sbj053
Koller, D., Belmonte, C., Lubomirov, R., Saiz-Rodríguez, M., Zubiaur, P., Román, M., Ochoa, D., Carcas, A., Wojnicz, A., & Abad-Santos, F. (2018). Effects of aripiprazole on pupillometric parameters related to pharmacokinetics and pharmacogenetics after single oral admin-

istration to healthy subjects. *Journal of Psychopharmacology, 32*(11), 1212–1222. https://doi.org/10.1177/0269881118798605

Kreis, I., Moritz, S., & Pfuhl, G. (2020). Objective versus subjective effort in schizophrenia. *Frontiers in Psychology, 11*(July), 1–8. https://doi.org/10.3389/fpsyg.2020.01469

Kreis, I., Zhang, L., Moritz, S., & Pfuhl, G. (2021). Spared performance but increased uncertainty in schizophrenia: Evidence from a probabilistic decision-making task. *Schizophrenia Research*. https://doi.org/10.1016/j.schres.2021.06.038

Kurtz, M. M., Gagen, E., Rocha, N. B. F., Machado, S., & Penn, D. L. (2016). Comprehensive treatments for social cognitive deficits in schizophrenia: A critical review and effect-size analysis of controlled studies. *Clinical Psychology Review, 43*, 80–89. https://doi.org/10.1016/j.cpr.2015.09.003

Leucht, S., Samara, M., Heres, S., Patel, M. X., Woods, S. W., & Davis, J. M. (2014). Dose equivalents for second-generation antipsychotics: The minimum effective dose method. *Schizophrenia Bulletin, 40*(2), 314–326. https://doi.org/10.1093/schbul/sbu001

Lidsky, A., Hakerem, G., & Sutton, S. (1971). Pupillary reactions to single light pulses in psychiatric patients and normals. *The Journal of Nervous and Mental Disease, 153*(4), 286–291. https://doi.org/10.1097/00005053-197110000-00007

Lincoln, T. M., Hartmann, M., Köther, U., & Moritz, S. (2015). Dealing with feeling: Specific emotion regulation skills predict responses to stress in psychosis. *Psychiatry Research, 228*(2), 216–222. https://doi.org/10.1016/j.psychres.2015.04.003

Loga, S., Curry, S., & Lader, M. (1981). Interaction of chlorpromazine and nortriptyline in patients with schizophrenia. *Clinical Pharmacokinetics, 6*(6), 454–462. https://doi.org/10.2165/00003088-198106060-00003

Luck, S. J., & Gold, J. M. (2008). The construct of attention in schizophrenia. *Biological Psychiatry, 64*(1), 34–39. https://doi.org/10.1016/j.biopsych.2008.02.014

Maia, T. V., & Frank, M. J. (2017). An integrative perspective on the role of dopamine in schizophrenia. *Biological Psychiatry, 81*(1), 52–66. https://doi.org/10.1016/j.biopsych.2016.05.021

Mäki-Marttunen, V., Andreassen, O. A., & Espeseth, T. (2020). The role of norepinephrine in the pathophysiology of schizophrenia. *Neuroscience and Biobehavioral Reviews, 118*, 298–314. https://doi.org/10.1016/j.neubiorev.2020.07.038

Martin, B. C., & Miller, L. S. (1998). Expenditures for treating schizophrenia: A population-based study of Georgia Medicaid recipients. *Schizophrenia Bulletin, 24*(3), 479–488. https://doi.org/10.1093/oxfordjournals.schbul.a033342

Mathôt, S. (2018). Pupillometry: Psychology, physiology, and function. *Journal of Cognition, 1*(1), 1–23. https://doi.org/10.5334/joc.18

May, P. R. A. (1947). Pupillary abnormalities in schizophrenia and during muscular effort. *Journal of Mental Science, 94*(394), 89–98. https://doi.org/10.1192/bjp.94.394.89

McCleery, A., & Nuechterlein, K. H. (2019). Cognitive impairment in psychotic illness: Prevalence, profile of impairment, developmental course, and treatment considerations. *Dialogues in Clinical Neuroscience, 21*(3), 239–248. https://doi.org/10.31887/DCNS.2019.21.3/amccleery

McGovern, J. E., Reddy, L. F., Reavis, E. A., & Green, M. F. (2020). Pupillary change on a cognitive effort task in schizophrenia: Associations with cognition and motivation. *International Journal of Psychophysiology, 155*(May), 1–7. https://doi.org/10.1016/j.ijpsycho.2020.05.003

Medalia, A., Saperstein, A. M., Hansen, M. C., & Lee, S. (2018). Personalised treatment for cognitive dysfunction in individuals with schizophrenia spectrum disorders. *Neuropsychological Rehabilitation, 28*(4), 602–613. https://doi.org/10.1080/09602011.2016.1189341

Millard, S. J., Bearden, C. E., Karlsgodt, K. H., & Sharpe, M. J. (2022). The prediction-error hypothesis of schizophrenia: New data point to circuit-specific changes in dopamine activity. *Neuropsychopharmacology, 47*, 628–640. https://doi.org/10.1038/s41386-021-01188-y

Minassian, A., Granholm, E., Verney, S., & Perry, W. (2004). Pupillary dilation to simple vs. complex tasks and its relationship to thought disturbance in schizophrenia patients. *International Journal of Psychophysiology, 52*(1), 53–62. https://doi.org/10.1016/j.ijpsycho.2003.12.008

Moran, E. K., Culbreth, A. J., Kandala, S., & Barch, D. M. (2019). From neuroimaging to daily functioning: A multimethod analysis of reward anticipation in people with schizophrenia. *Journal of Abnormal Psychology, 128*(7), 723–734. https://doi.org/10.1037/abn0000461

Morris, S. K., Granholm, E., Sarkin, A. J., & Jeste, D. V. (1997). Effects of schizophrenia and aging on pupillographic measures of working memory. *Schizophrenia Research, 27*(2–3), 119–128. https://doi.org/10.1016/S0920-9964(97)00065-0

Nieoullon, A. (2002). Dopamine and the regulation of cognition and attention. *Progress in Neurobiology, 67*(1), 53–83. https://doi.org/10.1016/S0301-0082(02)00011-4

Nuechterlein, K. H., & Dawson, M. E. (1984). Information processing and attentional functioning in the developmental course of schizophrenic disorders. *Schizophrenia Bulletin, 10*(2), 160–203. https://doi.org/10.1093/schbul/10.2.160

Nuechterlein, K. H., Asarnow, R. F., Subotnik, K. L., Fogelson, D. L., Ventura, J., Torquato, R. D., & Dawson, M. E. (1998). Neurocognitive vulnerability factors for schizophrenia: Convergence across genetic risk studies and longitudinal trait–state studies. In M. F. Lenzenweger & R. H. Dworkin (Eds.), Origins and development of schizophrenia: Advances in experimental psychopathology (pp. 299–327). *American Psychological Association.* https://doi.org/10.1037/10305-011

Okada, F., Kase, M., & Shintomi, Y. (1978). Pupillary abnormalities in schizophrenic patients during long-term administration of psychotropic drugs: Dissociation between light and near vision reactions. *Psychopharmacology, 58*(3), 235–240. https://doi.org/10.1007/BF00427385

Olfson, M., Gerhard, T., Huang, C., Crystal, S., & Stroup, T. S. (2015). Premature mortality among adults with schizophrenia in the United States. *JAMA Psychiatry, 72*(12), 1172–1181. https://doi.org/10.1001/jamapsychiatry.2015.1737

Palaniyappan, L., & Liddle, P. F. (2012). Does the salience network play a cardinal role in psychosis? An emerging hypothesis of insular dysfunction. *Journal of Psychiatry and Neuroscience, 37*(1), 17–27. https://doi.org/10.1503/jpn.100176

Palaniyappan, L., Mallikarjun, P., Joseph, V., White, T. P., & Liddle, P. F. (2011). Reality distortion is related to the structure of the salience network in schizophrenia. *Psychological Medicine, 41*(8), 1701–1708. https://doi.org/10.1017/S0033291710002205

Perry, W., & Braff, D. L. (1994). Information-processing deficits and thought disorder in schizophrenia. *American Journal of Psychiatry, 151*(3), 363–367. https://doi.org/10.1176/ajp.151.3.363

Prehn, K., Heekeren, H. R., & van der Meer, E. (2011). Influence of affective significance on different levels of processing using pupil dilation in an analogical reasoning task. *International Journal of Psychophysiology, 79*(2), 236–243. https://doi.org/10.1016/j.ijpsycho.2010.10.014

Prehn, K., Kazzer, P., Lischke, A., & Heinrichs, M. (2013). Effects of intranasal oxytocin on pupil dilation indicate increased salience of socioaffective stimuli. *Psychophysiology, 50*, 528–537. https://doi.org/10.1111/psyp.12042

Quigley, H., & MacCabe, J. H. (2019). The relationship between nicotine and psychosis. *Therapeutic Advances in Psychopharmacology, 9*, 204512531985996. https://doi.org/10.1177/2045125319859969

Reddy, L. F., Glynn, S. M., McGovern, J. E., Sugar, C. A., Reavis, E. A., & Green, M. F. (2023). A novel psychosocial intervention for motivational negative symptoms in schizophrenia: Combined motivational interviewing and CBT. *American Journal of Psychiatry, 180*(5), 367–376. https://doi.org/10.1176/appi.ajp.20220243

Reddy, L. F., Reavis, E. A., Wynn, J. K., & Green, M. F. (2018). Pupillary responses to a cognitive effort task in schizophrenia. *Schizophrenia Research, 199*, 53–57. https://doi.org/10.1016/j.schres.2018.03.005

Reichenberg, A., Harvey, P. D., Bowie, C. R., Mojtabai, R., Rabinowitz, J., Heaton, R. K., & Bromet, E. (2009). Neuropsychological function and dysfunction in schizophrenia and psychotic affective disorders. *Schizophrenia Bulletin, 35*(5), 1022–1029. https://doi.org/10.1093/schbul/sbn044

Riehle, M., Böhl, M. C., Pillny, M., & Lincoln, T. M. (2020). Efficacy of psychological treatments for patients with Schizophrenia and relevant negative symptoms: A meta-analysis. *Clinical Psychology in Europe, 2*(3), e2899. https://doi.org/10.32872/cpe.v2i3.2899

Robbins, T. W., & Arnsten, A. F. T. (2009). The neuropsychopharmacology of fronto-executive function: Monoaminergic modulation. *Annual Review of Neuroscience, 32*, 267–289. https://doi.org/10.1146/annurev.neuro.051508.135535
Rubin, L. S. (1962). Patterns of adrenergic-cholinergic imbalance in the functional psychoses. *Psychological Review, 69*, 501–519. https://doi.org/10.1037/h0046372
Rubin, L. S. (1970). Pupillary reflexes as objective indices of autonomic dysfunction in the differential diagnosis of schizophrenia and neurotic behavior. *Journal of Behavior Therapy and Experimental Psychiatry, 1*(3), 185–194. https://doi.org/10.1016/0005-7916(70)90001-7
Rubin, L. S., & Barry, T. J. (1972). The effect of the cold pressor test on pupillary reactivity of schizophrenics in remission. *Biological Psychiatry, 5*(2), 181–197.
Sadock, B. J., & Sadock, V. A. (Eds.). (2000). *Kaplan & Sadock's comprehensive textbook of psychiatry* ((7th ed.). ed.). Lippincott Williams and Wilkins Publishers.
Sakalis, G., Curry, S. H., Mould, G. P., & Lader, M. H. (1972). Physiologic and clinical effects of chlorpromazine and their relationship to plasma level. *Clinical Pharmacology & Therapeutics, 13*(6), 931–946. https://doi.org/10.1002/cpt1972136931
Sara, S. J., & Bouret, S. (2012). Orienting and reorienting: The locus coeruleus mediates cognition through arousal. *Neuron, 76*(1), 130–141. https://doi.org/10.1016/j.neuron.2012.09.011
Schaefer, J., Giangrande, E., Weinberger, D. R., & Dickinson, D. (2013). The global cognitive impairment in schizophrenia: Consistent over decades and around the world. *Schizophrenia Research, 150*(1), 42–50. https://doi.org/10.1016/j.schres.2013.07.009
Schlagenhauf, F., Huys, Q. J. M., Deserno, L., Rapp, M. A., Beck, A., Heinze, H. J., Dolan, R., & Heinz, A. (2014). Striatal dysfunction during reversal learning in unmedicated schizophrenia patients. *NeuroImage, 89*, 171–180. https://doi.org/10.1016/j.neuroimage.2013.11.034
Schlagenhauf, F., Sterzer, P., Schmack, K., Ballmaier, M., Rapp, M., Wrase, J., Juckel, G., Gallinat, J., & Heinz, A. (2009). Reward feedback alterations in unmedicated schizophrenia patients: Relevance for delusions. *Biological Psychiatry, 65*(12), 1032–1039. https://doi.org/10.1016/j.biopsych.2008.12.016
Schneider, M., Hathway, P., Leuchs, L., Sämann, P. G., Czisch, M., & Spoormaker, V. I. (2016). Spontaneous pupil dilations during the resting state are associated with activation of the salience network. *NeuroImage, 139*, 189–201. https://doi.org/10.1016/j.neuroimage.2016.06.011
Siegle, G. J., Granholm, E., Ingram, R. E., & Matt, G. E. (2001). Pupillary and reaction time measures of sustained processing of negative information in depression. *Biological Psychiatry, 49*(7), 624–636. https://doi.org/10.1016/S0006-3223(00)01024-6
Siegle, G. J., Price, R. B., Jones, N. P., Ghinassi, F., Painter, T., & Thase, M. E. (2014). You gotta work at it: Pupillary indices of task focus are prognostic for response to a neurocognitive intervention for rumination in depression. *Clinical Psychological Science, 2*(4), 455–471. https://doi.org/10.1177/2167702614536160
Siegle, G. J., Steinhauer, S. R., Friedman, E. S., Thompson, W. S., & Thase, M. E. (2011). Remission prognosis for cognitive therapy for recurrent depression using the pupil: Utility and neural correlates. *Biological Psychiatry, 69*(8), 726–733. https://doi.org/10.1016/j.biopsych.2010.12.041
Silverstein, S. M., Moghaddam, B., & Wykes, T. (2013). Schizophrenia: The nature of the problems and the need for evolution and synthesis in our approaches. In *Schizophrenia: Evolution and synthesis*. MIT Press. http://europepmc.org/abstract/MED/33886229
Silverstein, S. M., Spaulding, W. D., Menditto, A. A., Savitz, A., Liberman, R. P., Berten, S., & Starobin, H. (2009). Attention shaping: A reward-based learning method to enhance skills training outcomes in schizophrenia. *Schizophrenia Bulletin, 35*(1), 222–232. https://doi.org/10.1093/schbul/sbm150
Smolen, V. F., Murdock, H. R., Stoltman, W. P., Clevenger, J. W., Combs, L. W., & Williams, E. J. (1975). Pharmacological response data for comparative bioavailability studies of chlorpromazine oral dosage forms in humans: I. Pupilometry. *The Journal of Clinical Pharmacology, 15*(11–12), 734–751. https://doi.org/10.1002/j.1552-4604.1975.tb02338.x

Spohn, H. E., & Strauss, M. E. (1989). Relation of neuroleptic and anticholinergic medication to cognitive functions in schizophrenia. *Journal of Abnormal Psychology, 98*(4), 367–380. https://doi.org/10.1037//0021-843x.98.4.367

Steinhauer, S. R., Condray, R., & Pless, M. L. (2015). Pharmacological isolation of cognitive components influencing the pupillary light reflex. *Journal of Ophthalmology, 2015*, 179542. https://doi.org/10.1155/2015/179542

Steinhauer, S. R., Condray, R., van Kammen, D. P., Saito, H., Dougherty, G. G., & Kasparek, A. (1993). Cognitive dysfunction in schizophrenia is associated with disruption of sympathetically mediated autonomic pathways (conference abstract). *Schizophrenia Research, 9*(2–3), 167–168.

Steinhauer, S. R., & Hakerem, G. (1992). *The pupillary response in cognitive psychophysiology and schizophrenia* (Vol. 658, pp. 182–204). *New York Academy of Sciences*. http://www.wpic.pitt.edu/research/biometrics/Publications/BiometricsArchives PDF/832-1992 steinhauer%26hakerem.pdf

Steinhauer, S. R., Hakerem, G., & Spring, B. J. (1979). The pupillary response as a potential indicator of vulnerability to schizophrenia. *Psychopharmacology Bulletin, 15*(1), 44–45.

Steinhauer, S. R., Siegle, G. J., Condray, R., & Pless, M. (2004). Sympathetic and parasympathetic innervation of pupillary dilation during sustained processing. *International Journal of Psychophysiology, 52*(1), 77–86. https://doi.org/10.1016/j.ijpsycho.2003.12.005

Steinhauer, S. R., & Zubin, J. (1982). Vulnerability to schizophrenia: Information processing in the pupil and event-related potential. In E. Usdin & I. Hanin (Eds.), *Biological markers in psychiatry and neurology* (pp. 371–385). Pergamon Press.

Straube, E. R. (1982). Pupillometric, cardiac, electroderma reactivity of schizophrenic-patients under different stimulus conditions. *Psychophysiology, 19*(5), 589–590.

Strauss, G. P., Kappenman, E. S., Culbreth, A. J., Catalano, L. T., Lee, B. G., & Gold, J. M. (2013). Emotion regulation abnormalities in schizophrenia: Cognitive change strategies fail to decrease the neural response to unpleasant stimuli. *Schizophrenia Bulletin, 39*(4), 872–883. https://doi.org/10.1093/schbul/sbs186

Strauss, G. P., Kappenman, E. S., Culbreth, A. J., Catalano, L. T., Ossenfort, K. L., Lee, B. G., & Gold, J. M. (2015). Emotion regulation abnormalities in schizophrenia: Directed attention strategies fail to decrease the neurophysiological response to unpleasant stimuli. *Journal of Abnormal Psychology, 124*(2), 288–301. https://doi.org/10.1037/abn0000017

Strauss, G. P., Ossenfort, K. L., & Whearty, K. M. (2016). Reappraisal and distraction emotion regulation strategies are associated with distinct patterns of visual attention and differing levels of cognitive demand. *PLoS One, 11*(11), 1–18. https://doi.org/10.1371/journal.pone.0162290

Thakkar, K. N., Brascamp, J. W., Ghermezi, L., Fifer, K., Schall, J. D., & Park, S. (2018). Reduced pupil dilation during action preparation in schizophrenia. *International Journal of Psychophysiology, 128*, 111–118. https://doi.org/10.1016/j.ijpsycho.2018.03.012

Valentino, R. J., & Van Bockstaele, E. (2008). Convergent regulation of locus coeruleus activity as an adaptive response to stress. *European Journal of Pharmacology, 583*(2–3), 194–203. https://doi.org/10.1016/j.ejphar.2007.11.062

Van Der Meer, L., Swart, M., Van Der Velde, J., Pijnenborg, G., Wiersma, D., Bruggeman, R., & Aleman, A. (2014). Neural correlates of emotion regulation in patients with schizophrenia and non-affected siblings. *PLoS One, 9*(6), e99667. https://doi.org/10.1371/journal.pone.0099667

Vos, T., Barber, R. M., Bell, B., Bertozzi-Villa, A., Biryukov, S., Bolliger, I., Charlson, F., Davis, A., Degenhardt, L., Dicker, D., Duan, L., Erskine, H., Feigin, V. L., Ferrari, A. J., Fitzmaurice, C., Fleming, T., Graetz, N., Guinovart, C., Haagsma, J., et al. (2015). Global, regional, and national incidence, prevalence, and years lived with disability for 301 acute and chronic diseases and injuries in 188 countries, 1990-2013: A systematic analysis for the Global Burden of Disease Study 2013. *The Lancet, 386*(9995), 743–800. https://doi.org/10.1016/S0140-6736(15)60692-4

Waltz, J. A., Kasanova, Z., Ross, T. J., Salmeron, B. J., McMahon, R. P., Gold, J. M., & Stein, E. A. (2013). The roles of reward, default, and executive control networks in set-shifting

impairments in schizophrenia. *PLoS One, 8*(2), e57257. https://doi.org/10.1371/journal.pone.0057257
Wardhani, I. K., Mathôt, S., Boehler, C. N., & Laeng, B. (2020). Effects of nicotine on pupil size and performance during multiple-object tracking in non-nicotine users. *International Journal of Psychophysiology, 158*, 45–55. https://doi.org/10.1016/j.ijpsycho.2020.09.005
Westphal, A. (1907). Ueber ein im katatonischen Stupor beobachtetes Pupillenphänomen sowie Bemerkungen über die Pupillenstarre bei Hysterie. *Deutsche Medizintsche Wochenschrift, 33*(27), 1080–1084.
White, T. P., Joseph, V., Francis, S. T., & Liddle, P. F. (2010). Aberrant salience network (bilateral insula and anterior cingulate cortex) connectivity during information processing in schizophrenia. *Schizophrenia Research, 123*(2–3), 105–115. https://doi.org/10.1016/j.schres.2010.07.020
Woody, M. L., Vaughn-Coaxum, R. A., Siegle, G. J., & Price, R. B. (2020). Time course of pupillary response to threat words before and after attention bias modification for transdiagnostic anxiety disorders: A randomized controlled trial. *Brain and Behavior, 10*(8), 1–10. https://doi.org/10.1002/brb3.1664
Wynn, J. K., Green, M. F., Hellemann, G., Reavis, E. A., & Marder, S. R. (2019). A dose-finding study of oxytocin using neurophysiological measures of social processing. *Neuropsychopharmacology, 44*(2), 289–294. https://doi.org/10.1038/s41386-018-0165-y
Yamamoto, K., & Hornykiewicz, O. (2004). Proposal for a noradrenaline hypothesis of schizophrenia. *Progress in Neuro-Psychopharmacology and Biological Psychiatry, 28*(5), 913–922. https://doi.org/10.1016/j.pnpbp.2004.05.033
Zahn, T. P., Frith, C. D., & Steinhauer, S. R. (1991). Autonomic functioning in schizophrenia: Electrodermal activity, heart rate, pupillography. In *Handbook of schizophrenia: Vol. 5: Neurops* (pp. 185–224). http://www.wpic.pitt.edu/research/biometrics/Publications/Biometrics Archives PDF/821-1991 Zahn Frith Steinhauer.pdf
Zhang, L., Ai, H., Opmeer, E. M., Marsman, J. B. C., Van Der Meer, L., Ruhé, H. G., Aleman, A., & Van Tol, M. J. (2020). Distinct temporal brain dynamics in bipolar disorder and schizophrenia during emotion regulation. *Psychological Medicine, 50*(3), 413–421. https://doi.org/10.1017/S0033291719000217

Interpersonal Pupillometry: The Pupil as an Indicator of Emotion and Its Utility as a Social Cue

Daniella K. Cash, Laura A. Pazos, and Tiffany D. Russell

It is a bustling Saturday night, and you are sitting with your date for the evening at one of the best restaurants in town. As your eyes gaze across the candlelit table, they meet your date's. Their eyes are attentive, and their pupils are seemingly as large as the plates your meal arrived on. Are they enlarged because they want to get it on with you after the main course, or is this merely because of the dim, ambient lighting? Throughout the night, you will likely rely on numerous social cues to assess how well the date is going. For example, you might consider whether they are laughing at your jokes, whether they are making eye contact, or maybe if they are smiling. These are more explicit, conscious social cues, but you may also draw conclusions from cues you perceive unconsciously. Pupillary dilation may be one such cue you use to assess whether your date is a success or whether it is a hot mess. The current chapter evaluates not only the relationship between pupillary dilation and emotion but also examines how these changes in pupil size are interpreted by others.

As much as we would like, it is impossible to read the minds of other people. However, there are physiological and behavioral cues that people can rely on when engaging in interpersonal metacognition during social interactions. Among these cues, pupil size is one reliable marker that is readily available in nearly all forms of social communication. While well-known for its automatic response to changes in light (Gavas et al., 2017; Kawasaki, 1999), variations in pupillary size also serve as a reliable indicator of emotional valence, cognitive load, and attention (Ahern &

D. K. Cash (✉)
Department of Psychology & Philosophy, Sam Houston State University, Huntsville, TX, USA
e-mail: dkc025@shsu.edu

L. A. Pazos
Department of Psychology, University of Nevada Reno, Reno, NV, USA

T. D. Russell
Department of Psychology, University of North Dakota, Grand Forks, ND, USA

M. H. Papesh, S. D. Goldinger (eds.), *Modern Pupillometry*,
https://doi.org/10.1007/978-3-031-54896-3_11

Beatty, 1979; Beatty, 1982; Beatty & Kahneman, 1966; Bradshaw, 1967; Eldar et al., 2013; Goldwater, 1972; Hess & Polt, 1960, 1964; Hyönä et al., 1995; Kahneman, 1973; Laeng et al., 2012; Loewenfeld, 1993; Partala & Surakka, 2003). The automatic dilation and constriction of the pupil is a product of the sympathetic and parasympathetic nervous systems (Mathôt, 2018; Steinhauer et al., 2004). Two important muscles facilitate this process: the dilator and the sphincter. Pupillary dilation can be a product of increased activity in the sympathetic nervous system, which in turn increases activity in the dilator muscle, causing the pupil to dilate. Similarly, pupillary dilation can also be a product of lack of activation in the parasympathetic nervous system, decreasing activity in the sphincter, also resulting in pupillary dilation. Alternatively, the pupil constricts when increased activation in the parasympathetic nervous system stimulates the sphincter muscles, causing the pupil to contract (Kardon, 2005; Mathôt, 2018; McDougal & Gamlin, 2008). Despite both the sympathetic and parasympathetic nervous systems generally contributing to pupillary dilation and constriction, research has shown that pupillary changes as a function of emotional arousal are a product of increased activation in the sympathetic, rather than parasympathetic, nervous system (Bradley et al., 2008).

The evaluation of pupillary dilation as an index of emotional arousal first began in the 1960s through a series of studies conducted by Hess and colleagues. In one of the earliest studies examining emotional arousal and pupil size, Hess and Polt (1960) measured pupil sizes of male and female participants as they viewed a series of pictures. The pictures depicted partially nude males and females, babies, as well as pictures of mothers and babies together. The results from this seminal study revealed a 20% increase in pupil dilation compared to baseline pupil sizes when viewing partially nude photographs of the opposite sex. Additionally, women, but not men, also displayed an increase in dilation for the photographs containing babies. This led Hess and Polt (1960) to conclude that these pupillary responses were indicative of participants' emotional arousal and interest.

These findings were later expanded upon by Hess et al. (1965) to include a sample of homo- and heterosexual male participants. In line with expectations, pupillary dilation increased as a function of the gender of interest, with more pupillary dilation in homosexual participants to photographs of naked men compared to heterosexual participants who had more dilation when shown pictures of naked women. These findings have been further mirrored in the literature, with pupillary dilation being correlated with both self-reported sexual arousal and self-reported erections (Bernick et al., 1971; Hamel, 1974). Additionally, studies using heterosexual participants revealed an increase in pupil dilation for pictures where an opposite-sex model was wearing fewer as opposed to more clothes (Hamel, 1974; Nunnally et al., 1967) and increased pupil dilation for female participants who were ovulating when viewing pictures of their boyfriends (i.e., sexually relevant stimuli) as opposed to same-sex actresses (i.e., irrelevant sexual stimuli; Laeng & Falkenberg, 2007).

When considering the nature of these findings, it is necessary to make an important distinction regarding the definition of arousal. Generally, arousal is considered to be a heightened awareness that is independent of emotional valence (Kristjansson et al., 2009). Thus, according to this definition, arousal can be either positively or negatively valenced. Colloquially, arousal is also used interchangeably with sexual arousal (which would be positively valenced). The studies described earlier largely refer to the latter definition of arousal. This distinction is particularly important given the conflicting findings regarding the effect of emotional valence on pupil size. Originally, Hess and colleagues (e.g., Hess & Polt, 1960; Hess, 1972) posited that there was a continuum of pupil dilation such that more dilation was reserved for positively valenced stimuli, and more constriction was associated with negatively valenced stimuli. However, later research was unable to replicate these findings and noted several limitations such as small sample sizes, issues with the stimuli (e.g., color, luminance, contrast, limited stimuli trials), no statistical analyses, and limitations with measuring pupil dilation (Bradley et al., 2008; de Winter et al., 2021).

Other research examining pupillary dilation has further clarified this relationship. A substantial body of literature has accumulated suggesting that the notion that pupil size serves as a continuum of valence is inaccurate (although there are a few notable exceptions; e.g., Babiker et al., 2013; Kawai et al., 2013). For example, Loewenfeld (1966) found that other than varying light intensity, no other physical or psychological stimuli resulted in pupil constriction. Similarly, Janisse (1974) posited that negative stimuli do not produce pupil constriction and that earlier findings supporting this notion are a product of either the limited participant sample or the stimuli used. Janisse (1974) suggested that pupil size varies as a function of stimulus intensity such that there is maximal pupil dilation for maximally valenced (whether positive or negative) items, and minimal dilation for neutral items. Myriad research has also provided evidence suggesting that it is the degree of arousal, rather than valence, that impacts pupil size (Aboyoun & Dabbs, 1998; Bradley et al., 2008; de Winter et al., 2021; Henderson et al., 2014; Jin et al., 2015; Loewenfeld & Lowenstein, 1993; Kinner et al., 2017; Peinkhofer et al., 2019; Oliva & Anikin, 2018; Steinhauer & Hakerem, 1992; Steinhauer et al., 1983; Wang et al., 2018; Widmann et al., 2018). These arousal-based findings are not simply limited to visual stimuli but also extend to auditory stimuli (Gingras et al., 2015; Jin et al., 2015; Laeng et al., 2016; Oliva & Anikin, 2018; Widmann et al., 2018; Zekveld et al., 2018) and word stimuli (de Winter et al., 2021). Thus, the current literature suggests that it is arousal, rather than valence, that produces the pupillary dilation response.

In line with these findings, studies examining pupillary responses to stimuli designed to specifically elicit fear (Reinhard et al., 2006; Schaefer et al., 2014) and surprise (Einhäuser, 2017; Preuschoff et al., 2011; Reisenzein et al., 2019) as emotional responses have demonstrated similar patterns. More fear-evoking stimuli produce greater pupillary dilation compared to neutral stimuli (Reinhard et al., 2006; Schaefer et al., 2014). Despite some early work arguing that surprise should not be recognized as a true emotion (e.g., Lazarus, 1991; McDougall, 1908), more recent work has made a compelling case for its inclusion (Reisenzein et al., 2019) and, hence, warrants discussion in this chapter. Specifically, pupillary dilation has also been shown in response to unexpected events and violated predictions (Einhäuser, 2017). This pattern was initially termed the "probability effect" (Friedman et al., 1973) and predicted that larger pupillary dilations accompany the presentation of less probable stimuli, which has been substantiated within the surprise and uncertainty literature. These dilation increases can be seen in response to unexpected stimuli (Raisig et al., 2010; Kloosterman et al., 2015; Kamp & Donchin, 2015; Wetzel et al., 2016) and "surprise" regarding unpredicted outcomes (Preuschoff et al., 2011). Importantly, research has disentangled reward and uncertainty to reveal that the pupil dilates as a response to surprise and not the two former factors (Preuschoff et al., 2011).

One apparent exception to the general heuristic that increased arousal produces greater pupillary dilation is observed in relation to the role of disgust on pupil size. Despite most work focusing on emotions and pupil size revealing that these differences are indexed by an increase in dilation, rather than constriction, some studies examining disgust have found the inverse (Schienle et al., 2016; Ayzenberg et al., 2018). The reason for this discrepancy most likely lies in this response occurring as a product of parasympathetic, rather than sympathetic, nervous system activity (Calder et al., 2001; Murphy et al., 2003). Unlike fear, which evokes a sympathetic, "fight or flight" response, disgust is characterized as a response of the parasympathetic nervous system (de Jong et al., 2011; Levenson, 1988; van Overveld et al., 2009). The parasympathetic nervous system is termed as being responsible for "rest and digest" functions, which include decreasing heart rate, slowing cardiovascular functions, and sphincter muscle responses that innervate pupillary constriction. Thus, disgust appears to be an exception to the standard pupillary dilation response to more affective stimuli.

The literature regarding emotion on pupillary dilation largely portrays a consistent theme: more arousal produces more pupillary dilation. However, more clinically based work has also examined how pupillary responses might differ as a function of mood disorders. Research comparing clinically depressed to healthy patients revealed that depressed individuals displayed increased sustained pupillary dilation for emotionally charged information-processing tasks compared to individuals who were not depressed (Siegle et al., 2001, 2003; Steidtmann et al., 2010). Data from these studies posit the prolonged dilation to be a product of depressed individuals continuing to process negative information well after its presentation, particularly when that information is personally relevant. Other work has shown sustained amygdala activity in depressed individuals after viewing negative words that last beyond the negative event and persists into nonemotional trials (Siegle et al., 2002). These mood impairments can also be seen in those diagnosed with anxiety. Anxious individuals showed increased sustained pupil dilation rates to fearful faces compared to nonanxious individuals (Price et al., 2013). Despite the lack of clinical research on this topic, these studies suggest differential dilation patterns as a product of long-term mood disorders.

In sum, the literature is consistent in that more emotionally arousing stimuli result in increased pupillary dilation. Importantly, this response appears to be independent of the emotional valence of the stimuli used. While these data paint a comprehensive picture, an additional question is if and how these changes in pupil size are recognized and interpreted by others. Research examining pupillary contagion (or pupil mimicry) suggests that not only are people adept at recognizing these pupillary changes, but that they may unconsciously mirror them as well.

1 Pupillary Contagion as a Social Construct of Emotion

As described above in your date night scenario, pupil size does not occur in isolation and can be used as a social cue for third parties. In various social interactions, autonomic and nonverbal cues displayed by an individual can be mirrored by their conversational partner (Chartrand & van Baaren, 2009; Oberman et al., 2007). Pupillary contagion (also known as pupillary mimicry) represents one such form of imitation and has been suggested to be indicative of empathy and recognition of others' levels of arousal (Fawcett et al., 2016, 2022; Kret et al., 2015). Specifically, an individual's pupils will mirror the pupil size and changes of their conversational partner. Pupillary contagion has been demonstrated in adults (Carsten et al., 2019; Harrison et al., 2006; Hartley & Reed, 2022), infants (Aktar et al., 2020; Fawcett et al., 2016), and chimpanzees with increased dilation occurring within the same species (Kret et al., 2014). Thus, pupillary mimicry is posited to be either an innate or early developing mechanism that facilitates affective processing and social communication (Prochazkova & Kret, 2017). Additionally, this mimicry is more commonly observed as pupillary dilation as opposed to constriction (Aktar et al., 2020; Kret et al., 2014). In situations where pupillary constriction does occur, it seemingly is unrelated to social influences (Kret et al., 2015; Prochazkova et al., 2018a).

The first study to document the pupillary contagion phenomenon was conducted by Hess (1975), who observed that participants' pupils dilated in response to increased diameters of schematic pupils. Interestingly, this finding was only seen when two schematic eyes were presented, as opposed to one or three eyes, suggesting that it was the similarity to real eyes as opposed to the stimuli itself producing this effect. Harrison et al. (2006) also found evidence supportive of pupillary contagion, but only in response to sad, as opposed to happy, angry, or neutral expressions. This differential pupillary response may have been driven by other potentially confounding factors. Participants were able to independently scan the varying faces and choose which facial features they wanted to attend to. Given that people attend to various facial features when identifying different expressions (Ebner et al., 2011), it is likely that the amount of time evaluating the eye region varied as a function of emotional expression, making it unclear if these differences are truly indicative of differences in emotional valence. A later study by Kret et al. (2015) failed to observe any differences in pupillary contagion as a result of emotional facial expression. More recent data suggest that these pupillary responses are a product of pupil size rather than emotional expression (Carsten et al., 2019; Fawcett et al., 2022). Participants' pupil dilation mirrored pupils presented in static photographs independent of the emotional expression that was also presented in the image (although both studies note more pupillary dilation for angry facial expressions compared to any other emotional expression). To date, the current literature suggests that emotional expression is not necessary to evoke pupillary contagion and does not produce any marked differences in pupil size.

Researchers have attempted to evaluate individual differences that might produce differences in pupillary contagion. Some factors that have been evaluated include levels of empathy (Carsten et al., 2019; Harrison et al., 2006, 2007; Kang & Wheatley, 2017), gender (Axelsson & Fawcett, 2021; Fawcett et al., 2022; Kret & De Dreu, 2017), age (Fawcett et al., 2022; Hartley & Reed, 2022), socioeconomic status (SES; Fawcett et al., 2022), and psychosocial factors such as anxiety, depression, and sleep disturbances (Fawcett et al., 2022). Of the factors mentioned, only SES was found to impact pupillary contagion, with higher SES individuals displaying lower rates of pupillary contagion for neutral (but not emotionally expressive) eyes. These data are consistent with other work that has shown higher SES individuals consistently underperform on emotionally related cognitive tasks (e.g., Kraus et al., 2010). Fawcett and colleagues posit two potential explanations for these data. First, this relationship might be due to higher SES individuals needing to be less reliant on others, which may lower the utility and likelihood of engaging in perspective-taking and being empathetic. An alternative hypothesis is that people who are more emotionally considerate of others may be more likely to be drawn to careers in education or caring fields, which are also associated with lower salaries. Future research that better clarifies this relationship would benefit the literature.

Importantly, while some studies have found empirical evidence for pupillary contagion, others have argued pupillary contagion to be a product of the pupillary light response, rather than a social phenomenon (Derksen et al., 2018; Mathôt & Naber, 2018; although see Prochazkova et al., 2018b for a response to these criticisms). An additional consideration of the pupillary contagion literature involves the stimuli and tasks used to examine this phenomenon. As noted by Hartley and Reed (2022), the nature of these studies may limit their overall generalizability in two ways. The first criticism stems from the concern that the nature of the task itself may lend itself more readily to producing pupillary contagion. Across the pupil mimicry literature, the vast majority of studies utilize a task that directly orients attention to the face (e.g., Axelsson & Fawcett, 2021; Derksen et al., 2018, Experiment 1) or specifically to the pupils (e.g., Carsten et al., 2019; Kret et al., 2014, 2015; Prochazkova et al., 2018a, b). This particular attention to the face and/or eye region may enhance the likelihood of pupillary contagion relative to its occurrence in the real world. A secondary criticism is that the stimuli used may also serve to limit these findings' generalizability. In most studies examining pupillary contagion, participants are only shown a strip containing the eyes as opposed to a full facial image (e.g., Axelsson & Fawcett, 2021; Carsten et al., 2019; Fawcett et al., 2016, 2017; Hartley & Reed, 2022; Kret et al., 2014, 2015; Kret & De Dreu, 2017; Prochazkova et al., 2018a, b; Wehebrink et al., 2018). To date, only three studies have utilized full-face photographs without digitally altering the pupils (Derksen et al., 2018, Experiment 1; Harrison et al., 2006; Simms, 1967), and the results from these studies are inconsistent such that pupillary contagion either was not observed at all (Derksen et al., 2018, Experiment 1) or was partially observed in certain circumstances (Harrison et al., 2006; Simms, 1967). While the literature on pupillary contagion seems to currently suggest an independence from emotion, more research is needed to better understand this phenomenon and its generalizability.

2 Social Perception via Pupillary Dilation

The extant literature exploring pupillary responses to emotional stimuli seems to suggest that any changes in dilation are a product of increased arousal as opposed to specific emotional responses. Despite these findings implying that pupillary dilation may not be a helpful indicator of people's emotional states, that certainly doesn't mean they are not interpreted as such. In this way, people may be engaging in unconscious pupillometry in which pupil size serves as a cue for other interpersonal judgments. While this area of research remains in its nascent stages, research has provided some initial insight into how changes in pupil sizes are interpreted by others.

2.1 Inferring Emotions and Mental States from Pupil Size

Despite the literature largely suggesting that it is arousal, rather than valence, that results in pupillary dilation, there is limited research regarding how pupil size is actually interpreted by others with regard to their mental state. Of the research that exists, people have been shown to readily infer emotions from the eye region and that they do so unconsciously (Adams Jr & Kleck, 2005; Sadr et al., 2003; Leppänen et al., 2008). Additionally, there appears to be a consistent pattern with how pupil sizes are interpreted by others (Hess, 1975; Kret, 2018). An initial study examining perceived emotional state via pupil size was conducted by Hess (1975). Participants were provided line drawings of happy and angry faces where pupil size was absent from the drawing. Participants were then asked to complete the image by drawing in the absent pupils. There was a consistent pattern such that happy faces were more likely to be accompanied by larger pupils whereas angry faces were accompanied by smaller pupils, suggesting an implicit assumption regarding what pupillary cues accompany specific emotional states. Importantly, these assumptions are inconsistent with the literature that suggests emotional valence to be unrelated to pupillary dilation.

These findings were later expanded upon by Kret (2018), who found that this association between emotional states and pupil size is learned, as it is absent in children but develops and increases across the lifespan. Of interest is that the difference in emotional inferences between younger and older adults was largely driven by older participants being more likely to draw smaller pupils accompanying angry facial expressions compared to younger participants (a similar trend was observed for dilated pupils and happy expressions, but this comparison did not achieve statistical significance). These data replicate the finding that pupil sizes are used to discriminate between various emotional states, but that it is a cue that is learned over time.

In contrast, a series of studies conducted by Harrison et al. (2006, 2007) found that pupil size impacted perceived emotional state, but only for faces expressing sadness, as opposed to other emotions like fear, surprise, or disgust. Specifically, sad faces with smaller pupils were rated as being more negative and more emotionally intense than sad faces with larger pupil sizes. Interestingly, the extent to which participants utilized pupil size in their judgments was shown to be impacted by participants' degree of emotional empathy (Harrison et al., 2007).

The assumptions made regarding the meaning of changes in pupil sizes are not limited to only a target's perceived emotion but also as an indicator of the target's likes or dislikes (Mattavelli et al., 2022). A recent study by Mattavelli et al. (2022) had participants view targets who either gazed at the participant or a liked/disliked third party. Pupil size was also manipulated such that some targets had dilated or constricted pupils. When explicitly asked how the target felt about the person they were staring at (either the participant or third party), participants consistently indicated that targets with larger pupils felt more positively toward their gaze target than those with smaller pupils.

Clearly, people are willing to draw inferences regarding emotional valances (whether accurately or inaccurately) from pupillary dilations. One cue that has been shown to be correctly interpreted is how pupil size is perceived as an indicator of sexual arousal. Largely, prior research has documented that the pupil does dilate as a response to sexually arousing stimuli (Bernick et al., 1971; Hamel, 1974; Hess & Polt, 1960; Hess et al., 1965; Laeng & Falkenberg, 2007; Nunnally et al., 1967; although see Attard-Johnson et al., 2016; Scott et al., 1967; and Snowden et al., 2019 for more nuanced findings). A subsequent question to this finding is whether people can accurately use this cue when making judgments about another's sexual interest. To date, only one study has empirically evaluated how people interpret pupillary dilation as an indication of sexual arousal (Lick et al., 2016). The data from this study revealed that participants readily and accurately relied on pupillary dilation when evaluating others' sexual interest as participants rated individuals with dilated pupils as being more sexually aroused than those with constricted pupils. Taken as a whole, these data suggest that people are sensitive to pupillary dilations and use them as a cue for emotional attributions.

2.2 The Pupil as an Indicator of Perceived Attentional Allocation

Another interpersonal judgment that people can make is the amount of attentional resources others might be employing during social interactions. For example, during your date, you might assess whether your romantic interest is attentively engaged in the conversation. While eye gaze is more commonly used as an index of attentional allocation (see Tipper & Bayliss, 2011 for a review), pupil size may also serve as an indicator of attention. Recent research has shown that dilated pupils enter our own awareness faster than constricted pupils (Colombatto & Scholl, 2021), potentially serving to make this a salient cue. Prior research examining the relationship between pupil dilation and attentional allocation has shown pupil size to be a reliable indicator for activity within the locus coeruleus, which is heavily involved in the allocation and regulation of attention (Laeng et al., 2012), and that fluxes in pupil sizes are correlated with increased mind-wandering (Kang et al., 2014; Smallwood et al., 2011). To date, the only study examining pupillary dilation on perceived attention supported this notion such that faces with larger pupils were judged to be more attentive compared to their smaller pupil counterparts (Watier et al., 2017). Importantly, very few participants indicated having noticed the change in pupillary size, suggesting that these assessments are made unconsciously.

2.3 The Role of Pupillary Dilation in Impression Formation

The judgments that people may attempt to make about others based on pupil size need not be limited to emotional and mental states. In fact, the notion that people draw inferences about others' character based on pupil size is well documented in the literature (Hess, 1975; Kret & De Dreu, 2019; Kret et al., 2015; Weibel et al., 2010). For example, in the same Hess (1975) study described earlier, Hess found that women with dilated pupils were more likely to be described as friendly and warm compared to an identical photograph with constricted pupils. A later study by Kret and De Dreu (2019) replicated these findings in that participants rated individuals with dilated pupils as more friendly than those with constricted pupils. This general finding is in accordance with the notion that pupil dilation is often seen as indicative of interest, attention, or arousal (Granholm & Steinhauer, 2004). However, there is evidence that the effect of pupillary dilation on these judgments may vary as a function of gender. Pawling et al. (2017) found that women were rated as being more friendly when their pupils were dilated as opposed to constricted. This pattern was opposite for men such that men were rated more friendly when their pupils were constricted. This gender difference may be reflective of how these cues are interpreted. Specifically, if the women interpreted dilated pupils as indicative of either negative or unwanted sexual arousal, this may have resulted in the men being interpreted as threatening, rather than friendly (for similar findings, see Bliss-Moreau et al., 2008).

Given the fact that pupillary dilations are often inferred as a marker of friendliness, it is reasonable to expect this would translate to impact individuals' approach-avoidance responses. Prior work has shown that approach and avoidance responses are heavily reliant on valenced stimuli (see Laham et al., 2015 for a review), with people more likely to approach positively valenced and avoid negatively valenced stimuli. In this way, pupil size may serve as such an indicator. To date, only one study has tested this hypothesis (Brambilla et al., 2019). In line with this notion, Brambilla et al. (2019) found that participants were more likely to approach faces with dilated pupils and were conversely more likely to avoid faces with constricted pupils. These data suggest that pupillary dilations influence social interactions and the likelihood of approaching others.

2.4 The Reciprocal Relationship of Pupillary Dilation and Trust

An extension of the degree to which dilated pupils promote positive interpersonal judgments is how dilated pupils may serve to make others appear more trustworthy. Humans are social organisms and, as such, cooperation and trust are important cornerstones of society (De Dreu et al., 2010; Parks et al., 2013). Thus, determining whether others are trustworthy is a critical task to avoid mistakenly trusting deceptive or nefarious individuals (Axelrod & Hamilton, 1981; Cosmides & Tooby, 2005; Komorita & Parks, 1995). When making these trust judgments, people incorporate unconscious, nonverbal facial cues (Hatfield et al., 1994; Hess & Fischer, 2013), with one such readily utilized cue being pupillary dilation. A consistent finding in the literature by Kret and colleagues through multiple studies is that larger pupils evoke more trust in partners (Kret & De Dreu, 2017, 2019; Kret et al., 2015; Prochazkova et al., 2018a, b; Wehebrink et al., 2018). For example, Kret and De Dreu (2019) demonstrated that, in addition to other interpersonal judgments such as attractiveness and friendliness, participants viewed larger pupil dilation as being indicative of being more trustworthy. Pupillary dilation not only impacted participants' assessments of trustworthiness but also their behavior as participants were more likely to reciprocate and share financial gain with partners with dilated, as opposed to constricted, pupils. A later study by van Breen and colleagues (2018) also found that pupil dilations impacted participants' behaviors. Across three studies, participants made predictions regarding the outcome for coin tosses and were rewarded for correctly guessing the outcome. Participants also self-reported their accuracy scores, allowing them to be dishonest to procure a better reward for themselves. However, prior to reporting their accuracy, participants were shown a pair of eyes with pupils that were either dilated or constricted that supposedly belonged to their task partner. The data revealed that participants were less likely to be dishonest when their actions would harm a partner with dilated pupils. Thus, these data indicate that people readily utilize pupil dilation as an indicator of trust and base their behavioral decisions accordingly.

The impact of pupillary dilation on trust does not seem to be limited to only interpersonal judgments of trustworthiness. Research has also shown that an increased pupillary dilation in the observers' pupils serves to further enhance the degree of perceived trust for the target (Kret & De Dreu, 2017; Kret et al., 2015; Prochazkova et al., 2018a, b; Wehebrink et al., 2018) and suggests that pupillary contagion is an important mechanism in trust development. Behavioral mimicry generally serves to increase positive feelings and heightens similarity between organisms, which in turn would be expected to promote trust and cooperation (van Baaren et al., 2004). The first empirical investigation of pupillary contagion on increased trust was conducted by Kret et al. (2015), who found an increase in perceived trustworthiness when observers' pupils dilated to match the dilation of their partner. However, this relationship between trust and pupillary mimicry was not observed for out-group partners. Interestingly, pupillary contagion as it relates to trust seems to vary as a function of social contexts. While pupillary contagion has been shown to promote trust between individuals (Kang & Wheatley, 2017), competitive contexts have been shown to reduce its occurrence (van Breen et al., 2018). The reason for this is currently unclear, but there are two potential explanations. First, it is possible that being dishonest produces a substantial amount of arousal that outright inhibits pupillary contagion. A second potential explanation may be that pupillary mimicry is suppressed in situations where it would be detrimental to the observer. Given that pupillary contagion serves to promote trust and positive affect toward others, this may be an undesirable response when one is attempting to deceive. When the tasks did not harm the partner, the standard pupillary contagion effects were observed (van Breen et al., 2018). Research examining pupil mimicry and trust in clinical populations has revealed a nuanced relationship between the two for clinically depressed individuals. A study by Wehebrink et al. (2018) found that while patients with major depressive disorder (MDD) displayed similar rates of pupillary contagion compared to healthy participants, depressed patients interpreted such pupillary changes in their conversational partner to be indicative of them being less trustworthy compared to healthy individuals. These findings are important given that patients with MDD have been shown to trust others less and be less cooperative compared to healthy individuals (Clark et al., 2013; Kim et al., 2012; Lester & Gatto, 1990) and demonstrate a discrepancy in how pupil change is interpreted. Thus, these data suggest that not only are others' pupil sizes acting as indicators of trust, but that pupil mimicry may also serve as a feedback mechanism to further promote trust in certain situations.

2.5 *The Pupil as a Cue for Attraction and Interpersonal Liking*

The notion that pupil size serves as an indicator of attractiveness is certainly not novel. As early as the seventeenth century, Italian women had already noted the positive attributes of dilated pupils and attempted to induce pupillary dilation using belladonna eyedrops. Some of the research has lent credence to the idea that dilated pupils are viewed positively (e.g., Bull & Shead, 1979; Hess, 1975; Kret & De Dreu, 2019). A potential explanation for this preference may lie in the fact that pupil dilation is seen as a cue for sexual arousal (Aboyoun & Dabbs, 1998; Bernick et al., 1971; Hamel, 1974) or that it is seen as a sign of youth and reproductive fitness (Birren et al., 1950). The effects of pupillary dilation on perceived attractiveness have been shown to occur as early as puberty, which has been interpreted to signal its importance in reproductive strategies (Bull & Shead, 1979; Flade & Lindner, 1979; Hess, 1975; Tarrahian & Hicks, 1979; Tombs & Silverman, 2004). For example, Hess (1975) presented male participants with photographs of women where pupillary dilation varied between pictures. Despite none of the men reporting having noticed a difference between the photographs, the men consistently indicated that the woman with the dilated pupils was more attractive. Their judgments were not limited only to attractiveness, as they were more likely to use descriptors such as "more feminine" and "soft." In contrast, the women with smaller pupils were described as "selfish" or "cold." Hess noted that these differences in perception might stem from larger pupils being indicative of interest and, more specifically, sexual interest.

Despite these findings seeming somewhat intuitive, it is important to note that the literature regarding pupil size and perceived attractiveness has been mixed. In line with some of the earlier work mentioned, some of the more recent studies have also replicated the finding that increased pupil size is associated with an increase in perceived attractiveness (Demos et al., 2008; Gründl et al., 2012; Kret & De Dreu, 2019, Wiseman & Watt, 2010; but see Amemiya & Ohtomo, 2012; Carsten et al., 2019; Colombatto & Scholl, 2021; and Hagerman et al., 2017 for contrasting findings). In contrast, other researchers have noted that pupil dilation might be related, but not in as straightforward of a manner as initially proposed (da Costa Carvalho et al., 2019; Hicks et al., 1978; Tombs & Silverman, 2004; Tomlinson et al., 1978). For example, Tombs and Silverman (2004) observed gender differences with regard to pupil dilation preferences. When judging both male and female faces, responses from male participants indicated a linear relation between attractiveness and pupil size with faces showing more dilation being rated as more attractive. This was in contrast to the attractiveness ratings provided by female participants whose data indicated a bell curve where medium-sized pupils were rated as more attractive compared to large and small pupils. The authors interpreted this data as being indicative of the degree of preferred sexual interest, such that male participants preferred explicit, unrestricted sexual interest, while women largely preferred more moderate displays of interest. They posited that larger pupils might be indicative of the men being overly sexually aroused, jealous, or promiscuous. Lending further credence to this interpretation, Tombs and Silverman found that women who provided higher attraction ratings to more dilated pupils also indicated a preference for dating "bad boys." In contrast, da Costa Carvalho et al. (2019) found that men also display an inverted U preference in pupillary dilation. There is also evidence that these pupillary preferences in women change as a function of fertility levels (Caryl et al., 2009). Women who were in the follicular phase of their menstrual cycle (and, thus, more fertile) rated men with larger pupils as more attractive compared to women who were in their nonfertile phases. Interestingly, Caryl and colleagues did not find a relationship between a preference for larger pupils and "bad boy" traits as did Tombs and Silverman (2004). Given the conflicting and heavily varying findings across the literature, the exact relationship between pupillary dilation and perceived attractiveness currently remains ambiguous.

While the handful of studies examining pupil dilation and its link to perceived attractiveness have produced varied responses, studies focusing on general liking, rather than attractiveness, have provided a more consistent story regarding the role of pupil size. In three different studies, Stass and Willis (1967) had participants choose partners for an upcoming experimental task. In the first study, participants were told that the task would involve handling money and that participants should choose a trustworthy partner, whereas participants in experiments 2 and 3 were told that the task would involve intimate communication and that they should choose a partner that they enjoyed talking to. The partners that were available for selection varied in the degree of eye contact that they made as well as their pupillary dilation. Across all three studies, participants were more likely to choose task partners who made more eye contact and had dilated pupils. Interestingly, when asked why they chose their partner, only 11% of participants reported eye contact as a reason for their selection, with no participants mentioning pupil size. Instead, participants were more likely to provide responses suggesting they found their partner to be "friendly," "pleasant," and "nice" compared to the partner that they did not select.

A later study by Hagerman et al. (2017) denoted the important distinction between perceived attractiveness and liking in female participants. Overall, pupillary dilation did not impact ratings of perceived attractiveness for women when judging male faces with varying degrees of dilation. However, larger pupils were correlated with a stronger desire for friendship with the male face depicted. Finally, a study by Mattavelli et al. (2022) revealed an important interaction between how pupil size and gaze orientation (seen as an indicator of attentional allocation; Tomasello et al., 2007) impact the degree of liking between individuals. Prior to this study, the literature had ignored the potential interplay of the arousal-driven information provided by the pupil and the attentionally based information provided by gaze. Mattavelli et al. (2022) illustratively provided the example of mistakenly inferring someone with dilated pupils to be looking at you, only to quickly realize they are looking at a person behind you. Across two experiments, the data revealed that participants readily incorporated the emotionally driven information obtained via pupillary dilations and the attentionally indicative gaze orientation to make interpersonal liking judgments. Specifically, participants reported liking their partner more when partners with dilated pupils were gazing either at the participant or another liked individual. This was in contrast to when the partner with dilated pupils attended to a disliked individual. While the relationship between pupil size and attraction might be unclear, the current literature suggests that pupillary dilation does seem to be indicative of interpersonal liking.

3 Conclusion

Although the notion of being able to use pupil size as an indicator of how successful you are on the dating scene is innately appealing, the literature examining pupillary response as a function of emotional state suggests that this may not be something one can do effectively over a candlelit dinner. Despite this, people unconsciously attempt to use pupillary dilations not only to infer emotional state and sexual arousal but also a plethora of other interpersonal judgments. This review highlights an important discrepancy between what types of stimuli research has shown to produce pupillary dilation and what people expect to generate these responses. This review also underscores the need for additional research, using a wide variety of stimuli and experimental paradigms, to better understand the way pupillary dilation might influence interpersonal judgments. Because it is clear that people unconsciously engage in pupillometry in their daily lives, it is necessary to understand the accuracy and the extent of these judgments.

References

Aboyoun, D. C., & Dabbs, J. M. (1998). The Hess pupil dilation findings: Sex or novelty? *Social Behavior and Personality, 26*, 415–420. https://doi.org/10.2224/sbp.1998.26.4.415

Adams, R. B., Jr., & Kleck, R. E. (2005). Effects of direct and averted gaze on the perception of facially communicated emotion. *Emotion, 5*(1), 3–11. https://psycnet.apa.org/doi/10.1037/1528-3542.5.1.3

Ahern, S., & Beatty, J. (1979). Pupillary responses during information processing vary with Scholastic Aptitude Test scores. *Science, 205*(4412), 1289–1292. https://doi.org/10.1126/science.472746

Aktar, E., Raijmakers, M. E., & Kret, M. E. (2020). Pupil mimicry in infants and parents. *Cognition and Emotion, 34*(6), 1160–1170. https://doi.org/10.1080/02699931.2020.1732875

Amemiya, S., & Ohtomo, K. (2012). Effect of the observed pupil size on the amygdala of the beholders. *Social Cognitive and Affective Neuroscience, 7*(3), 332–341. https://doi.org/10.1093/scan/nsr013

Attard-Johnson, J., Bindemann, M., Ciardha, Ó., & C. (2016). Pupillary response as an age-specific measure of sexual interest. *Archives of Sexual Behavior, 45*(4), 855–870. https://doi.org/10.1007/s10508-015-0681-3

Axelrod, R., & Hamilton, W. D. (1981). The evolution of cooperation. *Science, 211*(4489), 1390–1396.

Axelsson, E. L., & Fawcett, C. (2021). Humans' pupillary contagion extends to cats and dogs. *Social Cognitive and Affective Neuroscience, 16*(1–2), 153–166. https://doi.org/10.1093/scan/nsaa138

Ayzenberg, V., Hickey, M. R., & Lourenco, S. F. (2018). Pupillometry reveals the physiological underpinnings of the aversion to holes. *PeerJ, 6*, e4185.

van Baaren, R. B., Holland, R. W., Kawakami, K., & Van Knippenberg, A. (2004). Mimicry and prosocial behavior. *Psychological Science, 15*(1), 71–74. https://doi.org/10.1111/j.0963-7214.2004.01501012.x

Babiker, A., Faye, I., & Malik, A. (2013, October). Pupillary behavior in positive and negative emotions. In *2013 IEEE international conference on signal and image processing applications* (pp. 379–383). IEEE.

Beatty, J. (1982). Task-evoked pupillary responses, processing load, and the structure of processing resources. *Psychological Bulletin, 91*(2), 276–292. https://doi.org/10.1037/0033-2909.91.2.276

Beatty, J., & Kahneman, D. (1966). Pupillary changes in two memory tasks. *Psychonomic Science, 5*(10), 371–372. https://doi.org/10.3758/BF03328444

Bernick, N., Kling, A., & Borowitz, G. (1971). Physiologic differentiation of sexual arousal and anxiety. *Psychosomatic Medicine, 33*, 341–352. https://psycnet.apa.org/doi/10.1097/00006842-197107000-00004

Birren, J. E., Casperson, R. C., & Botwinick, J. (1950). Age changes in pupil size. *Journal of Gerontology, 5*(3), 216–221. https://doi.org/10.1093/geronj/5.3.216

Bliss-Moreau, E., Barrett, L. F., & Wright, C. I. (2008). Individual differences in learning the affective value of others under minimal conditions. *Emotion, 8*(4), 479–493. https://doi.org/10.1037/1528-3542.8.4.479

Bradley, M. M., Miccoli, L., Escrig, M. A., & Lang, P. J. (2008). The pupil as a measure of emotional arousal and autonomic activation. *Psychophysiology, 45*(4), 602–607. https://doi.org/10.1111/j.1469-8986.2008.00654.x

Bradshaw, J. (1967). Pupil size as a measure of arousal during information processing. *Nature, 216*(5114), 515–516. https://doi.org/10.1038/216515a0

Brambilla, M., Biella, M., & Kret, M. E. (2019). Looking into your eyes: Observed pupil size influences approach-avoidance responses. *Cognition and Emotion, 33*(3), 616–622. https://doi.org/10.1080/02699931.2018.1472554

Bull, R., & Shead, G. (1979). Pupil dilation, sex of stimulus, and age and sex of observer. *Perceptual and Motor Skills, 49*(1), 27–30. https://doi.org/10.2466/pms.1979.49.1.27

Calder, A. J., Lawrence, A. D., & Young, A. W. (2001). Neuropsychology of fear and loathing. *Nature Reviews Neuroscience, 2*(5), 352–363.

Carsten, T., Desmet, C., Krebs, R. M., & Brass, M. (2019). Pupillary contagion is independent of the emotional expression of the face. *Emotion, 19*(8), 1343–1352. https://psycnet.apa.org/doi/10.1037/emo0000503

Caryl, P. G., Bean, J. E., Smallwood, E. B., Barron, J. C., Tully, L., & Allerhand, M. (2009). Women's preference for male pupil-size: Effects of conception risk, sociosexuality and relationship status. *Personality and Individual Differences, 46*(4), 503–508. https://doi.org/10.1016/j.paid.2008.11.024

Chartrand, T. L., & van Baaren, R. (2009). Human mimicry. *Advances in Experimental Social Psychology, 41*, 219–274. https://doi.org/10.1016/S0065-2601(08)00405-X

Clark, C. B., Thorne, C. B., Hardy, S., & Cropsey, K. L. (2013). Cooperation and depressive symptoms. *Journal of Affective Disorders, 150*(3), 1184–1187. https://doi.org/10.1016/j.jad.2013.05.011

Colombatto, C., & Scholl, B. J. (2021). Unconscious pupillometry: An effect of "attentional contagion" in the absence of visual awareness. *Journal of Experimental Psychology: General, 151*(2), 302–308. https://doi.org/10.1037/xge0000927

Cosmides, L., & Tooby, J. (2005). Neurocognitive adaptations designed for social exchange. In *The handbook of evolutionary psychology* (pp. 584–627). Wiley.

da Costa Carvalho, S. Q., de Andrade, M. J. O., & Dos Santos, N. A. (2019). Relationship between facial attractiveness and pupil diameter in young adults. *Psico, 50*(2), e30033. https://doi.org/10.15448/1980-8623.2019.2.30033

De Dreu, C. K., Greer, L. L., Handgraaf, M. J., Shalvi, S., Van Kleef, G. A., Baas, M., et al. (2010). The neuropeptide oxytocin regulates parochial altruism in intergroup conflict among humans. *Science, 328*(5984), 1408–1411. https://doi.org/10.1126/science.1189047

Demos, K. E., Kelley, W. M., Ryan, S. L., Davis, F. C., & Whalen, P. J. (2008). Human amygdala sensitivity to the pupil size of others. *Cerebral Cortex, 18*(12), 2729–2734. https://doi.org/10.1093/cercor/bhn034

Derksen, M., van Alphen, J., Schaap, S., Mathôt, S., & Naber, M. (2018). Pupil mimicry is the result of brightness perception of the iris and pupil. *Journal of Cognition, 1*(1), 32. https://doi.org/10.5334/joc.34

Ebner, N. C., He, Y. I., & Johnson, M. K. (2011). Age and emotion affect how we look at a face: Visual scan patterns differ for own-age versus other-age emotional faces. *Cognition & Emotion, 25*(6), 983–997. https://doi.org/10.1080/02699931.2010.540817

Einhäuser, W. (2017). The pupil as marker of cognitive processes. In Q. Zhao (Ed.), *Computational and cognitive neuroscience of vision*. Springer. https://doi.org/10.1007/978-981-10-0213-7_7

Eldar, E., Cohen, J. D., & Niv, Y. (2013). The effects of neural gain on attention and learning. *Nature Neuroscience, 16*(8), 1146–1153. https://doi.org/10.1038/nn.3428

Fawcett, C., Wesevich, V., & Gredebäck, G. (2016). Pupillary contagion in infancy: Evidence for spontaneous transfer of arousal. *Psychological Science, 27*(7), 997–1003. https://doi.org/10.1177/0956797616643924

Fawcett, C., Arslan, M., Falck-Ytter, T., Roeyers, H., & Gredebäck, G. (2017). Human eyes with dilated pupils induce pupillary contagion in infants. *Scientific Reports, 7*(1), 9601. https://doi.org/10.1038/s41598-017-08223-3

Fawcett, C., Nordenswan, E., Yrttiaho, S., Häikiö, T., Korja, R., Karlsson, L., et al. (2022). Individual differences in pupil dilation to others' emotional and neutral eyes with varying pupil sizes. *Cognition and Emotion, 36*(5), 928–942. https://doi.org/10.1080/02699931.2022.2073973

Flade, A., & Lindner, G. (1979). The role of the pupil's size in the perception of other persons. *Zeitschrift fur Experimentelle und Angewandte Psychologie, 26*(3), 436–447.

Friedman, D., Hakerem, G., Sutton, S., & Fleiss, J. L. (1973). Effect of stimulus uncertainty on the pupillary dilation response and the vertex evoked potential. *Electroencephalography and Clinical Neurophysiology, 34*(5), 475–484. https://doi.org/10.1016/0013-4694(73)90065-5

Gavas, R., Chatterjee, D., & Sinha, A. (2017, October). Estimation of cognitive load based on the pupil size dilation. In *In 2017 IEEE international conference on Systems, Man, and Cybernetics (SMC)* (pp. 1499–1504). IEEE. https://doi.org/10.1109/SMC.2017.8122826

Gingras, B., Marin, M. M., Puig-Waldmüller, E., & Fitch, W. T. (2015). The eye is listening: Music-induced arousal and individual differences predict pupillary responses. *Frontiers in Human Neuroscience, 9*, 619. https://doi.org/10.3389/fnhum.2015.00619

Goldwater, B. C. (1972). Psychological significance of pupillary movements. *Psychological Bulletin, 77*(5), 340–355. https://doi.org/10.1037/h0032456

Granholm, E. E., & Steinhauer, S. R. (2004). Pupillometric measures of cognitive and emotional processes. *International Journal of Psychophysiology, 52*(1), 1–6. https://doi.org/10.1016/j.ijpsycho.2003.12.001

Gründl, M., Knoll, S., Eisenmann-Klein, M., & Prantl, L. (2012). The blue-eyes stereotype: Do eye color, pupil diameter, and scleral color affect attractiveness? *Aesthetic Plastic Surgery, 36*(2), 234–240. https://doi.org/10.1007/s00266-011-9793-x

Hagerman, S., Woolard, Z., Anderson, K., Tatler, B. W., & Moore, F. R. (2017). Women's self-rated attraction to male faces does not correspond with physiological arousal. *Scientific Reports, 7*(1), 1–6. https://doi.org/10.1038/s41598-017-13812-3

Hamel, R. F. (1974). Female subjective and pupillary reaction to nude male and female figures. *The Journal of Psychology, 87*(2), 171–175. https://doi.org/10.1080/00223980.1974.9915687

Harrison, N. A., Singer, T., Rotshtein, P., Dolan, R. J., & Critchley, H. D. (2006). Pupillary contagion: Central mechanisms engaged in sadness processing. *Social Cognitive and Affective Neuroscience, 1*(1), 5–17. https://doi.org/10.1093/scan/nsl006

Harrison, N. A., Wilson, C. E., & Critchley, H. D. (2007). Processing of observed pupil size modulates perception of sadness and predicts empathy. *Emotion, 7*(4), 724–729.

Hartley, A. A., & Reed, C. L. (2022). Equivalent pupillary mimicry in younger and older adults. *Psychology and Aging, 37*(5), 604–613. Advance online publication. https://doi.org/10.1037/pag0000688

Hatfield, E., Cacioppo, J. T., & Rapson, R. L. (1994). Emotional contagion. *Current Directions in Psychological Science, 2*(3), 96–100.

Henderson, R. R., Bradley, M. M., & Lang, P. J. (2014). Modulation of the initial light reflex during affective picture viewing. *Psychophysiology, 51*(9), 815–818. https://doi.org/10.1111/psyp.12236

Hess, E. H. (1972). Pupillometrics. A method of studying mental, emotional, and sensory processes. In *Handbook of psychophysiology* (pp. 491–531). Holt, Rinehart and Winston.

Hess, E. H. (1975). The role of pupil size in communication. *Scientific American, 233*(5), 110–119. https://doi.org/10.1038/scientificamerican1175-110

Hess, U., & Fischer, A. (2013). Emotional mimicry as social regulation. *Personality and Social Psychology Review, 17*(2), 142–157. https://doi.org/10.1177/1088868312472607

Hess, E. H., & Polt, J. M. (1960). Pupil size as related to interest value of visual stimuli. *Science, 132*(3423), 349–350. https://doi.org/10.1126/science.132.3423.349

Hess, E. H., & Polt, J. M. (1964). Pupil size in relation to mental activity during simple problem-solving. *Science,143*(3611), 1190–1192.

Hess, E. H., Seltzer, A. L., & Shlien, J. M. (1965). Pupil response of hetero-and homosexual males to pictures of men and women: A pilot study. *Journal of Abnormal Psychology, 70*(3), 165–168. https://psycnet.apa.org/doi/10.1037/h0021978

Hicks, R. A., Pellegrini, R. J., & Tomlinson, N. (1978). Attributions of female college students to male photographs as a function of attractiveness and pupil size. *Perceptual and Motor Skills, 47*(3), 1265–1266. https://doi.org/10.2466/pms.1978.47.3f.1265

Hyönä, J., Tommola, J., & Alaja, A. M. (1995). Pupil dilation as a measure of processing load in simultaneous interpretation and other language tasks. *The Quarterly Journal of Experimental Psychology, 48*(3), 598–612. https://doi.org/10.1080/14640749508401407

Janisse, M. P. (1974). Pupil size, affect, and exposure frequency. *Social Behavior & Personality: An International Journal, 2*(2), 125–146. https://doi.org/10.2224/sbp.1974.2.2.125

Jin, A. B., Steding, L. H., & Webb, A. K. (2015). Reduced emotional and cardiovascular reactivity to emotionally evocative stimuli in major depressive disorder. *International Journal of Psychophysiology, 97*(1), 66–74. https://doi.org/10.1016/j.ijpsycho.2015.04.014

de Jong, P. J., van Overveld, M., & Peters, M. L. (2011). Sympathetic and parasympathetic responses to a core disgust video clip as a function of disgust propensity and disgust sensitivity. *Biological Psychology, 88*(2–3), 174–179.

Kahneman, D. (1973). *Attention and effort*. Prentice-Hall.

Kamp, S. M., & Donchin, E. (2015). ERP and pupil responses to deviance in an oddball paradigm. *Psychophysiology, 52*(4), 460–471. https://doi.org/10.1111/psyp.12378

Kang, O., & Wheatley, T. (2017). Pupil dilation patterns spontaneously synchronize across individuals during shared attention. *Journal of Experimental Psychology: General, 146*(4), 569–576. https://doi.org/10.1037/xge0000271

Kang, O. E., Huffer, K. E., & Wheatley, T. P. (2014). Pupil dilation dynamics track attention to high-level information. *PLoS One, 9*(8), e102463. https://doi.org/10.1371/journal.pone.0102463

Kardon, R. H. (2005). Anatomy and physiology of the autonomic nervous system. In N. R. Miller, N. J. Newman, V. Biousse, & J. B. Kerrison (Eds.), *Wash and Hoyt's clinical neuro-ophthalmology* (6th ed., pp. 649–714). Lippincott Williams & Wilkins.

Kraus, M. W., Côté, S., & Keltner, D. (2010). Social class, contextualism, and empathic accuracy. *Psychological Science, 21*(11), 1716–1723.

Kawai, S., Takano, H., & Nakamura, K. (2013). Pupil diameter variation in positive and negative emotions with visual stimulus. In *In 2013 IEEE international conference on systems, man, and cybernetics* (pp. 4179–4183). IEEE. https://doi.org/10.1109/SMC.2013.712

Kawasaki, A. (1999). Physiology, assessment, and disorders of the pupil. *Current Opinion in Ophthalmology, 10*(6), 394–400.

Kim, S. S., Chung, Y., Perry, M. J., Kawachi, I., & Subramanian, S. V. (2012). Association between interpersonal trust, reciprocity, and depression in South Korea: A prospective analysis. *PLoS One, 7*(1), e30602. https://doi.org/10.1371/journal.pone.0030602

Kinner, V. L., Kuchinke, L., Dierolf, A. M., Merz, C. J., Otto, T., & Wolf, O. T. (2017). What our eyes tell us about feelings: Tracking pupillary responses during emotion regulation processes. *Psychophysiology, 54*(4), 508–518. https://doi.org/10.1111/psyp.12816

Kloosterman, N. A., Meindertsma, T., van Loon, A. M., Lamme, V. A., Bonneh, Y. S., & Donner, T. H. (2015). Pupil size tracks perceptual content and surprise. *European Journal of Neuroscience, 41*(8), 1068–1078. https://doi.org/10.1111/ejn.12859

Komorita, S. S., & Parks, C. D. (1995). Interpersonal relations: Mixed-motive interaction. *Annual Review of Psychology, 46*(1), 183–207.

Kret, M. E. (2018). The role of pupil size in communication. Is there room for learning?. *Cognition and Emotion, 32*(5), 1139–1145.

Kret, M. E., & De Dreu, C. K. (2017). Pupil-mimicry conditions trust in partners: Moderation by oxytocin and group membership. *Proceedings of the Royal Society B: Biological Sciences, 284*(1850), 20162554. https://doi.org/10.1098/rspb.2016.2554

Kret, M. E., & De Dreu, C. K. (2019). The power of pupil size in establishing trust and reciprocity. *Journal of Experimental Psychology: General, 148*(8), 1299–1311. https://doi.org/10.1037/xge0000508

Kret, M. E., Tomonaga, M., & Matsuzawa, T. (2014). Chimpanzees and humans mimic pupil-size of conspecifics. *PLoS One, 9*(8), e104886. https://doi.org/10.1371/journal.pone.0104886

Kret, M. E., Fischer, A. H., & De Dreu, C. K. W. (2015). Pupil mimicry correlates with trust in in-group partners with dilating pupils. *Psychological Science, 26*(9), 1401–1410. https://doi.org/10.1177/0956797615588306

Kristjansson, S. D., Stern, J. A., Brown, T. B., & Rohrbaugh, J. W. (2009). Detecting phasic lapses in alertness using pupillometric measures. *Applied Ergonomics, 40*(6), 978–986.

Laeng, B., & Falkenberg, L. (2007). Women's pupillary responses to sexually significant others during the hormonal cycle. *Hormones and Behavior, 52*(4), 520–530. https://doi.org/10.1016/j.yhbeh.2007.07.013

Laeng, B., Sirois, S., & Gredebäck, G. (2012). Pupillometry: A window to the preconscious? *Perspectives on Psychological Science, 7*(1), 18–27. https://doi.org/10.1177/1745691611427305

Laeng, B., Eidet, L. M., Sulutvedt, U., & Panksepp, J. (2016). Music chills: The eye pupil as a mirror to music's soul. *Consciousness and Cognition, 44*, 161–178. https://doi.org/10.1016/j.concog.2016.07.009

Laham, S. M., Kashima, Y., Dix, J., & Wheeler, M. (2015). A meta-analysis of the facilitation of arm flexion and extension movements as a function of stimulus valence. *Cognition and Emotion, 29*(6), 1069–1090. https://doi.org/10.1080/02699931.2014.968096

Lazarus, R. S. (1991). Cognition and motivation in emotion. *American Psychologist, 46*(4), 352–367. https://doi.org/10.1037/0003-066X.46.4.352

Leppänen, J. M., Hietanen, J. K., & Koskinen, K. (2008). Differential early ERPs to fearful versus neutral facial expressions: A response to the salience of the eyes? *Biological Psychology, 78*, 150–158. https://doi.org/10.1016/j.biopsycho.2008.02.002

Lester, D., & Gatto, J. L. (1990). Interpersonal trust, depression, and suicidal ideation in teenagers. *Psychological Reports, 67*(3), 786–786.

Levenson, R. W. (1988). Emotion and the autonomic nervous system: A prospectus for research on autonomic specificity. In *Social psychophysiology: Theory and clinical applications*. Wiley.

Lick, D. J., Cortland, C. I., & Johnson, K. L. (2016). The pupils are the windows to sexuality: Pupil dilation as a visual cue to others' sexual interest. *Evolution and Human Behavior, 37*(2), 117–124. https://doi.org/10.1016/j.evolhumbehav.2015.09.004

Loewenfeld, I. E. (1966). Pupil size. *Survey of Opthalmology, 11*, 291–294.

Loewenfeld, I. E. (1993). *The pupil: Anatomy, physiology, and clinical applications* (Vol. 2). Iowa State University Press.

Loewenfeld, I. E., & Lowenstein, O. (1993). *The pupil: Anatomy, physiology, and clinical applications*. Iowa State University Press/Wayne State University Press.

Mathôt, S. (2018). Pupillometry: Psychology, physiology, and function. *Journal of Cognition, 1*(1), 16. https://doi.org/10.5334/joc.18

Mathôt, S., & Naber, M. (2018). There is no evidence that pupil mimicry is a social phenomenon. *Proceedings of the National Academy of Sciences, 115*(50), E11565.

Mattavelli, S., Brambilla, M., & Kret, M. E. (2022). It is written in the eyes: Inferences from pupil size and gaze orientation shape interpersonal liking. *Social Cognition, 40*(1), 88–106. https://doi.org/10.1521/soco.2022.40.1.88

McDougal, D. H., & Gamlin, P. D. R. (2008). Pupillary control pathways. In R. H. Masland & T. Albright (Eds.), *The senses: A comprehensive reference* (pp. 521–536). Academic. https://doi.org/10.1016/B978-012370880-9.00282-6

McDougall, W. (1908). The principal instincts and the primary emotions of man. In *An introduction to social psychology* (pp. 45–89). Methuen & Co.

Murphy, F. C., Nimmo-Smith, I. A. N., & Lawrence, A. D. (2003). Functional neuroanatomy of emotions: A meta-analysis. *Cognitive, Affective, & Behavioral Neuroscience, 3*(3), 207–233.

Nunnally, J. C., Knott, P. D., Duchnowski, A., & Parker, R. (1967). Pupillary response as a general measure of activation. *Perception & Psychophysics, 2*, 149–155. https://doi.org/10.3758/BF03210310

Oberman, L. M., Winkielman, P., & Ramachandran, V. S. (2007). Face to face: Blocking facial mimicry can selectively impair recognition of emotional expressions. *Social Neuroscience, 2*(3–4), 167–178. https://doi.org/10.1080/17470910701391943

Oliva, M., & Anikin, A. (2018). Pupil dilation reflects the time course of emotion recognition in human vocalizations. *Scientific Reports, 8*(1), 1–10. https://doi.org/10.1038/s41598-018-23265-x

van Overveld, W. J. M., de Jong, P. J., & Peters, M. L. (2009). Digestive and cardiovascular responses to core and animal-reminder disgust. *Biological Psychology, 80*(2), 149–157.

Parks, C. D., Joireman, J., & Van Lange, P. A. (2013). Cooperation, trust, and antagonism: How public goods are promoted. *Psychological Science in the Public Interest, 14*(3), 119–165. https://doi.org/10.1177/1529100612474436

Partala, T., & Surakka, V. (2003). Pupil size variation as an indication of affective processing. *International Journal of Human-Computer Studies, 59*(1–2), 185–198. https://doi.org/10.1016/S1071-5819(03)00017-X

Pawling, R., Kirkham, A. J., Tipper, S. P., & Over, H. (2017). Memory for incidentally perceived social cues: Effects on person judgment. *British Journal of Psychology, 108*(1), 169–190. https://doi.org/10.1111/bjop.12182

Peinkhofer, C., Knudsen, G. M., Moretti, R., & Kondziella, D. (2019). Cortical modulation of pupillary function: Systematic review. *PeerJ, 7*, e6882.

Preuschoff, K., Hart, B. M., & Einhauser, W. (2011). Pupil dilation signals surprise: Evidence for noradrenaline's role in decision making. *Frontiers in Neuroscience, 5*, 115. https://doi.org/10.3389/fnins.2011.00115

Price, R. B., Siegle, G. J., Silk, J. S., Ladouceur, C., McFarland, A., Dahl, R. E., & Ryan, N. D. (2013). Sustained neural alterations in anxious youth performing an attentional bias task: A pupilometry study. *Depression and Anxiety, 30*(1), 22–30. https://doi.org/10.1002/da.21966

Prochazkova, E., & Kret, M. E. (2017). Connecting minds and sharing emotions through mimicry: A neurocognitive model of emotional contagion. *Neuroscience & Biobehavioral Reviews, 80*, 99–114. https://doi.org/10.1016/j.neubiorev.2017.05.013

Prochazkova, E., Prochazkova, L., Giffin, M. R., Scholte, H. S., De Dreu, C. K., & Kret, M. E. (2018a). Pupil mimicry promotes trust through the theory-of-mind network. *Proceedings of the National Academy of Sciences, 115*(31), E7265–E7274. https://doi.org/10.1073/pnas.1803916115

Prochazkova, E., Prochazkova, L., Giffin, M. R., Scholte, H. S., De Dreu, C. K., & Kret, M. E. (2018b). Reply to Mathôt and Naber: Neuroimaging shows that pupil mimicry is a social phenomenon. *Proceedings of the National Academy of Sciences, 115*(50), E11566–E11567. https://doi.org/10.1073/pnas.1815545115

Raisig, S., Welke, T., Hagendorf, H., & van der Meer, E. (2010). I spy with my little eye: Detection of temporal violations in event sequences and the pupillary response. *International Journal of Psychophysiology, 76*(1), 1–8. https://doi.org/10.1016/j.ijpsycho.2010.01.006

Reinhard, G., Lachnit, H., & König, S. (2006). Tracking stimulus processing in Pavlovian pupillary conditioning. *Psychophysiology, 43*(1), 73–83. https://doi.org/10.1111/j.1469-8986.2006.00374.x

Reisenzein, R., Horstmann, G., & Schützwohl, A. (2019). The cognitive-evolutionary model of surprise: A review of the evidence. *Topics in Cognitive Science, 11*(1), 50–74. https://doi.org/10.1111/tops.12292

Sadr, J., Jarudi, I., & Sinha, P. (2003). The role of eyebrows in face recognition. *Perception, 32*(3), 285–293. https://doi.org/10.1068/p5027

Schaefer, H. S., Larson, C. L., Davidson, R. J., & Coan, J. A. (2014). Brain, body, and cognition: Neural, physiological and self-report correlates of phobic and normative fear. *Biological Psychology, 98*, 59–69. https://doi.org/10.1016/j.biopsycho.2013.12.011

Schienle, A., Übel, S., Gremsl, A., Schöngassner, F., & Körner, C. (2016). Disgust proneness and the perception of disgust-evoking pictures. *Journal of Psychophysiology, 30*(3), 124–129. https://doi.org/10.1027/0269-8803/a000162

Scott, T. R., Wells, W. H., Wood, D. Z., & Morgan, D. I. (1967). Pupillary response and sexual interest re-examined. *Journal of Clinical Psychology, 23*, 433–438. https://doi.org/10.1002/1097-4679(196710)

Siegle, G. J., Granholm, E., Ingram, R. E., & Matt, G. E. (2001). Pupillary and reaction time measures of sustained processing of negative information in depression. *Biological Psychiatry, 49*(7), 624–636. https://doi.org/10.1016/S0006-3223(00)01024-6

Siegle, G. J., Steinhauer, S. R., Thase, M. E., Stenger, V. A., & Carter, C. S. (2002). Can't shake that feeling: Event-related fMRI assessment of sustained amygdala activity in response to emotional information in depressed individuals. *Biological Psychiatry, 51*(9), 693–707.

Siegle, G. J., Steinhauer, S. R., Carter, C. S., Ramel, W., & Thase, M. E. (2003). Do the seconds turn into hours? Relationships between sustained pupil dilation in response to emotional information and self-reported rumination. *Cognitive Therapy and Research, 27*(3), 365–382. https://doi.org/10.1023/A:1023974602357

Simms, T. M. (1967). Pupillary response of male and female subjects to pupillary difference in male and female picture stimuli. *Perception & Psychophysics, 2*, 553–555.

Smallwood, J., Brown, K. S., Tipper, C., Giesbrecht, B., Franklin, M. S., Mrazek, M. D., et al. (2011). Pupillometric evidence for the decoupling of attention from perceptual input during offline thought. *PLoS One, 6*(3), e18298.

Snowden, R. J., McKinnon, A., Fitoussi, J., & Gray, N. S. (2019). Pupillary responses to static images of men and women: A possible measure of sexual interest?. *The Journal of Sex Research, 56*(1), 74–84.

Stass, J. W., & Willis, F. N. (1967). Eye contact, pupil dilation, and personal preference. *Psychonomic Science, 7*(10), 375–376. https://doi.org/10.3758/BF03331131

Steidtmann, D., Ingram, R. E., & Siegle, G. J. (2010). Pupil response to negative emotional information in individuals at-risk for depression. *Cognition and Emotion, 24*(3), 480–496. https://doi.org/10.1080/02699930902738897

Steinhauer, S. R., & Hakerem, G. (1992). The pupillary response in cognitive psychophysiology and schizophrenia. *Annals of the New York Academy of Sciences, 658*(1), 182–204.

Steinhauer, S. R., Boller, F., Zubin, J., & Pearlman, S. (1983). Pupillary dilation to emotional visual stimuli revisited. *Psychophysiology, 20*, 472.

Steinhauer, S. R., Siegle, G. J., Condray, R., & Pless, M. (2004). Sympathetic and parasympathetic innervation of pupillary dilation during sustained processing. *International Journal of Psychophysiology, 52*(1), 77–86. https://doi.org/10.1016/j.ijpsycho.2003.12.005

Tarrahian, G. A., & Hicks, R. A. (1979). Attribution of pupil size as a function of facial valence and age in American and Persian children. *Journal of Cross-Cultural Psychology, 10*(2), 243–249. https://doi.org/10.1177/0022022179102010

Tipper, S. P., & Bayliss, A. P. (2011). The impact of social gaze perception on attention. In A. Calder, G. Rhodes, M. Johnson, & J. Haxby (Eds.), *Handbook of face perception* (pp. 551–570). Oxford University Press.

Tomasello, M., Hare, B., Lehmann, H., & Call, J. (2007). Reliance on head versus eyes in the gaze following of great apes and human infants: The cooperative eye hypothesis. *Journal of Human Evolution, 52*(3), 314–320. https://doi.org/10.1016/j.jhevol.2006.10.001

Tombs, S., & Silverman, I. (2004). Pupillometry: A sexual selection approach. *Evolution and Human Behavior, 25*(4), 221–228. https://doi.org/10.1016/j.evolhumbehav.2004.05.001

Tomlinson, N., Hicks, R. A., & Pellegrini, R. J. (1978). Attributions of female college students to variations in pupil size. *Bulletin of the Psychonomic Society, 12*(6), 477–478. https://doi.org/10.3758/BF03329742

Van Breen, J. A., De Dreu, C. K., & Kret, M. E. (2018). Pupil to pupil: The effect of a partner's pupil size on (dis)honest behavior. *Journal of Experimental Social Psychology, 74*, 231–245.

Wang, C. A., Baird, T., Huang, J., Coutinho, J. D., Brien, D. C., & Munoz, D. P. (2018). Arousal effects on pupil size, heart rate, and skin conductance in an emotional face task. *Frontiers in Neurology, 9*, 1029. https://doi.org/10.3389/fneur.2018.01029

Watier, N., Healy, C., & Armstrong, H. (2017). Effects of photo-depicted pupil diameter on judgments of others' attentiveness and on facial recognition memory. *Perceptual and Motor Skills, 124*(2), 366–379. https://doi.org/10.1177/0031512516685191

Wehebrink, K. S., Koelkebeck, K., Piest, S., de Dreu, C. K., & Kret, M. E. (2018). Pupil mimicry and trust–Implication for depression. *Journal of Psychiatric Research, 97*, 70–76. https://doi.org/10.1016/j.jpsychires.2017.11.007

Weibel, D., Stricker, D., Wissmath, B., & Mast, F. W. (2010). How socially relevant visual characteristics of avatars influence impression formation. *Journal of Media Psychology, 22*(1), 37–43. https://doi.org/10.1027/1864-1105/a000005

Wetzel, N., Buttelmann, D., Schieler, A., & Widmann, A. (2016). Infant and adult pupil dilation in response to unexpected sounds. *Developmental Psychobiology, 58*(3), 382–392. https://doi.org/10.1002/dev.21377

Widmann, A., Schröger, E., & Wetzel, N. (2018). Emotion lies in the eye of the listener: Emotional arousal to novel sounds is reflected in the sympathetic contribution to the pupil dilation response and the P3. *Biological Psychology, 133*, 10–17. https://doi.org/10.1016/j.biopsycho.2018.01.010

de Winter, J. C. F., Petermeijer, S. M., Kooijman, L., & Dodou, D. (2021). Replicating five pupillometry studies of Eckhard Hess. *International Journal of Psychophysiology, 165*, 145–205. https://doi.org/10.1016/j.ijpsycho.2021.03.003

Wiseman, R., & Watt, C. (2010). Judging a book by its cover: The unconscious influence of pupil size on consumer choice. *Perception, 39*(10), 1417–1419. https://doi.org/10.1068/p6834

Zekveld, A. A., Koelewijn, T., & Kramer, S. E. (2018). The pupil dilation response to auditory stimuli: Current state of knowledge. *Trends in Hearing, 22*, 2331216518777174. https://doi.org/10.1177/2331216518777174

Using Pupillometry in the Detection of Deception

Anne E. Cook, Andrew Potts, Monika Kupcova, Douglas J. Hacker, Dan J. Woltz, and John C. Kircher

Pupillometry has a long history in cognitive psychology and psychophysiology. Hess and Polt's (1960, 1964) finding that increases in pupil dilation correspond to increases in cognitive effort and emotional arousal led to applications of pupillometry in several contexts. Examples include digit string transformation (Kahneman & Beatty, 1966); mental arithmetic (Ahern & Beatty, 1979; Bradshaw, 1968; Schaefer et al., 1968); reading (Just & Carpenter, 1993; Schluroff, 1982); complex learning tasks (van Gerven et al., 2002; Zheng & Cook, 2012); lexical translation (Hyona et al., 1995); and important to the topic of this chapter, the detection of deception (Baker et al., 1992; Bradley & Janisse, 1981; Dionisio et al., 2001; Heilveil, 1976; Webb et al., 2009a, b). The goal of this chapter is to describe our own work using pupillometry in the detection of deception using the ocular-motor deception test (ODT).

1 The Origin of the Ocular-Motor Deception Test (ODT)

Based on previous findings that reading strategies can measurably alter reading behaviors (e.g., Hyona & Nurminen, 2006), and the previously cited findings that cognitive effort leads to increased pupil dilation, two members of our research team—John Kircher, a psychophysiologist with expertise in the detection of deception, and Doug Hacker, an educational psychologist with expertise in reading, writing, and metacognition—speculated on whether an individual's reading behaviors and pupil diameter would be diagnostic of deception during a computer-based test. They raised this question with Dan Woltz, a cognitive psychologist with expertise in

A. E. Cook (✉) · A. Potts · M. Kupcova · D. J. Hacker · D. J. Woltz · J. C. Kircher
Department of Educational Psychology, University of Utah, Salt Lake City, UT, USA
e-mail: anne.cook@utah.edu

M. H. Papesh, S. D. Goldinger (eds.), *Modern Pupillometry*,
https://doi.org/10.1007/978-3-031-54896-3_12

working memory, measurement, and methodology, and Anne Cook, a cognitive psychologist with expertise in reading and eye tracking technology. This research team of four initiated what turned out to be an 18-year research agenda pursuing answers to the question of how ocular-motor measures can be used to detect deception. During this time, approximately eight laboratory studies and five field validation studies have been conducted to develop, test, and validate what eventually became known as the ocular-motor deception test (ODT).

In this chapter, we first will discuss four underlying assumptions that serve as the psychological and physiological bases for the ODT. We then outline the basic procedures of the standard ODT, including the various measures that are used. The bulk of the chapter will then focus on our recent research that has extended the uses of the ODT, which include testing multiple issues, developing an auditory version of the ODT, and developing a directed-lie test. We end the chapter with an examination of possible countermeasures that might be used to compromise the results of the ODT.

2 Underlying Assumptions of the ODT

The ODT is an automated, computer-based, psychophysiological test for detecting deception in screening environments. Its success in discriminating between truthful and deceptive people rests on four assumptions: Deception is cognitively taxing, emotionally arousing, and it requires self-regulation and the self-motivation to generate and maintain deception.

Deception Is Cognitively Taxing Previous research has provided strong support for the hypothesis that being deceptive is more cognitively demanding than being truthful (Johnson et al., 2005; Kircher, 1981; Lubow & Fein, 1996; Raskin, 1979; Seymour et al., 2000; Vendemia et al., 2005; Vrij, 2008; Vrij et al., 2007, 2009; Walczyk et al., 2003, 2009). When responding deceptively to a statement, whether presented in text or orally, a person must first fully comprehend the statement. Greater complexity of the semantic content and syntactic structure of the statement will require greater cognitive resources for comprehension. Once the deceptive person comprehends the statement, he or she must evaluate its relationship with truth, inhibit the truth, and produce a falsehood that counters that truth. The deceptive person also must commit the falsehood to memory so that subsequent responses maintain consistency with the falsehood. In addition, the deceptive person may allocate further cognitive resources to thoughts concerning the possible negative consequences he or she may incur should that consistency be broken and he or she is revealed to be a liar.

For a deceptive individual, these multiple sources of cognitive load and the demands placed on working memory should result in large increases in pupil dilation (Ahern & Beatty, 1979; Beatty & Wagoner, 1978: Colman & Paivio, 1970; Hess & Polt, 1964; Kahneman & Beatty, 1966). In contrast, a truthful person would face none of these cognitive demands other than those imposed from comprehending the statement, and therefore their pupillary responses should be relatively small by comparison. In this context, therefore, task-evoked pupillary responses would serve as a psychophysiological index of deception.

Deception Is Emotionally Arousing From nearly the outset of interest in pupillary responses, there has been interest in the effects of emotional arousal on pupil dilations. Hess and Polt's (1960) seminal study provided clear evidence that pupillary dilation occurred in response to positive and negative stimuli. Subsequent research has provided substantial support for the influence of emotion on pupil dilation (e.g., Bayer et al., 2011; Bradley et al., 2008; Partala et al., 2000; Vo et al., 2008).

The emotions aroused by the negative threatening consequences of being deceptive and possibly being caught in a deception have been shown to contribute to pupil dilation (Dionisio et al., 2001). Compared to truthful individuals, deceptive individuals are more concerned about failing a test of deception and should show greater pupil dilation. The results of Bradley and Janisse (1981), Heilveil (1976), and Webb et al. (2009a, b) supported this claim. It is important to recognize that the increased pupil dilation caused by emotion is likely confounded with dilation caused by cognitive effort. However, the multiple causes for dilation should actually serve as an asset in a psychophysiological test of deception, in which concealing deception is both cognitively demanding and emotionally arousing.

Deception Requires Self-Regulation Although self-regulation has been studied mainly in the context of self-regulated learning, the theoretical underpinnings of self-regulation are applicable to many contexts; moreover, the context of the ODT can arguably be considered a learning environment. Self-regulation is the extent to which a person is metacognitively, motivationally, and behaviorally active in his or her own learning (Zimmerman, 2008). The metacognitive component of self-regulation consists, in part, of a person's ability to monitor and control his or her cognitive and affective states and processes (Flavell, 1979; Hacker, 1997). As mentioned earlier, creating and maintaining deception can be cognitively taxing, requiring comprehension of potentially incriminating statements, evaluating their relationship with truth, fabricating falsehoods, and maintaining consistency with the falsehood. To avoid detection of the deception, we assume that the deceptive individual must metacognitively monitor each of these cognitive processes and the emotional reactions they may have about creating the falsehood and potentially being caught. Based on the input from monitoring, the deceptive person must then control these cognitive and affective processes to further maintain the deception.

Metacognitive monitoring and control add to the cognitive load a person must accommodate for successful learning (Fischer et al., 2008; Schwonke, 2015; Sweller & Paas, 2017). People frequently have difficulty with monitoring and control of their learning (Azevedo, 2002), and the more conscious the deceptive person becomes of these difficulties, emotional reactions to being caught in a lie are likely to rise. The metacognitive load imposed by self-regulation can be beneficial, negligible, or detrimental (Schwonke, 2015). It can be beneficial in that the increased cognitive load can act as a cue for the deceptive person to increase his or her efforts to maintain the deception. It can be negligible in that the monitoring and control of cognitive and affective processes has no impact on the deceptive person. Or it can be detrimental in that the increased cognitive load may cause a working memory breakdown, leading to obvious inconsistencies in the person's deceptions.

Deception Requires Sufficient Self-Motivation to Maintain the Deception We must assume that a person who successfully engages in a deception is sufficiently motivated to do so, an assumption supported by Webb (2008). Success at generating a falsehood and maintaining that falsehood depends on the person's self-motivation to engage in relevant metacognitive processes necessary to self-regulate their cognitive and affective processes.

Self-motivation has been linked to four social-cognitive constructs: self-efficacy, outcome expectations, task interest, and goal orientation (Zimmerman & Moylan, 2009). Self-efficacy refers to a person's beliefs about their capabilities to perform at designated levels (Zimmerman et al., 1992). Deceptive individuals must believe that they are capable of generating and maintaining a plausible deception. As their beliefs grow stronger, so does the likelihood that they will engage in deceptive activities, persist in those activities, and self-monitor those activities (Bouffard-Bouchard et al., 1991). Outcome expectancies refer to beliefs about the ultimate goal of one's performance (Zimmerman & Moylan, 2009). The deceptive person must engage in a deception, believing that he or she will succeed at convincing someone that he or she is telling the truth. Task interest is a cognitive and affective predisposition to engage in a particular activity (Deci & Ryan, 1985; Eccles & Wigfield, 2002; Hidi & Renninger, 2006). Because both truthful and deceptive individuals are intent on appearing truthful, the assumption is that they would have strong interest in engaging in a task that would result in their appearance of innocence. Finally, goal orientation involves a person's beliefs or feelings about the purpose for engaging in a task. A person can attempt to accomplish a task because the task is intrinsically meaningful to them. Alternatively, a person can attempt to accomplish a task because of extrinsic rewards received upon success (Zimmerman & Moylan, 2009). For a deceptive person, success at appearing truthful can be both intrinsically and extrinsically meaningful: Intrinsically meaningful because he or she has personally succeeded at appearing truthful, and extrinsically meaningful because the deception has allowed him or her access to some desired goal (e.g., gaining employment). In sum, self-motivation to engage successfully in generating and maintaining a deception results from a person's beliefs that he or she is capable of deceiving, will succeed at deceiving, is strongly interested in deceiving, and has both an intrinsic as well as extrinsic orientation to appear truthful.

3 The Basic Procedures of the Standard ODT

The ODT incorporates these basic assumptions about deception to predict that deceptive individuals' reactions and responses to questions in which they must conceal their deception will result in different patterns of reading behaviors and pupillary responses than those produced by truthful individuals. In laboratory studies on the ODT, we derive ground truth by assigning participants to deceptive or truthful conditions; deceptive individuals are instructed to engage in mock crimes, whereas truthful individuals are only told about the crimes. Both groups of individuals report to the lab, where they are given the ODT by an examiner who is blind to their experimental condition. All participants are paid for their participation and are promised a monetary bonus if they appear innocent on the ODT. The ODT is a computer-based screening test in which multiple true/false statements regarding an examinee's possible involvement in illicit activities appear one at a time in the center of a computer monitor. Prior to taking the test, the examinee is told to answer each statement quickly and accurately or they will fail the test. As an examinee reads the statement and responds by pressing one of two keys, corresponding to either true or false, their eye movements and pupil diameter are continually measured (at 60 Hz) with an eye tracker that is located at the bottom of the computer monitor. In addition to the behavioral measures of response time and error rates, fixation-based measures of reading behavior are extracted from the eye movement record. A subset of pupillary, response time, and reading measures is used in a logistic regression equation to compute the probability of deception, which then classifies the examinee as truthful or deceptive (Kircher, 2018).

The ODT design was influenced by the Relevant Comparison Test (RCT), which was originally proposed by Kircher et al. (2012) as a new polygraph test format to screen travelers for trafficking drugs or transporting explosives at ports of entry (Kircher, 2018). The RCT makes use of two types of statements concerning relevant issues (R1 and R2), such as use of illegal drugs (R1) and statements about espionage (R2). The R1 issue is the issue of interest for the test, and the R2 issue serves as a control, which usually has a very low probability of occurrence. The differences in reactions to the two types of statement are used to determine whether the examinee was truthful or deceptive to either of the issues. In addition, neutral statements (e.g., I am reading this on a day that is not Sunday), which require a low cognitive load, are placed among the relevant statements to allow the examinee recovery from responding to the more cognitively demanding relevant statements. All of the statements are repeated across multiple blocks (usually five). By signal averaging across multiple presentations of a statement, the measures to be used in the logistic regression analysis are calculated.

To classify examinees as truthful or deceptive, we calculate a contrast for each R1-R2 feature for each block of test statements. Relying on our underlying assumptions, our examinations of these contrasts have shown that a deceptive examinee who is motivated to appear truthful expends greater cognitive effort and has greater emotional involvement on R1 statements compared to R2 statements, as indicated by greater pupil dilation compared to truthful examinees. In addition, deceptive examinees, on average, take longer than truthful examinees to respond to the statements, suggesting that they are metacognitively monitoring and controlling their reading and responses to a greater extent than truthful examinees. This metacognitive load serves to exacerbate the cognitive demands on the deceptive examinee, thereby enhancing the pupillary response. However, when focusing on the relevant R1 statements compared to the control R2 statements, deceptive examinees make fewer fixations and spend less time reading and re-reading when compared to truthful examinees. In addition, the deceptive examinee has shorter response times on relevant statements when compared to control statements. Recalling the directions that examinees receive prior to testing that they are to answer quickly and accurately, we believe deceptive examinees try to respond rapidly when they are deceptive to avoid detection but the cognitive effort and emotional arousal involved in this effort ultimately give them away (Kircher, 2018).

These are the general procedures we have followed across multiple laboratory and field validation studies, although we have made various changes to some of our studies in order to investigate alternative methods. From our laboratory and field studies, the standard ODT procedures have produced, on average, between 80% and 86% accuracy (Kircher, 2018). We believe that these high accuracy rates warrant further investigations and modalities of the ODT, which will be described in the subsequent sections.

4 Continuing Research on the ODT

4.1 Investigating Deception on Multiple Issues

Psychophysiological detection of deception tests can be divided into two categories: screening tests and specific-incident tests. A number of characteristics differentiate the two tests. First, screening examinations cover several different topics that can have varying degrees of overlap with one another, whereas a specific-incident examination focuses on a single issue. Second, screening test questions are more general in scope than specific-incident test questions. For example, a question on a screening test might be, “Have you ever committed a serious crime?”, whereas a question on a specific-incident test might be, “Did you break into the National Bank last night?” The time period is broader for screening examinations than specific-incident examinations and could refer to an examinee’s entire lifetime or a window of time, such as 5 years. Generally, the relevant issues covered in screening tests are less well-defined than those in specific-incident tests, and examinees are likely to be less clear in their own mind about their truthfulness. Consequently, the accuracy of screening exams is likely to be lower than that of specific-incident exams (American Polygraph Association, 2011; Nelson, 2015; NRC, 2003). There is less research on screening examinations than on specific-incident testing (Handler et al., 2009; Krapohl, 2002).

Because screening tests cover multiple issues, accuracy can be assessed at the level of the person and at the level of the topic. At the level of the person, accuracy refers to the test's ability to correctly identify an individual as deceptive or truthful. To be classified as truthful, an examinee must appear truthful to all of the issues on the test. If it appears the examinee is deceptive to any one of the issues, they are classified as deceptive. Topic-level accuracy refers to the test's ability to accurately identify which of several issues were answered deceptively.

Multiple studies have evaluated the validity of polygraph screening examinations. The literature supports the idea that the polygraph can classify individuals as deceptive or truthful on multiple-issue screening examinations with approximately 85% accuracy, but accuracy drops to approximately 65% when attempts are made to identify the specific issue(s) that elicited a deceptive response (American Polygraph Association, 2011; Correa & Adams, 1981; Department of Defense Polygraph Institute Research Division Staff, 1997; Honts & Amato, 2007). Because the topic-level accuracy is relatively low, Nelson (2015) advised that polygraph practitioners not report decisions for individual topics.

The shortcomings of the polygraph in screening applications led the National Research Council to advocate for new screening technologies (2003). In 2012, Cook, Hacker, Webb, Osher, Kristjansson, Woltz, and Kircher introduced the ODT for use in a screening environment. As mentioned previously, the ODT is based on the RCT but only tests examinees on two issues. To expand on that work, we recently conducted a mock crime experiment to evaluate a new, multiple-issue screening protocol called the Multiple Comparison Test (MCT; Potts et al., 2020). The MCT covered four mock crimes. Three of four crimes served as relevant issues, and the fourth crime was used as a comparison issue. The three relevant issues included the theft of $20 from a secretary, the theft of a VISA gift card from a wallet, and the vandalism of a parking kiosk. The comparison issue was filing a false police report. The comparison issue provided a baseline of nondeceptive behavior. Reactions to the relevant issues were compared to the comparison issue to assess the relative magnitude of the examinees' reactions to the relevant issues. If reactions to the relevant issue were significantly greater than reactions to the comparison issue, an examinee was classified as deceptive on that particular issue.

One hundred and eighty participants were recruited from the University of Utah and surrounding community. Sixty participants were assigned to the innocent condition (*innocent*), 60 participants were assigned to steal $20 (*cash*), and 60 participants were assigned to steal $20 and the VISA gift card (*cash + card*). Participants were told some guilty participants vandalized a parking kiosk or filed a false police report, but no one actually committed those crimes.

After participants carried out their instructions, they were given the MCT. The MCT contained 12 True/False statements about each of the four topics (48 items), and the set of 48 items was repeated in different orders across five blocks. The statements appeared one at a time, in the center of the computer screen. To the right of the statement, the words "True" and "False" appeared in separate boxes. Instructions and test items appeared in black characters on a gray background to minimize fluctuations in illumination that might have affected the pupil. Participants answered true or false by pressing the left arrow key or the right arrow key on the keyboard, respectively. There was a 600 ms pause after an examinee's response until the next item appeared. Between each of the five blocks of test statements, examinees solved ten simple arithmetic problems. The purpose of the intervening task was to clear examinees' working memory of the test statements (Hacker et al., 2014). A Tobii 4C remote eye tracker affixed to a computer monitor recorded eye movements and pupil diameter at 60 Hz. Viewing was binocular, but analyses used data only from the right eye. Although the eye tracker allowed for free head movement, a chin rest was used to keep the participant's head still. Participants were told they should respond quickly and accurately to the statements or they would fail the test.

The measures analyzed for the MCT were the same as those analyzed in previous studies (Cook et al., 2012; Hacker et al., 2014; Kircher, 2018). For each of the measures, RMANOVA provided an omnibus test of the Guilt X Statement Type interaction to determine if the three groups reacted differently to the four types of statements. To test if the groups reacted in the manner consistent with the four-factor theory, three planned comparisons were conducted when the omnibus test was significant. One contrast was the difference between cash and false report scores. *Innocent* participants were expected to show no significant difference between cash and false report items, whereas both groups of guilty participants were expected to react more strongly to cash than false report statements. Another contrast was the difference between cash and card scores. The final contrast was the difference between card and false report scores. *Cash* participants were expected to show a significant difference between cash and card statements but no difference between card and false report statements, whereas *cash + card* participants were expected to show no difference between cash and card statements and a significant difference between card and false report items. The first comparison tested for the expected difference between guilty and *innocent* groups. The remaining two comparisons tested for expected differences between the two guilty groups.

The difference between reactions to cash and false report statements distinguished between truthful and deceptive participants on 8 of 11 outcome measures. The mean partial η^2 effect size for significant contrasts was 0.16. Previous ODT research showed a mean effect size of 0.16 (Kircher, 2018), suggesting that the MCT was as effective as the RCT in discriminating between deceptive and truthful participants. The mean effect size for the difference between guilty groups on the cash-card contrasts was also 0.16. This indicates that the differences between the two guilty groups were as great as the differences between guilty and *innocent* groups. Another contrast indicated that differences between relevant and comparison issues were smaller for *cash + card* than for *cash* participants. Habituation might account for the difference between guilty groups, since cash participants lied to questions about one of four relevant topics, whereas *cash + card* participants lied to questions about two of four relevant topics.

Of the eight significant outcome measures, response time, number of fixations, pupil diameter (PD) peak amplitude, PD area, and PD level were most diagnostic of deception and were used to develop a logistic regression model. A four-fold cross-validation was performed to assess the accuracy of the model. Participants were randomly partitioned into four subgroups (folds) such that each subgroup contained 45 participants: 15 *innocent*, 15 *cash*, and 15 *cash + card*. A logistic regression model was trained on three subgroups and validated on the fourth. Each subgroup was used three times in a training model and once as the validation sample. For each of the four subgroups, all possible combinations of the five variables were analyzed in R (R Core Team, 2017) with the package *leaps* (Lumley, 2020) to determine the best model in terms of accuracy and parsimony. The analyses indicated that number of fixations and PD level best predicted guilt for all partitions.

For each validation, accuracy was obtained at the person level and at the topic level. Participants were classified as innocent if deception was not indicated to any of the three relevant issues. The mean person-level classification accuracy was 85.5%: accuracy was 83.3% for *innocent* participants and 86.6% for guilty participants. The logistic model correctly classified 91.7% of *cash* participants and 81.4% of *cash + card* participants. Mean accuracy at the level of individual topics was 87.6%: 86.1% for cash statements, 81.1% for card statements, and 95.6% for vandalism statements. ROC curves were generated for cash classifications and card classifications. The area under the ROC curve was 0.92 for cash classifications and.85 for card classifications.

The results of our initial study with the MCT demonstrated that the ODT can be expanded to use for screening on multiple issues, and that measures of pupil diameter are strong predictors in the distinction between innocent and deceptive individuals, as well as between different levels or degrees of deception (single vs. multiple issues). In the next section, we describe an extension of the MCT in which decisions about participant guilt are weighted more heavily on pupillary reactions to test statements.

4.2 Creation of an Auditory MCT

Since the ODT involves reading and responding to items on a computer screen, there is legitimate concern that the ODT is less accurate for individuals with limited reading skills. If comprehension is poor when reading ODT items, the impact of item content on reading or pupillary responses is likely to be much reduced. Furthermore, fixation patterns during reading differ for skilled and unskilled readers under normal comprehension tasks (e.g., Rayner et al., 2012; Wei, 2019); these differences may be enhanced in a more complex comprehension task like the ODT. Finally, pupillary change in unskilled readers during the ODT may reflect the extra cognitive effort required to decode and comprehend the text in addition to concealing deceptive responses. This could potentially obscure pupil responses associated with deception relative to skilled readers. For these reasons, we have investigated the effectiveness of an auditory version of the ODT.

Despite the perceived need for the auditory presentation of items for poor reading populations, the loss of reading fixation measures presents a major challenge to the potential accuracy of our approach to deception detection. As described in the previous sections, individuals' reading behaviors differ as a function of whether they are deceptive to the items they are reading. All else being equal across visual and aural processing of items, the loss of this information in the analysis model should reduce the overall accuracy of an auditory ODT. However, we also considered the possibility that deception effects on pupil change could be greater in aural presentation for several reasons, which ideally would offset the loss of the reading behavior measures.

Recent evidence suggests there are minimal differences in semantic representations acquired from reading and listening; in fine-grained fMRI analysis of cortical activity during reading and listening to the same narratives, Deniz et al. (2019) concluded that the representation of semantic information is largely independent of the encoding modality. This suggests that any working memory or cognitive resource differences between reading and listening result from differences in the encoding processes. If reading requires fewer attentional resources than listening, it could reflect the availability of both visual (orthographic) and auditory (phonemic) information being available temporally in working memory. This dual coding of new information is expected to facilitate initial memory representations (Baddeley, 2007; Paivio, 1990). In addition, reading is self-paced with opportunity to pause and reread when comprehension is challenged. The pace of encoding during listening is typically beyond the control of the listener, and there is no visual representation to facilitate immediate memory.

If listening is associated with increased cognitive demand on encoding information compared to reading, why might we expect pupil change to be a better measure of deception during listening? First, it is conceivable that the inability to pause input or re-listen to previous content could increase the effort required to process input well enough to successfully appear truthful, and this increased effort would be reflected in pupil dilation measures. Related to this, Klingner et al. (2011) compared the magnitude of pupil dilation in several cognitive tasks under visual and aural stimulus presentation while non-cognitive influences on pupil dilation were controlled. On two tasks representing working memory (mental multiplication and digit span), performance accuracy was somewhat better when stimuli were presented visually. This comports with dual code theories of working memory (Baddeley, 2007; Paivio, 1990) and parallels the claim that listening can place greater demands on cognitive resources than reading. Of primary importance to our interests, Klingner et al. showed that pupillary change during the working memory tasks was similar across modalities in its onset, temporal pattern, and duration, but the magnitude of the change was greater for auditory stimulus presentation. In their mental multiplication task, mean dilation was 0.16 mm (SD = 0.13) in the visual condition and 0.35 mm (SD = 0.11) in the aural condition. In their digit recall task, mean dilation was 0.24 mm (SD = 0.17) for the visual condition and 0.44 mm (SD = 0.22) in the aural condition. The numeric working memory tasks used in Klingner et al.'s research differ considerably from language comprehension tasks, especially from a verbal test requiring participants to appear truthful in responding to questions. Nevertheless, the findings were consistent across several memory tasks and suggest that auditory items create a greater cognitive load than visual items. The question of interest was whether an increased cognitive load in listening to ODT items would intensify effort needed by guilty individuals to respond deceptively, and whether such effort would be reflected in greater pupil differences with innocent subjects.

Beyond the cognitive processing differences inherent in reading and listening versions of the ODT, we considered how auditory presentation could offer some advantage in controlling critical content in order to maximize pupil measurement. In the reading ODT, it has been standard practice to vary item structure as much as possible. This variability includes the location within items of critical information needed for a correct response. Such variability presumably prevents individuals from developing a simple effort-reducing strategy to locate the key information if it is always found in a particular location. One consequence of this variability is that some items will contain the critical information initially (e.g., *I am innocent...*), possibly allowing for an easy response decision—even before determining the relevant issue (e.g., … of *stealing the $20*). Auditory item presentation precludes strategies to systematically scan ahead for specific content, so variability in item structure no longer serves the same purpose. In fact, after extensive pilot testing, we settled on a degree of item consistency rather than variability in an attempt to maximize our ability to measure pupil response over the timeframe of item presentation. Following an orienting stimulus on the display, each auditory item began with a phrase clearly identifying the issue (e.g., *Regarding the theft of the credit card…*). The critical information needed for determining the correct response is then presented at the end of the item (e.g., *…I did not do it*). We demonstrated in a series of pilot experiments that the initial presentation of the relevant issue serves to promote an anticipatory pupil dilation response that differs for guilty and innocent participants. The phrasing of the final clause is varied, so participants must attend to this wording carefully to determine a response that reflects innocence. However, the final decision process is always performed with knowledge of the issue given the initial phrase. We also pilot tested several item types that varied in the complexity of the final clause but opted for the simplest wording given the already increased working memory load associated with listening comprehension. Further research could attempt to identify the level of linguistic complexity that optimizes the pupil response associated with deception.

Another issue faced in the development of an auditory ODT task was a pragmatic one. Accurate pupil measurement requires participants to fixate their eyes consistently on the computer display. The requirement is satisfied inherently in a reading task, but other mechanisms must be employed to ensure a consistent gaze forward when item content is presented auditorily. In the ODT, initial attention cues are presented prior to the onset of each item, and participants are prevented from responding until a signal appears at the end of the item presentation. Given the standard instruction to respond quickly and accurately in order to appear innocent, the visual cue as to when a response is accepted helps to keep eye gaze focused on the display.

Auditory ODT Study 1 A small laboratory study was conducted on an initial version of the auditory ODT with 42 participants randomly assigned to guilty and innocent conditions in a mock crime experiment, in which participants either stole \$20 from a secretary's purse (guilty) or were innocent. The experimental procedure was consistent with previous studies other than the auditory presentation and the consistent item structure described here. In addition, there was a brief visual presentation of the item topic presented on the display prior to the audio recording of the item. This was intended to promote an anticipation response that differed by guilt condition and also focus attention on the display for pupil recording. We used an RCT-like design for this study; R1 was stealing \$20 from a secretary's purse, R2 was stealing an iPhone, the neutral items pertained to whether the participant was sitting or standing during the test. Item length was equivalent across issues and audio files were created by text to speech software. There were eight items (half true and half false) per issue, and the 24 items were repeated across five blocks.

Given concerns about pupil information loss from participants looking away from the display, we analyzed the frequency of missing pupil data. For the five blocks, the average proportion of lost pupil data for the 42 participants was 0.085 ($SD = 0.035$). This was deemed an acceptable rate of loss.

The average standardized pupil diameter wave forms were measured from 2 s before item response to 2 s after response (participant response is at time interval 20 in these figures). Innocent participants had a slightly stronger pupil reaction to R2 compared to R1 items, which reflects experimental instructions intended to emphasize the importance of the R2 issue. In contrast, the guilty participants had a stronger response to R1 items. These differences persisted from the topic introduction, aural processing of the item content, response decision, and into the subsequent item. The contrast of R1-R2 standardized pupil diameter correlated with guilt, $r = 0.64$.

Consistent with previous ODT findings, innocent and guilty participants differed in their response times to R1 and R2 items. In the auditory ODT, response time is measured from the end of the audio file to the response. Guilty participants on average took longer to respond but were relatively faster to R1items in an apparent effort to appear innocent. The contrast of R1-R2 standardized response time correlated with guilt, $r = -0.52$.

We predicted guilt from a logistic model with only two variables: R1-R2 contrast of standardized PD values and the R1-R2 contrast of standardized RT values. Despite the absence of reading measures available when items are presented in text, accuracy was high; the model correctly predicted 90.5% of innocent participants and 85.7% of guilty participants, for an overall accuracy of 88%. However, this was a relatively small sample, and a larger test study is necessary to evaluate the accuracy of the auditory presentation method.

Auditory ODT Study 2 A second, larger study on an auditory ODT was conducted at Converus, Inc. (Converus, 2020). This study employed an auditory ODT similar to that described in Study 1, but it included multiple relevant issues (MCT). A mock crime experiment was conducted similar to previous research with four relevant issues: theft of cash, theft of a gift card, theft of a cell phone, and theft of a headlamp.

One hundred and eighty participants recruited from the local community were assigned to one of three conditions. In one guilty condition, participants stole \$20 from a secretary (n = 74). In a different guilty condition, participants performed two mock crimes: stealing \$20 from a secretary and stealing a gift card from a wallet (n = 55). In the third condition (innocent), participants did not perform either mock crime (n = 51). All participants were informed of the four crimes and that some subjects in the experiment performed one or more of them. As in previous experiments, participants were paid for their participation and offered a monetary bonus if they appeared innocent on the ODT. A k-fold validation procedure of logistic regression analysis predicting guilty versus innocent responses resulted in an average classification accuracy of 81%. The accuracy was similar for questions answered truthfully (82%) and questions answered deceptively (80%).

In comparing these results from those reported in the previous section on the MCT, it is evident that the auditory MCT had slightly lower classification accuracy than the conventional MCT. Furthermore, the classification accuracy in the Converus study was lower than that of Study 1 on the auditory ODT, although the number of relevant issues differed. Given the larger sample size of the Converus auditory ODT study, it is reasonable to conclude that ODT accuracy is slightly reduced due to the loss of reading behavior measures and the demand for different WM processes required in listening. However, the accuracy of the auditory ODT versions tested still appears sufficient for use in populations that have limited reading skills and emphasizes the utility of the pupillary response in discriminating between deceptive and innocent individuals. Further research is needed to evaluate our assumption that an auditory ODT will produce greater accuracy in deception detection than a reading ODT when used in a population with limited reading ability. As discussed earlier, the various versions of the ODT were developed for use in screening contexts. The next section will discuss adaptation of the ODT to specific-incident investigations.

4.3 Directed Lie Test

RCT and MCT formats were designed for pre-employment screening and periodic testing of existing employees. These tests cover several relevant issues, and decisions are based on comparisons of reactions to the various relevant issues. This type of test usually is inappropriate for a specific-incident criminal investigation, in which all of the relevant questions address the same issue. Specific-incident polygraph techniques base decisions on within-subject comparisons of reactions to relevant and comparison questions. The type of comparison question used in the test distinguishes among the different polygraph techniques. The Relevant-Irrelevant technique compares reactions to relevant and irrelevant questions (Horowitz et al., 1997). The Probable-Lie Test compares reactions to relevant and probable-lie questions (Raskin & Kircher, 2014). The Concealed Information Test uses a multiple-choice format to compare reactions to target and non-target alternatives (Verschuere & Ben-Shakhar, 2011).

During the Vietnam War, U.S. military intelligence developed another polygraph technique known as the Directed Lie Test (DLT; Fuse, 1982). Directed lie questions ask about inappropriate acts in the person's life (e.g., "In your entire life, have you ever told a lie?"). Even though everyone has lied at some point in their life, the examiner instructs the examinee to lie to the directed lie questions. The examiner tells the examinee that reactions to directed lie questions establish a baseline physiological response for deception during the test, and it is important that they react appropriately when they lie to the directed lie questions or they will fail the test. These statements draw attention to the directed lie questions. The DLT predicts that an innocent person who answers the relevant questions truthfully will attend more to the directed lie questions out of concern they might fail the test. The test predicts that the directed lie questions will pose the greatest threat to innocent examinees, and that will cause innocent examinees to react more strongly to the directed lie questions than to the relevant questions. In contrast, guilty people answer the relevant questions deceptively. Because the relevant questions go to the heart of the matter under investigation, guilty examinees are expected to be more concerned about relevant questions and react more strongly to them than to the directed lie questions.

Polygraph research supports use of the DLT for criminal investigation and screening (Ad Hoc Committee on Polygraph Techniques, 2011; Bell et al., 2008; Honts & Reavy, 2015). In their meta-analysis, the APA (2011) reported that the mean accuracy of decisions was 84.4%, with an inconclusive rate of 8.8%. Accuracy of decisions was 84.5% in Bell et al. experiment and 82.2% in the Honts and Reavy study, with inconclusive rates of 13.5% and 13.6%, respectively.

Raskin (personal communication, March 2018) suggested that the directed lie format might provide a way to conduct specific-incident ODTs. We ran a mock crime experiment to explore that possibility. To obtain an ODT that addressed a single relevant issue, we began with the RCT and replaced questions about one of the two relevant issues with directed-lie questions.

Participants were offered $30 pay and a possible $30 bonus if they passed the test. They were randomly assigned to treatment conditions. Twenty-four participants were guilty of taking $20 from a secretary's desk, and 25 participants were innocent of the theft. Five participants were dropped because they failed to follow instructions or did not meet criteria for participation. Participants were given a pretest to familiarize them with the testing procedures and ensure they understood the instructions to lie to directed-lie questions. The test contained six True/False statements about the theft of the $20, six directed-lie statements, and 13 arithmetic problems. The set of test statements was repeated three times.

In contrast to the RCT format, we presented different types of test statements in a consistent, predictable order (arithmetic problem, directed lie, arithmetic problem, relevant; arithmetic problem, directed lie, arithmetic problem, relevant; etc.), although we varied statements of a given type over repetitions. The computer presented statements visually as text on the screen and aurally with a synthesized voice.

Finally, unlike previous ODT studies, we did not encourage participants to answer as quickly as possible, nor did we inform them that slow response times were characteristic of deception.

Accuracy of the DLT in a Laboratory Experiment Analysis of the behavioral and physiological measures reported previously for the RCT revealed significant effects of guilt on PD area (partial $\eta^2 = 0.26$), PD amplitude (partial $\eta^2 = 0.28$), and PD level (partial $\eta^2 = 31$; Cook et al., 2012). Although the DLT produced large effects on PD measures, they were largely redundant (mean intercorrelation = 0.68), and there were no significant effects on measures of reading (number of fixations, first pass, re-reading) or behavior (response time, errors). No requirement to answer quickly might be responsible for the lack of effects on reading and behavior.

Logistic regression analysis classified 19 of 25 innocent (76.0%) and 16 of 19 of guilty participants correctly (84.2%). Unweighted mean accuracy was 80.1%, and area under the ROC curve was 0.810 (Green & Swets, 1966). The sample was too small for N-fold validation. Instead, we created ten random subsets of 67% of cases to develop a logistic regression model and used it to classify the remaining 33% of cases. The 33% of cases in the ten hold-out samples provided ten unbiased estimates of the model's diagnostic validity in the population. Percent correct classifications in the holdout samples ranged from 50% to 87.5% for innocent participants and from 62.5% to 100% for guilty participants. On average, the logistic models correctly classified 76.4% of innocent and 83.4% of guilty participants in the hold-out samples. Unweighted mean accuracy on cross-validation was 79.9%.

Accuracy of the DLT in a Field Study In addition to the laboratory experiment, we obtained cases from actual field examinations. Thirty-one confirmed truthful tests concerning illegal drug use were obtained from students at a religious university. We assumed the base rate of deception was near zero and all who took the DLT were truthful. To the extent that our assumption was wrong and deceptive cases were included in the sample of truthful cases, observed accuracy was negatively biased. Another confirmed truthful case was obtained when the police later reported that their investigation revealed no crime had been committed. The last case was obtained when someone other than the examinee confessed to the crime. The resulting sample contained 34 confirm truthful cases.

Confirmed deceptive cases were obtained from field examiners that met the following criteria:

- The examinee was given an ocular-motor DLT,
- The examinee confessed to the crime, or incontrovertible physical evidence was obtained subsequent to the test, and
- The examinee had not been informed of the outcome of the DLT prior to their confession.

Efforts were made to exclude cases where a confession or investigation to uncover physical evidence was the result of a deceptive outcome on the DLT. In such cases, the criterion (deceptive status) would be contaminated by the deceptive test outcome, inflating estimates of test validity artificially (Iacono, 2008). Forty cases were included in the confirmed deceptive sample where case notes indicated that the confessions and physical evidence were obtained independently of the test outcome.

Logistic regression correctly classified 32 of 34 confirmed truthful examinees (94.1%) and 35 of 40 confirmed deceptive examinees (87.5%). Mean unweighted accuracy was 90.9%, and area under the ROC curve was 0.931. The sampling procedures described above for the laboratory experiment were used to cross-validate the logistic regression model. Percent correct classifications in holdout samples ranged from 70% to 100% for truthful cases and from 66.7% to 100% for deceptive cases. On average, 93.8% of truthful and 82.9% of deceptive examinees were classified correctly. Unweighted mean accuracy on cross-validation was 88.3%.

Accuracy was somewhat higher in the field study than in the laboratory experiment. Although the difference might be due to sampling error, there is reason to expect accuracy to be higher in the field than the laboratory. We give participants in laboratory experiments instructions on two occasions. When they arrive for the experiment, we instruct some participants to commit a mock crime and lie about it. We instruct other participants to answer truthfully about the crime on the test. Subsequently, when they take the test, the computer instructs participants to lie to the directed lie questions and then makes them practice lying to those questions before the test. At this point, guilty participants might become confused by their responsibilities as a participant and may become more focused on following instructions than the crime. In a field setting, guilty people are not likely to be confused by their task or lose sight of the importance of questions about the crime. Accuracy may be greater in the field than the laboratory because field examinees are not as likely to be confused by the task as participants in a laboratory experiment.

Although the results from field examinations are encouraging, they may be biased. Deceptive cases were referred by the service providers who conducted the tests rather than being sampled at random from the population of deceptive field suspects. We encouraged reports of post-ODT confessions and physical evidence regardless of the outcome of the test, but it is possible that providers were selective in the cases they reported. Our validity estimates were biased to the extent that service providers under- or over-reported false negative errors. False negative errors occurred when the examinees passed the DLT and subsequently confessed or evidence was uncovered to incriminate them.

Current evidence of ocular-motor DLT validity is positive but tenuous. There are reasons to believe accuracy observed in the laboratory may be lower than that in the field, and the accuracy observed in the present sample of field cases may be higher. The best point estimate of the accuracy of the ocular-motor DLT may be obtained by pooling the cases across laboratory and field settings (N = 118). Accuracy on cross-validation was 86.4% for the 59 innocent cases and 83.1% for the 59 deceptive cases. Mean accuracy was 84.7%. The 95% confidence interval for percent correct ranged from 77.0% to 90.7%. Based on the available data, we would not expect mean accuracy of the DLT in the field to fall below 77% or exceed 91%.

We find it very promising that the use of pupillary responses in the ODT for screening purposes can also be extended to specific incident examinations. The studies described so far report on our investigations of factors designed to facilitate the detection of deception with the ODT. The next describes behaviors on the part of examinees that may thwart these efforts.

5 Countermeasures

A person who undergoes a deception detection test is usually highly motivated to pass it, no matter the circumstances (criminal investigation, employment screening, laboratory study, etc.), since examinees typically benefit more from passing a deception test than from failing it. Some people deliberately attempt to increase their chances of passing the test by means of countermeasures. It is paramount that researchers examine whether there are any strategies that participants may use to successfully defeat the ODT.

Previous research has shown that countermeasures jeopardize the classification accuracy of deception detection methods, including the polygraph (e.g. Honts et al., 1987, 1994; Peth et al., 2016) or fMRI based tests (Ganis et al., 2011). In a study on the effectiveness of pupil diameter in a Comparison Question Test (CQT), Webb et al. (2009b) suggested that further research is necessary to examine whether pupil diameter is resistant to countermeasures, and the need to determine whether countermeasures may pose a threat to the ODT accuracy was acknowledged by the authors of the test as well (Cook et al., 2012).

Types of Countermeasures Available research on methods that counter other psychophysiological tests for deception may help us identify possible threats to the ODT. Although deception detection techniques may differ in important ways, there are implications for all techniques to the extent that the same cognitive and emotional processes underlie them. Twyman et al. (2013) distinguished between *physical* and *mental* countermeasures. In his review, Honts (1987) classified polygraph countermeasures as either *general state countermeasures* or *specific point countermeasures*. General state countermeasures are attempts to alter the examinee's psychological and/ or physiological state throughout the test. The most common general state countermeasures would be drug use, exercising to exhaustion, or relaxation techniques. Specific point countermeasures, on the other hand, are used by the examinee to change psychological and/or physiological state during a specific period of the examination. Honts (2014) concluded that general state countermeasures have little effect on accuracy of the Concealed Information Test (CIT) or Comparison Question Tests (CQT) polygraph tests, whereas training in the application of specific point countermeasures may pose a threat to polygraph test validity. Contrary to that, Elaad and Ben Shakkar (1991) argued that efforts to dissociate from the tests' content using relaxing thoughts could reduce physiological responses through decreased attention to the relevant items. The results of their experiments indicated that continuous dissociation throughout a CIT was associated with lower differential responsivity between target and nontarget alternatives, whereas the item-specific dissociation tended to make detection more likely. The authors noted that mental dissociation could be used without external manifestations which would make it difficult to detect.

Potential threats to the validity of the ODT. Since changes in pupil size are not unique to deception, it is possible that strategies unobservable to the examiner may be used to defeat an ocular-motor deception test. Though the ODT was originally based on the cognitive workload hypothesis (Hacker et al., 2014), it remains unclear to what extent the pupil dilation during deception is influenced by emotional arousal. Therefore, either cognition- or emotion-based countermeasures may affect pupillary reactions and threaten the validity of the ODT. These strategies could be physical or mental and primarily cognitive or emotional.

Physical Countermeasures Twyman et al. (2013) examined countermeasures in a deception detection study in which in addition to other measures, they measured pupillary response to target items in the CIT. Twyman et al. hypothesized that if participants self-inflicted pain such as biting the tongue during nontarget items in the CIT, this would decrease the difference between pupil reactions to target and nontarget items and thereby increase the number of false negative errors. Contrary to their predictions, "pupil dilation was the strongest effect among those investigated and appeared to be the most resilient to countermeasures. The pupil dilation resulting from the orienting response was strong, and there was no decrease in this effect when mental distraction or pain was used" (pp. 12–13).

Other studies concerned with physical countermeasures in polygraph testing, however, show that performing simple physical actions, such as biting the tongue or pressing the toes to the floor, can help defeat polygraph tests by increasing physiological reactions to comparison questions. These findings demonstrate that in emotional reaction-based methods to detect deception, physical countermeasures are effective and hard to detect (e.g., Honts et al., 1985, 1987, 1994; Peth et al., 2016). Further research is needed to determine whether physical countermeasures pose a threat to the ODT.

Mental Countermeasures Walczyk et al. (2013) suggested that if a lie is well rehearsed, "lie construction" during deception becomes "lie recall," which requires less mental effort. They found that a brief rehearsal of lies lowered cognitive load, as indexed by response time, below that of unrehearsed lies. The authors noted that rehearsal must be considered a countermeasure for cognitive load-based lie detection methods.

One of the strongest indicators of deceit in the ODT is pupil dilation. To appear innocent, it is possible that examinees may attempt to engage in cognitively demanding tasks or induce emotional arousal while control items are presented to produce pupil reactions to control items that are as large as those to relevant items. Twyman et al. (2013) instructed participants in their mental countermeasures group to count backwards by seven from a three-digit number. There was no effect on pupil dilation when mental distraction was used. However, when the counting backwards countermeasure was used in a CQT polygraph, decision accuracy was lower than accuracy for the guilty control group. Moreover, none of the guilty examinees who used mental countermeasures were identified by the examiner (Honts et al., 1994).

Peth et al. (2016) administered CIT to their participants and tested for effects of mental and physical countermeasures on number of fixations, mean fixation duration, eye blinks, and other psychophysiological measures. For the mental countermeasures group, pictures of an unpleasant scene (i.e., a car accident, a street fight, a dentist treatment) were rehearsed and later recalled during the presentation of nontargets. Physical countermeasure participants were instructed to move their toes during nontarget item presentations. Generally, mental countermeasures were less effective than physical countermeasures. However, mental countermeasures reduced the difference between target and nontarget items on number of fixations, fixation duration, and number of eyeblinks.

Suchotzki et al. (2017) conducted a meta-analysis on response time countermeasures to four deception detection techniques. They found that the ability to defeat the test requires knowledge about the test principles and/or clear instructions on how to strategically manipulate the responses. The effect of information on a CQT was examined by Rovner (1986) in an experiment with participants divided into three treatment groups: the control group received no special treatment, the informed group obtained specific, detailed information about the CQT, and the information + practice group received information and then was given two practice polygraph tests. Decision accuracy was 95% for both the control and information groups but dropped to 71% for the information + practice group. Guilty participants who were given information and practice were better able to defeat the test than guilty control participants.

ODT Countermeasures Study A recent study at Masaryk University followed the mock crime protocol described in previous studies of the ODT (e.g., Patnaik et al., 2016). A total of 80 participants (40 females) recruited from the general population were randomly assigned to one of four cells in a 2 × 2 design. Half of the participants were guilty and half were innocent of the mock crime. Prior to the test, half of the participants in each group were provided with a "How to defeat the ODT" document that contained information on eye-tracking, pupil dilation, cognitive effort during lying, and suggestions on how to defeat the ODT. The remaining participants were uninformed. Measures of pupil diameter, response time, response accuracy, and reading behaviors (number of fixations, first-pass duration) were obtained for each statement and repetition of the test items. Consistent with Rovner's (1986) results for the CQT, there was no reliable evidence that information about the ODT and suggestions about how to defeat it moderated the diagnostic validity of the measured features. Whether practice in the use of specific countermeasures would reduce the accuracy of the ODT is an avenue for further research. However, it is promising that even participants who were given information about the ODT were not able to develop countermeasures to successfully defeat it.

6 Conclusion

We have made clear in the preceding sections that measurement of pupil diameter is a critical variable in the ODT. Across multiple laboratory and field experiments, pupillary response remains one of the strongest indicators of deception. When measures of pupil diameter are combined with behavioral responses in the ODT, or with other psychophysiological measures in polygraph (e.g., Webb et al., 2009b), the result is a highly diagnostic tool that can be used across a broad range of screening and criminal investigation scenarios, and with a wide range of participants. Early research has demonstrated that the use of this multi-method technique is relatively resistant to effects of countermeasures. However, continued examination of potential countermeasures is certainly a ripe topic for future research. In addition, expanding our focus on potential contexts and variations of the ODT, and the role of the pupillary response in deception, will benefit both basic research on pupillometry as well as more applied domains.

References

Ad Hoc Committee on Polygraph Techniques. (2011). Meta-analytic survey of criterion accuracy of validated polygraph techniques. *Polygraph, 40*, 193–305.

Ahern, S. K., & Beatty, J. (1979). Pupillary responses during information processing vary with Scholastic Aptitude Test scores. *Science, 205*, 1289–1292.

American Polygraph Association. (2011). Meta-analytic survey of criterion accuracy of validated polygraph techniques. *Polygraph, 40*, 196–305.

Azevedo, R. (2002). Beyond intelligent tutoring systems: Using computers as METAcognitive tools to enhance learning? *Instructional Science, 30*, 31–45.

Baddeley, A. (2007). *Working memory, thought, and action* (Vol. 45). Oxford University Press.

Baker, L., Goldstein, R., & Stern, J. A. (1992). *Saccadic eye movements in deception: Report DoDPI92-R-003*. Department of Defense Polygraph Institute.

Bayer, M., Sommer, W., & Schacht, A. (2011). Emotional words impact the mind but not the body: Evidence from pupillary responses. *Psychophysiology, 48*, 1533–1561.

Beatty, J., & Wagoner, B. L. (1978). Pupillometric signs of brain activation vary with level of cognitive processing. *Science, 199*, 1216–1218.

Bell, B. G., Kircher, J. C., & Bernhardt, P. C. (2008). New measures improve the accuracy of the directed-lie test when detecting deception using a mock crime. *Physiology & Behavior, 94*, 331–340.

Bouffard-Bouchard, T., Parent, S., & Larivee, S. (1991). Influence of self-efficacy on self-regulation and performance among junior and senior high-school age students. *International Journal of Behavioral Development, 14*, 153–164.

Bradley, M. T., & Janisse, M. P. (1981). Accuracy demonstrations, threat, and the detection of deception: Cardiovascular, electrodermal, and pupillary measures. *Psychophysiology, 18*, 307–315.

Bradley, M. T., Micolli, L., Escrig, M. A., & Lang, P. J. (2008). The pupil as a measure of emotional arousal and autonomic activation. *Psychophysiology, 45*, 602–607.

Bradshaw, J. L. (1968). Pupil size and problem solving. *Quarterly Journal of Experimental Psychology, 20*, 116–122.

Colman, F., & Paivio, A. (1970). Pupillary dilation and mediation processes during paired association learning. *Canadian Journal of Psychology, 24*, 261–270.

Converus. (2020). *EyeDetect audio multi-issue comparison test (AMCT): Development and validation summary*. Converus, Inc. https://converus.com/wp-content/uploads/2020/10/EyeDetect-Audio-MCT-Dev-and-Validation-Summary-20200318.pdf

Cook, A. E., Hacker, D. J., Webb, A. K., Osher, D., Kristjansson, S. D., Woltz, D. J., & Kircher, J. C. (2012). Lyin' eyes: Ocular-motor measures of reading reveal deception. *Journal of Experimental Psychology. Applied, 18*, 301–313.

Correa, E. I., & Adams, H. E. (1981). The validity of the preemployment polygraph examination and the effects of motivation. *Polygraph, 10*, 143–155.

Deci, E. L., & Ryan, R. M. (1985). Cognitive evaluation theory. In *Intrinsic motivation and self-determination in human behavior. Perspectives in social psychology* (pp. 43–85). Springer.

Deniz, F., Nunez-Elizalde, A. O., Huth, A. G., & Gallant, J. L. (2019). The representation of semantic information across human cerebral cortex during listening versus reading is invariant to stimulus modality. *Journal of Neuroscience, 39*, 7722–7736.

Department of Defense Polygraph Institute Research Division Staff. (1997). A comparison of psychophysiological detection of deception accuracy rates obtained using the counterintelligence scope polygraph and the test for espionage and sabotage question formats. *Polygraph, 26*, 79–106.

Dionisio, D. P., Granholm, E., Hillix, W. A., & Perrine, W. F. (2001). Differentiation of deception using pupillary responses as an index of cognitive processing. *Psychophysiology, 38*, 205–211.

Eccles, J. S., & Wigfield, A. (2002). Motivational beliefs, values, and goals. *Annual Review of Psychology, 53*, 109–132.

Elaad, E., & Ben-Shakhar, G. (1991). Effects of mental countermeasures on psychophysiological detection in the guilty knowledge test. International journal of psychophysiology. *Official Journal of the International Organization of Psychophysiology, 11*, 99–108.

Fischer, P., Greitemeyer, T., & Frey, D. (2008). Self-regulation and selective exposure: The impact of depleted self-regulation resources on confirmatory information processing. *Journal of Personality and Social Psychology, 94*, 382–395.

Flavell, J. H. (1979). Metacognition and cognitive monitoring: A new area of cognitive developmental inquiry. *American Psychologist, 34*, 906–911.

Fuse, L. S. (1982). *Directed-lie control testing technique*. Paper presented at the Seminar of the Federal Interagency Polygraph Committee, Federal Bureau of Investigation Academy.

Ganis, G., Rosenfeld, J., Meixner, J., Kievit, R., & Schendan, H. (2011). Lying in the scanner: Covert countermeasures disrupt deception detection by functional magnetic resonance imaging. *NeuroImage, 55*, 312–319.

Green, D. M., & Swets, J. A. (1966). *Signal detection theory and psychophysics*. Wiley.

Hacker, D. J. (1997). Comprehension monitoring of written discourse across early-to-middle adolescence. *Reading and Writing: An Interdisciplinary Journal, 9*, 207–240.

Hacker, D. J., Kuhlman, B. B., Kircher, J. C., Cook, A. E., & Woltz, D. J. (2014). Detecting deception using ocular metrics during reading. In D. C. Raskin, C. R. Honts, & J. C. Kircher (Eds.), *Credibility assessment: Scientific research and applications* (pp. 159–216). Academic.

Handler, M. D., Honts, C. R., Krapohl, D. J., Nelson, R., & Griffin, S. (2009). Integration of pre-employment polygraph screening into the police selection process. *Journal of Police and Criminal Psychology, 24*, 69–86.

Heilveil, I. (1976). Deception and pupil size. *Journal of Clinical Psychology, 32*, 675–676.

Hess, E. H., & Polt, J. M. (1960). Pupil size as related to interest value of visual stimuli. *Science, 132*, 349–350.

Hess, E. H., & Polt, J. M. (1964). Pupil size in relation to mental activity during simple problem-solving. *Science, 143*, 1190–1192.

Hidi, S., & Renninger, K. A. (2006). The four-phase model of interest development. *Educational Psychologist, 41*, 111–127.

Honts, C. R. (1987). Interpreting research on polygraph countermeasures. *Journal of Police Science and Administration, 15*, 204–209.

Honts, C. R. (2014). Countermeasures and credibility assessment. In D. C. Raskin, C. R. Honts, & J. C. Kircher (Eds.), *Credibility assessment: Scientific research and applications* (pp. 133–155). Academic Press.

Honts, C. R., & Amato, S. (2007). Automation of a screening polygraph test increases accuracy. *Psychology, Crime & Law, 13*, 187–199.

Honts, C. R., & Reavy, R. (2015). The comparison question polygraph test: A contrast of methods and scoring. *Physiology & Behavior, 143*, 15–26.

Honts, C. R., Hodes, R. L., & Raskin, D. C. (1985). Effects of physical countermeasures on the physiological detection of deception. *Journal of Applied Psychology, 70*(1), 177–187.

Honts, C. R., Raskin, D. C., & Kircher, J. C. (1987). Effects of physical countermeasures and their electromyographic detection during polygraph tests for deception. *Journal of Psychophysiology, 1*, 241–247.

Honts, C. R., Raskin, D. C., & Kircher, J. C. (1994). Mental and physical countermeasures reduce the accuracy of polygraph tests. *The Journal of Applied Psychology, 79*, 252–259.

Horowitz, S. W., Kircher, J. C., Honts, C. R., & Raskin, D. C. (1997). The role of control questions in physiological detection of deception. *Psychophysiology, 34*, 108–115.

Hyönä, J., & Nurminen, A. M. (2006). Do adult readers know how they read? Evidence from eye movement patterns and verbal reports. *British Journal of Psychology, 97*, 31–50.

Hyönä, J., Tommola, J., & Alaja, A. M. (1995). Pupil dilation as a measure of processing load in simultaneous interpretation and other language tasks. *The Quarterly Journal of Experimental Psychology, 48A*, 598–612.

Iacono, W. G. (2008). Accuracy of polygraph techniques: Problems using confessions to determine ground truth. *Physiology & Behavior, 95*, 24–26.

Johnson, R., Jr., Barnhardt, J., & Zhu, J. (2005). Differential effects of practice on the executive processes used for truthful and deceptive responses: An event-related brain potential study. *Cognitive Brain Research, 24*, 386–404.

Just, M. A., & Carpenter, P. A. (1993). The intensity of dimension of thought: Pupillometric indices of sentence processing. *Canadian Journal of Experimental Psychology, 47*, 310–339.

Kahneman, D., & Beatty, J. (1966). Pupil diameter and load on memory. *Science, 154*, 1583–1585.

Kircher, J. C. (1981). *Psychophysiological processes in the detection of deception.* Doctoral Dissertation, University of Utah.

Kircher, J. C. (2018). Ocular-motor deception test. In J. P. Rosenfeld (Ed.), *Detecting concealed information and deception* (pp. 187–212). Elsevier.

Kircher, J. C., Kristjansson, S., Gardner, M. K., & Webb, A. K. (2012). Human and computer decision-making in the psychophysiological detection of deception. *Polygraph, 41*, 77–126.

Klingner, J., Tversky, B., & Hanrahan, P. (2011). Effects of visual and verbal presentation on cognitive load in vigilance, memory, and arithmetic tasks. *Psychophysiology, 48*, 323–332.

Krapohl, D. J. (2002). The polygraph in personnel screening. In M. Kleiner (Ed.), *Handbook of polygraph testing* (pp. 217–236). Academic Press.

Lubow, R. E., & Fein, O. (1996). Pupillary size in response to a visual guilty knowledge test: New technique for the detection of deception. *Journal of Experimental Psychology: Applied, 2*, 164–177.

Lumley, T. (2020). *Regression subset selection* (3.1).

National Research Council. (2003). *The polygraph and lie detection.* National Academies Press.

Nelson, R. (2015). Scientific basis for polygraph testing. *Polygraph, 44*, 28–61.

Paivio, A. (1990). *Mental representations: A dual coding approach.* Oxford University Press.

Partala, T., Jokiniemi, M., & Surakka, V. (2000). Pupillary responses to emotionally provocative stimuli. In *Proceedings of the 2000 symposium on eye tracking research and applications* (pp. 123–129). New York, NY.

Patnaik, P., Woltz, D. J., Hacker, D. J., Cook, A. E., Ramm, M. L., Webb, A. K., & Kircher, J. C. (2016). Generalizability of an ocular-motor test for deception to a Mexican population. *International Journal of Applied Psychology, 6*, 1–9.

Peth, J., Suchotzki, K., & Gamer, M. (2016). Influence of countermeasures on the validity of the Concealed Information Test. *Psychophysiology, 53*, 1429–1440.

Potts, A. C., Kircher, J. C., Webb, A. K., Cook, A. E., Hacker, D. J., Son, S., & Woltz, D. J. (2020). *1, 2, 3 crimes you're out: Ocular-motor methods for detecting deception in a multiple-issue screening protocol*. University of Utah.

R Core Team. (2017). *R: A language and environment for statistical computing. (3.4.2)*. R Foundation for Statistical Computing. http://www.R-project.org/

Raskin, D. C. (1979). Orienting and defensive reflexes in the detection of deception. In H. D. Kimmel, B. H. van Olst, & J. F. Orlebeke (Eds.), *The orienting reflex in humans* (pp. 587–605). Erlbaum.

Raskin, D. C., & Kircher, J. C. (2014). Validity of polygraph techniques and decision methods. In D. C. Raskin, C. R. Honts, & J. C. Kircher (Eds.), *Credibility assessment: Scientific research and applications* (pp. 63–129). Elsevier.

Rayner, K., Pollatsek, A., Ashby, J., & Clifton, C., Jr. (2012). *The psychology of reading* (2nd ed.). New York, NY.

Rovner, L. I. (1986). The accuracy of physiological detection of deception for subjects with prior knowledge. *Polygraph, 15*, 1–39.

Schaefer, T., Jr., Ferguson, J. B., Klein, J. A., & Rawson, E. B. (1968). Pupillary responses during mental activities. *Psychonomic Science, 14*, 137–138.

Schluroff, M. (1982). Pupil responses to grammatical complexity of sentences. *Brain and Language, 17*, 133–145.

Schwonke, R. (2015). Metacognitive load–Useful, or extraneous concept? Metacognitive and self-regulatory demands in computer-based learning. *Journal of Educational Technology & Society, 18*, 172–184.

Seymour, T. L., Seifert, C. M., Shafto, M. G., & Mosmann, A. L. (2000). Using response time measures to assess "guilty knowledge". *Journal of Applied Psychology, 85*, 30–37.

Suchotzki, K., Verschuere, B., Van Bockstaele, B., Ben-Shakhar, G., & Crombez, G. (2017). Lying takes time: A meta-analysis on reaction time measures of deception. *Psychological Bulletin, 143*, 428–453.

Sweller, J., & Paas, F. (2017). Should self-regulated learning be integrated with cognitive load theory? A commentary. *Learning and Instruction, 51*, 85–89.

Twyman, N., Schuetzler, R., Proudfoot, J. G., & Elkins, A. C. (2013). A systems approach to countermeasures in credibility assessment interviews. In *Proceedings of the international conference on information systems (ICIS 2013): Reshaping society through information systems design* (pp. 265–283). Association for Information Systems.

Van Gerven, P. W. M., Paas, F. G. W. C., Van Merriënboer, J. J. G., & Schmidt, H. G. (2002). Cognitive load theory and aging: Effects of worked examples on training efficiency. *Learning and Instruction, 12*, 87–105.

Vendemia, J. M. C., Buzan, R. F., & Green, E. P. (2005). Practice effects, workload, and reaction time in deception. *American Journal of Psychology, 118*, 413–429.

Verschuere, B., & Ben-Shakhar, G. (2011). Theory of the concealed information test. In B. Verschuere, G. Ben-Shahkar, & E. Meijer (Eds.), *Memory detection: Theory and application of the concealed information test* (pp. 128–148). Cambridge University Press.

Vo, M. L. H., Jacobs, A. M., Kuchinke, L., Hofmann, M., Conrad, M., Schacht, A., & Hutzler, F. (2008). The coupling of emotion and cognition in the eye: Introducing the pupil old/new effect. *Psychophysiology, 45*, 130–140.

Vrij, A. (2008). *Detecting lies and deceit: Pitfalls and opportunities* (2nd ed.). Wiley.

Vrij, A., Mann, S., Kristen, S., & Fisher, R. P. (2007). Cues to deception and ability to detect lies as a function of police interview styles. *Law and Human Behavior, 31*, 599–518.

Vrij, A., Leal, S., Granhag, P. A., Mann, S., Fisher, R. P., Hillman, J., & Sperry, K. (2009). Outsmarting the liars: The benefit of asking unanticipated questions. *Law and Human Behavior, 33*, 159–166.

Walczyk, J. J., Roper, K. S., Seemann, E., & Humphrey, A. M. (2003). Cognitive mechanisms underlying lying to questions: Response time as a cue to deception. *Applied Cognitive Psychology, 17*, 755–774.

Walczyk, J. J., Mahoney, K. T., Doverspike, D., & Griffith-Ross, D. A. (2009). Cognitive lie detection: Response time and consistency of answers as cues to deception. *Journal of Business and Psychology, 24*, 33–49.

Walczyk, J. J., Grifith, D. A., Yates, R., Visconte, S., & Simoneaux, B. (2013). Eye movements and other cognitive cues to rehearsed and unrehearsed deception when interrogated about a mock crime. *Applied Psychology in Criminal Justice, 9*, 1–23.

Webb, A. K. (2008). *Effects of motivation and item difficulty on oculomotor and behavioral measures of deception*. University of Utah.

Webb, A. K., Hacker, D. J., Osher, D., Cook, A. E., Woltz, D. J., Kristjansson, S., & Kircher, J. C. (2009a). Eye movements and pupil size reveal deception in computer administered questionnaires. In D. D. Schmorrow, I. V. Estabrooke, & M. Grootjen (Eds.), *Foundations of augmented cognition: Neuroergonomics and operational neuroscience* (pp. 553–562). Springer.

Webb, A. K., Honts, C. R., Kircher, J. C., Bernhardt, P. C., & Cook, A. E. (2009b). Effectiveness of pupil diameter in a probable-lie comparison question test for deception. *Legal and Criminal Psychology, 14*, 279–292.

Wei, W. (2019). *The relationship between reading speed and reading skill in reading narrative inconsistencies: Evidence from eye movements*. University of Utah.

Zheng, R., & Cook, A. (2012). Solving complex problems: A convergent approach to cognitive load measurement. *British Journal of Educational Technology, 43*, 233–246.

Zimmerman, B. J. (2008). Investigating self-regulation and motivation: Historical background, methodological developments, and future prospects. *American Educational Research Journal, 45*, 166–183.

Zimmerman, B. J., & Moylan, A. R. (2009). Self-regulation: Where metacognition and motivation intersect. In *Handbook of metacognition in education* (pp. 311–328). Routledge.

Zimmerman, B. J., Bandura, A., & Martinez-Pons, M. (1992). Self-motivation for academic attainment: The role of self-efficacy beliefs and personal goal setting. *American Educational Research Journal, 29*, 663–676.

Part III
Practical Guides

Methodological Aspects of Pupillometry

Bruno Laeng and Sebastiaan Mathôt

1 Methodological Aspects of Pupillometry

1.1 The Long History of Observing Pupil Size

Anecdotal knowledge about pupillary changes dates back centuries (Loewenfeld, 1993); for example, approximately three to six centuries ago, women already used the extract of the plant Belladonna to artificially enlarge the pupils to appear both more interesting and interested. However, understanding the causes and effects of pupil responses and the scientific measurement of pupil size in psychology and medical research are quite recent developments (e.g., Wilhelm et al., 1999).

Neurology was one of the first disciplines to use changes in pupil size to identify specific pathological processes in the nervous system by observing, often with the naked eye, large-scale automatic changes in pupil diameter when the eye was stimulated by light (i.e., the pupil light reflex). Currently, changes in pupil size are still commonly used by doctors as well as by the police to establish if a person may be under the influence of some specific drug (e.g., opioids dramatically constrict, while amphetamine dilates, the eye pupils; Rodney et al., 1983).

Psychology discovered pupillometry in the early 1960 in the USA and it was originally based on simply filming the eye at close range and then deriving pupil diameters from the projection, frame by frame, of the film (see the section on Equipment below). However, at about the same time, eye tracking became a central method within the psychological sciences and, despite the label "eye tracker" (ET)

B. Laeng (✉)
Department of Psychology, University of Oslo, Oslo, Norway
e-mail: bruno.laeng@psykologi.uio.no

S. Mathôt
Faculty and Behavioral and Social Sciences, Experimental Psychology, University of Groningen, Groningon, Netherlands
e-mail: s.mathot@cogsci.nl

M. H. Papesh, S. D. Goldinger (eds.), *Modern Pupillometry*,
https://doi.org/10.1007/978-3-031-54896-3_13

apparently identifies the monitoring of gaze (e.g., saccades and fixations) as its principal function, precise pupillary data can be obtained nowadays when using a commercial infrared eye tracker.

1.2 What Is Measured with Pupillometry?

With current ET technology, it is possible to register with high spatial and temporal precision the movements of the pupils and to reconstruct precisely both gaze direction and the actual eye's opening diameter. It is important to understand from the outset that the main adaptive goal of controlling the aperture of the eye is the optimal adjustment of the optical image (Loewenfeld, 1993; Mathôt, 2020) to the level of electromagnetic radiation entering the eye at any time point (Pirenne & Denton, 1952).

However, research conducted during the last 60 years within the field of psychology has established that the eye pupil dilates (i.e., expands its diameter) proportionally to the degree of mental workload. In an influential study, published in 1960 by the American psychologist Eckard Hess, it became clear that with each added step in difficulty of arithmetic problems, the pupils' diameter also increased in steps. In the same years, the psychologist Daniel Kahneman (1973) systematically applied pupillometry to the study of a key latent cognitive variable that he labeled "mental effort," showing the diameter of the pupil increases proportionally to the amount of cognitive load (e.g., number of items held in working memory). Kahneman made a distinction between the "selective" aspect of attention (what we attend to) and the "intensive" aspect of attention (how intensely we attend, which is related to the level of arousal). In eye tracking, this distinction corresponds to measuring the direction of the gaze (which captures, at least, "overt" selective attention) versus measuring the diameter of the eye pupil, which reflects ongoing arousal levels.

We know now that the pupil can reflect a variety of ongoing conscious mental states like affective processes (e.g., Rieger & Savin-Williams, 2012), mental images (e.g., Laeng & Sulutvedt, 2014), optical illusions (e.g., Laeng & Endestad, 2012; Laeng, Nabil & Kitaoka, 2022), and cognitive control processes (e.g., Einhäuser et al., 2008; Chiew & Braver, 2013). However, the method of pupillometry is also capable of revealing cognitive processes under the threshold or at the "fringe" of consciousness, like perception-in-the-making, subconscious decisions, and unfelt emotions (for a review: Laeng et al., 2012). Hence, monitoring pupillary responses is useful also because, differently from eye movements, they are difficult to control voluntarily (e.g., Loewenfeld, 1993; Ekman et al., 2008; Laeng & Sulutvedt, 2014; Ehlers et al., 2016). From early in life, we possess a partially voluntary control of gaze such that we can control where we direct overt attention. Moreover, we can suppress looking directly at something and hide the true focus of our attention. In contrast, the adjustments of the pupil appear the unconscious expression of ongoing mental states and seem to be fully automatic (Weiskrantz et al., 1999; Tamietto et al., 2009). Borrowing a term from biological game theory, the pupil provides an "honest signal" of internal states (Laeng & Alnæs, 2019).

1.3 This Chapter

In the following sections, we will focus on how we can apply pupillometry optimally to psychological research or to the more general goal of testing scientific hypotheses and obtaining valid results. We will first survey the existing types of equipment that allow the measurement of the pupil diameters. We will then discuss how a pupillometry experiment should be designed and then set up in the controlled environment of the laboratory. We will discuss pupillometry as a quantitative method that provides measurements on an interval scale (typically millimeters) that directly describes change in a real-world physical dimension (i.e., the size of the eye opening). Hence, statistical tests on quantitative data are used typically to compare results in each of the conditions of the experiment and infer relationships between experimental conditions and behavior or physiology. Since there is no fixed pupil size value (e.g., anything greater than 0.3 mm from baseline) that counts per se as a valid or meaningful pupillary signal (e.g., of increased mental effort or arousal), all pupillary changes must be compared through inferential statistical methods to compare pupil size deviations from a baseline or control condition while considering individual differences. We will also discuss a central methodological issue in pupillometry consisting in the control and removal of potential artifacts when measuring the pupil in specific experimental situations. As we shall see, one needs to consider a variety of things and potential problems when doing a pupillometry experiment. However, many aspects are no different from hypothesis testing in any other field of empirical science.

2 Equipment

2.1 The Naked Eye

The easiest way to "measure" pupil size is with the naked eye: by looking at someone's pupils to see whether and how they respond. This is routinely done by medical doctors when they administer the swinging-flashlight test, a common neurological test that was apparently first described by Levatin (1959). During this test, the doctor uses a portable flashlight to illuminate each of the patient's eyes in alternation by literally swinging the flashlight between both eyes. Normally, this procedure results in relatively stable pupils, because pupil responses are consensual (i.e., both pupils respond equally to illumination of both eyes), and therefore it doesn't matter whether a single eye is illuminated constantly or whether illumination alternates between both eyes. However, systematic dilation or constriction of both pupils in response to swings of the flashlight indicates an afferent deficit; that is, this result indicates that visual input from one eye is not processed normally. Furthermore, asymmetry in the response of the pupils (i.e., a lack of the normal consensual response) indicates an efferent deficit; that is, this result indicates that the motor response of one of the pupils is impaired.

The reason that the swinging-flashlight test can be administered without specialized equipment is that the pupil light response is large, often corresponding to changes of several millimeters of diameter; therefore, deficits of the pupil light response, even subtle ones, are readily observable with the naked eye. However, pupil responses to cognitive processes, such as dilation in response to arousal, are far smaller, generally corresponding to changes of less than a millimeter of diameter. With training, people can learn to detect even such tiny pupil responses with above-chance performance (Naber & Nakayama, 2013); presumably, this is one of the techniques that mentalists use to "read minds." However, the usefulness of the naked eye as a measurement tool for cognitive pupillometry research is very limited.

2.2 *Photography*

Irene Loewenfeld, one of history's most influential pupillometry researchers, started her career as a technician in the laboratory of Otto Lowenstein (Thompson & Kardon, 2006). One of her tasks was to manually determine pupil size from photos of eyes of cats, rabbits, and human participants (see e.g., Fig. 2 from Loewenfeld, 1958). As already mentioned above, manual scoring of photos (or video frames) was also the approach taken by Kahneman and Beatty (1966) and Hess and Polt (1960) in their early studies on the effects of cognitive processes on pupil size.

Pupil size can be very accurately determined from photos. However, the clear downside of this method is the amount of work involved in manually scoring each photo. According to modern standards (see also the section below on *Sample size and number of trials*), an experiment with sufficient statistical power has around 1600 observations per condition, for example from 40 participants with 40 trials each per condition (Brysbaert & Stevens, 2018). A contemporary replication of Kahneman and Beatty's (1966) study on the effect of working-memory load on pupil size would therefore require manual scoring of 200,000 photos (1600 observations × 5 conditions × 25 photos per trial). For this reason, photography is no longer a viable option for cognitive pupillometry research.

2.3 Automated Pupillometers

Automated pupillometers are devices that, in one way or another, automatically quantify pupil size. The first automated pupillometer appears to have been designed by Lowenstein and Loewenfeld (1958; see also Lowenstein, 1954 for an early, less successful attempt), and essentially measured the amount of near-infrared light that reflected off the eye (see also Loweth, 1973 for an early pupillometer that works based on a similar principle). Currently, there are many kinds of automated pupillometers. Most of these are commercial products, typically handheld, that provide a more reliable way to assess neurological impairments as compared to the traditional swinging flashlight test described above (Phillips et al., 2019).

Many automated pupillometers are all-in-one solutions for neurological testing that handle stimulus presentation (i.e., they flash a light); measurement (i.e., they record pupil size over time), and data analysis (i.e., they automatically provide several standard metrics, such as maximum constriction velocity and response latency). For cognitive pupillometry research, automated pupillometers are of limited use, because their convenience in medical settings comes at the expense of a lack of flexibility.

2.4 Video-Based Near-Infrared Eye Trackers

Many contemporary eye trackers, such as the EyeLink (SR Research) and GazePoint (GazePoint Research), consist of a near-infrared video camera combined with one or more near-infrared light-emitting diodes (LEDs) that illuminate the eye. The pupil corresponds to the darkest part of the camera image, whereas the reflections of the LEDs on the cornea correspond to the brightest part. After a calibration procedure, the vector between the center of the pupil and the center of the corneal reflection(s) can then be used to derive gaze position.

The primary use of eye trackers is to measure gaze position, but they also measure pupil size. The details vary from device to device. For example, the EyeLink appears to[1] simply measure pupil size as the number of pixels that fall below a certain brightness threshold and are therefore considered part of the pupil; this results in a measure of pupil size in arbitrary units that does not directly translate to millimeters of diameter (but see Mathôt & Ivanov, 2019 for an example of how these units can be converted to mm). Other eye trackers, such as the Tobii (Tobii AB), also measure the distance between the device and the head of the participant, which allows the device to express pupil size directly in millimeters of diameter.

[1] The implementation details of commercial eye trackers are generally proprietary and not disclosed. Therefore, some of the information presented here is based on an educated guess of how the devices work based on their behavior.

Most eye trackers provide extensive software development kits (SDKs) that allow users to build experiments that interact with the eye tracker in sophisticated ways. For example, users can send messages to the eye tracker to log when certain events occur during the experiment, and users can request real-time information about eye position and pupil size to implement experiments in which participants can, for example, respond through eye movements (so-called gaze-contingent experiments).

Because of their ease of use, flexibility, and the fact that most psychological laboratories are equipped with them, video-based near-infrared eye trackers are the most widely used devices for cognitive pupillometry research.

2.5 Experimental Design

Typically, pupillometry experiments that appear in psychology journals are based on the "standard method" of science, in which empirical evidence is gathered to decide for or against competing hypotheses or theoretical accounts. Psychological experiments are typically considered part of the social sciences where there exists a variety of experimental approaches. Although any single experiment may never "prove" anything once-and-for-all, its empirical findings can contribute to add support to the hypothesis or contradict it, although many experiments turn out to be just inconclusive. Standard experiments have "treatment" and "control" conditions, and participants (typically human individuals) are either randomly assigned to a condition, as in between-subjects designs, or subjected to all conditions, as in within-subject designs. The goal is to assess empirically whether there is an average treatment effect or a difference in outcomes between treatment and control conditions. In the standard method, the control condition differs from the experimental condition in just one aspect (e.g., the absence or presence of a particular stimulus). Human participants receive similar instructions in both conditions to have similar expectations about the outcome of the experiment. The contrast between the two conditions gives meaning and facilitates interpretation of the data.

An experiment can be exploratory, without an a priori expectation about the results; but even in this case, the experiment must be conducted according to an optimal design that is free from artifacts (i.e., systematic errors or biases). Artifacts should be isolated or measured, taken into consideration when analyzing data and, if possible, eliminated. This often means that eye position and visual input need to be as constant as possible across conditions (see also the section on Preprocessing below).

As pointed out by Rosnow and Rosenthal (1997, p. 3) in their critical review about behavioral research: "Artifacts are not simply inconsequential effects in a research design; they may actually jeopardize the validity of the researcher's inferences from his or her results." Moreover, sometimes, "the same conditions discounted as 'nuisance variables' at one time may later be exploited as variables of interest in their own right." An example of an artifact, within the domain of pupillometry, that may be considered as a relevant variable per se is the use of "eye-blinks"—which disrupt pupil measurements by occluding it—as the variable of interest (e.g., Siegle et al., 2008; Smilek et al., 2010). Similarly, the pre-stimulus or pre-experiment baseline, originally intended to reduce the amount of measurement noise by discounting "uninteresting" factors that affect pupil size, has been used as a possible index of cognitive capacity (e.g., Unsworth & Robison, 2017; Unsworth et al., 2019; Aminihajibashi et al., 2019, 2020).

All scientific inquiry is subject to error and possible confounding factors can disrupt the outcomes of the experiment and in turn the interpretation of the results. Indeed, all measurements of human behavior or physiology vary all the time, also in identical conditions. Human experiments will always contain random variation (or "noise") during data collection. Simply averaging the measurements is the preferred way in psychology to "average out" random (instead of systematic) errors, so that random overestimates and underestimates from a true value will cancel out.

2.6 *Choosing Your Conditions*

If you want to contrast different conditions with one another, these need to be chosen carefully such that pupil size is only affected by the manipulation(s) of interest. In clinical or pharmacological studies, the manipulation is often an intervention, and the goal is to assess how this intervention affects pupil size; for example, we recently tested the effect of nicotine on pupil size during an attentional task (Wardhani et al., 2020). In such studies, the design should include a control condition or a control group that is hypothesized to be free of the effect of interventions (Pearl, 2009). In the case of our nicotine study, the control condition consisted of a separate session during which participants received a placebo pill. These are often referred to as the base levels of the independent variable/s. In another kind of design, which is typical for cognitive pupillometry, the manipulation consists of two or more experimental conditions, and the goal is to assess how pupil size differs between these conditions.

To facilitate exposition, we use as illustration an actual experiment in visual cognition by Mathôt, Grainger, and Strijkers ('Pupillary Responses to Words That Convey a Sense of Brightness or Darkness, 2017), in which words referring to bright or dark objects or events were presented to student participants while their pupils' size was concomitantly registered by a video-based near-infrared eye tracker. Here, the experimental contrast of interest was that between brightness- and darkness-related words.

2.7 The Theoretical Account

In principle, all empirical investigations originate from some theory that ideally is declared explicitly. Mathôt and colleagues' (2017) study is a paradigmatic psychological experiment using pupillometry where a theoretical account frames the study, hypotheses are developed from it, and given some specific conditions (i.e., variables manipulated, experimental set-up, stimuli) in which the experiment is conducted, specific predictions are made and subsequently put to test (statistically) against the obtained data. This study's theoretical framework originates from current accounts of language comprehension where a word is thought to trigger a process of internal mental "simulations" that, at least, contribute to representing the meaning of the word (reviewed in Zwaan, 2014). Such evoked simulations are not posited to be just episodes of mental imagery (e.g., we read or hear the word "sun" and we generate a mental image of a strong source of light up above) but also simulations of what one would do in the presence of the referent. For example, when exposed to strong sunlight, our body reacts to avoid being dazzled, among which a highly adaptive reaction is that the eye pupil automatically constricts to reduce the amount of light entering the eye. Such bodily adjustments or preparatory motions can contribute strongly to the "embodied" meaning of a word.

2.8 The Hypotheses

Hypotheses are ideally framed as "if… then" statements, which often imply a causal relationship between an intervention or independent variable (event or stimulus that is experimentally manipulated) and an outcome or dependent variable (what is measured) which, in pupillometry studies, is pupil size. Its interpretation is facilitated by collecting simultaneously other behavioral measures like accuracy or response times. In our example, "if" word comprehension involves bodily simulations, "then" the semantic brightness of words should affect pupil size.

2.9 The Predictions

Hypotheses and predictions are not identical, though commonly they are thought to be synonyms to be used interchangeably. A hypothesis is an explanation for something happening. Hypotheses are testable and use precise but abstract, theoretical, language. Predictions are guesses about the future that are scientific (as opposed to the predictions of prophets) since they can be inferred from the hypotheses and the experimental situation. Hence, they are educated and testable guesses. A prediction uses concrete and precise language pertaining to observable phenomena in a specific situation. In our example, "bright" words (e.g., sun) should trigger small but observable (with an eye tracker) pupil constrictions as compared to "dark" words (e.g., night). Hence, a scientific prediction is what should happen in this specific experiment if the hypothesis is true. In a well-thought and well-designed experiment, we also formulate predictions about what should happen if the hypothesis is false (based on competitive theoretical accounts). In the present example, an alternative hypothesis is that the pupil response, given its reliance on the noradrenergic arousal system of the brain, will simply reflect the emotional meaning of the word. For example, bright things referred to by "bright" words are experienced as strongly positive ("sun" and "light" are good things). However, intense and positive experiences are known to be associated also with dilations of the pupil (a manifestation of increased arousal), therefore predicting the opposite outcome of the original, theoretically derived, hypothesis.

2.10 Stimulus Selection and Presentation

Experiments are rarely based on a single stimulus, and what one presents in an experiment is several members of a class of stimuli (e.g., faces, a class of words). Sets of stimuli from different conditions should be matched on other features that they may possess that might potentially exert effects on the outcome that are unwanted or artifacts. In pupillometry, the strongest effect on pupil size is light itself (or its lack) which can increase or decrease the pupil diameter in proportion to the level of illumination in the environment and the lightness (reflectance) or brightness (emitted by a source) of objects. In fact, cognitive effects on pupils tend to be quite small (e.g., less than 0.5 mm). The pupil diameter of an adult is 3 or 4 mm at rest with an average range of 2.5–5 mm in typical room illumination conditions, but its response to light levels can range from 1 to 8 mm (MacLachlan & Howland, 2002) and in total darkness is approximately 6.5 mm with an average of normal pupils size ranging from 5 to 8.5 mm. Hence, external stimuli varying in light energy have an effect on the pupil diameter that is several orders of magnitude greater than the effect of internal mental states.

Hence, a very important methodological aspect of pupillometry studies is that during the experiment, illumination in the room and screen brightness are kept constant and changes of luminance across stimuli are minimized or made equal. Making stimuli equiluminant is challenging when using visual stimuli, but an easy approach is to benefit from graphic software like Adobe Photoshop which allows to change the mean brightness levels of an image (e.g., by opening the "histogram" window and adjusting brightness/contrast level in each image to normalize it to a desired average pixels' brightness (e.g., mean = 150). If equalizing brightness levels would interfere with feature (e.g., color or contrast) of the image, a solution is to present right before each target stimulus a luminance-matched baseline image for a brief period of a few hundreds of ms (e.g., Mathôt et al., 2017) to 1 s (e.g., Laeng & Sulutvedt, 2014). One should keep in mind that long baselines have the disadvantage of being susceptible to pupil-size fluctuations during the baseline period (e.g., eyeblinks), but short baseline periods have the disadvantage of being susceptible to recording noise (e.g., Mathôt et al., 2018).

Pupil responses to internal states differ from those to external stimuli like light flashes in velocity and they can be considered relatively slow, with dilations being generally slower than constrictions. Moreover, also for the pupillary light reflex, the diameter of the pupil tends to return to "baseline" after a couple of seconds from the stimulus offset (Bergamin et al., 2003; Barbur, 2004). Depending on the cognitive task, pupil dilations can peak as early as 1.5 s and return to baseline around 2 s from onset (e.g., in the Stroop task; Laeng et al., 2011) but with some tasks dilations may peak around 3 s and relax at 5 s (e.g., when doing additions and multiplications: Jainta & Baccino, 2010). This state of things implies that, in a well-designed experiment, trials should be relatively slow-paced and spaced, with variable inter-trial intervals, so that the full pupil response in a single trial can be captured in its complete waveform, and so that there are no carry-over effects from one trial to the next.

Self-paced experiments, during which participants take their own time before initiating the next trial, have the advantage that participants are likely to be in the right state of attentive readiness for the next trial, thus reducing the likelihood that a stimulus appears during a mind-wandering episode (cf., Unsworth & Robison, 2018; Pelagatti et al., 2020). Self-pacing an experiment also has the effect of reducing eyeblinks during the trial, since the participant tends to blink within the inter-trial interval.

The disadvantage of self-paced experiments is that participants have to press a key on each trial, which may disrupt the flow of the experiment. Eye trackers often provide an intermediate, semi-self-paced solution in the form of an automated drift-correction procedure, which pauses the experiment until a stable fixation is detected. This allows the participant to momentarily pause the experiment after each trial by blinking or closing the eyes, while avoiding the need to press a key to initiate each trial.

In our illustration from Mathôt and colleagues' 2017 study, a word was presented in the center of the screen for 3 s or auditorily, while looking at a gray background and each trial started with an automated drift correction. Most eye-tracking software allows recording only from the onset to the offset of specific events, making it possible to obtain appropriately sized data files without lengthy recordings of the pupil during the irrelevant inter-trial intervals.

2.11 Sample Size and Number of Trials

Effects of cognitive and affective manipulations on pupil size tend to be small and one can expect effect sizes that are at best small-intermediate (e.g., Cohen's d = 0.4). According to principled statistical power calculation, with 80% power, one would need over 50 participants in order to compare two within-participants conditions. A look at the pupillometry literature will reveal that in practice the great majority of studies have used smaller sample sizes than N = 50. Most studies have based their sample size simply on the N of previous studies, occasionally increasing sample size of two to three times as many participants per experiment (e.g., as in Mathôt and colleagues' 2017 study). However, taking into account also between-subjects factors or interactions, hundreds of participants would be needed without running underpowered studies, which is rarely seen in pupillometry studies (a notable exception – with N = 325 – is a study on the effect of sexual arousal on the pupil by Rieger & Savin-Williams, 2012).

An alternative statistical approach to a power analysis for sample size can be based on Bayesian statistics by using a sequential Bayes factor design, as recommended by Mani et al. (2021). In practice, one can begin collecting data from a small group of 20 participants and, if a set level of inference threshold (e.g., for H1: BF = 5 and for H0: BF = 1/5) is not reached, then a sequential Bayes analysis is initiated (e.g., using JASP) that examines sequentially, after the test of each additional participant, the evidence for H1/H0. As soon as the Bayesian factors cross the predefined inference level and remains stable for another three to four participants, data collection can be terminated, and a conclusion reached for either H1 or H0. The downside of sequential sampling, Bayesian or otherwise, is that on average it requires even more participants, because a more stringent criterion needs to be adopted to compensate for the fact that multiple statistical tests are being performed.

The number of trials within a condition has also relevance for obtaining robust data given that with eye tracking you often lose trials due to mistracking or artifacts. Some researchers have provided rule-of-thumb sample sizes of at least 15–25 trials (Winn et al., 2018). In our illustration study by Mathôt and colleagues (2017), about 30 words per condition were selected, as a compromise between having well-matched stimuli and having a reasonable number of observations per participant and condition.

Nevertheless, a survey of the literature will reveal a wide variety of number of trials and participants, starting with the minimalist but classic studies of Hess and Polt (1960, 1964), both published in *Science* and often considered to have kick-started the pupillometry method in psychology; these had only 4/5 trials and 5/6 participants (in 1960 and 1964, respectively). Note that the 1964 study by Hess and Polt, which showed increasing enlargement of the pupil with task difficulty (or "effort" *a la* Kahneman), had effects corresponding to an unusually "very large" Cohen's d = 1.85.

2.12 Pupillometry Experiments with Other Modalities than Vision or Outside the Laboratory

Most pupillometry experiments up to now have used visual stimuli. However, there exists also a sizable literature on auditory experiments, including the early classic experiment by Kahneman and Beatty (1966) on working memory, which were auditory in the sense that they used the auditory modality for presenting stimuli in their study of internal processes like memory, reasoning, and decision-making. Often findings do not differ depending on whether stimuli are either presented visually or auditorily, as in the illustration study of Mathôt and colleagues (2017), where words were either presented on screen or spoken through loudspeakers. Several studies have specifically targeted listening effort in sentence comprehension (Winn et al., 2018) or even during music listening (Endestad et al., 2020; Bishop et al., 2021).

On the one hand, non-visual studies circumvent several of the artifacts that we discussed earlier, since during auditory presentations one can keep the brightness and contrast presented on screen absolutely constant at all times. On the other hand, auditory stimuli may evolve over significantly longer period of time (e.g., minutes, as when playing a musical piece; e.g., Laeng et al., 2016) than the brief presentation (typically a few seconds) of pictures or single words. With long stimulus presentations, there is the tendency for the mind to begin to wander, as testified by several studies on reading text showing clear changes in pupil size (e.g., Franklin et al., 2013; Unsworth & Robison, 2018). Finally, the low-level properties of an auditory stimulus, such as its loudness or pitch, likely also affect pupil size, although in ways that are less predictable than the effect of brightness.

Pupillometry studies on the chemical senses like smell and taste remain sparse (e.g., Hess & Polt, 1966; Aguillon-Hernandez et al., 2015). Similarly, there are few pupillometry studies on touch (e.g., Ellingsen et al., 2013) but several on pain (e.g., Aissou et al., 2012; Connelly et al., 2014) or on sexual pleasure (e.g., Whipple et al., 1992; Rieger & Savin-Williams, 2012).

Only in recent years, pupillometry has been adapted to studies with electroencephalography (EEG; e.g., Miles et al., 2017) and functional magnetic resonance imaging (fMRI; e.g. DiNuzzo et al., 2019), since eye tracking can be adapted to the magnetic environment. Moreover, eye tracking has been adapted as well as within virtual reality environments (by using eye-tracking goggles; e.g., Juvrud et al., 2018; Liu et al., 2020), and even applied in semi-ecological conditions (e.g., during golf putting on artificial greens: Carnegie et al., 2020; in musicians during a real concert: Bishop et al., 2021; during visual search of books on real bookshelves: Zhou et al., 2020). Of course, field measurements in real life situations, especially where pupillometry could be a very useful application (e.g., while driving a vehicle to measure attentive states), encounter the challenge of discounting enormous variations in illumination in natural conditions, which calls for future methodological solutions.

3 Preprocessing

This section will focus on the steps involved in going from the "raw" pupil data, as stored by the eye tracker during recording, to a data visualization that fits the research question. As a concrete example, we will use data from Mathôt et al. (2019), which is a preprint that describes a replication of the study by Mathôt et al. (2017) on brightness- and darkness-related words; the reason for using data from this preprint, rather than from the original study, is that the preprint is accompanied by a more user-friendly analysis script that readers of this article can refer to, and that includes code to generate all figures from this section (see https://osf.io/qd3vp/).

3.1 The Raw Signal

Eye tracking is a field without standardization; this means that each brand of eye tracker stores data in its own idiosyncratic format. However, differences in data format aside, eye-tracking data corresponds to three time series, that is, three measures that are recorded over time: an x-coordinate, a y-coordinate, and pupil size (Fig. 1). Many eye trackers also maintain an "event log" that consists of short, time-stamped text messages; these messages indicate when particular events occurred (e.g., the start of a trial, the onset of a stimulus), store experimental variables (e.g., experimental conditions or participant responses), and store various kinds of other information (e.g., region-of-interest coordinates for the display, calibration results).

3.1.1 Dealing with Blinks

Eyeblinks result in both artifactual and real changes in pupil size. The artifactual changes are clearly visible in Fig. 1a. Specifically, at the start of an eyeblink, the eye lid rapidly closes while progressively covering more of the pupil surface. Most eye trackers record this as a sharp, rapid drop in pupil size (see Figs. 1a and 2a). While the eyes remain closed, pupil size is generally recorded as 0. When the blink ends, the eye lid rapidly re-opens while progressively revealing more of the pupil surface; most eye trackers record this as a sharp, rapid increase in pupil size.

There are various ways to deal with blinks in pupil data. One approach is to remove all trials on which a blink occurred. The advantage of this approach is that it's relatively straightforward to do, although it does require a way to reliably detect blinks in the signal (Hershman et al., 2018). The disadvantage of this method is that most people blink so frequently that many trials will have to be removed.

A second approach is to remove only those samples that are affected by blinks (rather than removing entire trials). This means replacing all the 0 values (assuming that missing data is indicated by 0 values) by *nan*, or not-a-number, values, so that they are subsequently ignored. Importantly, to deal with the sharp drops and rises in pupil size before and after blinks, a margin of a few samples before and after each blink should be removed as well. The disadvantage of removing missing data is that it may result in step-like changes in pupil size after averaging; these step-like changes correspond to individual trials temporarily "dropping out" because of a blink, which can cause a sudden change in mean pupil size.

A third approach is to reconstruct pupil size during blinks through an interpolation algorithm, such as linear interpolation, which draws a straight line from the start until the end of the blink, again with a margin of a few samples, or cubic-spline

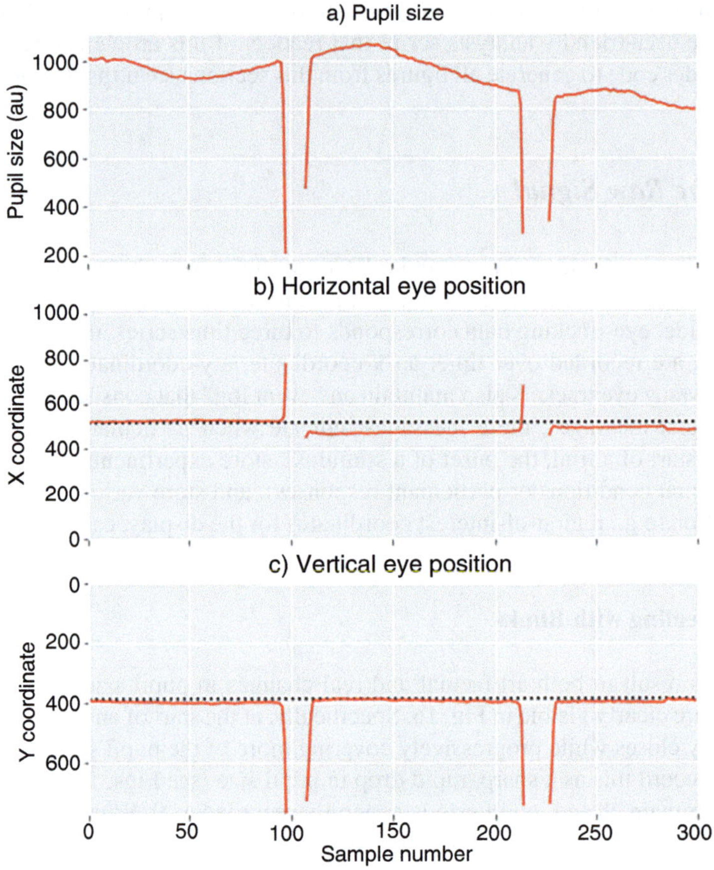

Fig. 1 Raw eye-tracking data from a single trial recorded at 100 Hz from Mathôt et al. (2019). (**a**) Pupil size (in arbitrary units) as a function of sample number (i.e., one observation for each 10 ms). The sharp drops in pupil size, followed and preceded by missing data, correspond to eye blinks, which distort pupil-size recordings due to closing of the eyelid. (**b**) Horizontal eye position (in display pixels), with the display center indicated by the dotted line and rightwards plotted up. (**c**) Vertical eye position (in display pixels), with the display center indicated by the dotted line and upwards plotted up. The distorting effect of eye blinks is clearly visible in the eye position as well

interpolation, which mimics smooth pupil-size changes based on four points around the period of the blink (Mathôt, 2013). The main advantage of interpolation is that it results in a smooth signal after averaging. For the example study, we opted to reconstruct blinks using cubic-spline interpolation (Fig. 2a).

Even when blinks have been properly dealt with using one of the methods described above, a few important considerations remain. First, blinks are followed by a pupil constriction (Knapen et al., 2016); this is not an artifact in the sense that this constriction is real, presumably resulting from changes in visual input due to the brief closing of the eye (Mathôt, 2020). Second, blink rate is affected by various factors, such as fatigue (Stern et al., 1994). This means that differences in blink rate

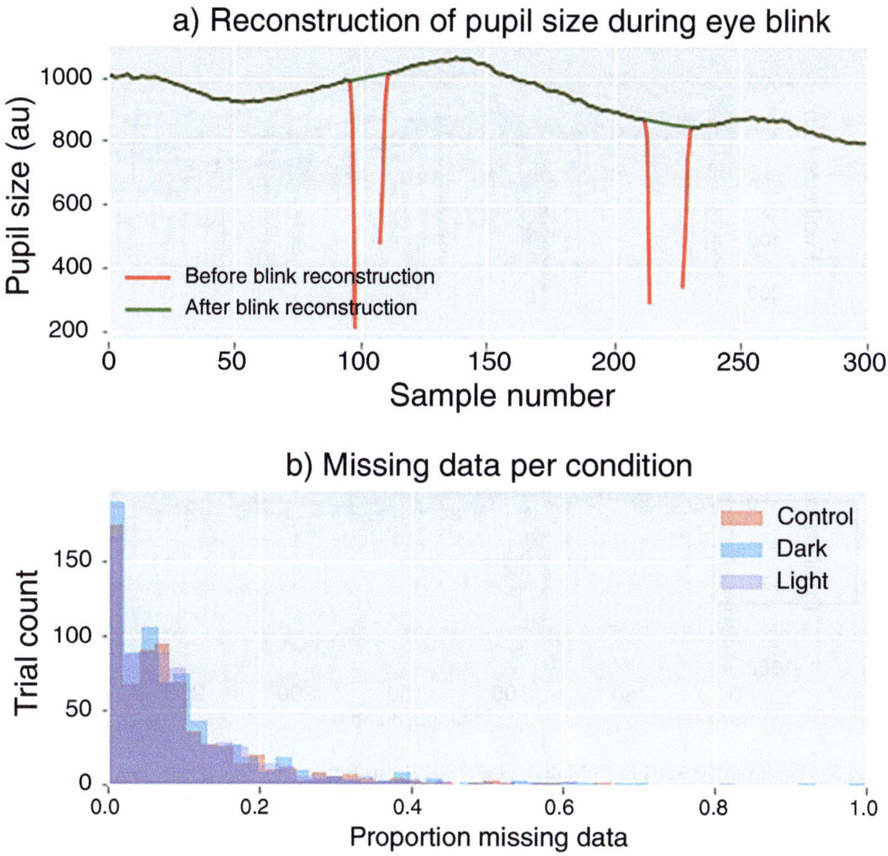

Fig. 2 (**a**) Pupil size before (red line) and after (green line) blink reconstruction through cubic-spline interpolation. Data from the same trial as shown in Fig. 1. This panel only shows blink reconstruction, not any other forms of preprocessing. (**b**) Distribution of missing data, expressed as proportion of nan-samples per trial, for the different experimental conditions from Mathôt et al. (2019)

between conditions can directly result in blink-mediated differences in mean pupil size. Therefore, for our example study, we ensured that there was an approximately equal number of blinks in each condition by plotting a histogram with the proportion of missing data for each trial separately for each condition (Fig. 3a).

3.1.2 Dealing with Changes in Eye Position

Changes in eye position also result in both artifactual and real changes in recorded pupil size. The artifactual changes are related to the angle at which the eye-tracking camera records the pupil. Imagine that the participant looks straight into the lens; in this case, the camera has a perfect frontal view of the pupil, which therefore appears

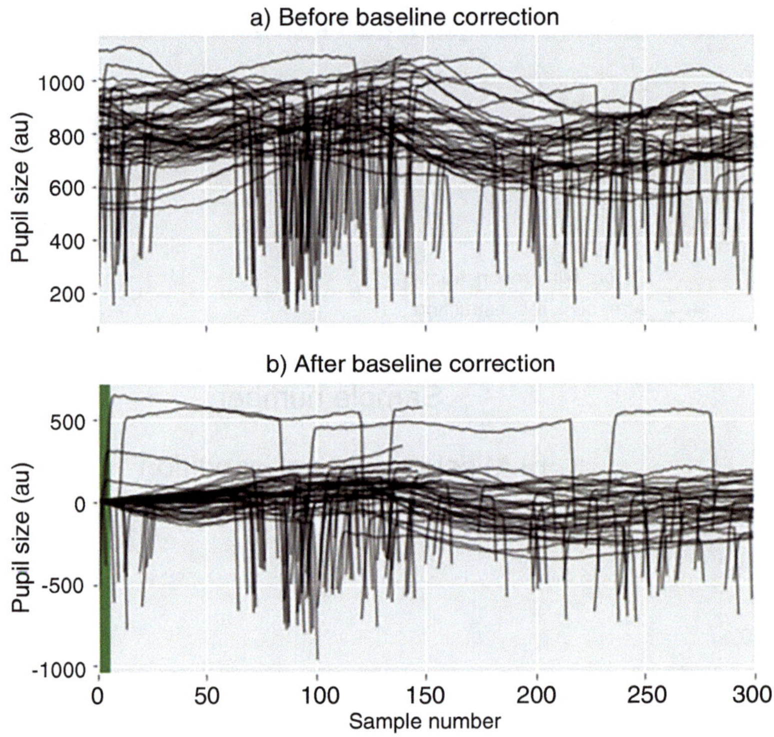

Fig. 3 Data from 50 trials from Mathôt et al. (2019). (**a**) Before baseline correction, pupil size at the onset of the relevant stimulus is different on each trial. (**b**) After baseline correction, pupil size is always 0 at the onset of the relevant stimulus. For these trials, the first five samples of the signal, indicated in green, were used as the baseline period. This panel only shows baseline correction, not any other forms of preprocessing

as a circle. Now imagine that, as is the case in many eye-tracking setups, the participant looks at a display positioned several tens of centimeters above the camera; in this case, the camera records the pupil from a slight angle, and the pupil therefore appears as an ellipse. Crucially, given constant pupil size, the surface of this ellipse will be slightly smaller than that of the circle; for many eye trackers, this translates to an artifactual decrease in recorded pupil size, a phenomenon referred to as the pupil foreshortening error (PFE).

Since the PFE is predictable and depends solely on the geometry of the recording set-up, a correction can be made for it. This correction usually takes the form of a linear model that predicts real pupil size as a function of horizontal eye position, vertical eye position, and recorded pupil size (Brisson et al., 2013). An alternative is to create a (non-linear) model that incorporates the relative positions of the eye tracker and the participant's eyes, and similarly uses this to predict real pupil size as a function of eye position and recorded pupil size (Gagl et al., 2011; Hayes & Petrov, 2016).

However, even after applying a correction for the PFE, pupil size still changes as a function of eye position in at least two ways. First, eye movements are followed by a pupil constriction (Knapen et al., 2016; Mathôt et al., 2015; Zuber et al., 1966), presumably—and just as with blinks—as a result of changes in visual input (Mathôt, 2020). This means that differences in the number of eye movements between conditions can directly result in differences in pupil size between conditions. Second, pupil size can actually depend on eye position; for example, Mathôt et al. (2015) observed that participants' pupils were slightly larger when participants fixated the top of the display, as compared to the bottom. This may have been related to the fact that, in the specific setup used in that study, fixating the top of the display was slightly less comfortable than fixating the bottom.

In our example study, eye movements played at most a minor role because the only stimuli on the display were centrally presented; therefore, we did not apply any PFE correction. Nevertheless, even in this design, participants made small eye movements to inspect different parts of the presented words, and they occasionally made larger eye movements towards objects in the laboratory.

In summary, eye movements can induce artifactual changes in recorded pupil size through the pupil foreshortening error (PFE), and real changes in pupil size through post-saccadic pupil constriction and real eye-position-dependent changes in pupil size.

3.2 Baseline Correction

The size of the pupil changes over time with slow, seemingly random fluctuations that are sometimes referred to as hippus or pupillary unrest. Because of this, in a trial-based experiment in which stimuli are presented many times over, pupil size at the onset of stimulus presentation changes from trial to trial. These fluctuations are a source of noise that reduce statistical power when you are interested in stimulus-evoked pupil responses, in which case you are interested in *changes* in pupil size rather than in absolute pupil size.

To reduce the effect of these trial-to-trial fluctuations in pupil size, you can, instead of looking at absolute pupil size, look at the *difference* in pupil size relative to a baseline period; here, the baseline period is usually a short moment before the presentation of the relevant stimulus. This is baseline correction (reviewed in Mathôt et al., 2018). Practically speaking, baseline correction entails calculating, for each trial separately, the mean (or median) pupil size during the baseline period and subtracting the resulting value from the entire pupil-size time series.

When selecting a baseline period, two considerations are important. First, the baseline period should itself not be affected by any experimental manipulations. Second, the baseline period should be right before the period of interest. In our example study, we applied baseline correction to pupil size during the presentation of the word stimulus, using the first 50 ms of that same period as the baseline period (Fig. 3). This is appropriate, because the pupil response has a response latency of about 200–250 ms (e.g., Ellis, 1981); therefore, pupil size during the first 50 ms after the onset of the word is not (yet) affected by the word itself, and can safely be used as a baseline period.

When applying baseline correction, artifacts during the baseline period can strongly impact the entire trial (Mathôt et al., 2018). For example, if a blink occurred during the baseline period of some trial, and if this blink was for some reason not properly reconstructed or removed, then baseline pupil size will be very small for that trial; consequently, baseline-corrected pupil size will be very large. Because of this, baseline correction can introduce, rather than remove, noise if there are many trials with artifacts during the baseline period. To ensure that this doesn't happen, it is good practice to plot a histogram of baseline pupil sizes and remove those trials on which baseline pupil size was unrealistically small or large. In our example study, we first converted pupil size from arbitrary units to millimeters of diameter (Mathôt & Ivanov, 2019), and then excluded trials in which baseline pupil size was less than 2 mm or more than 6.5 mm. These cut-offs were based on a visual inspection of the histogram shown in Fig. 4.

3.3 Choosing a Summary Statistic

For the purpose of data analysis, pupil size generally needs to be reduced to a measure that lends itself to statistical analysis. Phrased differently, you need to derive a summary statistic from your pupil-size data, and this summary statistic is subsequently used as the dependent measure in a statistical analysis.

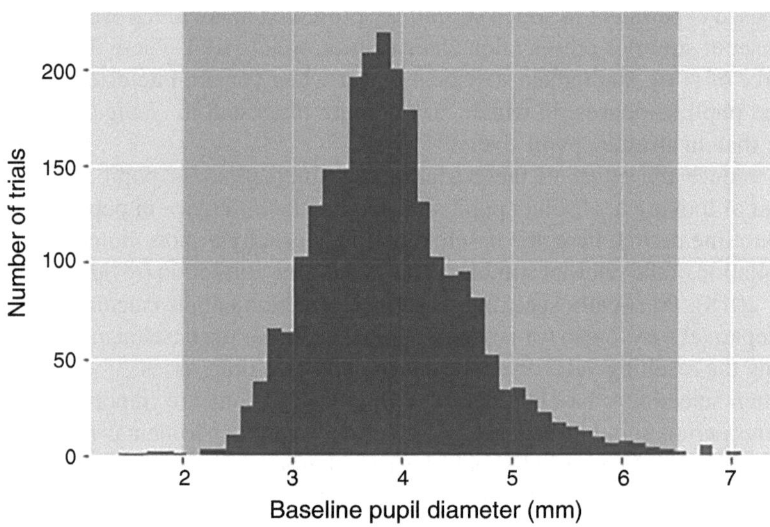

Fig. 4 A histogram of baseline pupil sizes for individual trials. (Data from Mathôt et al. (2019))

One common approach is to determine a single value for each trial. Most commonly, this single value is mean pupil size (or pupil change, when baseline correction is applied) during a period of interest; for example, Laeng and Sulutvedt (2014) analyzed mean pupil size during 5 s of mental imagery, when investigating whether mental imagery of bright environments results in pupil constriction compared to imagery of dark environments. However, there are a number of other measures that can also be used, such as peak pupil dilation (i.e., the maximum, rather than mean, pupil size); a more in-depth discussion of such measures can be found in the chapter by Reilly et al. (2019).

Another common approach is to analyze pupil size during each time point separately. This is the approach that we took for our example study, effectively resulting in 300 dependent measures, one for each 10 ms time for the 3000 ms window of word presentation. We then analyzed each of these samples with a separate statistical test. The advantage of this approach is that it's fairly easy to do. The disadvantage is that these many separate statistical tests need to be controlled for multiple comparisons. A crude form to approximate this is to take a minimum window during which the effect of interest must meet some criterion; specifically, for our example study, we specified that the t value for the difference in pupil size between brightness- and darkness-related words, as determined with a linear mixed-effects model, needed to exceed 1.96 for at least 200 ms. A criterion like this is convenient and can be reasonable for data that is smooth and does not span a long period, but, importantly, it is not a formal method for multiple-comparison correction. Better alternatives are cluster-based permutation testing (reviewed in Sassenhagen & Draschkow, 2019), and a recently proposed "flattened average" approach (Bowman et al., 2020).

3.4 Data Visualization

To effectively communicate results, data visualization is important. Just as there are many ways to reduce pupil-size data to a summary statistic, there are many ways to summarize pupil-size data for the purpose of visualization. In fact, both are related, because the way that data are visualized should ideally match the way that it has been analyzed statistically.

In our example study, we chose to visualize the result as a line plot in which each line corresponds to mean pupil size as a function of time for the three different word categories (Fig. 5). This preserves the time dimension in the visualization, which matches our analysis approach of conducting separate tests for each 10 ms window, as described above.

To illustrate the importance of preprocessing for data quality, Fig. 5a shows the results without any preprocessing applied; Fig. 5b shows the results after the following preprocessing steps in this order: blink reconstruction, conversion from arbitrary units to millimeters of diameter, baseline correction, and removal of trials with exceedingly small (<2 mm) or large (>6.5 mm) baselines. Only after preprocessing,

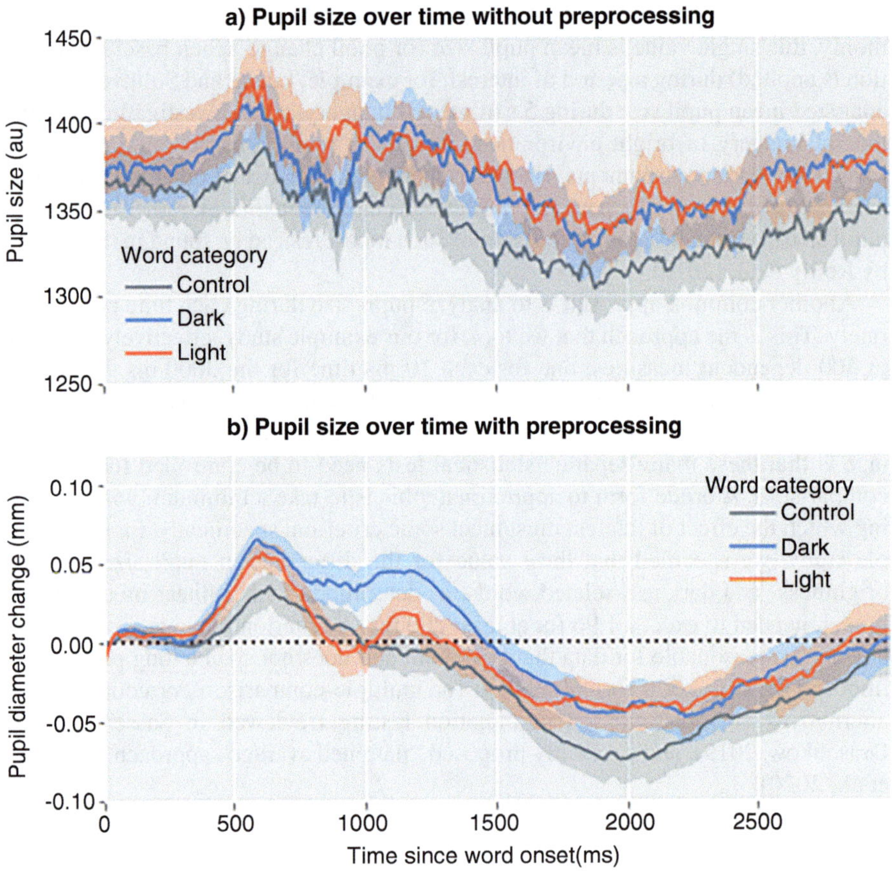

Fig. 5 A common way to visualize pupil-size data is through a line plot that shows plot pupil size over time, separately for each experimental condition. (**a**) The results from Mathôt et al. (2017) without any preprocessing. (**b**) The same results with preprocessing as described in the main text. Shadings around the lines indicate the grand standard error

the difference in pupil size in response to darkness-related and brightness-related words is apparent.

Figure 5 also highlights the importance of selecting a proper experimental contrast (see the section *Choosing your conditions* above). In the case of our example experiment, the relevant contrast is that between brightness- and darkness-related words, which were matched based on various properties, such as lexical frequency and visual intensity. The control words served as filler items that served to make it less obvious to participants that the key manipulation in the experiment was the semantic brightness of words; these control words were not matched to the words of interest, and consequently the associated pupil response is markedly different, characterized by an overall more pronounced constriction. Therefore, the brightness- and darkness-related words should be compared to each other, rather than to the

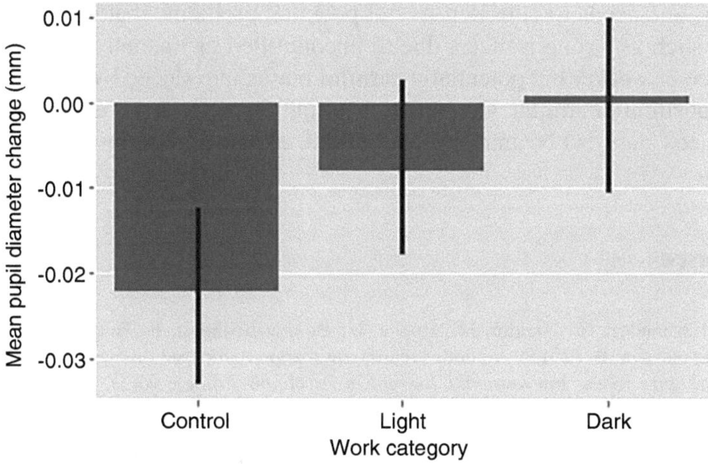

Fig. 6 Another common way to visualize pupil-size data is to present mean pupil size for different conditions in a bar plot or line plot. This figure shows the same data as Fig. 4b, after averaging over the full three seconds of word presentation

control words. (A related question is whether the results for the control words should be shown at all. Here we chose to show them for transparency, at a slight risk of confusing the reader.)

Another way to visualize pupil-size data is to plot mean pupil size averaged over time as a bar plot or line plot, where each bar or point represents a different condition (Fig. 6). The main advantage of this visualization approach is that it allows for the results of multifactorial designs to be visualized intuitively without overwhelming the reader with too much information. Another advantage is that such plots lend themselves well to also including information about individual subjects, for example in the form of a so-called violin plot. The disadvantage of this approach is that the time course of the effect is no longer visible, whereas in pupil-size experiments time is often a relevant dimension.

4 Conclusion

Pupillometry is a psychophysiological method of research that can be conducted with a variety of existing types of equipment, although the most common is the eye tracker. It is a quantitative method that provides measurements on an interval scale of the size of the eye opening. As for any other psychophysiological measurements, its validity depends on well-controlled and well-designed experiments, on the use of appropriate signal processing procedures to reduce noise and artifacts, and on the use of inferential statistical methods to compare pupil size deviations from other conditions, including baseline or control conditions. Thus, doing a pupillometry experiment is no different from hypothesis testing in any field of empirical science.

However, one needs to consider several potential problems that are specific to the method, such as strong artifacts due to uncontrolled or uncontrollable changes in luminance or weaker but potentially harmful biases introduced by gaze movements and the position of stimuli. Given that the pupil diameter is a continuous measure, pupillometry data can be analyzed and visualized as a time series variable.

References

Aguillon-Hernandez, N., Naudin, M., Roché, L., Bonnet-Brilhault, F., Belzung, C., Martineau, J., & Atanasova, B. (2015). An odor identification approach based on event-related pupil dilation and gaze focus. *International Journal of Psychophysiology, 96*(3), 201–209. https://doi.org/10.1016/j.ijpsycho.2015.03.009

Aissou, M., Snauwaert, A., Dupuis, C., Atchabahian, A., Aubrun, F., & Beassier, M. (2012). Objective assessment of the immediate postoperative analgesia using pupillary reflex measurement: A prospective and observational study. *Anesthesiology, 116*, 1006–1012.

Aminihajibashi, S., Hagen, T., Folldal, M. D., Laeng, B., & Espeseth, T. (2019). Individual differences in resting-state pupil size: Evidence for association between working memory capacity and pupil size variability. *International Journal of Psychophysiology, 140*(1–7), 1. https://doi.org/10.1016/j.ijpsycho.2019.03.007

Aminihajibashi, S., Hagen, T., Andreassen, O. A., Laeng, B., & Espeseth, T. (2020). The effects of cognitive abilities and task demands on tonic and phasic pupil sizes. *Biological Psychology, 156*, 107945. https://doi.org/10.1016/j.biopsycho.2020.107945

Barbur, J. (2004). Learning from the pupil: Studies of basic mechanisms and clinical applications. *The Visual Neurosciences, 1*, 641–656.

Bergamin, O., Zimmerman, M. B., & Kardon, R. H. (2003). Pupil light reflex in normal and diseased eyes: Diagnosis of visual dysfunction using waveform partitioning. *Ophthalmology, 110*(1), 106–114. https://doi.org/10.1016/s0161-6420(02)01445-8

Bishop, L., Jensenius, A. R., & Laeng, B. (2021). Musical and bodily predictors of mental effort in string quartet music. *Frontiers in Psychology, 12*, 653021. https://doi.org/10.3389/fpsyg.2021.653021

Bowman, H., Brooks, J. L., Hajilou, O., Zoumpoulaki, A., & Litvak, V. (2020). Breaking the circularity in circular analyses: Simulations and formal treatment of the flattened average approach. *PLoS Computational Biology, 16*(11), e1008286. https://doi.org/10.1371/journal.pcbi.1008286

Brisson, J., Mainville, M., Mailloux, D., Beaulieu, C., Serres, J., & Sirois, S. (2013). Pupil diameter measurement errors as a function of gaze direction in corneal reflection eyetrackers. *Behavior Research Methods, 45*(4), 1322–1331.

Brysbaert, M., & Stevens, M. (2018). Power analysis and effect size in mixed effects models: A tutorial. *Journal of Cognition, 1*(1), 9. https://doi.org/10.5334/joc.10. PMID: 31517183; PMCID: PMC6646942.

Carnegie, E., Marchant, D., Towers, S., & Ellison, P. (2020). Beyond visual fixations and gaze behavior: Using pupillometry to examine the mechanisms in the planning and motor performance of a golf putt. *Human Movement Science, 71*, 102622. https://doi.org/10.1016/j.humov.2020.102622

Chiew, K. S., & Braver, T. S. (2013). Temporal dynamics of motivation–cognitive control interactions revealed by high-resolution pupillometry. *Frontiers in Psychology, 4*, 15. https://doi.org/10.3389/fpsyg.2013.00015

Connelly, M. A., Brown, J. T., Kearns, G. L., Anderson, R. A., St Peter, S. D., & Neville, K. A. (2014). Pupillometry: A non-invasive technique for pain assessment in paediatric patients. *Archives of Disease in Childhood, 99*(12), 1125–1131. https://doi.org/10.1136/archdischild-2014-306286

DiNuzzo, M., Mascali, D., Moraschi, M., Bussu, G., Maugeri, L., Mangini, F., Fratini, M., & Giove, F. (2019). Brain networks underlying Eye's pupil dynamics. *Frontiers in Neuroscience, 13*, 965. https://doi.org/10.3389/fnins.2019.00965

Ehlers, J., Strauch, C., Georgi, J., & Huckauf, A. (2016). Pupil size changes as an active Information Channel for biofeedback applications. *Applied Psychophysiology and Biofeedback, 41*, 331–339. https://doi.org/10.1007/s10484-016-9335-z

Einhäuser, W., Stout, J., Koch, C., & Carter, O. (2008). Pupil dilation reflects perceptual selection and predicts subsequent stability in perceptual rivalry. *Proceedings of the National Academy of Sciences of the United States of America, 105*(5), 1704–1709. https://doi.org/10.1073/pnas.0707727105

Ekman, I., Poikola, A., Mäkäräinen, M., Takala, T., & Hämäläinen, P. (2008). Voluntary pupil size change as control in eyes only interaction. In *Proceedings of the 2008 symposium on eye tracking research & applications* (pp. 115–118). ACM.

Ellingsen, D. M., Wessberg, J., Chelnokova, O., Olausson, H., Laeng, B., & Leknes, S. (2013). In touch with your emotions: Oxytocin and touch change social impressions. *Psychoneuroendocrinology, 39*, 11–20. https://doi.org/10.1016/j.psyneuen.2013.09.017

Ellis, C. J. (1981). The pupillary light reflex in normal subjects. *British Journal of Ophthalmology, 65*(11), 754–759. https://doi.org/10.1136/bjo.65.11.754

Endestad, T., Godøy, R. I., Sneve, M. H., Hagen, T., Bochynska, A., & Laeng, B. (2020). Mental effort when playing, listening, and imagining music in one Pianist's eyes and brain. *Frontiers in Human Neuroscience, 14*, 576888. https://doi.org/10.3389/fnhum.2020.576888

Franklin, M. S., Broadway, J. M., Mrazek, M. D., Smallwood, J., & Schooler, J. W. (2013). Window to the wandering mind: Pupillometry of spontaneous thought while reading. *Quarterly Journal of Experimental Psychology, 66*(12), 2289–2294. https://doi.org/10.1080/17470218.2013.858170

Gagl, B., Hawelka, S., & Hutzler, F. (2011). Systematic influence of gaze position on pupil size measurement: Analysis and correction. *Behavior Research Methods, 43*(4), 1171–1181. https://doi.org/10.3758/s13428-011-0109-5

Hayes, T. R., & Petrov, A. A. (2016). Mapping and correcting the influence of gaze position on pupil size measurements. *Behavior Research Methods, 48*(2), 510–527. https://doi.org/10.3758/s13428-015-0588-x

Hershman, R., Henik, A., & Cohen, N. (2018). A novel blink detection method based on pupillometry noise. *Behavior Research Methods, 50*(1), 107–114. https://doi.org/10.3758/s13428-017-1008-1

Hess, E. H., & Polt, J. M. (1960). Pupil size as related to interest value of visual stimuli. *Science, 132*, 349–350.

Hess, E. H., & Polt, J. M. (1964). Pupil size in relation to mental activity during simple problem solving. *Science, 140*, 1190–1192. https://doi.org/10.1126/science.143.3611.1190

Hess, E. H., & Polt, J. M. (1966). Changes in pupil size as a measure of taste difference. *Perceptual and Motor Skills, 23*(2), 451–455.

Jainta, S., & Baccino, T. (2010). Analyzing the pupil response due to increased cognitive demand: An independent component analysis study. *International Journal of Psychophysiology, 77*(1), 1–7. https://doi.org/10.1016/j.ijpsycho.2010.03.008

Juvrud, J., Gredebäck, G., Åhs, F., Lerin, N., Nyström, P., Kastrati, G., & Rosén, J. (2018). The immersive virtual reality lab: Possibilities for remote experimental manipulations of autonomic activity on a large scale. *Frontiers in Neuroscience, 12*. https://www.frontiersin.org/article/10.3389/fnins.2018.00305

Kahneman, D. (1973). *Attention and effort*. Prentice Hall.

Kahneman, D., & Beatty, J. (1966). Pupil diameter and load on memory. *Science, 154*, 1583–1585. http://www.jstor.org/stable/1720478

Knapen, T., de Gee, J. W., Brascamp, J., Nuiten, S., Hoppenbrouwers, S., & Theeuwes, J. (2016). Cognitive and ocular factors jointly determine pupil responses under equiluminance. *PLoS One, 11*(5), e0155574. https://doi.org/10.1371/journal.pone.0155574

Laeng, B., & Endestad, T. (2012). Bright illusions reduce the eye's pupil. *Proceedings of the National Academy of Sciences USA, 109*, 2162–2167.

Laeng, B., & Sulutvedt, U. (2014). The eye pupil adjusts to imaginary light. *Psychological Science, 25*(1), 188–197. https://doi.org/10.1177/0956797613503556

Laeng, B., Ørbo, M., Holmlund, T., & Miozzo, M. (2011). Pupillary Stroop effects. *Cognitive Processing, 12*, 13–21.

Laeng, B., Sirois, S., & Gredebäck, G. (2012). Pupillometry: A window to the preconscious? *Perspectives on Psychological Science, 7*, 18–27.

Laeng, B., & Alnæs, D. (2019). Pupillometry. In Klein, Christoph & Ettinger, Ulrich (Ed.), Eye Movement Research. Springer Nature.

Laeng, B., Eidet, L. M., Sulutvedt, U., & Panksepp, J. (2016). Music chills: The eye pupil as a mirror to music's soul. *Consciousness and Cognition, 44*, 161–178. https://doi.org/10.1016/j.concog.2016.07.009

Laeng, B., Nabil, S., & Kitaoka, A. (2022). The Eye Pupil Adjusts to Illusorily Expanding Holes. *Frontiers in Human Neuroscience, 16*, 877249. https://doi.org/10.3389/fnhum.2022.877249

Levatin, P. (1959). Pupillary escape in disease of the retina or optic nerve. *Archives of Ophthalmology, 62*, 768–799.

Liu, H., Laeng, B., & Czajkowski, N. O. (2020). Does stereopsis improve face identification? A study using a virtual reality display with integrated eye-tracking and pupillometry. *Acta Psychologica, 210*, 103142. https://doi.org/10.1016/j.actpsy.2020.103142

Loewenfeld. (1958). Mechanisms of reflex dilatation of the pupil. Historical reviewand experimental analysis. *Documenta Ophthalmologica, 12*, 185.

Loewenfeld, I. (1993). *The pupil: Anatomy, physiology, and clinical applications*. Wayne State University Press.

Lowenstein, O. (1954). Clinical pupillary symptoms in lesions of the optic nerve, optic chiasm, and optic tract. *AMA Archives of Ophthalmology 52*(3), 385–403. https://doi.org/10.1001/archopht.1954.00920050387006. PMID: 13188481.

Lowenstein, O., & Loewenfeld, I. E. (1958). Electronic pupillography. A new instrument and some clinical applications. *AMA Archives of Ophthalmology, 59*, 352.

Loweth, H. A. (1973). A device for continuous monitoring of pupil dilation. *Behavior Research Methods & Instrumentation, 5*, 473–474. https://doi.org/10.3758/BF03200237

MacLachlan, C., & Howland, H. C. (2002). Normal values and standard deviations for pupil diameter and interpupillary distance in subjects aged 1 month to 19 years. *Ophthalmic and Physiological Optics, 22(3)*, 175–182. https://doi.org/10.1046/j.1475-1313.2002.00023.x

Mani, N., Schreiner, M. S., Brase, J., Köhler, K., Strassen, K., Postin, D., & Schultze, T. (2021). Sequential Bayes factor designs in developmental research: Studies on early word learning. *Developmental Science, 24*(4), e13097. https://doi.org/10.1111/desc.13097

Mathôt, S. (2013). A simple way to reconstruct pupil size during eye blinks (Version1). *Figshare*. https://doi.org/10.6084/m9.figshare.688001.v1

Mathôt, S. (2020). Tuning the senses: How the pupil shapes vision at the earliest stage. *Annual Review of Vision Science, 6*, 433–451.

Mathôt, S., & Ivanov, Y. (2019). The effect of pupil size and peripheral brightness on detection and discrimination performance. *PeerJ, 7*, e8220. https://doi.org/10.7717/peerj.8220

Mathôt, S., Melmi, J. B., & Castet, E. (2015). Intrasaccadic perception triggers pupillary constriction. *PeerJ, 3*(e1150), 1–16. https://doi.org/10.7717/peerj.1150

Mathôt, S., Grainger, J., & Strijkers, K. (2017). Pupillary responses to words that convey a sense of brightness or darkness. *Psychological Science, 28*(8), 1116–1124. https://doi.org/10.1177/0956797617702699. Epub 2017 Jun 14. PMID: 28613135; PMCID: PMC5549816

Mathôt, S., Fabius, J., Heusden, E. V., & der Stigchel, S. V. (2018). Safe and sensible preprocessing and baseline correction of pupil-size data. *Behavior Research Methods, 4*, 1–13. https://doi.org/10.3758/s13428-017-1007-2

Mathôt, S., Sundermann, L., & Rijn, H. van. (2019). The effect of semantic brightness on pupil size: A replication with Dutch words. *bioRxiv*. https://doi.org/10.1101/689265.

Miles, K., McMahon, C., Boisvert, I., Ibrahim, R., de Lissa, P., Graham, P., & Lyxell, B. (2017). Objective assessment of listening effort: Coregistration of Pupillometry and EEG. *Trends Hear, 21*, 2331216517706396. https://doi.org/10.1177/2331216517706396

Naber, M., & Nakayama, K. (2013). Pupil responses to high-level image content. *Journal of Vision, 13*(6), 7. http://www.journalofvision.org/content/13/6/7.full

Pearl, J. (2009). Causal inference in statistics. *Statistics Surveys, 3*, 96–146. https://doi.org/10.1214/09-SS057

Pelagatti, C., Binda, P., & Vannucci, M. (2020). A closer look at the timecourse of mind wandering: Pupillary responses and behaviour. *PLoS One, 15*(4), e0226792. https://doi.org/10.1371/journal.pone.0226792

Pirenne, M. H., & Denton, E. J. (1952). Accuracy and sensitivity of the human eye. *Nature, 170*(4338), 1039–1042.

Phillips, S. S., Mueller, C. M., Nogueira, R. G., & Khalifa, Y. M. (2019). A systematic review assessing the current state of automated pupillometry in the NeuroICU Neurocrit. Care, pp. 1–20.

Reilly, J., Kelly, A., Kim, S. H., Jett, S., & Zuckerman, B. (2019). The human task-evoked pupillary response function is linear: Implications for baseline response scaling in pupillometry. *Behavior Research Methods, 51(2)*, 865–878. https://doi.org/10.3758/s13428-018-1134-4

Rieger, G., & Savin-Williams, R. C. (2012). The eyes have it: Sex and sexual orientation differences in pupil dilation patterns. *PLoS One, 7*(8), e40256. https://doi.org/10.1371/journal.pone.0040256

Rodney B., Murray, M. W., Adler, A., & Korczyn D. (1983). *The Pupillary Effects of Oploids Life Sciences, 33*(6), 495–509. https://doi.org/10.1016/0024-3205(83)90123-6

Rosnow, R., & Rosenthal, R. (1997). *People studying people: Artifacts and ethics in behavioral research.* W. H. Freeman and Company.

Sassenhagen, J., & Draschkow, D. (2019). Cluster-based permutation tests of MEG/EEG data do not establish significance of effect latency or location. *Psychophysiology, 56*(6), e13335.

Siegle, G. J., Ichikawa, N., Steinhauer, S. (2008). Blink before and after you think: blinks occur prior to and following cognitive load indexed by pupillary responses. *Psychophysiology, 45*(5), 679–687. https://doi.org/10.1111/j.1469-8986.2008.00681.x. Epub 2008 Jul 24. PMID: 18665867.

Smilek, D., Carriere, J. S. A., & Cheyne, J. A. (2010). Out of mind, out of sight: Eye blinking as indicator and embodiment of mind wandering. *Psychological Science, 21*, 786–789.

Stern, J. A., Boyer, D., & Schroeder, D. (1994). Blink rate: A possible measure of fatigue. *Human Factors, 36*(2), 285–297. https://doi.org/10.1177/001872089403600209

Tamietto, M., Castelli, L., Vighetti, S., Perozzo, P., Geminiani, G., Weiskrantz, L., & de Gelder, B. (2009). Unseen facial and bodily expressions trigger fast emotional reactions. *Proceedings of the National Academy of Sciences of the USA, 106*(42), 17661–1766. https://doi.org/10.1073/pnas.0908994106. Epub 2009 Oct 5. PMID: 19805044; PMCID: PMC2764895.

Thompson, H. S., & Kardon, R. H. (2006). The Argyll Robertson pupil. *Journal of Neuro-Ophthalmology, 26*(2), 134–138. https://doi.org/10.1097/01.wno.0000222971.09745.91

Unsworth, N., & Robison, M. K. (2017). The importance of arousal for variation in working memory capacity and attention control: A latent variable pupillometry study. *Journal of Experimental Psychology: Learning, Memory, and Cognition, 43*(12), 1962–1987. https://doi.org/10.1037/xlm0000421

Unsworth, N., & Robison, M. K. (2018). Tracking arousal state and mind wandering with pupillometry. *Cognitive, Affective, & Behavioral Neuroscience, 18*(4), 638–664. https://doi.org/10.3758/s13415-018-0594-4

Unsworth, N., Robison, M. K., & Miller, A. L. (2019). Individual differences in baseline oculometrics: Examining in baseline pupil diameter, spontaneous eye blink rate, and fixation stability. *Cognitive, Affective, & Behavioral Neuroscience, 20*, 1074. https://doi.org/10.3758/s13415-019-00709-z

Wardhani, I. K., Mathôt, S., Boehler, C. N., & Laeng, B. (2020). Effects of nicotine on pupil size and performance during multiple-object tracking in non-nicotine users. *International Journal of Psychophysiology, 158*, 45–55. https://doi.org/10.1016/j.ijpsycho.2020.09.005. Epub Oct 17. PMID: 33080291.

Weiskrantz, L., Cowey, A., & Barbur, J. L. (1999). Differential pupillary constriction and awareness in the absence of striate cortex. *Brain, 122*, 1533–1538.

Wilhelm, B., Wilhelm, H., & Lüdtke, H. (1999). Pupillography: Principles and applications in basic and clinical research. In J. Kuhlmann & M. Böttcher (Eds.), (pp. 1–10). München, Germany: Zuckscjwerdt Verlag.

Whipple, B., Ogden, G., & Komisaruk, B. R. (1992). Physiological correlates of imagery-induced orgasm in women. *Archives of Sexual Behavior, 21*, 121–133.

Winn, M. B., Wendt, D., Koelewijn, T., & Kuchinsky, S. E. (2018). Best practices and advice for using Pupillometry to measure listening effort: An introduction for those who want to get started. *Trends in Hearing, 22*, 2331216518800869. https://doi.org/10.1177/2331216518800869

Zhou, C., Lorist, M. M., & Mathôt, S. (2020). *Eye movements in real-life search are guided by task-irrelevant working-memory content* (p. 2020.05.18.101410). https://doi.org/10.1101/2020.05.18.101410

Zuber, B. L., Stark, L., & Lorber, M. (1966). Saccadic suppression of the pupillary light reflex. *Experimental Neurology, 14*(3), 351–370. https://doi.org/10.1016/0014-4886(66)90087-2

Zwaan, R. A. (2014). Embodiment and language comprehension: Reframing the discussion. *Trends in Cognitive Sciences, 18*(5), 229–234. https://doi.org/10.1016/j.tics.2014.02.008

A Primer on Design and Data Analysis for Cognitive Pupillometry

Jamie Reilly, Bonnie Zuckerman, and Alexandra E. Kelly

1 Introduction

This chapter is focused on cognitive pupillometry, a set of methods used for contrasting small perturbations in the surface area of the pupil evoked by cognitive demands. Cognitive pupillometry historically demanded expertise in optics, psychophysics, and engineering. Scientists were once compelled to construct their own elaborate systems for presenting stimuli and measuring pupil responses using analog film capture with sampling rates of about five images per second. Many of these early systems involved jury-rigged arrays of pulleys, mirrors, winches, and bite-bars. In contrast, infrared eyetracking systems are capable of remote recording with sampling rates exceeding 1000 Hz. Much of the hardware used in modern pupillometry is commercially available and guided by intuitive point-and-click graphical user interfaces. In addition, researchers now have access to a variety of open source software packages dedicated to automated processing and statistical analyses of pupillometry data. These advances have made pupillometry a widely accessible tool whose popularity has extended beyond esoteric corners of psychophysics to a much broader range of applied cognitive science. Figure 1 illustrates this trend of exponential growth in citation counts for pupillometry in PubMed indexed articles across the interval 1960–2020.

The original version of the chapter has been revised. A correction to this chapter can be found at https://doi.org/10.1007/978-3-031-54896-3_16

J. Reilly (✉) · B. Zuckerman
Eleanor M. Saffran Center for Cognitive Neuroscience, Philadelphia, PA, USA

Department of Communication Sciences and Disorders, Temple University, Philadelphia, PA, USA
e-mail: Jamie_reilly@temple.edu

A. E. Kelly
Department of Psychology, Drexel University, Philadelphia, PA, USA

M. H. Papesh, S. D. Goldinger (eds.), *Modern Pupillometry*,
https://doi.org/10.1007/978-3-031-54896-3_14

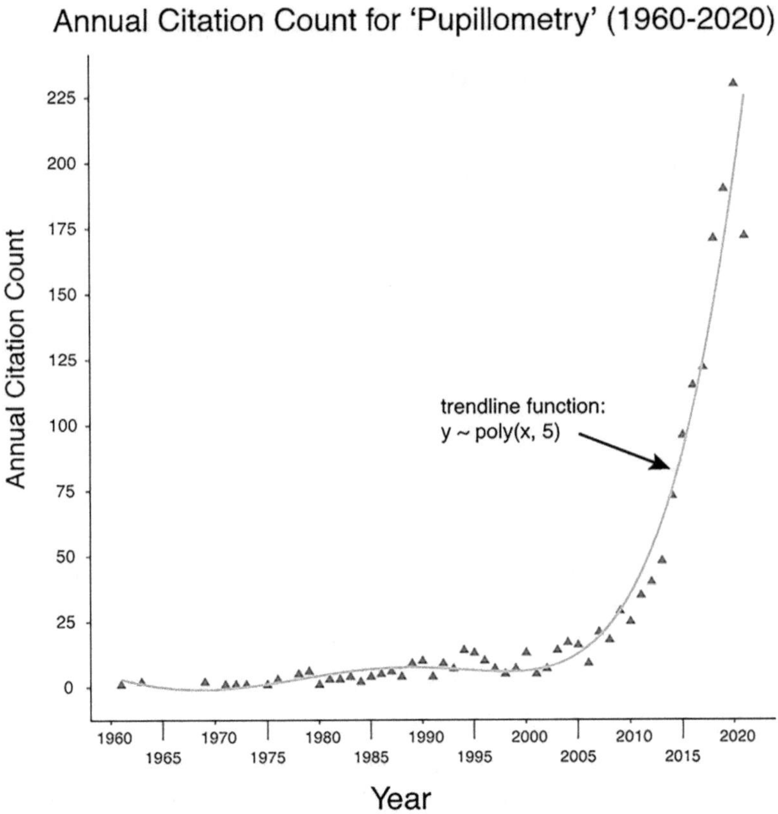

Fig. 1 PubMed citation counts by year for pupillometry-related search terms

Researchers unfamiliar with pupillometry may hold naïve biases about the complexity of this measurement tool. After all, pupillary data reflect the dynamics of a single channel fluctuating relatively slowly over time. How hard can that be? It may be true that collecting and analyzing raw pupillometry data is not particularly challenging. However, executing a valid and reliable cognitive pupillometry study is quite difficult. Our aims here are to introduce methodological challenges and identify solutions to common pitfalls in experimental design, execution, and analysis of this multifaceted neurobiological signal.

1.1 Physiological and Behavioral Indices of Pupillary Response Functions

The morphology of the pupil and the robustness of its response to light are well-established markers of neurological and ocular pathology. For example, fixed and dilated pupils are a symptom of brainstem dysfunction incurred in severe head trauma and disorders of consciousness (Hoffmann et al., 2012; Jennett & Teasdale,

1977; Marmarou et al., 2007). Pinpoint pupils may indicate acute opiate intoxication or chemical pesticide exposure (Davies et al., 1975; Larson, 2008; Rengstorff, 1994; Rollins et al., 2014), and acute anisochoria (i.e., asymmetric pupil size) may suggest the presence of a unilateral brain tumor or glaucoma (Lam et al., 1987). These conditions highlight the utility of clinical pupillometry as a tool for inferring disease states within the eye(s) and/or the brain that guide differential diagnosis and medical management.[1] Clinical pupillometry typically involves measurement of macroscale features such as the shape of the pupil or its responsiveness to light. Many of these characteristics are observable with the naked eye or using simple handheld magnification (e.g., ophthalmoscope). In the chapter to follow, we will describe challenges involved in measuring a far more subtle pupillary response.

Eckard Hess and James Polt (1960) introduced the English-speaking scientific community to a new neurophysiological response function. In this study, the authors filmed the pupils of six adult men and women as they viewed five photographs comprised of (1) neutral landscape, (2) baby, (3) mother + baby, (4) partially nude female, (5) partially nude male. The remarkable finding was that small fluctuations in pupil size (assessed by % change) were evoked by "interest value" of the stimuli. Female participants showed the highest peak dilation when viewing the partially nude male photo, whereas male participants showed the opposite pattern. Crucially, Hess and Polt controlled illuminance during the experiment. This allowed the authors to isolate a response evoked by cognitive factors rather than light reflexes. Hess and Polt (1960) touted the far-reaching implications of this biological signal, and a star was born.

Throughout the latter half of the twentieth century, pupillometry research has gradually shifted focus from more nebulous constructs such as "interest level" to better operationalized variables such as physiological arousal (Beatty, 1982; Bradley et al., 2008; Nassar et al., 2012; Peysakhovich et al., 2017). We have since learned that the human pupil dilates in response to a vast range of cognitive and perceptual challenges, including memory encoding and retrieval (Goldinger & Papesh, 2012, 2013; Papesh et al., 2012; Papesh & Goldinger, 2015), effortful listening while perceiving speech in noise (Kuchinsky et al., 2014; Van Engen & McLaughlin, 2018; Zekveld et al., 2010, 2011, 2014; Zekveld & Kramer, 2014), and difficulty manipulations during mental arithmetic (Causse et al., 2017). In fact, Tryon (1975) compiled a non-exhaustive list of 23 sources of pupil variation with the caveat that many more sources of variability lurk beneath the surface.

1.2 *Cognitive Pupillometry*

Pupillometry experienced a boom throughout the latter half of the twentieth century. Beatty (1982) coined the term task-evoked pupillary response or TEPR as a descriptive label for this particular response function. Many researchers have since adopted this nomenclature, and "TEPR" remains in common use today. Nevertheless, the

[1] For reviews of clinical pupillometry and applications, see Barbur (2004), Bremner (2009)

"TEPR" distinction has not been met with universal acceptance. The "task-evoked" component of the TEPR implies a discrete, exogeneous demand such as detecting a target word or solving a math problem. Yet, TEPR-like activation can also be observed in the absence of an external stimulus. Here we will collectively refer to such pupillary responses using the broader distinction of *cognitive pupillometry.*

Within the past 20 years, many disciplines including psychophysics, cognitive psychology, and cognitive neuroscience adopted pupillometry as a go-to tool. The promise of this technique was that non-invasive measurement of the surface of the eye could yield a proxy measure of brain activity within noradrenergic brainstem nuclei such as the locus coeruleus (LC) that modulate tonic and phasic arousal (Aston-Jones & Cohen, 2005; Gilzenrat et al., 2010; Mathôt, 2018; Wang & Munoz, 2015). Coupling between the pupil and LC has previously been demonstrated within macaque using combinations of both invasive and non-invasive neurophysiological recording techniques (Aston-Jones & Cohen, 2005; Joshi et al., 2016). Recent efforts have leveraged multimodal human neuroimaging with simultaneous pupil recording to elucidate the functional architecture of the ascending arousal network (Elman et al., 2017; Murphy et al., 2014; Wainstein et al., 2021).

Cognitive pupillometry experiments typically employ event-related designs wherein stimuli from two or more conditions are interspersed at jittered intervals. Consider, for example, a hypothetical experimental design where a psycholinguist wishes to test her hypothesis that verbs have higher processing demands than nouns in the context of background noise. She plans to test this hypothesis via a lexical decision experiment where participants hear nouns, verbs, nonwords, and filler words in the context of interfering background noise as pupil size is continuously recorded. When she inspects the raw data, she will see a noisy, possibly non-stationary time series (i.e., rising or falling baselines) littered with missing and other complex artifacts. Figure 2 illustrates a raw pupillary time series continuously recorded from the left eye of a single subject in our own laboratory. In this particular example, the participant heard stimuli corresponding to two conditions. One condition reflected true statements about the world (e.g., Paris is the capitol of France), whereas the other condition involved false statements about the world matched in length to the true statements (e.g., Paris is the capitol of Italy). Our aim in this ongoing study is to contrast pupil responses for true and false statements as a potential screening tool for language comprehension in severe brain injury. Figure 2 gives the reader a sneak peek at what raw pupillometry data look like sampled over a single 6 min session for one participant.[2]

There is great heterogeneity across pupillometry studies as to what constitutes an event and how to window its subsequent pupil response. In a continuous pupillometry study, pupil dilation is modeled over an extended response interval such as during mental arithmetic (Hess & Polt, 1964; Klingner et al., 2011) or during creative problem solving tasks (Bradshaw, 1968; de Rooij et al., 2018). In contrast, discrete pupillometry involves time-locking a pupil response to an infinitesimally brief stimulus. This discrete approach is analogous to deconvolution techniques for analyzing hemodynamic response function (HRF) (Gitelman et al., 2003). In ERP and fMRI time series analyses, the stimuli are typically treated as delta functions.

[2] Raw data in pupillometry are typically a hot mess (see Fig. 2).

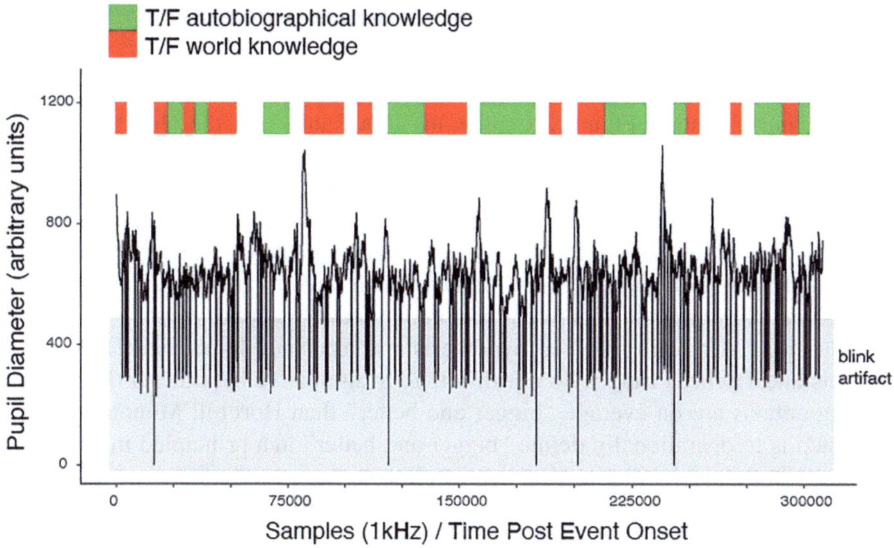

Fig. 2 Raw single-subject data in an event-related pupillometry study*Note*: These data reflect a session of a single-subject raw pupil time series sampled from the participant's left eye via an Eyelink 1000 Plus eyetracker (1000 Hz sampling rate). The color bar reflects points in the time series corresponding to event onsets within two experimental conditions. The experimental conditions involve true or false statements about the world versus the self

All subsequent activation is time-locked to that particular event onset/offset. This method has allowed researchers to develop mathematical basis functions for many biological signals. For example, the canonical HRF has a characteristic wave peaking about 6 s after event onset/offset followed by a slow decay to baseline after about 16 s (Handwerker et al., 2004).

Much remains to be learned about the pupil response function including how its parameters are impacted by individual differences. However, some of its global features have gained mainstream acceptance as canonical in nature (Denison et al., 2020). In a typical pupillometry experiment, researchers contrast pupil change per unit time normalized to a 0 mm baseline pupil size. When evoked change in pupil size (Y) is plotted against time (X), a canonical pupil response function bears resemblance to a time-compressed HRF or a slightly positively skewed mountain.

This chapter will focus primarily on optimizing study designs to evoke discrete pupillary response functions. Essentially, the analysis goal of cognitive pupillometry involves determining whether two or more composite mountains differ from each other. In the section to follow, we describe complexities, obstacles, and assumptions involved in the technical challenge of contrasting two composite mountains.

1.3 The Two Mountain Problem: A Measurement Metaphor

Consider a scenario where you live on a two-dimensional planet with two prominent mountain ranges. Figure 3 depicts the Ibis Range and the Hornbill Range. Mountain climbers flock to the Hornbill Range, whereas the Ibis Range is less frequented. Neither mountain range has been mapped, nor is it feasible to catalogue the entire population of mountains. Ibis Tourism Adventures, Inc. has hired you to substantiate their claim that the Ibis Range has "bigger and better" mountains than the Hornbill Range. Your budget for undertaking this investigation is enormous, and your client expects a full scientific report in 6 months. Where to begin?

You might start with a formal hypothesis test. Your null hypothesis (H_0) is that the Ibis and Hornbill ranges do not differ. Your alternative hypothesis (H_1) is that Ibis mountains are on average "bigger and better" than Hornbill Mountains. Your next step is to operationally define "bigger and better" in a principled manner that promotes both falsifiability and replication. You settle upon the following measurement parameters for "bigger and better" illustrated in Fig. 4:

1. *Base-to-Peak Altitude*: "Height" of the mountain at its tallest point normalized for the initial altitude of the mountain (i.e., raw peak altitude minus raw initial altitude) (y_{max}).
2. *Horizontal Distance to Peak*: Horizontal distance from the origin (0,0) to the point on the x-axis corresponding to the mountain's peak altitude (x_{obs} at y_{max}).
3. *Horizontal Distance to Baseline*: Horizontal distance from the initial rise of the mountain (0,0) until it descends to baseline or alternatively plateaus.
4. *Average Altitude*: The average "height" of the mountain across all x-values normalized for initial starting altitude (i.e., raw peak altitude minus raw starting altitude).
5. *Area Under the Curve*: Total approximate area of the mountain from initial rise to return to baseline.

At this point in your investigation, you established a guiding hypothesis, operationalized an abstract construct (e.g., "bigger and better"), and identified a set of

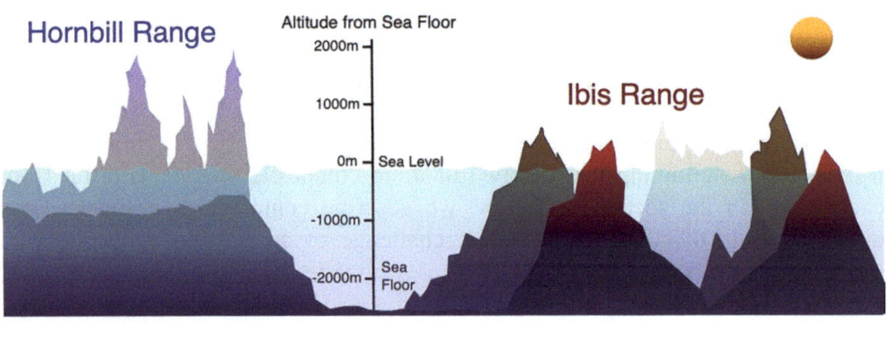

Fig. 3 The Two Mountain Problem

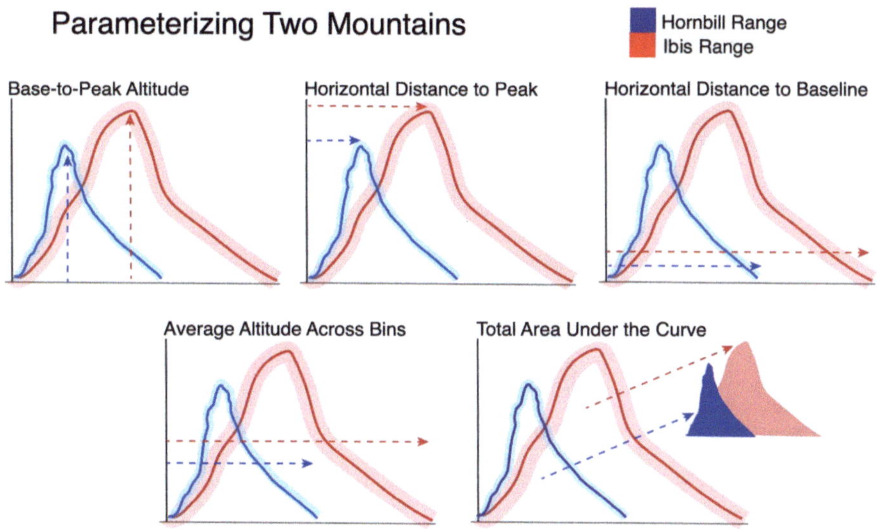

Fig. 4 How to measure a mountain

objectively measurable variables. You next devise sampling procedures that will yield accurate and unbiased estimates of the mountain ranges. GPS-equipped drones and remote submersibles are soon dispatched to the respective mountain ranges to sample altitude (y-axis) and global position (x-axis) at many points. Your hope is that extensive sampling will promote excellent source reconstruction and that your margin of measurement error will be small. However, these hopes are soon dashed when you learn of a high rate of equipment loss from winds, snow squalls, and giant squids. Your recovered data are punctuated by strange artifacts and missingness. What next?

Your next step involves data cleaning with the goal of removing outliers and imputing missing observations. Once these preprocessing steps are compete, you are now prepared to statistically evaluate your original hypothesis. You run statistical tests that are sensitive to autocorrelation (e.g., one point on a mountain is not independent of the previous point) and overfitting and interpret the results. Your data processing pipeline is illustrated in Fig. 5.

Despite being over 80% submerged, the Ibis Mountains have a higher base-to-peak altitude and greater total area than Hornbill Mountains. In contrast, Hornbill Mountains have steeper slopes. You report these findings to your client, and your job is done. Cognitive pupillometry shares many similarities with the Two Mountain Problem. When modeled as an event-related design, researchers typically examine relative change in pupil size evoked by discrete events. Each event in a pupillometry experiment corresponds to an individual mountain in the Two Mountain Problem. Many events together comprise an experimental condition.

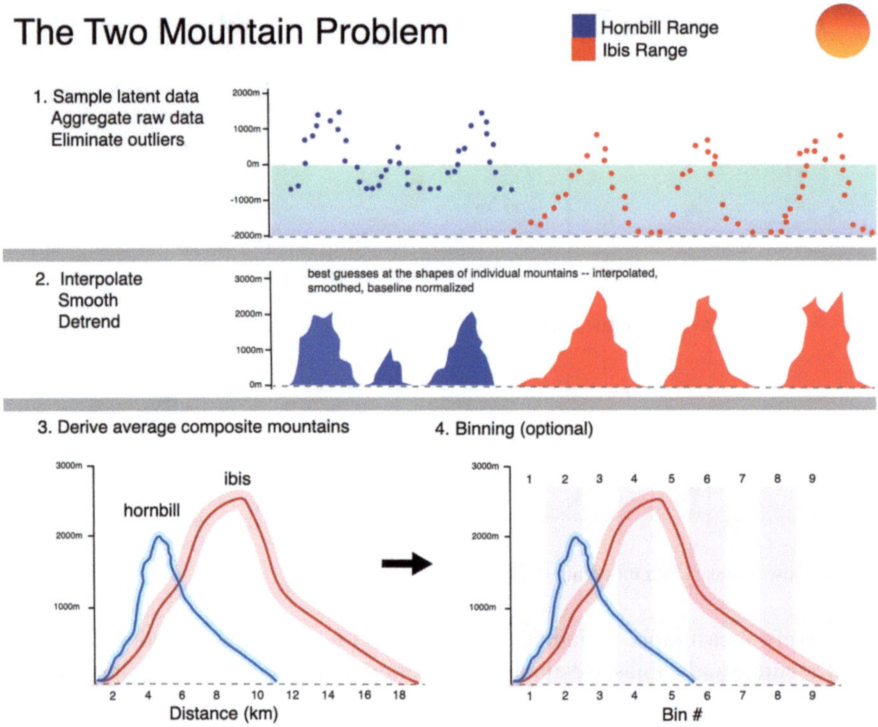

Fig. 5 A data processing pipeline for the Two Mountain Problem

1.4 Expertise and Technical Specialization in Pupillometry

The engineering challenges of pupillometry are immense. A machine must detect near microscopic (~0.1 mm) fluctuations of a slightly oblong disk simultaneously rotating within an eye and translating within a head. Hess and Polt (1960) evaluated changes in pupil size using an analog film camera to photograph the pupil less than five times per second. The authors then projected the images onto a screen and manually measured their dimensions. Pupillometry has since evolved an ever more sophisticated suite of hardware and software solutions for data acquisition. In addition, a number of inspiring teams of researchers have independently developed comprehensive open source pupillometry processing pipelines such as GazeR (Geller et al., 2019), CHAP (Hershman et al., 2019), and pupillometryR (Forbes, 2020). Each of these programs is free, well-supported, and transparent with respect to their processing and artifact correction algorithms.

Pupillometry has reached a level of sophistication where it is impractical if not impossible to master all of the nuances of the constituent processes (e.g., digital signal processing, optics, artifact detection). The good news is that you don't have to. Inexpensive eyetracking hardware, supportive user communities, and open source software have made pupillometry a widely accessible tool. Nevertheless, the benefits of automation are counterbalanced by a number of pitfalls. There are a great many ways to conduct a "bad" pupillometry study (Loewenfeld & Lowenstein, 1993), even with advanced measurement and automated processing tools. Pupillometry is a methodological minefield with project-sinking danger at all stages from design through experiment execution and analysis to the global interpretation of results. An experimenter might spend months meticulously matching stimuli for illuminance and visual complexity only to overlook key individual differences (e.g., fatigue, substance abuse, motivation) or task-correlated covariates (e.g., blinks). Many of these sources of error are difficult to detect and some are impossible to retroactively correct. Biased data or faulty analyses can in turn support spurious conclusions. To follow, we focus on common threats to validity during design (before), execution (during), and analysis (after) of your pupillometry study.

2 Before Your Pupillometry Study

2.1 *Before: Theoretical Considerations*

Consider content and construct validity when debating the use of pupillometry as your dependent measure. Validity is typically regarded as a psychometric dimension in the development of behavioral scales and assessments. Yet, a critical evaluation of validity can also be a useful exercise in weighing the value of pupillometry as a source of evidence for your particular research question. Construct validity is the extent to which a particular tool accurately measures the latent concept it is intended to measure (Cronbach & Meehl, 1955; Sireci, 1998). For example, a wrongheaded emotion researcher might operationally define an abstract concept such as "happiness" as "the number of steps a person takes in a day." This operational definition lends itself to reliable objective measurement, but it has no construct validity. In contrast, content validity is the extent to which an instrument measures all components of a latent construct. A critical evaluation of validity compels answers to two key questions:

(a) What exactly do we hope to measure?
(b) What exactly does pupillometry measure?

Construct validity has special relevance for pupillometry. When a researcher seeks to assess cognitive load, for example, she should provide a clear operational definition of this construct. A chronic lack of falsifiable and reproducible operational definitions has been a major shortcoming in pupillometry. Researchers have shifted the goalposts over time as to what the cognitive pupil response indexes (e.g., interest level, cognitive load, cognitive control, mental effort, mental workload, mental activity, resource allocation, attention changes, executive functioning, effortful mental activity, phasic arousal, adaptive gain, attentional change). One possibility is that the pupil responds selectively to each of these constructs and that these orthogonal signals all bottleneck at the pupil. A more plausible alternative hypothesis is that many of these constructs load on the same latent factor, i.e., physiological arousal. Thus, the pupillary system might be agnostic to subtle distinctions between cognitive load and cognitive control but instead fuels the metabolic demands required by each. In any case, researchers must consider the multifactorial nature of this neurobiological signal and exercise appropriate caution regarding the certainty of their inferences.

Pupillometry is typically used for confirmatory hypothesis testing. For example, in a study of listening effort and speech intelligibility, Zekveld and Kramer (2014) hypothesized that, "pupil dilation would be largest at medium intelligibility levels, and smaller in both easy conditions and in extremely difficult listening conditions resulting in cognitive overload." In this particular design, the null hypothesis (H_0) is that pupil dilation will not differ across the listening conditions. The importance of a clearly stated set of falsifiable hypotheses cannot be overstated. In the Zekveld and Kramer (2014) study, the independent variable was speech intelligibility, and the authors manipulated speech intelligibility by modulating the signal-to-noise ratio. From a philosophical standpoint, this is an outstanding manipulation because it is both content-valid and construct-valid.

Rigorous design in pupillometry should include formal hypothesis tests and validity checks. In an ideal world, experiment planning should also include principled stopping rules and well-justified sample size estimates. Power estimation remains a lacuna in pupillometry, however. Heterogeneity in measurement scales across different studies (e.g., pixels, arbitrary units, mm) and the lack of an extensive database of age-stratified norms have left researchers with few options other than to plan sample sizes by mimicking previous studies.

2.2 *Before: Methodological Considerations*

Numerous trait- and state-level individual differences moderate the cognitive pupil response. Trait variables known to impact pupillometry tend to remain relatively stable or evolve slowly over time (e.g., age, attention deficit disorder, memory span). In contrast, state factors involve behaviors which rapidly change in response to specific conditions (e.g., exaggerated startle reflexes to sudden sounds, agitation after consuming caffeine, migraine headache, etc.). Individual differences pose unique threats to validity especially when studying clinical populations whose levels of arousal, fatigue, motor coordination, affect, motivation, and other neurocognitive abilities fluctuate throughout the course of a day or on idiosyncratic medication dosage schedules. Thus, conducting a valid pupillometry study among special populations requires exhaustive consideration of a wide range of etiology-specific anatomical and neurobehavioral characteristics.[3] To follow, we focus on broad considerations for conducting cognitive pupillometry among neurotypical adults.

2.2.1 Before: Considering Trait-Level Individual Differences

Trait-level individual differences are sometimes the focus of pupillometry. For example, Tsukahura and colleagues reported significant correlations between baseline pupil size and fluid intelligence (a trait-level factor) (Tsukahara et al., 2016; Tsukahara & Engle, 2021). However, individual differences in trait-level variability are more often considered as "noise" or obstacles toward the aim of making clean inferences about a particular experimental manipulation. A common method of controlling for trait-level differences involves specifying inclusion/exclusion criteria.

Age is trait-level variable known to influence the pupillary response. The levator muscles within our eyelids may stretch and weaken with age causing our eyelids to droop (Finsterer, 2003; Friedman, 2005). This phenomenon, known as ptosis, can present a conundrum for eyetracking especially for researchers who are unaware of it. In our own experience, eyetrackers tend to omit many observations and often misattribute ptosis to blink artifact. Correcting for ptosis during an experiment might involve simple steps such as recording from the less droopy eye or asking participants to open their eyes as widely as possible.

Another trait-level factor associated with aging is smaller baseline pupil size (Kim et al., 2000; Morris et al., 1997; Van Gerven et al., 2004). It remains an open question whether the amplitude of evoked pupil responses from this reduced baseline is dampened in aging. Some studies have reported higher task-evoked dilation for older adults (Piquado et al., 2010), whereas others have reported the opposite phenomenon (Gerven et al., 2004), an issue which also depends on the statistical approach taken (McLaughlin et al., 2022). In any case, researchers should be cautious when conducting between-subjects contrasts such as younger vs. older adults because their canonical pupil response functions likely differ in ways that are not yet fully understood.

Many pupillometry studies among neurotypical adults use a common set of exclusion criteria for trait-level factors. These include the presence of sensory deficits (e.g., sensorineural hearing loss), a history of neurodevelopmental disorders (e.g., dyslexia, specific language impairment, attention deficit disorder), ocular disease or trauma (e.g., cataracts), and neurological disorders (e.g., traumatic brain injury, stroke). Pupillometry researchers typically do not control for handedness as a proxy measure for language lateralization since it is assumed that the cognitive pupillary response is coupled across both eyes. Some researchers do, however, control for bilingualism. This decision tends to be motivated (however implicitly) by

[3] For recent representative pupillometric studies in neuropsychiatric disorders, see Burley et al. (2019), Kries et al. (2021), and Schneider et al. (2018).

the need to control for the additional processing demands imposed by language proficiency and/or code switching.

Researchers interested in testing neurotypical participants have several options for confirming roughly normal global cognition. The first option involves asking participants to identify whether they have a history of learning disability or neurological disorder. It is also helpful to ask participants if they are currently experiencing difficulties in memory, language, or concentration. Although self-report can be expedient, it has limitations with respect to sensitivity and specificity. Participants might be unaware that they are experiencing declines in cognition (e.g., mild cognitive impairment), or they might fear stigma of disclosing an impairment.

A more rigorous alternative to self-report involves formally assessing global cognition using a standardized neuropsychological screening tool such as the Montreal Cognitive Assessment (MoCA) (Nasreddine et al., 2005) or the Mini Mental State Examination (MMSE) (Folstein et al., 1975). There are advantages and disadvantages to confirming normal cognition using a standardized measure. First, specificity is problematic for the MoCA because this particular measure tends to misclassify older African American adults as cognitively impaired with high false positive rates (Rossetti et al., 2017; Zahodne et al., 2017). As a consequence, your well-intended screening tool could introduce cultural and/or racial bias by screening out particular group(s) of people whose inclusion is essential for promoting representativeness.

Another ethical consideration involves follow-up for people who fail a cognitive screen. A common and significant fear in middle-aged and older-adults is memory loss and the onset of dementia (Kessler et al., 2012). Experimenters rarely consider the impact that being disqualified from a study might cause. Researchers should plan for this contingency and encourage participants to discuss subjective memory complaints with their primary care provider. In addition, we have found it helpful to provide people with concrete resources such as contact information for helplines (e.g., Alzheimer's Association) and university clinics where they might receive free or low-cost treatment.[4]

2.2.2 Before: Considering State-Level Individual Differences

A dedicated pupillometry researcher must consider many state-level individual differences. People experience normal fluctuations in mood, arousal, and motivation throughout the day. These daily fluctuations are nested within longer cycles of maturation playing out over months and years. Some of this variability is predictable. For example, a pupillometry study conducted among college students at exam time is likely to involve elevated stress, sleeplessness, fatigue, and anxiety that could contaminate your results. Similarly, time of testing may play a significant role in the

[4] This is important! Please contact the first author (reillyj@temple.edu) if he can assist with your efforts.

performance of people with insomnia. Although it is impossible to assess all possible trait-level factors, a useful strategy is to consider factors that impact a person's ability to intensely concentrate for a prolonged duration (e.g., alertness, fatigue, motivation, drug effects).

Wakefulness is an important consideration for planning the length and timing of a pupillometry study. Drowsiness produces an abnormal pattern of pupil oscillation known as hippus (Lüdtke et al., 1998; Wilhelm et al., 1998). Moreover, sleep-deprived participants tend to demonstrate decreased pupil diameters (Morad et al., 2000) and diminished pupillary light reflexes (Lowenstein et al., 1963) due to diminished sympathetic and parasympathetic innervation of the pupillary muscles. Possible methods for counteracting fatigue include structuring relatively brief testing sessions along with regular interaction from the experimenters. This performance boost might at least in part be attributable to the "good-subject effect," the tendency for a person to perform better when seeking tacit approval from an observer (Nichols & Maner, 2008).

Many of us respond to unpleasant states by self-medicating. For example, people might counter headache pain with ibuprofen, fatigue with caffeine, and intermittent allergy symptoms with antihistamines. Each of these states (e.g., pain, fatigue, dry eyes) that originally motivated us to medicate can influence pupillary responses in isolation. In addition, drugs commonly used to treat these states also impact the pupil response, either through direct or indirect pathways. One example of a direct effect on pupillary behavior involves the action of anticholinergic medications such as diphenhydramine (a common ingredient in Benadryl and Nyquil) on the sphincter muscles of the iris resulting in dilation (mydriasis) (Harris et al., 1946; Jaanus, 1992). The same anticholinergic drug may produce indirect effects on the pupil response by inducing drowsiness. Thus, people who take common antihistamines to cope with allergy symptoms before a pupillometry study may experience unintended side effects such as drowsiness and over-dilated pupils. Other "red flags" for state-induced pharmacological effects in pupillometry include opiate or ethanol intoxication (Larson, 2008).

The range of pharmacological effects on nervous system function are substantial. Banning a drug such as caffeine solves one problem (e.g., no participants consumed caffeine) while creating an arguably worse problem (e.g., some participants experience early symptoms of withdrawal) (O'Shea & Moran, 2018). There is no clear guidance on how to approach the issue of pharmacological effects in pupillometry. Many researchers pursue a more pragmatic approach to drug exclusions. Participants are typically excluded if they are sedated and/or exhibiting altered mental status at the time of testing regardless of the drug or psychological/neurological etiology.

Other state-level factors are linked to stimulus anticipation and test anxiety. Some of these factors can be directly observed. For example, people may bite their lips, blink, or produce extraneous motor movements correlated with the onset of events. Some of these behaviors may be ameliorated by interacting with participants and adding a sufficient number of feedback-based training trials prior to conducting the true experiment. In our own experience, asking participants if they have questions or concerns, and how we can make them feel comfortable tends to put them at ease. These interactions and frequent check-ins can in turn result in better retention and higher data quality.

A final consideration that is often overlooked is how the task-evoked pupil response changes over the course of an experiment. Figure 6 shows model fits to pupil response throughout an experiment (D. McLaughlin, personal communication). The change from baseline is biggest in the beginning of the experiment, and smallest at the end. This change may reflect fatigue, habituation to the experimental paradigm, changes in tonic pupil response, or some combination of these factors. Regardless of the underlying cause(s), from a practical perspective, a response that diminishes over time is a challenge for studies in which changes in cognitive effort are of interest (for example, a perceptual learning or training study). Equally distributing experimental conditions throughout the experiment is also important, so that time is not confounded with experiment condition (if Condition A always comes first, and Condition B always last, a smaller response in Condition B might be due to changes in effort *or* simply because it occurs later, and later responses are smaller). Statistically accounting for time may provide better model fits and greater sensitivity to effects of interest.

2.2.3 Before: Considering Environmental Factors

"Let there be well-controlled light" should be the mantra of pupillometry. The pupillary system is highly photosensitive, and the dynamic range of pupillary dilation and constriction in response to light is hundreds of times larger than the magnitude of the cognitive pupil response. As such, uncontrolled luminance in the testing room and variable luminance contours within the stimuli themselves can easily eclipse the cognitive pupil response.

Recording pupillary data in extremely dark or intensely bright lighting conditions presents substantial challenges. Most pupillometry studies do not assess the full dynamic range of pupillary dilation and constriction. Instead, a more common laboratory setup involves testing participants who are seated at a computer monitor under fluorescent lighting. Winn et al. (2018) recommended maintaining ambient light intensities between 10 and 200 lux. Handheld light meters are inexpensive but

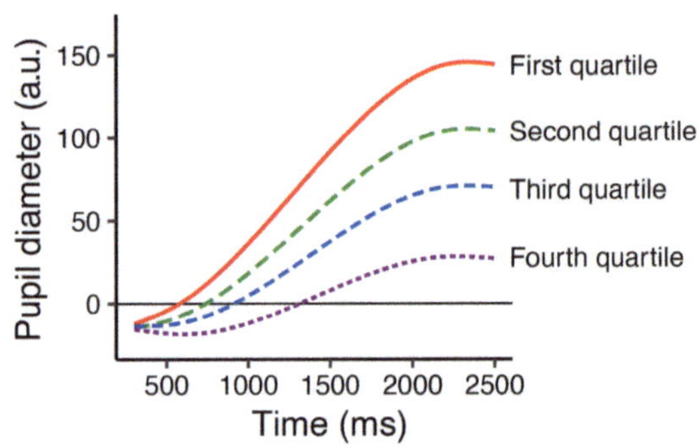

Fig. 6 Changes in task-evoked pupil response over the course of an experiment

essential tools for ensuring that light intensities fall within acceptable ranges. In addition to a moderate range of intensity, researchers should strive for relatively static lighting conditions over the course of an experiment. Testing in a windowless room can help attenuate idiosyncratic light and shadow. In our own lab, we test participants in a sound attenuated booth with blackout film covering the single window. Once the door to the sound booth is sealed, the only ambient light source is the computer monitor.

In addition to ambient lighting in the testing room, it is essential to control for systematic differences in the stimuli. For example, photographs of objects in one condition might simply be darker than another condition (e.g., snowy scenes vs. forest scenes). In this particular example, an experimenter interested in executive demands would almost certainly expect to encounter higher evoked pupil dilation when people view photographs of dark forests. However, they could *not* attribute this response to the variable of interest (executive function). We can only be sure that the participants' pupils dilated more to forests because forest scenes tend to be darker than snowy scenes. Tools such as the SHINE toolbox for MATLAB (Willenbockel et al., 2010) are helpful for matching visual stimuli on luminance and other relevant variables (e.g., complexity).

Another method of controlling for systematic differences in luminance across two stimulus conditions is to avoid visual presentation altogether. Participants in our own pupillometry research hear stimuli (e.g., pure tones, curse words) while viewing a static gray background against which they view a central fixation cross (Reilly et al., 2018, 2020). Auditory presentation circumvents some visual confounds (luminance, accommodation), but it is not a panacea. Sudden loud, novel, and/or frightening stimuli are capable of triggering startle responses (Davis, 1984). Startle reflexes tend to be conserved across other mammals, and they are thought to represent an adaptation for promoting rapid escape (withdrawal) while protecting the eyes and neck (Geyer & Swerdlow, 1998) Moreover, control of central visual fixation remains crucial even in an auditory-only paradigm.

Researchers must also consider idiosyncratic factors unique to their own laboratories when planning their studies. Computer hardware and software constraints (e.g., sound/graphic cards, headphones, RAM) often impact the reliability of stimulus timing, buffering, and presentation. Anti-virus programs, firewalls, and automatic software updates can initiate unpredictably, resulting in heartbreaking data loss. In a busy lab with several different ongoing projects, experimenters often change software settings or switch peripherals (button boxes, joysticks). Participants themselves may be tempted to touch or adjust the monitor or infrared eyetracking sensor. Of course there is no way to plan for every possible contingency. Steps such as disabling computer updates and conducting pre-experiment diagnostics during pilot testing are integral for identifying and preventing many such issues.

2.2.4 Before: The Importance of Baselines

Pupillometry researchers are typically most interested in absolute change relative to some baseline. In a non-stationary time series, the baseline amplitude changes over time. Therefore, it is necessary to correct for the starting amplitude of each trial in order to eventually contrast the magnitude of evoked change from a uniform standard (0 mm). Researchers have historically applied many different baseline correction techniques including non-linear corrections (% change) which assess relative change. Our laboratory empirically tested whether the pupil response scales linearly or non-linearly from different baseline amplitudes (Reilly et al., 2018). We manipulated baseline pupil amplitude by having participants complete tasks with comparable executive demands (e.g., counting pure tones) while situated in either dark or bright light. We discovered that the pupil tends to scale linearly across different baselines and that subtractive scaling is warranted (e.g., Change—Baseline) (see also Mathôt et al., 2018).

Event-related pupillometry is almost always concerned with relative change in pupil amplitude from a neutral baseline. It is crucial to include a sufficient number of trials with their own baselines. A baseline period during an inter-trial interval should have no task demands. Its raison d'etre is to allow the pupil to settle into a steady state before the next event is initiated. In our own research, we tend to jitter inter-trial intervals so that event timing is slightly eccentric. During these intervals, participants view a static screen with a centrally positioned attention fixation cross. We later extract a 500 ms window of pupil size immediately preceding each event for use as that event's unique baseline, a point we will revisit in the analysis section later in the chapter. When designing your own study, mind your baselines.

2.2.5 Before: Be Prepared and Make Principled Decisions

Technicians must cope with numerous demands during an experiment, including monitoring the participant's wellbeing, troubleshooting technical issues, and ensuring that the experiment is proceeding as planned. Optimizing and standardizing eyetracker settings (e.g., sampling rate, which eye to sample from) beforehand will minimize the number of free parameters that will go wrong on testing day. Do not rely on the default settings of your eyetracker. Principled choices must be made before your study.

An often neglected step in planning involves ensuring that your measurement scale promotes replication. Some eyetrackers record pupil dimensions using arbitrary units or pixels, whereas other systems report pupil size using a standard metric scale (mm or mm^2). Please report pupil dimensions (diameter or area) in metric units. Since the pupil does not appear to scale non-linearly, measurements of % change from an arbitrary baseline (e.g., 8753 arbitrary units or 7653 pixels) are neither reproducible, nor particularly interpretable (Reilly et al., 2018). In contrast, reporting evoked change in millimeters facilitates replication efforts across any eyetracking system.[5]

Another fundamental consideration is how researchers should report pupil size. Studies perhaps most often report evoked pupil change in terms of pupil diameter. The validity of pupil diameter is premised on the assumption that the human eye is roughly spherical and that on-center measurement of the pupil approximates a flat disc. This assumption justifies a simple transformation between diameter and surface area ($\pi \times d$). In reality, the human eye is slightly aspherical (Binda et al., 2013; Laeng et al., 2011).

Another principled decision involves determining an appropriate sampling rate. Hess and Polt (1960) sampled about four times a second, whereas many eyetracking systems today sample in excess of 1000 Hz. There is no consensus as to what constitutes either an optimal sampling rate or thresholds for reliable minima/maxima. A general principle borrowed from digital signal processing is that higher sampling rates yield better source reconstruction. Nevertheless, a sampling rate of 1000 Hz may be unnecessarily high for reconstructing the pupillary response function. Higher sampling rates are not always "better" in that oversampling can incur costs both in terms of data proliferation and model overfitting. Researchers must balance these costs/benefits in selecting a sampling rate. *Option 1* (not recommended) is to rely on the default sampling rate for your eyetracker and hope for the best. *Option 2* is to reduce the default sampling rate and acquire data at a more plausible rate of biological change (e.g., 1000 Hz to 250 Hz). *Option 3* is to downsample the raw data after acquisition through procedures such as binning. *Option 4* is to retain all of the original data (e.g., 3000 observations per 3000 ms event) and analyze evoked change using growth curve modeling, being cautious to avoid overfitting (Mirman, 2014).

3 During Your Pupillometry Study

You have settled on a sampling rate and measurement scale that will promote replication. You have successfully screened and consented your participant who is comfortably positioned in a chinrest. You have cycled through your custom "pre-flight" checklist and ensured that all eyetracking settings are accurate as planned. It's go time! This section will cover considerations for ensuring high quality data collection during a pupillometry study.

[5] Conversion from arbitrary units or pixels to mm can be achieved by measuring a known reference such as an artificial eyeball. Our laboratory had no idea where to acquire an artificial eyeball, so we recorded diameter (in arbitrary units) of several black dots printed on cardstock and taped to eyeglasses at a fixed distance of 60 cm from the sensor. Prior to tracking the dots, we confirmed their diameters (e.g., 3 mm, 4 mm, 5 mm, 6 mm) using a measurement caliper. See your eyetracker's documentation for a recommended conversion technique. Sometimes this is a matter of simply checking a box for mm as preferred output.

3.1 During: Minimize Movement Artifacts

Analog film capture during the early days of pupillometry required rigid and prolonged head stabilization. Many remote infrared eyetrackers today are capable of recording data without any head stabilization whatsoever. Advances in motion compensation have yielded unprecedented flexibility in testing populations for whom head stabilization or restraint is impossible. Nevertheless, remote eyetracking is not without cost in terms of sacrificing data quality. Researchers must continue to minimize head motion as much as possible. An ophthalmological chinrest seems to represent a reasonable compromise between no head stabilization and *A Clockwork Orange* style restraint system. Many chinrests clamp to the side of a desk or table. If times are hard, you can even 3D print your own chinrest (Murphy, 2019) https://github.com/nimh-nif/SCNI_Toolbar/wiki/RestEasy:-An-open-source-chin-rest-for-human-psychophysics-experiments.

Gaze tracking studies are typically designed to analyze patterns of visual fixation and saccade dynamics as our eyes move across a screen. These large eye movements are antithetical to pupillometry where design must minimize pupil foreshortening artifacts. That is, the most precise measurement of the pupil occurs while our eyes remain centrally fixated. Stimuli presented at larger visual angles tend to result in unreliable estimates of pupil size because the pupil is no longer spherical at oblique angles. Although several corrections have been proposed (Gagl et al., 2011; Brisson et al., 2013; Hayes & Petrov, 2016), there are also practical steps to minimizing peripheral eye movements. Visual stimuli should be presented centrally and within a narrow visual angle. Allow participants frequent breaks and ensure that they are engaged with the task (i.e., mind wandering = gaze wandering).

3.2 During: Minimize Fatigue and Maximize Engagement

Many of us find our own research fascinating. Fewer of us consider how boring our experiments might be for participants. Boringness is not a trivial consideration for data quality in pupillometry due to the pupillary system's sensitivity to an interaction between cognitive effort, arousal, and reward motivation (Aston-Jones & Cohen, 2005; Gilzenrat et al., 2010). Boring and repetitive tasks that tax sustained attention can induce fatigue and ultimately task disengagement (Granholm et al., 1996). When participants disengage from a demanding task and enter a resting state, they typically show a corresponding reduction (or absence) of task-evoked pupil dilation (Franklin et al., 2013).

Participants disengage for many reasons during an experiment. A common cause is when task demands exceed cognitive capacities. One of the earliest and most robust confirmations of this phenomenon was reported for memory encoding in immediate serial recall of digits (Beatty & Kahneman, 1966). Several studies in the subsequent decades have replicated this effect (Johnson et al., 2014), confirming that when memory load, as indexed by list length, is manipulated, participants show systematically larger pupil dilation during encoding as list lengths grow (e.g., to a length of approximately 7 digits) until maximum span of immediate recall (e.g., N = 9 digits) is surpassed. That is, once participants disengage from the problem, pupil amplitude ceases to increase. A researcher who only considers the raw data might come to the erroneous conclusion that challenging math problems are no more demanding than simple math problems.

Another possible cause of disengagement is motivation. If there is no intrinsic reward to the participant, she might not care enough to invest her full effort. When participants are left alone in a testing room and exposed to long runs of boring stimuli, one can hardly blame them. Regardless of intrinsic motivation, keep sessions brief and provide ample breaks. For projects that require multiple testing sessions for the same participant dispersed over different days, consider scheduling each visit for roughly the same time if possible (see also Veneman et al., 2013). This will reduce potential confounds associated with normal fluctuations in arousal people experience throughout the day.

4 After Your Pupillometry Study: What Next?

Meticulous planning and solid execution have gifted you with useable data. However, even under the best circumstances, your data will require substantial preprocessing before statistical contrasts are possible. The section to follow covers steps in transforming a noisy raw time series into a series of smoothed, aggregated mountains. Data cleaning and analysis in pupillometry represents a dynamic area of methods development. To follow, we focus on some of the major considerations in developing a customized data processing and analysis pipeline.

4.1 After: Data Inspection and Outlier Identification

Inspect your raw pupillometry data. Then inspect it again. Your output should include event codes (e.g., trial number), timestamps, and pupil size data at a bare minimum. Since you are most likely human, your ability to detect subtle trends within hundreds of thousands of numbers is probably limited. Plot your raw data as an uncorrected time series. Visual inspection of each session and every participant is essential for identifying both global and local artifacts.

One advantage that pupillometry researchers have is that the pupil's approximate dynamic range has known anatomical constraints. The human pupil diameter varies between approximately 2.0 and 9.0 mm in extremely bright and dark lighting conditions (Loewenfeld & Lowenstein, 1993; Wang & Munoz, 2015). Observations falling outside of this dynamic range are anatomically impossible and must, therefore, reflect artifact. These thresholds provide benchmarks for blink detection. When the eyelid briefly occludes the pupil during a blink, eyetrackers typically record pupil size as rapidly dropping to 0 mm. An abnormally high rate of constriction coupled with complete occlusion together indicates the presence of a blink (Hershman et al., 2018, 2019). Innovative methods of blink correction involve both detection of the blink event and the removal of "pathological" observations preceding and following the blink. Blinks are not the only cause of signal dropout. People sneeze, cough, and look away from the infrared sensor when distracted. Each of these artifacts produce gaps in an otherwise continuous time series. In the next section, we outline several techniques for filling these gaps.

4.2 After: Random vs. Systematic Data Missingness

Consider yet another scenario. You are an exterminator hired by a statistician to evaluate termite damage on a staircase. You must supply a damage report, but your client is also curious whether the damage is systematic or random. Your pest removal training has not prepared you well for this moment, but you reason that a random termite attack would result in a roughly uniform distribution of termite damage across the entire staircase. In contrast, a systematic attack would involve focal or concentrated attacks on specific subsections of the staircase. Staircases and pupillary time series both exemplify systems that can tolerate more random than systematic missingness. Pupillometry researchers must accordingly consider both the source and extent of randomness within their data.

Intermittent coughing, sneezing, or unscheduled fire alarms likely constitute more random than systematic error. In a well-powered design, random missingness should not bias one condition over another because random noise is absorbed evenly across all experimental conditions. Similarly, the structural integrity of a staircase can survive more diffuse than focal (systematic) termite damage. Random signal dropout does, however, have limits. When termites decimate enough wood, the entire staircase is susceptible to collapse. Similarly, pupillary time series that require extensive imputation are subject to bias. There is no universal standard for a threshold of random data loss in pupillometry.

Non-random or systematic missingness can be especially difficult to control. Two common systematic artifacts include mental imagery and blinks. When people are engaged in spatial problem solving and/or visual imagery, their eye movements and pupil dynamics are both impacted (Grant & Spivey, 2003; Just & Carpenter, 1985; Laeng & Sulutvedt, 2014; Mathôt et al., 2017; Thomas & Lleras, 2007; Zavagno et al., 2017). When people engage in mental rotation or other complex working memory tasks, they experience changes in pupil size along with idiosyncratic eye movements. These perturbations are likely the result of perceptual simulation (i.e., visual imagery) rather than cognitive load. Researchers should be aware of these confounding effects when considering inferential validity.

Blinks represent another complex artifact in pupillometry. The frequency and duration of blinks are strongly correlated with task demands (Siegle et al., 2008) and blink rate provides a complementary index of cognitive load in its own right (Chen & Epps, 2014; Recarte et al., 2008). Although a well-designed pupillometry study should seek to minimize blink artifact, blinks are inevitable. In the section to follow, we discuss methods for correcting blinks and considerations for preventing blinks.

4.3 Artifact Detection

Even after careful planning and controlling for systematic error, your raw pupillary data will require cleaning (Mathôt, 2018, Mathôt & Vilotijević, 2022). As individual laboratories gain experience with cognitive pupillometry, many develop their own custom cleaning pipelines. Although these approaches differ in some respects such as the order of operations, they also share commonalities including artifact detection, imputation of missing values, and baseline correction.

One of the first steps in cleaning a raw pupillary time series involves filtering impossibly high and low values for raw pupil size and rate of change (i.e., acceleration). A simple approach to gross artifact rejection involves applying a bandpass filter to the raw pupil time series using known anatomical constraints on the dynamic range of the human pupil. For example, a bandpass filter with lower and upper bounds of 2 and 9 mm respectively will omit both impossibly low and high observations outside the known dynamic range of the pupil. Mind the distinction between missing values and zero when applying filters or other artifact rejection procedures. When participants close their eyes or look away from an infrared sensor, eyetrackers often register a rapid drop in pupil size to 0 mm. Since no pupil has a 0 mm diameter, impossible observations must be omitted. Marking these data as missing (e.g., NA) is necessary for data imputation.

Some pupillometry artifacts are more subtle and will likely survive a bandpass filter. These artifacts necessitate special detection and correction techniques. Blinks compromise data acquisition both while the eyelid has completely occluded the pupil (all 0 mm observations) but also during the intervals surrounding the blink when the eyelid is closing and re-opening. A simple blink artifact rejection requires both replacing 0 mm observations with NA but also replacing leading and following observations within a temporal window surrounding the blink (Geller et al., 2019). Blinks represent one such case where researchers have proposed artifact detection algorithms (Hershman et al., 2018; Kret & Sjak-Shie, 2019). The idea behind this correction approach is that the blink and the intervals surrounding the blink can be identified through a change in dilation speed (Kret & Sjak-Shie, 2019) or monotonic pattern (Hershman et al., 2018). Blink correction is built in to programs such as GazeR (Geller et al., 2019), CHAP (Hershman et al., 2019).

Recall that pupil measurement can deform when the eye rotates at oblique angles (Brisson et al., 2013; Gagl et al., 2011; Hayes & Petrov, 2016). These pupil foreshortening artifacts are not always detectable upon visual inspection of a raw pupil time series. Although complex mathematical corrections exist for off-center gaze, stimulus arrays should minimize visual angle as much as possible.

4.4 Making Things Whole Again: Data Imputation

Artifact rejection produces gaps in an otherwise continuous univariate time series. The purpose of imputation is to replace these missing values with reasonable estimates of what might have been. Pupillometry as a univariate time series represents a unique challenge for imputation. Since pupillary data are autocorrelated, the most accurate estimate of any missing observation can be derived by looking to its neighbors. For example, my best forecast of the high temperature in Philadelphia tomorrow is derived by the high temperature in Philadelphia today ± a small fudge factor. Single imputation techniques employed in randomly sampled datasets (e.g., replacing missing values with sample means or medians) are inappropriate for estimating missing values because of non-independence between successive observations. Pupillometry researchers typically approach the imputation problem by applying various forms of interpolation (e.g., linear, spline) over their datasets.

The general idea of interpolation (e.g. linear, cubic, spline) is that gaps in a noncontinuous time series are estimated by "connecting" the datapoints which endcap each side of the gap. The most simple type of interpolation (i.e., linear) involves plotting a straight line between endpoints and distributing all missing point estimates equidistantly along the line.

4.5 Smoothing Pupillary Time Series

There exist various smoothing algorithms, but all share the same overarching purpose. Smoothing reduces point-to-point variability across a time series. In our own work, we smooth the data using a simple moving average (window size = 5). That is, each new observation is averaged with n observations surrounding it to yield a new time series. Larger window sizes produce smoother time series. Thus, what starts as a jagged but continuous mountain range should appear more rounded after smoothing. Although there are no recommended minimum/maximum thresholds for window size (n), researchers should exercise caution in specifying larger windows that obscure meaningful variability.

4.6 Baseline Correction and Event Extraction

One of the most common pupillometry designs involves comparing absolute differences in evoked pupil dilation within one or more conditions. Since pupil responses are typically initiated from different baseline amplitudes (e.g., 3 mm vs. 6 mm), it is necessary to normalize evoked change from a uniform standard (0 mm). Subtractive baseline correction is typically implemented with the rationale that baseline pupil size varies within and between individuals, effectively constituting non-stationary time series.

Our own research has shown that the pupil likely scales linearly independent of the baseline amplitude (Reilly et al., 2019). This property of the pupil response is best modeled using a linear scaling technique (i.e., subtractive correction). We typically conduct subtractive baseline correction by computing an average pupil diameter over the 500 ms neutral interval preceding an event. We subtract this median baseline pupil size from each subsequent observation extending outward from the event onset (0 ms) to a specified window of 3000 ms. This procedure normalizes each event to its own baseline, allowing flexibility in baseline pupil size over a session. If you apply dynamic baseline correction using this method, there is no need for detrending because all events are normalized to the same starting amplitude (0 mm).

During the final stages of preprocessing, you will extract all baseline-corrected events from the time series. You must specify the duration over which you will window the pupil response function. The event window must be of a sufficient duration to characterize the rise and fall of the pupil response. However, the event window must not be so long that it captures rest or anticipation of the next subsequent stimulus. Event duration is typically constrained both by stimulus duration and the nature of the canonical pupil response.

4.7 Data Analysis: Contrasting Two Mountains

The past decade has seen advances in the sophistication, power, and complexity of pupillometry data analysis. The vices and virtues of these particular statistical models (e.g., generalized additive mixed models) are beyond the scope of this primer. Nevertheless, pupillometry does pose challenges for any statistical model you ultimately pursue. We address some of these issues to follow.

We have identified five parameters that characterize a two-dimensional mountain (see Fig. 3). These include peak amplitude, time-to-peak, sustained amplitude, area under the curve, and time-to-decay. Researchers face a dilemma in selecting which parameter(s) to contrast because no comprehensive neurobiological model of these parameters exists. As such, we have a limited understanding of how the canonical pupil response function might shift under different neural and behavioral challenges across the lifespan. Reciprocal relationships between brainstem activation, arousal, and pupil dilation have motivated an intense focus on peak amplitude and time-to-peak as parameters of interest in pupillometry. Nevertheless, great variability exists in how peak amplitude is derived (Tun et al., 2009) and indeed whether the pupil response might better be modeled using a two-peak solution with an early peak at approximately 600 ms and a late peak at 1200 ms (Steinhauer & Hakerem, 1992).

5 Concluding Remarks, Unanswered Questions, Future Directions

A proliferation of low cost eyetracking hardware along with a supportive user community has seeded exponential growth in the popularity of pupillometry. In turn, pupillometry has recently generated remarkable insights into cognition, consciousness, and mental imagery (Laeng et al., 2012; Laeng & Sulutvedt, 2014; Mathôt et al., 2015, 2017; Zavagno et al., 2017). The advent of automated software applications has made pupillometry accessible to a wide range of disciplines not typically steeped in psychophysiology. Our goal in this primer is to highlight complexities of measuring this biological signal for non-experts.

There are many factors to consider if your aim is to conduct a rigorous pupillometry study. Newcomers to this technique might be frustrated to learn that there are no uniform best practices for design, execution, and analysis of pupillometry, although this does appear to be changing (Winn et al., 2018). Pupillometry often reflects a long-term commitment to learning the nuances of this technique over time. Our own laboratory has stumbled upon many of the pitfalls discussed here through trial-and-error over about a decade. A preventable error as simple as a technician switching eyetracker settings between studies has resulted in data loss for our lab on several occasions. We have also learned firsthand the necessity to monitor fatigue and to double check stimulus delays and response logging. Checklists are essential. Develop your own and share them with other researchers.

In conclusion, much remains to be learned about pupillometry. New methods of design and time series analysis hold promise for improving the rigor of pupillometry. Yet, the field remains limited by an anemic history of replication (but see de Winter et al., 2021) and the lack of age-stratified norms against which effect sizes might be derived. A deeper understanding of the neural and neuromuscular substrates of the cognitive pupil response function is essential for understanding which cognitive processes are indexed by specific parameters and how these parameters might be selectively perturbed.

Acknowledgements We thank Jason Geller, Jonathan Peelle, Drew McLaughlin, and members of the Concepts and Cognition Laboratory at Temple University. We are grateful to all of the scientists who have developed open source software applications for pupillometry. Your immeasurable support represents the best spirit of science. This work was supported in part by a grant from the US National Institute on Deafness and Other Communication Disorders (NIH/NIDCD R01 DC0103063)

References

Aston-Jones, G., & Cohen, J. D. (2005). An integrative theory of locus coeruleus-norepinephrine function: Adaptive gain and optimal performance. *Annual Review of Neuroscience, 28*, 403–450.

Barbur, J. L. (2004). Learning from the pupil-studies of basic mechanisms and clinical applications. *The Visual Neurosciences, 1*, 641–656.

Beatty, J. (1982). Task-evoked pupillary responses, processing load, and the structure of processing resources. *Psychological Bulletin, 91*(2), 276–292.

Beatty, J., & Kahneman, D. (1966). Pupillary changes in two memory tasks. *Psychonomic Science, 5*(10), 371–372.

Binda, P., Pereverzeva, M., & Murray, S. O. (2013). Attention to bright surfaces enhances the pupillary light reflex. *Journal of Neuroscience, 33*(5), 2199–2204. https://doi.org/10.1523/JNEUROSCI.3440-12.2013

Bradley, M. M., Miccoli, L., Escrig, M. A., & Lang, P. J. (2008). The pupil as a measure of emotional arousal and autonomic activation. *Psychophysiology, 45*(4), 602–607.

Bradshaw, J. L. (1968). Pupil size and problem solving. *Quarterly Journal of Experimental Psychology, 20*(2), 116–122. https://doi.org/10.1080/14640746808400139

Bremner, F. (2009). Pupil evaluation as a test for autonomic disorders. *Clinical Autonomic Research, 19*(2), 88–101. https://doi.org/10.1007/s10286-009-0515-2

Brisson, J., Mainville, M., Mailloux, D., Beaulieu, C., Serres, J., & Sirois, S. (2013). Pupil diameter measurement errors as a function of gaze direction in corneal reflection eyetrackers. *Behavior Research Methods, 45*(4), 1322–1331. https://doi.org/10/f5kq7b

Burley, D. T., Gray, N. S., & Snowden, R. J. (2019). Emotional modulation of the pupil response in psychopathy. *Personality Disorders: Theory, Research, and Treatment, 10*(4), 365. https://doi.org/10/ggrjqp

Causse, M., Peysakhovich, V., & Mandrick, K. (2017). Eliciting sustained mental effort using the Toulouse N-back task: Prefrontal cortex and pupillary responses. In *Advances in Neuroergonomics and cognitive engineering* (pp. 185–193). Springer.

Chen, S., & Epps, J. (2014). Using task-induced pupil diameter and blink rate to infer cognitive load. *Human–Computer Interaction, 29*(4), 390–413. https://doi.org/10/ghs95n

Cronbach, L. J., & Meehl, P. E. (1955). Construct validity in psychological tests. *Psychological Bulletin, 52*(4), 281. https://doi.org/10/dcsjjf

Davies, J. E., Barquet, A., Freed, V. H., Haque, R., Morgade, C., Sonneborn, R. E., & Vaclavek, C. (1975). Human pesticide poisonings by a fat-soluble organophosphate insecticide. *Archives of Environmental Health: An International Journal, 30*(12), 608–613. https://doi.org/10/gmhnzh

Davis, M. (1984). The mammalian startle response. In *Neural mechanisms of startle behavior* (pp. 287–351). Springer.

de Rooij, A., Vromans, R. D., & Dekker, M. (2018). Noradrenergic modulation of creativity: Evidence from Pupillometry. *Creativity Research Journal, 30*(4), 339–351. https://doi.org/10.1080/10400419.2018.1530533

de Winter, J. C. F., Petermeijer, S. M., Kooijman, L., & Dodou, D. (2021). Replicating five pupillometry studies of Eckhard Hess. *International Journal of Psychophysiology, 165*, 145–205. https://doi.org/10/gmhjhb

Denison, R. N., Parker, J. A., & Carrasco, M. (2020). Modeling pupil responses to rapid sequential events. *Behavior Research Methods, 52*(5), 1991–2007. https://doi.org/10/gg4zs2

Elman, J. A., Panizzon, M. S., Hagler, D. J., Eyler, L. T., Granholm, E. L., Fennema-Notestine, C., Lyons, M. J., McEvoy, L. K., Franz, C. E., Dale, A. M., & Kremen, W. S. (2017). Task-evoked pupil dilation and BOLD variance as indicators of locus coeruleus dysfunction. *Cortex, 97*, 60–69. https://doi.org/10.1016/j.cortex.2017.09.025

Finsterer, J. (2003). Ptosis: Causes, presentation, and management. *Aesthetic Plastic Surgery, 27*(3), 193–204. https://doi.org/10/bkph2x

Folstein, M. F., Folstein, S. E., & McHugh, P. R. (1975). Mini-mental state: A practical method for grading the state of patients for the clinician. *Journal of Psychiatric Research, 12*, 189–198.

Forbes, S. (2020). PupillometryR: An R package for preparing and analysing pupillometry data. *Journal of Open Source Software, 5*(50), 2285. https://doi.org/10/gg2qmh

Franklin, M. S., Broadway, J. M., Mrazek, M. D., Smallwood, J., & Schooler, J. W. (2013). *Window to the wandering mind: Pupillometry of spontaneous thought while reading*. SAGE.

Friedman, O. (2005). Changes associated with the aging face. *Facial Plastic Surgery Clinics, 13*(3), 371–380. https://doi.org/10/c9d2xm

Gagl, B., Hawelka, S., & Hutzler, F. (2011). Systematic influence of gaze position on pupil size measurement: Analysis and correction. *Behavior Research Methods, 43*(4), 1171–1181. https://doi.org/10/c75xwk

Geller, J., Winn, M., Mahr, T., & Mirman, D. (2019). GazeR: A package for processing gaze position and pupil size data. *PsyArXiv*. https://doi.org/10.31234/osf.io/gvcxb

Gerven, P. W. M. V., Paas, F., Merriënboer, J. J. G. V., & Schmidt, H. G. (2004). Memory load and the cognitive pupillary response in aging. *Psychophysiology, 41*(2), 167–174. https://doi.org/10.1111/j.1469-8986.2003.00148.x

Geyer, M. A., & Swerdlow, N. R. (1998). Measurement of startle response, prepulse inhibition, and habituation. *Current Protocols in Neuroscience, 3*(1), 8–7. https://doi.org/10/fqf37m

Gilzenrat, M. S., Nieuwenhuis, S., Jepma, M., & Cohen, J. D. (2010). Pupil diameter tracks changes in control state predicted by the adaptive gain theory of locus coeruleus function. *Cognitive, Affective, & Behavioral Neuroscience, 10*(2), 252–269.

Gitelman, D. R., Penny, W. D., Ashburner, J., & Friston, K. J. (2003). Modeling regional and psychophysiologic interactions in fMRI: The importance of hemodynamic deconvolution. *NeuroImage, 19*(1), 200–207. https://doi.org/10/bc8ct8

Goldinger, S. D., & Papesh, M. H. (2012). Pupil dilation reflects the creation and retrieval of memories. *Current Directions in Psychological Science, 21*(2), 90–95.

Goldinger, S. D., & Papesh, M. H. (2013). Recollection is fast and easy: Pupillometric studies of face memory. In *Psychology of learning and motivation* (Vol. 59, pp. 191–222). Elsevier.

Granholm, E., Asarnow, R. F., Sarkin, A. J., & Dykes, K. L. (1996). Pupillary responses index cognitive resource limitations. *Psychophysiology, 33*(4), 457–461.

Grant, E. R., & Spivey, M. J. (2003). Eye movements and problem solving: Guiding attention guides thought. *Psychological Science, 14*(5), 462–466. https://doi.org/10/d97qck

Handwerker, D. A., Ollinger, J. M., & D'Esposito, M. (2004). Variation of BOLD hemodynamic responses across subjects and brain regions and their effects on statistical analyses. *NeuroImage, 21*(4), 1639–1651. https://doi.org/10/dv6qcs

Harris, R., McGavack, T. H., & Elias, H. (1946). The nature of the action of dimethylaminoethyl benzhydryl ether hydrochloride (Benadryl): Effects upon the human eye. *The Journal of Laboratory and Clinical Medicine, 31*(10), 1148–1152. https://doi.org/10.5555/uri:pii:0022214346901564

Hayes, T. R., & Petrov, A. A. (2016). Mapping and correcting the influence of gaze position on pupil size measurements. *Behavior Research Methods, 48*(2), 510–527.

Hershman, R., Henik, A., & Cohen, N. (2018). A novel blink detection method based on pupillometry noise. *Behavior Research Methods, 50*(1), 107–114.

Hershman, R., Henik, A., & Cohen, N. (2019). CHAP: Open-source software for processing and analyzing pupillometry data. *Behavior Research Methods., 51*, 1059. https://doi.org/10.3758/s13428-018-01190-1

Hess, E. H., & Polt, J. M. (1960). Pupil size as related to interest value of visual stimuli. *Science, 132*(3423), 349–350.

Hess, E. H., & Polt, J. M. (1964). Pupil size in relation to mental activity during simple problem-solving. *Science, 143*(3611), 1190–1192. https://doi.org/10/fb3kx4

Hoffmann, M., Lefering, R., Rueger, J. M., Kolb, J. P., Izbicki, J. R., Ruecker, A. H., Rupprecht, M., Lehmann, W., & Surgery, T. R. (2012). Pupil evaluation in addition to Glasgow coma scale components in prediction of traumatic brain injury and mortality. *British Journal of Surgery, 99*(S1), 122–130.

Jaanus, S. D. (1992). Ocular side effects of selected systemic drugs. *Optometry Clinics, 2*(4), 73–96.

Jennett, B., & Teasdale, G. (1977). Aspects of coma after severe head injury. *The Lancet, 309*(8017), 878–881.

Johnson, E. L., Miller Singley, A. T., Peckham, A. D., Johnson, S. L., & Bunge, S. A. (2014). Task-evoked pupillometry provides a window into the development of short-term memory capacity. *Frontiers in Psychology, 5*, 218. https://doi.org/10/ghs96d

Joshi, S., Li, Y., Kalwani, R. M., & Gold, J. I. (2016). Relationships between pupil diameter and neuronal activity in the locus coeruleus, colliculi, and cingulate cortex. *Neuron, 89*(1), 221–234.

Just, M. A., & Carpenter, P. A. (1985). Cognitive coordinate systems: Accounts of mental rotation and individual differences in spatial ability. *Psychological Review, 92*(2), 137. https://doi.org/10/b8rk8j

Kessler, E.-M., Bowen, C. E., Baer, M., Froelich, L., & Wahl, H.-W. (2012). Dementia worry: A psychological examination of an unexplored phenomenon. *European Journal of Ageing, 9*(4), 275–284. https://doi.org/10/f4cn3v

Kim, M., Beversdorf, D. Q., & Heilman, K. M. (2000). Arousal response with aging: Pupillographic study. *Journal of the International Neuropsychological Society, 6*(3), 348–350. https://doi.org/10/c6x3gf

Klingner, J., Tversky, B., & Hanrahan, P. (2011). Effects of visual and verbal presentation on cognitive load in vigilance, memory, and arithmetic tasks. *Psychophysiology, 48*(3), 323–332. https://doi.org/10/fqzf3m

Kreis, I., Zhang, L., Moritz, S., & Pfuhl, G. (2021). Spared performance but increased uncertainty in schizophrenia: Evidence from a probabilistic decision-making task. *Schizophrenia Research*. https://doi.org/10/gmgb5x

Kret, M. E., & Sjak-Shie, E. E. (2019). Preprocessing pupil size data: Guidelines and code. *Behavior Research Methods, 51*(3), 1336–1342. https://doi.org/10/gf5ssx

Kuchinsky, S. E., Ahlstrom, J. B., Cute, S. L., Humes, L. E., Dubno, J. R., & Eckert, M. A. (2014). Speech-perception training for older adults with hearing loss impacts word recognition and effort. *Psychophysiology, 51*(10), 1046–1057. https://doi.org/10/f6kpcp

Laeng, B., & Sulutvedt, U. (2014). The eye pupil adjusts to imaginary light. *Psychological Science, 25*(1), 188–197. https://doi.org/10.1177/0956797613503556

Laeng, B., Orbo, M., Holmlund, T., & Miozzo, M. (2011). *Pupillary Stroop effects. Cognitive Processes, 12*(1), 13–21. https://doi.org/10.1007/s10339-010-0370-z

Laeng, B., Sirois, S., & Gredeback, G. (2012). Pupillometry: A window to the preconscious? *Perspectives on Psychological Science, 7*(1), 18–27. https://doi.org/10.1177/1745691611427305

Lam, B. L., Thompson, H. S., & Corbett, J. J. (1987). The prevalence of simple anisocoria. *American Journal of Ophthalmology, 104*(1), 69–73. https://doi.org/10/gmhpbt

Larson, M. D. (2008). Mechanism of opioid-induced pupillary effects. *Clinical Neurophysiology, 119*(6), 1358–1364. https://doi.org/10/chkw8r

Loewenfeld, I. E., & Lowenstein, O. (1993). *The pupil: Anatomy, physiology, and clinical applications* (Vol. 2). Wiley-Blackwell.

Lowenstein, O., Feinberg, R., & Loewenfeld, I. E. (1963). *Pupillary movements during acute and chronic fatigue: A new test for the objective evaluation of tiredness* (Vol. 65). Federal Aviation Agency, Office of Aviation Medicine.

Lüdtke, H., Wilhelm, B., Adler, M., Schaeffel, F., & Wilhelm, H. (1998). Mathematical procedures in data recording and processing of pupillary fatigue waves. *Vision Research, 38*(19), 2889–2896. https://doi.org/10/d8fwmf

Marmarou, A., Lu, J., Butcher, I., McHugh, G. S., Murray, G. D., Steyerberg, E. W., Mushkudiani, N. A., Choi, S., & Maas, A. I. (2007). Prognostic value of the Glasgow coma scale and pupil reactivity in traumatic brain injury assessed pre-hospital and on enrollment: An IMPACT analysis. *Journal of Neurotrauma, 24*(2), 270–280.

Mathôt, S. (2018). Pupillometry: Psychology, physiology, and function. *Journal of Cognition, 1*(1) https://doi.org/10/gfkmbc, 16.

Mathôt, S., & Vilotijević, A. (2022). Methods in cognitive pupillometry: Design, preprocessing, and statistical analysis. In *Behavior research methods*. Advance online publication. https://doi.org/10.3758/s13428-022-01957-7

Mathôt, S., Melmi, J.-B., Van der Linden, L., & Van der Stigchel, S. (2015). The mind-writing pupil: Near-perfect decoding of visual attention with pupillometry. *Journal of Vision, 15*(12), 176. https://doi.org/10.1167/15.12.176

Mathôt, S., Grainger, J., & Strijkers, K. (2017). Pupillary responses to words that convey a sense of brightness or darkness. *Psychological Science, 0956797617702699*, 1116.

Mathôt, S., Fabius, J., Van Heusden, E., & Van der Stigchel, S. (2018). Safe and sensible baseline correction of pupil-size data. *Behavior Research Methods, 50*(1), 94–106.

McLaughlin, D. J., Zink, M. E., Gaunt, L., Spehar, B., Van Engen, K. J., Sommers, M. S., & Peelle, J. E. (2022). Pupillometry reveals cognitive demands of lexical competition during spoken word recognition in young and older adults. *Psychonomic Bulletin & Review, 29*(1), 268–280. https://doi.org/10.3758/s13423-021-01991-0

Mirman, D. (2014). *Growth curve analysis and visualization using R* (1st ed.). Chapman & Hall.

Morad, Y., Lemberg, H., Yofe, N., & Dagan, Y. (2000). Pupillography as an objective indicator of fatigue. *Current Eye Research, 21*(1), 535–542. https://doi.org/10/bqc9tn

Morris, S. K., Granholm, E., Sarkin, A. J., & Jeste, D. V. (1997). Effects of schizophrenia and aging on pupillographic measures of working memory. *Schizophrenia Research, 27*(2–3), 119–128. https://doi.org/10/bk3fzf

Murphy, P. R., O'Connell, R. G., O'Sullivan, M., Robertson, I. H., & Balsters, J. H. (2014). Pupil diameter covaries with BOLD activity in human locus coeruleus. *Human Brain Mapping, 35*(8), 4140–4154. https://doi.org/10.1002/hbm.22466

Nasreddine, Z. S., Phillips, N. A., Badirian, V., Charbonneau, S., Whitehead, V., Collin, I., Cummings, J. L., & Chertkow, H. (2005). The Montreal cognitive assessment, MoCA: A brief screening tool for mild cognitive impairment. *Journal of the American Geriatrics Society, 53*(4), 695–699.

Nassar, M. R., Rumsey, K. M., Wilson, R. C., Parikh, K., Heasly, B., & Gold, J. I. (2012). Rational regulation of learning dynamics by pupil-linked arousal systems. *Nature Neuroscience, 15*(7), 1040–1046. https://doi.org/10.1038/nn.3130

Nichols, A. L., & Maner, J. K. (2008). The good-subject effect: Investigating participant demand characteristics. *The Journal of General Psychology, 135*(2), 151–166. https://doi.org/10/c2gcm5

O'Shea, H., & Moran, A. (2018). To go or not to go? Pupillometry elucidates inhibitory mechanisms in motor imagery. *Journal of Cognitive Psychology, 30*(4), 466–483. https://doi.org/10/gmgb5v

Papesh, M. H., & Goldinger, S. D. (2015). Pupillometry and memory: External signals of metacognitive control. In *Handbook of biobehavioral approaches to self-regulation* (pp. 125–139). Springer. https://doi.org/10.1007/978-1-4939-1236-0_9

Papesh, M. H., Goldinger, S. D., & Hout, M. C. (2012). Memory strength and specificity revealed by pupillometry. *International Journal of Psychophysiology, 83*(1), 56–64. https://doi.org/10/bw977q

Peysakhovich, V., Vachon, F., & Dehais, F. (2017). The impact of luminance on tonic and phasic pupillary responses to sustained cognitive load. *International Journal of Psychophysiology, 112*, 40–45.

Piquado, T., Isaacowitz, D., & Wingfield, A. (2010). Pupillometry as a measure of cognitive effort in younger and older adults. *Psychophysiology, 47*(3), 560–569. https://doi.org/10.1111/j.1469-8986.2009.00947.x

Recarte, M. Á., Pérez, E., Conchillo, Á., & Nunes, L. M. (2008). Mental workload and visual impairment: Differences between pupil, blink, and subjective rating. *The Spanish Journal of Psychology, 11*(2), 374–385. https://doi.org/10/gmjb35

Reilly, J., Kelly, A., Kim, S. H., Jett, S., & Zuckerman, B. (2018). The human task-evoked pupillary response function is linear: Implications for baseline response scaling in pupillometry. *Behavior Research Methods*, 1–14.

Reilly, J., Zuckerman, B., Kelly, A., Flurie, M., & Rao, S. (2020). Neuromodulation of cursing in American English: A combined tDCS and pupillometry study. *Brain and Language, 206*, 104791. https://doi.org/10.1016/j.bandl.2020.104791

Rengstorff, R. H. (1994). Vision and ocular changes following accidental exposure to organophosphates. *Journal of Applied Toxicology, 14*(2), 115–118. https://doi.org/10/dnv74r

Rollins, M. D., Feiner, J. R., Lee, J. M., Shah, S., & Larson, M. (2014). Pupillary effects of high-dose opioid quantified with infrared pupillometry. *Anesthesiology, 121*(5), 1037–1044. https://doi.org/10/gmhnzg

Rossetti, H. C., Lacritz, L. H., Hynan, L. S., Cullum, C. M., Van Wright, A., & Weiner, M. F. (2017). Montreal cognitive assessment performance among community-dwelling African Americans. *Archives of Clinical Neuropsychology, 32*(2), 238–244. https://doi.org/10.1093/arclin/acw095

Schneider, M., Leuchs, L., Czisch, M., Sämann, P. G., & Spoormaker, V. I. (2018). Disentangling reward anticipation with simultaneous pupillometry/fMRI. *NeuroImage, 178*, 11–22. https://doi.org/10/gdxffm

Siegle, G. J., Ichikawa, N., & Steinhauer, S. (2008). Blink before and after you think: Blinks occur prior to and following cognitive load indexed by pupillary responses. *Psychophysiology, 45*(5), 679–687. https://doi.org/10/bc3fc3

Sireci, S. G. (1998). The construct of content validity. *Social Indicators Research, 45*(1), 83–117. https://doi.org/10/bx7dg3

Steinhauer, S. R., & Hakerem, G. (1992). The pupillary response in cognitive psychophysiology and schizophrenia. *Annals of the New York Academy of Sciences, 658*(1), 182–204.

Thomas, L. E., & Lleras, A. (2007). Moving eyes and moving thought: On the spatial compatibility between eye movements and cognition. *Psychonomic Bulletin & Review, 14*(4), 663–668. https://doi.org/10/bzs8j6

Tryon, W. W. (1975). Pupillometry: A survey of sources of variation. *Psychophysiology, 12*(1), 90–93.

Tsukahara, J. S., & Engle, R. W. (2021). Is baseline pupil size related to cognitive ability? Yes (under proper lighting conditions). *Cognition, 211*, 104643. https://doi.org/10/gjcgv3

Tsukahara, J. S., Harrison, T. L., & Engle, R. W. (2016). The relationship between baseline pupil size and intelligence. *Cognitive Psychology, 91*, 109–123. https://doi.org/10.1016/j.cogpsych.2016.10.001

Tun, P. A., McCoy, S., & Wingfield, A. (2009). Aging, hearing acuity, and the attentional costs of effortful listening. *Psychology and Aging, 24*(3), 761–766. https://doi.org/10.1037/a0014802

Van Engen, K. J., & McLaughlin, D. J. (2018). Eyes and ears: Using eye tracking and pupillometry to understand challenges to speech recognition. *Hearing Research, 369*, 56–66. https://doi.org/10/gfqg8t

Veneman, C. E., Gordon-Salant, S., Matthews, L. J., & Dubno, J. R. (2013). Age and measurement time-of-day effects on speech recognition in noise. *Ear and Hearing, 34*(3), 288–299. https://doi.org/10/gmjcs9

Wainstein, G., Rojas-Libano, D., Medel, V., Alnæs, D., Kolskår, K. K., Endestad, T., Laeng, B., Ossandon, T., Crossley, N., Matar, E., & Shine, J. M. (2021). The ascending arousal system promotes optimal performance through meso-scale network integration in a visuospatial attentional task. *Network Neuroscience, 5*, 1–32. https://doi.org/10/gmw45x

Wang, C.-A., & Munoz, D. P. (2015). A circuit for pupil orienting responses: Implications for cognitive modulation of pupil size. *Current Opinion in Neurobiology, 33*, 134–140.

Wilhelm, B., Wilhelm, H., Lüdtke, H., Streicher, P., & Adler, M. (1998). Pupillographic assessment of sleepiness in sleep-deprived healthy subjects. *Sleep, 21*(3), 258–265.

Willenbockel, V., Sadr, J., Fiset, D., Horne, G. O., Gosselin, F., & Tanaka, J. W. (2010). Controlling low-level image properties: The SHINE toolbox. *Behavior Research Methods, 42*(3), 671–684. https://doi.org/10/dxddqm

Winn, M. B., Wendt, D., Koelewijn, T., & Kuchinsky, S. E. (2018). Best practices and advice for using pupillometry to measure listening effort: An introduction for those who want to get started. *Trends in Hearing, 22*, 2331216518800869.

Zahodne, L. B., Manly, J. J., Smith, J., Seeman, T., & Lachman, M. E. (2017). Socioeconomic, health, and psychosocial mediators of racial disparities in cognition in early, middle, and late adulthood. *Psychology and Aging, 32*(2), 118–130. https://doi.org/10.1037/pag0000154

Zavagno, D., Tommasi, L., & Laeng, B. (2017). The eye Pupil's response to static and dynamic illusions of luminosity and darkness. *I-Perception, 8*(4), 2041669517717754. https://doi.org/10.1177/2041669517717754

Zekveld, A. A., & Kramer, S. E. (2014). Cognitive processing load across a wide range of listening conditions: Insights from pupillometry. *Psychophysiology, 51*(3), 277–284.

Zekveld, A. A., Kramer, S. E., & Festen, J. M. (2010). Pupil response as an indication of effortful listening: The influence of sentence intelligibility. *Ear and Hearing, 31*(4), 480–490. https://doi.org/10/cx648q

Zekveld, A. A., Kramer, S. E., & Festen, J. M. (2011). Cognitive load during speech perception in noise: The influence of age, hearing loss, and cognition on the pupil response. *Ear and Hearing, 32*(4), 498–510. https://doi.org/10/cfgzkx

Zekveld, A. A., Heslenfeld, D. J., Johnsrude, I. S., Versfeld, N. J., & Kramer, S. E. (2014). The eye as a window to the listening brain: Neural correlates of pupil size as a measure of cognitive listening load. *NeuroImage, 101c*, 76–86. https://doi.org/10.1016/j.neuroimage.2014.06.069

Processing and Analyzing of Pupillometry Data

Ronen Hershman, Dalit Milshtein, and Avishai Henik

1 Introduction

In simple words, we can describe the pupil as a hole located in the center of the iris of the eye. This special hole functions as a leaky sensory gate, controlling the light striking the retina. The changes in illumination can provoke pupil dilations of more than double its size and respectively, regulate the amount of light entering the eye (MacLachlan & Howland, 2002). This process is known as the pupillary light reflex (PLR). Increasing or decreasing pupil diameter is controlled by the dilator and sphincter muscles that are innervated by sympathetic and parasympathetic action (Beatty & Lucero-Wagoner, 2000; Loewenfeld, 1958). Therefore, the stimulation of the dilator and the inhibition of the constrictor can change pupillary dynamics as well. While the sympathetic branch enlarges the pupil by directly innervating the

[1] The LC-NE system is one of several brainstem neuromodulatory nuclei, situated in the dorsal pons, with widely distributed, ascending projections to the neocortex, and is the sole source of cortical norepinephrine (NE).

[2] *flip* function returns the same array, but with the order of the elements reversed.

[3] CHAP calculates peak and dip parameters by using the mean pupil size curve. Users who are interested in analysis of these parameters by using trial-by-trial values may use the processed data created for each participant during the group analysis (trials_data.csv).

We wish to thank Ms. Desiree Meloul for her helpful comments and useful input on this article.

R. Hershman (✉)
Department of Psychology, University of Innsbruck, Innsbruck, Austria
e-mail: ronen.hershman@uibk.ac.at

D. Milshtein
Department of Psychology, University of Haifa, Haifa, Israel

A. Henik
Department of Psychology and The Zelman Center for Brain Science, Ben-Gurion University of the Negev, Beer-Sheva, Israel

M. H. Papesh, S. D. Goldinger (eds.), *Modern Pupillometry*,
https://doi.org/10.1007/978-3-031-54896-3_15

dilator muscles, inhibition of the Edinger–Westphal nucleus (i.e., the parasympathetic branch) causes relaxation of the sphincter muscles and as a result, the pupil dilates (Steinhauer et al., 2004).

The pupil not only constricts in response to light and dilates in response to darkness but it also has been found to be an effective indicator of various cognitive and affective processes (Goldwater, 1972; Kahneman, 1973; Loewenfeld, 1993) (for recent reviews see Laeng et al., 2012; Mathôt, 2018; van der Wel & van Steenbergen, 2018). Importantly, although pupil dilations occur spontaneously and are difficult to control voluntarily, intentional stimulation—cognitive and emotional as well—can provoke pupillary responses (e.g., emotional imagery; Henderson et al., 2018). The extent to which cognitive and emotional processing activity is uniquely associated with the sympathetic or parasympathetic pathway, as well as its reflection in pupillometry, is the subject of ongoing research (for a recent review see Joshi & Gold, 2020). A wide corpus of findings has indicated the pivotal role that the brain's locus coeruleus–noradrenergic (LC–NE) system[1] plays in pupillary regulation (Gilzenrat et al., 2010; Joshi et al., 2016; Murphy et al., 2011, 2014). The LC–NE system has been tied mainly to general functions such as arousal, slow-wave sleep, and memory. But recently, several specific hypotheses have been proposed, including the LC–NE system's role in the cognitive control of adaptive behavior (Aston-Jones & Cohen, 2005; McGinley et al., 2015). As a result, pupil size may provide a very simple tracking device for moment-to-moment LC neural activity (Costa & Rudebeck, 2016). Thus, these proposals have significantly expanded the range of cognitive processes that have implications for pupillometry.

In addition to the above, pupillometry has several other characteristics that can make it an attractive research tool for various cognitive domains. Akin to other physiological measurements (e.g., heart rate, blood pressure, galvanic skin response), pupillometry can provide temporal inspections of cognitive processes and thus provide a better understanding of their dynamics along a timeline. Furthermore, pupil response latencies (i.e., 200 ms and above) may help in pointing out changes, which may be elusive when using other means of measurement (Ellis, 1981). That is, differences among conditions could be indicated for relatively long intervals. Pupillometry's flexibility and the relatively few physical limitations it imposes on participants make it convenient for studies with preverbal or just-verbal populations also (Hepach & Westermann, 2016). All these advantages have been known for over half a century. However, until recently, pupillometry did not become a common measurement, mainly due to some serious difficulties largely in the data processing stage. Nevertheless, technological and computational developments in the last decade have promoted the use of this study tool (Forbes, 2020; Geller et al., 2020; Hershman et al., 2019). That is, pupillometry has increasingly become useful in various domains such as high-order visual processing (Einhauser et al., 2008; Kloosterman et al., 2015; Naber & Nakayama, 2013; Sapir et al., 2021), attention (Geva et al., 2013), cognitive control (Hershman et al., 2020, 2021, 2022; van der Wel & van Steenbergen, 2018), mental workload (Klingner et al., 2011), surprise (Braem et al., 2015), memory (Goldinger & Papesh, 2012; Otero et al., 2011), arousal and affective processing (Bradley et al., 2008; Cohen et al., 2015), imagery and mental manipulation (Mathôt et al., 2017), and movement preparation (Wang et al., 2015, 2016). Pupillometry is also frequently used in ophthalmology studies focusing on pupil physiology, as well as various eye diseases (Binda et al., 2017; Feigl et al., 2012; Kostic et al., 2016). Several methodological guides for pupillometry data acquisition and processing have been published (Kret & Sjak-Shie, 2019; Lemercier et al., 2014; Mathôt & Vilotijević, 2022; Winn et al., 2018). However, the field of pupillometry is changing rapidly and hence, updated uniform norms for data preparation and analysis beyond specific domains (e.g., hearing, emotions) are required. In the following sections, we will present steps required to perform this task, the different methods available to researchers, and the pros and cons of each. In addition, we include instructions for CHAP (open-source software written in MATLAB, Hershman et al., 2019) for carrying out the data preparation and some of the presented analyses.

2 Pre-processing

2.1 *Artifacts*

Analysis of pupillometry data requires considering several artifacts that may skew the results. Therefore, those artifacts should be addressed before the data processing stage. Eyeblinks (i.e., spontaneous momentary coverage of the eyeball by the eyelid) as well as eye and head movements are the most common artifacts. These artifacts will be discussed in the following sections.

2.1.1 Eyeblinks

Covering the eyeball by an eye blink prevents eye-tracking devices from detecting the pupils properly. As a result, incorrect reports of reduction in pupil size are obtained, although no actual change occurs (see Fig. 1 for a visual example). Normal subjects blink about every 2.5 s and for clinical subjects, the frequency may rise to every 2 s (Karson, 1983). Although eyeblinks may occur randomly during an experimental trial, and be uncorrelated with stimulus timing landmarks, they add noise that may cover up an actual effect. However, sometimes eyeblink occurrences can be affected directly by a task's characteristics such as stimulus presentation, the relevance of the stimulus, or a behavioral response (Fukuda, 2001; Siegle et al., 2008; Stern et al., 1984). In the above circumstances, the apparent systematic decrease of pupil diameter in the relevant time-windows might be misinterpreted as an actual pupillary effect.

In light of the above, eyeblink data should be excluded from the raw data. Replacing the blink data samples with other values is the preferred solution, usually by using interpolation algorithms (e.g., linear interpolation (Sirois & Jackson, 2012) or cubic interpolation (Mathôt, 2013)). The repaired values (according to the chosen interpolation approach) fill the gap between each two data points representing an eyeblink's onset and offset.

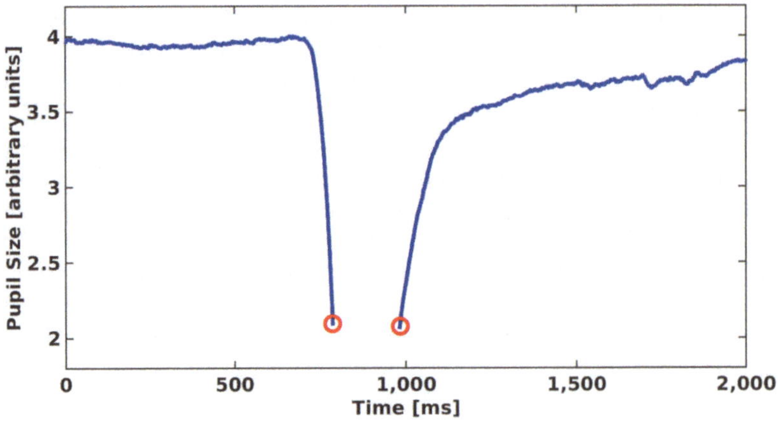

Fig. 1 A typical eye blink (sampling rate = 1000 Hz). The beginning and ending of the eye blink are characterized by sharp slopes (an artifact that is created by the closing and opening of the eyelid). The red circles represent the onset and the offset of a time-window in which no information about the pupils exists

Importantly, the detection of the specific time points between which the eyeblinks occur is not a trivial matter. In most eye-tracking devices used in research, no inherent indications of these actual time-points can be found. Moreover, most often, the eye-tracking devices define blink onset and offset as the onset and the offset of the time-window that does not contain any information about the pupils (i.e., when the eyelid completely covers the pupils). However, the eyelid movement is not instantaneous, that is, the duration between the blink onset and the first missing values, as well the duration between the missing values and the blink offset, is larger than zero ms. Thus, blink-related data starts a bit before the first missing sample and ends a bit after the last missing sample (see Fig. 1 for a visual example). Accordingly, blinks are characterized by a sharp decrement in pupil size during the start of the blink (Caffier et al., 2003). Hence, detecting clear starting and ending points for the blink window can be a significant difficulty.

Several methods have been suggested for dealing with eyeblink detection before interpolation of problematic values. The most naïve approach determines the nearest time-point before and after the missing values as the eyeblink onset and offset, respectively (see the red circles in Fig. 1). Inspection of Fig. 1 indicates that this approach (that takes into account samples that occurred during the time the eyes were closed and reopened) may entail including incorrect values in the statistical analysis. Including confidence intervals before and after these values (e.g., 100–200 ms before and after the time-window of the missing values) can be used also for an improved approach (de Gee et al., 2014; Nyström et al., 2016; Satterthwaite et al., 2007; Wierda et al., 2012). Applying this approach may result in replacing valid values with interpolated ones and therefore might distort the results.

The movement of the eyelid during the eyeblink is faster than any physiological natural change in pupil size that may be reflected in cognitive and perceptual processes. This characterization can be used in an advanced approach that looks for rapid changes in pupil dilation;, namely, defining blink onset and offset points based on the time-points when the velocity of changes in pupil size crosses a threshold (Mathôt, 2013). However, the velocity of pupil size changes during the eyeblinks is not a constant, neither across participants, nor across trials. Therefore, applying this approach may lead to too many misidentifications, resulting in incorrect values.

An additional approach is based on the typical noise produced by an error measurement that characterizes the commonly used eye-tracking devices. In eye-tracking devices, the image processing of the data received from the camera sensor produces an inherent noise. However, in the time period between closing and reopening of the eyelids during eyeblinks, this noise is relatively eliminated. That is, the strong signal recorded during the rapid movements of the eyelid is stronger than the inherent noise. By identifying the moment in time in which the noise signal vanishes and re-emerges, eyeblink onset and offset can be defined more accurately (Hershman et al., 2018) (Practice Box 1).

Practice Box 1

CHAP allows to users to detect and to reconstruct blink data. Once the data file is loaded, the user is able to define how to interpolate the data (red field) by using linear or cubic interpolation. In addition, the user can decide to keep the blink data as is and not interpolate it (CHAP's default interpolation is linear interpolation).

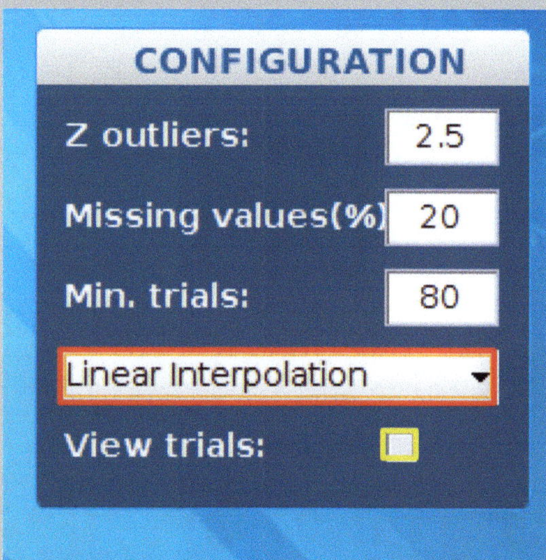

Choosing the *View trials* option (yellow field) will present the trials one by one. The data will be presented with and without the blink corrections. In addition, the detected blinks' onset and offset (as detected by Hershman et al. (2018) algorithm) will be indicated in the presented figures. Clicking on the space bar will present the next trial.

2.1.2 Eye Movements

The pupils not only constantly change in size but also continuously change position in the eye socket. An eye movement caused by a change in direction of gaze is self-evident, but even during attempted fixation to maintain high visual acuity, small eye movements appear (Leigh & Zee, 2015). In fact, eye movements are a necessary and essential condition for keeping the retina steadily pointed at the environment. This eye movement repertoire may become a serious artifact in pupillometry study. Specifically, incorrect detection (due to eye or head movements) may result in an incorrect calculation and misleading reporting of the pupil size. Moreover, changing the position of the pupil relative to the eye-tracker device may cause a distortion of perspective. That is, different locations of the pupil within the eyeball are perceived by the device as different shapes. A pupil located in the center of the eyeball, just in front of the eye-tracking device, is perceived as a circle. However, if the pupil is tilted to the side (relative to the eye-tracker), the shape perceived in the device lens is an ellipse. The more a pupil is tilted, the narrower the ellipse that is detected, and a smaller pupil size might be recorded (see Fig. 2 for a visual example).

Unlike the ease of detecting eyeblinks by inspecting pupillometry data only, the eye movement artifact is more complicated to handle and repair. Eye-tracking devices detect the pupil by using image processing algorithms. Several approaches

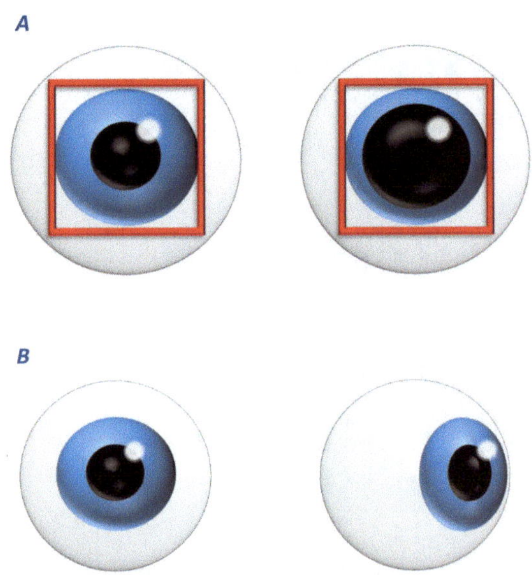

Fig. 2 Examples of potential distortion/bias in the observation of the pupils by an eye tracker device. (**a**) Difference due to physiological variability. (**b**) Observed difference due to variance in the position of the pupils relative to the device. On the left, the pupil is at the front of the eyeball and therefore in the front of the device. Hence, the shape of observed pupil is a circle. On the right, the pupil is in the periphery (here at the left of the eyeball) of the camera. Therefore, the shape of the observed pupil is an ellipse

have been suggested for dealing with eye movement artifacts. Exclusion of trials with eye movements based on the saccade amplitude (i.e., the distance from the required fixation point) is the most common approach (Binda et al., 2013; Tkacz-Domb & Yeshurun, 2018). Advanced mathematical models provide formulas that correct the pupil size to the actual pupil size (Brisson et al., 2013; Gagl et al., 2011; Hayes & Petrov, 2016). However, in most cases, these models are tailored to specific devices and require appropriate preparation in advance. A better approach could be reducing the research set fragility to the eye movement artifact at the procedure design stage; that is, use of a procedure that does not require eye movements and encourages participants to look at one fixed point. For example, instead of presentation of full sentences that require eye movements along the text, the researchers may consider a serial presentation (i.e., word-by-word presentation). This method eliminates the need for eye movements during reading. Therefore, sensitivity to this artifact may be decreased as well (see Exp. 2 in Hershman et al., 2023).

2.2 *Outlier Samples*

Due to technical limitations of the eye-tracking devices in image processing ability, occasionally extremely large or small values that cannot be associated with actual physiological activity of the pupil may be reported. These values may be defined as outlier samples. The outlier values are not associated with task characteristics, and they may occur randomly during the experimental trials. Similar to the artifacts described above (i.e., eyeblinks and eye movements), these samples may produce noise, which cover an actual effect. However, unlike the well-defined eyeblink pattern (in terms of changes in pupil size), no systematic pattern may characterize these outlier samples. Dealing with outlier samples is done at three levels: the sample level, trial level, and participant level. In the first stage, outlier exclusion is the most recommended method for dealing with the outlier samples (Akdoğan et al., 2016; Cohen et al., 2015; Wainstein et al., 2017). Specifically, in each trial (or any other well-defined time-window), samples with extremely large or small values should be excluded by using a well-defined threshold (e.g., 2.5 Z-scores). After the exclusion of the problematic values, similar to dealing with eyeblinks, interpolation (e.g., linear interpolation) between the first and last known points is required (see Fig. 3 for visual example).

The interpolation method is suitable for bridging the gap between two adjacent time points (e.g., eyeblink onset and offset). However, excessive interpolations can cast doubt on the validity of the data in representing the actual results, that is, the creation of non-reliable data for the analysis. Thus, a recorded time-window consisting mainly of missing values (due to noise or any other artifacts) needs to be excluded from the analyzed data. Accordingly, in the second stage of handling outliers, problematic trials consisting of too many outlier samples need to be excluded (Chiew & Braver, 2013; Koelewijn et al., 2012; Zekveld et al., 2014; Zekveld & Kramer, 2014). The exclusion of a trial can be done by a well-defined threshold

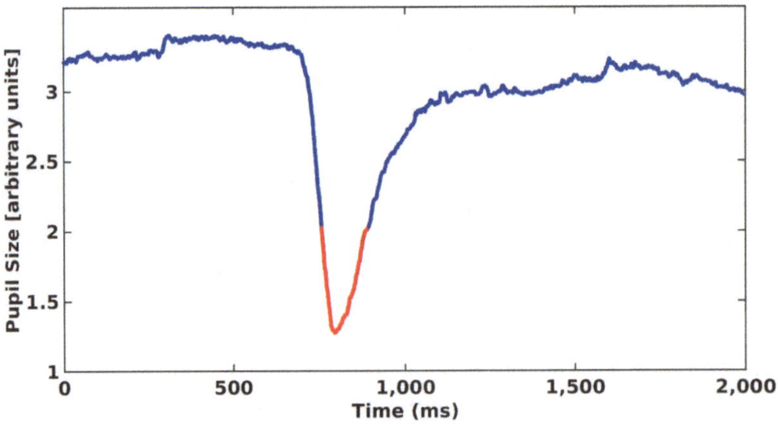

Fig. 3 An example of outlier samples that should be excluded from the analysis. Here, the combined curve represents the given pupil size as a function of time. The blue curve represents the samples that have valid values (inside the range between 2.5 above and below the mean of the trial), and the red curve represents invalid samples that should be excluded from the analysis

based on the required time-window. Note that setting an unfitted threshold may distort the results. On one hand, a stringent threshold produces clean and informative data. However, it may cause exclusion of too many outlier trials and reduce the statistical power. On the other hand, an indeterminate threshold produces few outlier trials. Nevertheless, these trials may not reflect the actual data. We highly recommend using a threshold of at most 20% of missing values, although researchers may use a less stringent criterion but no more than 50% of missing values.

In the third stage of the outlier handling process, the researcher may choose a minimum threshold for trials that will allow for statistical analysis. Importantly, trials may be excluded due to behavioral responses (e.g., slow or wrong responses) as well. This may result in lack of a sufficient number of trials in some investigated conditions for several participants. In this case, participants who do not have enough trials for each investigated condition need to be excluded from the analysis. The threshold for excluding samples and trials is not dependent on the experimental procedure employed. However, the threshold for participants' exclusion is derived from the eye-tracking device quality (i.e., the better the quality of the device, the smaller number of trials that are required per condition) and from the task itself. Currently, there is no clear rule of thumb indicating a specific threshold for each experimental procedure or tracking device. It is therefore necessary to examine the subjects of each experiment on their own merits. That is, if the data are too noisy, more trials are probably needed (Practice Box 2).

Practice Box 2

CHAP allows users to define three criteria for exclusion:

1. Exclusion of outlier samples by using a Z-score.
2. Exclusion of trials with too many outlier samples.
3. Exclusion of a participant who doesn't have a minimal number valid (pupillometry) trials per condition.

Once the data file is loaded, the user is able to define the relevant criteria. *Z outliers* (red field) will define the Z-score that will be used as the threshold for sample exclusions (CHAP's default value is a Z-score of 2.5). *Missing values (%)* (yellow field) will define the maximum percent of missing values (caused by eyeblinks and other artifacts) in a trial that will cause the trial to be excluded (CHAP's default value is 20%). *Min. trials* (green field) will define the minimum number of valid (pupillometry) trials that is required for each condition. Participants with a smaller number of trials than the chosen number in one or more of the investigated conditions will be excluded from analysis (CHAP's default value is 80 trials).

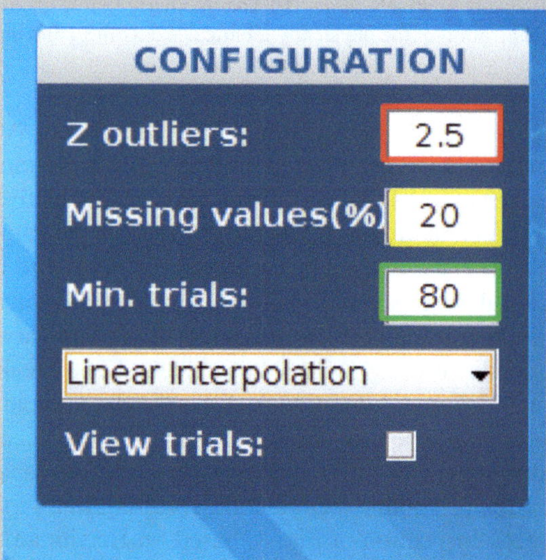

3 Data Processing

3.1 Data Normalization

Eye-tracking devices are based on image processing of the data from the camera sensors, that is, converting pixel groups into a number that represents the pupil size. Unless advance converting to millimeters is applied, the unit of the provided values

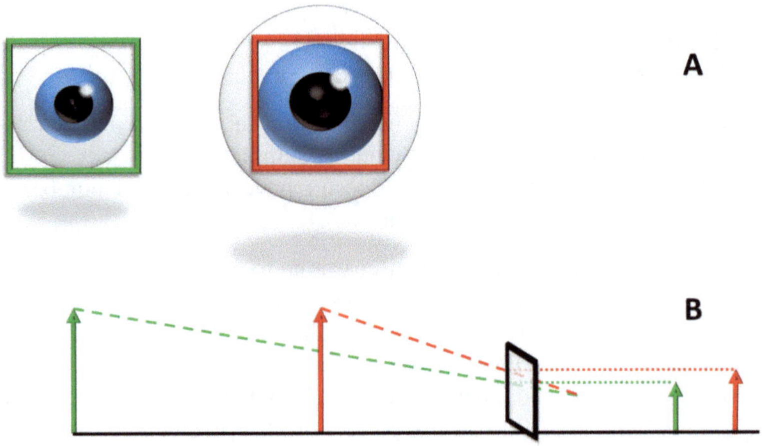

Fig. 4 An example of the observed pupils by the eye-tracking devices in the case of different distances of the eyes from the eye-tracking device. (**a**) The observed pupil size in the eye tracker. On the left (the green frame), the eyes and the pupils are relatively far from the camera. In this case, the observed pupil size is relatively small. On the right (the red frame), the eyes are close to the camera. Therefore, the observed pupil size is relatively large. (**b**) The geometry that leads to the different values. On the left is the actual position of the physically identical sized object and on the right is the observed size

is arbitrary. Many eye-tracking devices (e.g., EyeLink, The Eye Tribe, ASL, Gazepoint, and VPIXX) do not provide pupil size in millimeters. Importantly, expressing pupil size values by arbitrary units (e.g., number of pixels) is sensitive to biases caused by the position of the camera and the distance between the camera and the participant. That is, the shorter the distances, the larger the pupil size that will be observed by the eye tracker device (see Fig. 4 for a visual demonstration). However, the distance between the subjects and an eye-tracking device can vary, as can the distance under different conditions (e.g., when the subject takes a break or changes physical position between different conditions). Accordingly, comparing pupil size across participants, and across experimental blocks and trials, can become a major obstacle. In particular, different scales may lead to high scattering (as can be reflected by a large standard error or confidence interval), resulting in a type II error (i.e., rejection of a true effect). Inspection of Fig. 5, based on Hershman and Henik's (2019) study, demonstrates relatively large scattering when analyzing pupillometry data without any normalization. Although the means of the conditions imply meaningful differences between the investigated conditions, the scatter is too wide for accepting this hypothesis. On the other hand, the same data, analyzed after a normalization procedure, yield completely different results (see Fig. 6). In line with the above, for avoiding a type II error, normalization of the pupil size values should be done before performing the analysis.

There are several approaches for normalizing pupillometry data. The most reliable approach starts with converting arbitrary units of each participant into

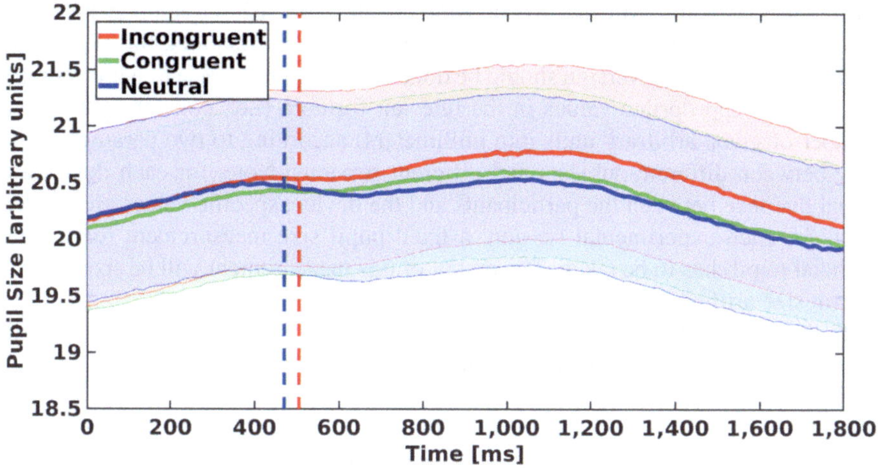

Fig. 5 Mean pupil size as a function of time in a two-color Stroop task run on 19 participants (Exp. 1 in Hershman & Henik, 2019). Participants were instructed to respond to the color of a word and to ignore its meaning. The task consisted of congruent, incongruent, and neutral trials (e.g., congruent: BLUE, incongruent: BLUE, and neutral: XXXX). 0 represents the time of the Stroop stimulus onset, and the vertical lines close to 500 ms represent mean participant response. The shaded areas represent 1 standard error from the mean. Here the data were averaged as is, without any normalization

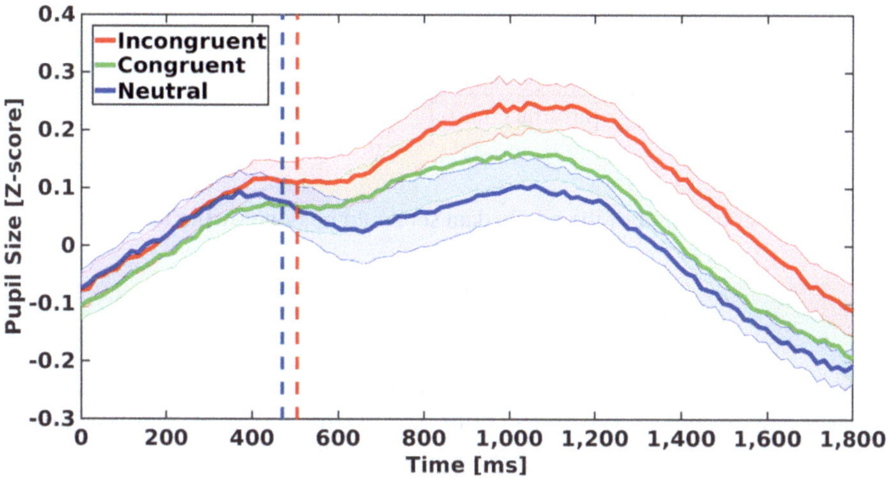

Fig. 6 Mean pupil size as a function of time in a two-color Stroop task identical to the data presented in Fig. 5. 0 represents the time of the Stroop stimulus onset, and the vertical lines close to 500 ms represent mean participant response. The shaded areas represent 1 standard error from the mean. Here the data was averaged after Z-score normalization

millimeters (Binda & Murray, 2015a; Binda et al., 2014; Blaser et al., 2014; Siegle et al., 2004). While a few devices provide the data in millimeters, for most of the devices, a post hoc conversion should be done. To this end, the researcher is required to place the appropriate values in the relevant formula (i.e., converting an actual number of given arbitrary units into millimeters) according to two parameters that vary between different subjects and different sessions. Note, for each device and actual distance between the participants and the device, specific values are needed. First, for each experimental session, a fixed pupil size measurement (e.g., of an artificial pupil) has to be taken. The output of this measurement will be size of well-known size artificial pupil (AP) in arbitrary units.

In the second stage, in the case of measurement of the pupil diameter, the ratio between the size in arbitrary units and the actual size in millimeters provides the key for the total conversion:

$$\text{key} = \frac{\text{AP}_{\text{diameter}}\left[\text{mm}\right]}{\text{AP}_{\text{diameter}}\left[\text{arbitrary}_{\text{units}}\right]}.$$

Then, the conversion of the whole data set could be done by the formula:

$$\text{pupil\_diameter}\left[\text{mm}\right] = \text{pupil\_diameter}\left[\text{arbitrary\_units}\right] \bullet \text{key}$$

In the case of measurement of the pupil area, the key would be defined by the following formula:

$$\text{key} = \frac{\text{AP}_{\text{diameter}}\left[\text{mm}\right]}{\text{AP}_{\text{diameter}}\left[\text{arbitrary units}\right]} = \frac{\text{AP}_{\text{diameter}}\left[\text{mm}\right]}{2\sqrt{\dfrac{\text{AP}_{\text{area}}\left[\text{arbitrary units}\right]}{\pi}}}$$

Then the conversion of the whole data set could be done by the formula:

$$\text{pupil}_{\text{diameter}}\left[\text{mm}\right] = 2\sqrt{\frac{\text{pupil}_{\text{area}}\left[\text{arbitrary units}\right]}{\pi}} \cdot \text{key}.$$

Normalization by converting data into millimeters requires preparation in advance and must be done before or after each recording session. This type of normalization is highly recommended because it provides the actual required parameter—the pupil size. However, sometimes it might be difficult to use this normalization (e.g., when the recording of an artificial pupil is difficult to carry out). In that case, other normalization approaches should be used. One of the most popular normalization approaches is Z-scores (Cohen et al., 2015; Einhauser et al., 2008; Kang & Wheatley, 2015; Wainstein et al., 2017). In this approach, the mean (μ) and the standard error (σ) across the whole experiment are calculated for each participant (after the artifacts are resolved). Next, for each pupillometry value (x), the Z-score is recalculated using the formula:

$$Z_x = \frac{x - \mu}{\sigma}$$

Inspection of Fig. 6 illustrates a meaningful influence of the normalization on the results; namely, not only the means of the conditions indicate meaningful differences between the investigated conditions, but also the scattering is relatively small. Therefore, a type II error may be avoided (Practice Box 3).

Practice Box 3

Once the data for each condition has been selected and presented, CHAP allows users to normalize data by converting the data into Z-scores (red field). By using this approach, the Z-score will be calculated by using the mean and the standard error pupil size across the whole experiment.

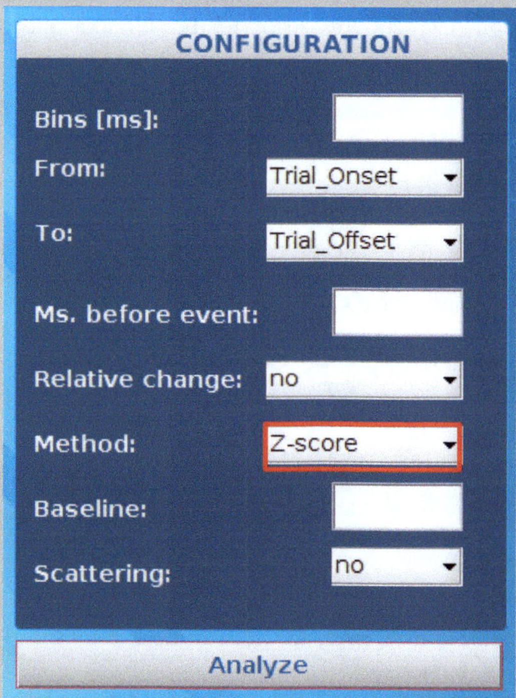

A second useful normalization method is based on calculation of the relative changes in pupil dilation compared over a well-defined time-window. That is, all the measured samples are divided by a representative parameter (e.g., mean or median)

from a chosen time-window (Binda et al., 2014; Kloosterman et al., 2015; Mathôt et al., 2013; Snell et al., 2018). Unlike the normalization methods described above (i.e., millimeters and Z-scores), the relative changes method can detect differences between conditions regardless of the starting point pupil size (further explanation can be found in the section below discussing baseline correction). Importantly, this method converts measured values into units of pupil dilation/constriction that cannot be used for measuring absolute pupil size (Sirois & Jackson, 2012) (Practice Box 4).

Practice Box 4

Once the data for each condition has been selected and presented, CHAP allows users to normalize data by presenting relative changes compared to a well-defined baseline. First, the event of interest should be selected. In our example, we are interested in the stimulus onset. Therefore, time 0 will be define as the stimulus onset moment (red field). Then, we will define how the relative change (yellow field) in the data should be presented; in this particular example, the relative change is in percentage. After that, we will define the baseline to be the mean pupil size 200 ms prior to the stimulus onset (green field).

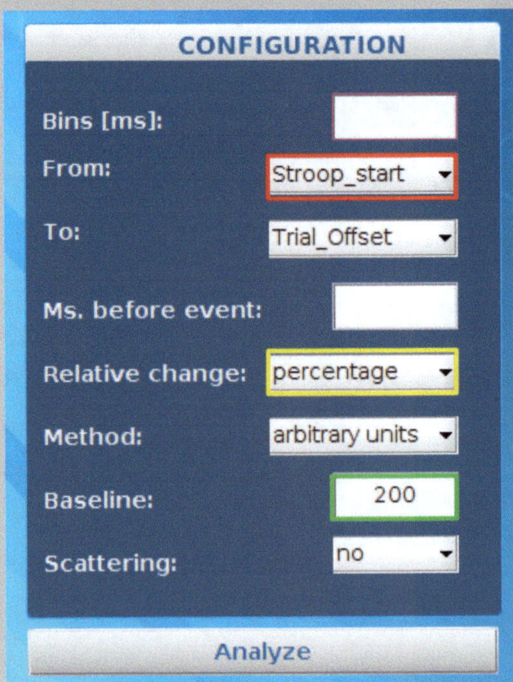

3.2 Baseline Correction

In typical pupillometry experiments, the relevant data are changes in pupil size (i.e., dilation and constriction) resulting from the relevant experimental manipulations. If there is no difference in pupil size when the stimulus onset occurs, the changes can be represented in absolute values. However, this requirement might be often violated, due to differences between conditions or other artifacts that can affect pupil size between trials. In order to solve this problem, a baseline correction procedure is proposed, that is, a transformation of the data from an absolute size to relative change size from a baseline value. However, it has been accepted that baseline correction may help detect changes in pupil size between conditions that do not necessarily result from the manipulation applied. For instance, inspection of Fig. 6 may reveal evident differences between congruent trials (i.e., green curve) and neutral trials (i.e., blue curve) prior to stimulus onset for about 500 ms. It goes without saying that in this case, it is difficult to attribute the observed change in pupil size to the experimental manipulation. Baseline correction may clarify this issue. As described above, pupil size can be displayed as changes relative to a well-defined time-window. In other words, the information provided is not about the absolute size of the pupil at a specific time point, but the relative change following the experimental manipulation regardless of the actual pupil size at the initial reference point. Inspection of Fig. 7 demonstrates that applying baseline corrections (i.e., comparing the change values instead of the absolute values) eliminated the apparent differences observed prior to the stimulus onset.

Two optional methods for performing baseline correction are typically employed by researchers: *subtractive baseline correction*, and *divisive baseline correction*. In

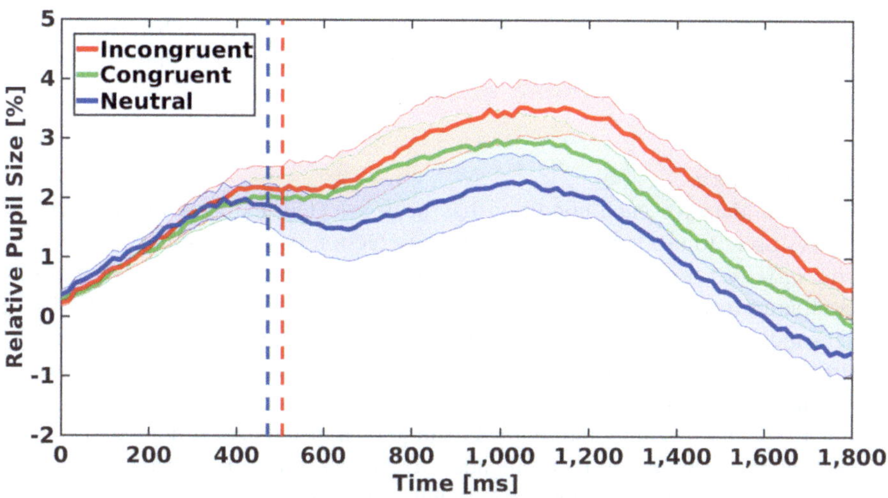

Fig. 7 Mean pupil size as a function of time in a two-color Stroop task identical to the data presented in Figs. 5 and 6. 0 represents the time of the Stroop stimulus onset, and the vertical lines close to 500 ms represent mean participant response. The shaded areas represent 1 standard error from the mean. Here the data was calculated relative (divided baseline correction) to the mean pupil size in the time-window of 200 ms before the manipulation onset

both methods, determining the baseline value can be based on a representative parameter (e.g., mean (Blaser et al., 2014) or median (Mathôt et al., 2018)) from a selected time-window. The time-window duration can vary between a few milliseconds (e.g., the first sample of the trial) up to 1000 ms (Höfle et al., 2008). Short time-windows may include relatively homogeneous values. Long time-windows are likely to be more heterogeneous, revealing the pupil size changing pattern during the baseline period. In the *subtractive baseline correction* (Binda et al., 2014; Binda & Murray, 2015a; Steinhauer et al., 2004), each measured sample—whether obtained from the eye-tracking device or after normalization—is subtracted from the baseline. In the *divisive baseline correction* (Kloosterman et al., 2015; Mathôt et al., 2013; Snell et al., 2018), each measured sample is divided by the baseline. Accordingly, the values of the samples convert to ratios.

There are two optional methods to calculate and present the relative changes. The methods differ from each other in the definition of the reference point (i.e., the baseline). In the first method, identity of the actual pupil size and the baseline value is represented by 0 or 0% (Hershman & Henik, 2019; Kloosterman et al., 2015; Preuschoff et al., 2011). Accordingly, −0.5 (or −50%) represents decreasing pupil size by half its size relative to baseline value. Similarly, a value of 1.0 (or 100%) represents an increase of the pupil size to double relative to baseline value (see Fig. 7). Alternatively, instead of using zero as baseline, 1.0 (or 100%) may represent equality between the actual pupil size and the baseline value (Mathôt et al., 2013; Snell et al., 2018; Tkacz-Domb & Yeshurun, 2018). Therefore, decreasing pupil size by half its size relative to the baseline value is represented by 0.5 (or 50%) and 2.0 (or 200%) represents doubling the pupil size relative to the baseline value.

Subtractive baseline correction uses the measured units. Hence, in the case of non-normalized (arbitrary) units, the mean pupil size across trials (and across participants) may take into account extreme values. The mean is influenced by extreme values and therefore, it may indicate false effects. As such, researchers who prefer the *subtractive baseline correction* need to make sure that the data units are reliable (Practice Box 5).

Practice Box 5

Once the data for each condition has been selected and presented, CHAP allows users to use subtractive baseline correction (see instructions for *divisive baseline correction* in Practice 4) data by presenting relative changes compared to a well-defined baseline. First, the event of interest should be selected. In our example, we are interested in the stimulus onset. Therefore, time 0 will be define as the stimulus onset moment (red field). Then, we will define how relative change (yellow field) in the data should be presented; in this particular example, the relative change is in difference. If the data is not normalized, it is highly recommended to use a Z-score instead of arbitrary units (green field). After that, we will define the baseline to be the mean pupil size 200 ms prior to the stimulus onset (purple field).

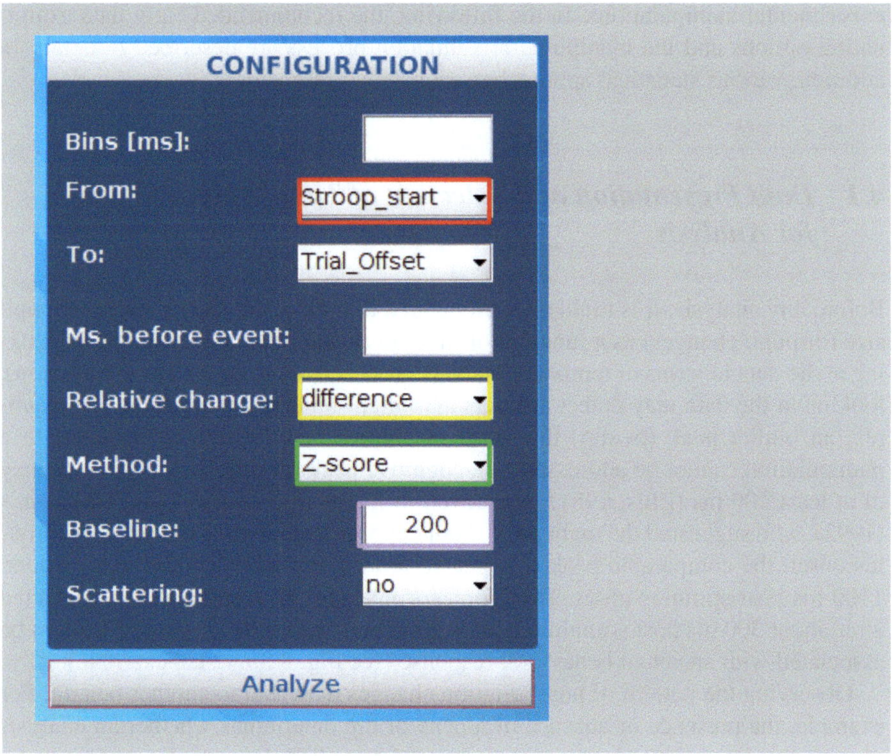

4 Data Analysis

In contemporary research one can find different statistical approaches for analyzing pupil size data. These methods differ not only in the technical and computational aspects but also in the theoretical assumptions they reflect. The most common analysis approach is based on calculating a single value by collapsing multiple samples over a selected time-window. This group of methods includes comparing between averages (Binda et al., 2014; Laeng et al., 2011), peak amplitudes (Binda & Murray, 2015b; Hess & Polt, 1964; Schmidtke, 2014), dip amplitudes (Bitsios et al., 2004; Henderson et al., 2014; Steinhauer et al., 2000), peak latency (Hess & Polt, 1964; Koelewijn et al., 2015; Laeng et al., 2011; Schmidtke, 2014), and dip latency (Bitsios et al., 2004; Henderson et al., 2014; Steinhauer et al., 2000). The temporal approach suggests a different perspective, namely, analyzing pupil changes over time (Hershman & Henik, 2019, 2020; Hoeks & Levelt, 1993; Kuchinsky et al., 2013; Sirois & Brisson, 2014). Regardless of the chosen statistical approach, it is advisable to examine the data in advance, in particular using its visualization along the timeline. Such an inspection can aid in choosing (or rejection) of a specific analyzing approach, in locating outliers, and in supporting the validation of experimental manipulations. In the following, the recommended early data examination options and the manipulation validation process are described in detail. In addition, various statistical approaches are briefly presented.

4.1 Data Presentation and Selection of Time-Window for Analysis

Before any analysis, it is highly recommended to look at the data in terms of pupil size (or pupil changes) as a function of time. In addition to the advantages of looking at the data in terms of temporal analysis (as described at the end of this section), looking at the data may detect artifacts that might influence the results. For example, an outlier peak (or dip) that does not represent a behavioral response to a manipulation cannot be addressed as a cognitive peak (or dip). Pupils have latency of at least 200 ms (Ellis, 1981). Moreover, in line with Steinhauer and Hakerem's (1992), who suggested the main peak should appear after about 1100 ms post stimulus onset, the comparable peak is expected around this time-window (i.e., around 1100 ms post stimulus onset. Therefore, for instance, an amplitude peak detected after about 300 ms post stimulus onset is apparently an outlier because it cannot be associated with an actual behavioral response (see Fig. 8 for visual example).

Observing the pattern of pupil dilation changes may lead to another benefit. For example, the presence or absence of a peak or dip determines whether an analysis of peak (or dip) amplitude or latency should be applied. Importantly, absence of a

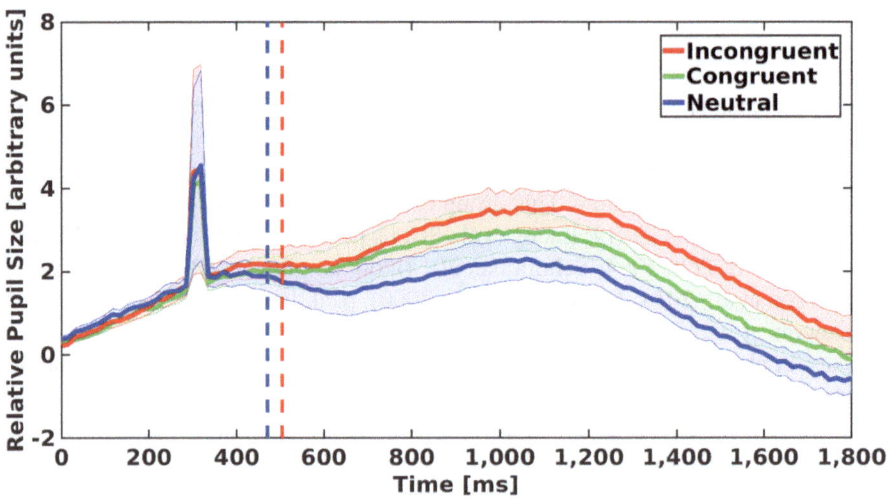

Fig. 8 Manipulated mean pupil size as a function of time in a two-color Stroop task (Hershman & Henik, 2019). 0 represents the time of the Stroop stimulus onset, and the vertical lines close to 500 ms represent mean participant response. The shaded areas represent 1 standard error from the mean. Here the data have an incorrect peak after about 300 ms post stimulus onset

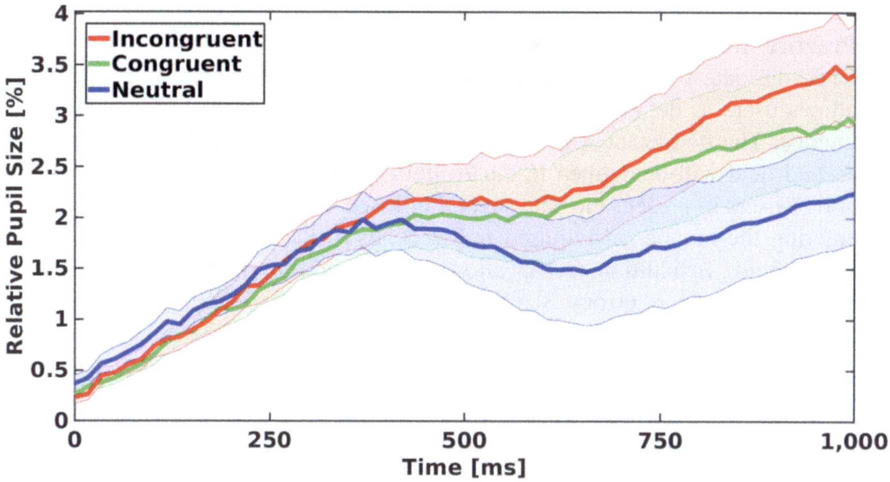

Fig. 9 Mean pupil size as a function of time in a two-color Stroop task (cut data from Exp. 1 in Hershman & Henik, 2019). 0 represents the time of the Stroop stimulus onset, and the vertical lines close to 500 ms represent mean participant response. The shaded areas represent 1 standard error from the mean. Here the data has no global peak

global peak characterizes findings from diverse cognitive and emotional pupillometry studies (van Rij et al., 2019). In these studies, the pupils' incremental change continues until the end of the chosen time-window, as is clearly demonstrated in continuous discourse or reading tasks (Kaakinen & Simola, 2020; Schluroff et al., 1986; Vogelzang et al., 2016). In cases like these, analysis of peak (or dip) amplitude would not be beneficial (due to absence of extreme amplitudes or due to continuous dilation throughout the time-window). For example, in Fig. 9, there are three conditions that led to pupil dilation. Obviously, in this example, it is possible that the selected time-window was not appropriate. Still, if the selection of the time-window is valid, one cannot apply an analysis of peak (or dip) amplitude or latency. This is because pupil dilations continue until the end of the chosen time-window.

Sometimes the visual presentation may help researchers to clarify their findings. For instance, inspection of Fig. 7 suggests that the differences between the conditions become prominent from about 500 ms post stimulus onset to the end of the trial. No such differences can be found at the beginning of the trial (i.e., 0–500 ms). Such observations can clarify why a comparison between different conditions based on a full time-window may not yield clear differences. Specifically, a full time-window (including the time period of the first 500 ms) may hide the effects and lead to type II errors (see discussion in the temporal analysis approach section).

While looking at the pupil size (or changes) as a function of time may provide clearer information about the differences and the similarities between the conditions, the scattering of the data may indicate the validity of the effects. For example, Fig. 10 implies a meaningful difference between the conditions between 0 and 250 ms post stimulus onset). However, these differences are negligible, given the wide overlapping of the scattering (i.e., standard error from the mean) (Practice Box 6).

Practice Box 6
Once the data for each condition has been selected and presented, CHAP allows users to define the time-window of interest by using two events of interest. The first event will be defined in the *From* field (red field) and the second field will be defined in the *To* field (yellow field). Moreover, CHAP allows users to present the data several milliseconds before the first event by entering the number of required milliseconds in the *MS. before event field* (green field). In addition, CHAP allows users to present scattering around the mean (e.g., standard error—SE) by choosing the required scattering option in the *Scattering* field (purple field).

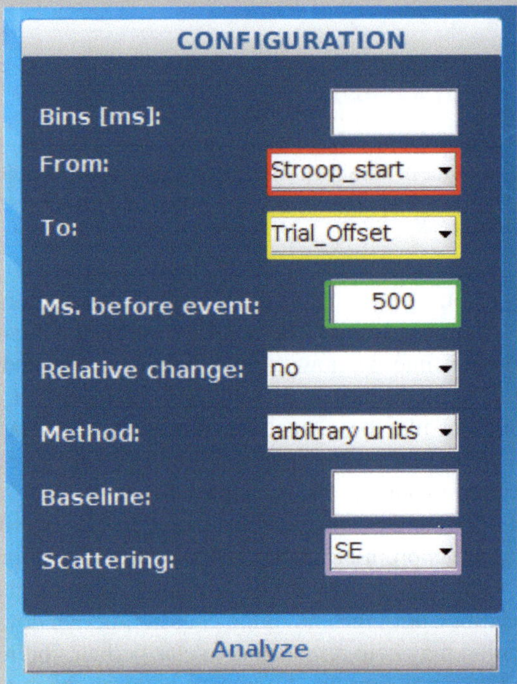

4.2 Manipulation Check

In general, pupil size is compared between different manipulations. The manipulation could be visual (Kloosterman et al., 2015; Partala & Surakka, 2003; van Steenbergen & Band, 2013), auditory (Koelewijn et al., 2015; Winn et al., 2015; Zekveld & Kramer, 2014), or imaginary (Henderson et al., 2018). In these cases, the assumption is that the manipulation causes the differences in pupil sizes. Therefore, any observed differences after the manipulation onset would be associated with the manipulation itself. To be sure that the differences are due to the manipulation itself, an analysis of the time-windows before the manipulation onset should be done.

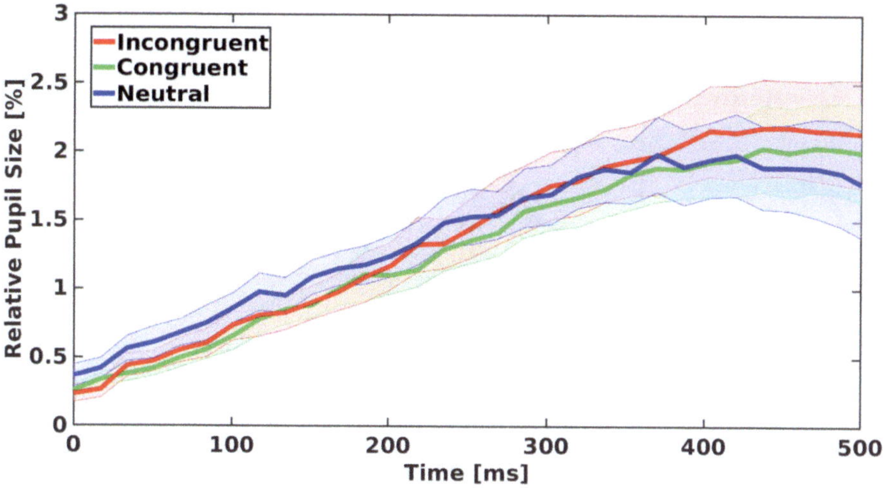

Fig. 10 Mean pupil size as a function of time in a two-color Stroop task (Hershman & Henik, 2019). 0 represents the time of the Stroop stimulus onset, and the vertical lines close to 500 ms represent mean participant response. The shaded areas represent 1 standard error from the mean. Here the data include only the first 500 ms post-stimulus onset

Figure 11 presents a reliable effect. In this figure, we manipulated the values of the pupil dilation in the neutral condition before the manipulation onset (data(1 : 30) = data (1 : 30) + 0.03 × flip(1 : 30);)[2]. As is reflected in the figure, it seems that after about 450 ms post manipulation onset, the pupil size of the neutral condition (the blue curve) deviates from those of both the congruent and the incongruent conditions. Ostensibly, it seems that the manipulation has a strong influence on the pupil size. However, the question about the validity of the manipulation (in terms of pupil size) cannot be answered by these results only. As is reflected from the pattern of the pupil dilation before the manipulation onset, there are differences between the conditions *before* the manipulation onset. Hence, it will be difficult to conclude that the manipulation itself led to the results at 450 ms post stimulus onset. Moreover, in this scenario, the analysis itself should be re-considered.

5 Statistical Methods

5.1 The Collapsing to Singular Value Methods

Although pupillometry is essentially a temporal measure, the most common approach for analyzing pupil change data is based on collapsing multi-samples to a singular value across a chosen time-window (for detailed reviews see Attard-Johnson et al., 2019; Hershman et al., 2023). There are several variants for this

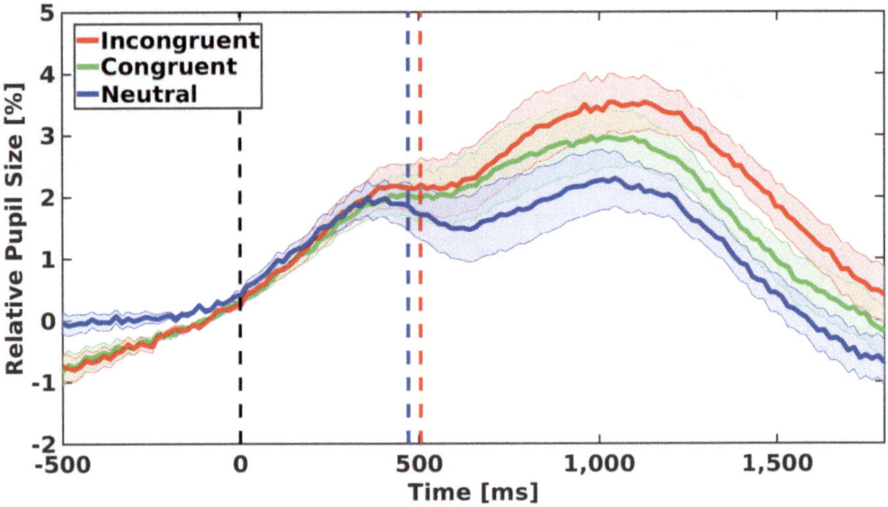

Fig. 11 *Manipulated* mean pupil size as a function of time in a two-color Stroop task (Hershman & Henik, 2019). 0 represents the time of the Stroop stimulus onset, and the vertical lines close to 500 ms represent mean participant response. The shaded areas represent 1 standard error from the mean. Here there is smaller pupil dilation in the neutral condition compared to both congruent and incongruent conditions at 450 ms post stimulus onset. However, in the present figure of manipulated pupil size, there are also differences between the observed pupil dilation between the conditions before the manipulation onset

approach. Calculating the average of all samples (depending on the size of the time-window and on the sampling rate) and then comparing the averages of the various experimental conditions is the most frequently used technique (see example in the introduction of the *Data analysis* section). Importantly, it has been suggested in previous studies that mean pupil dilation is correlated with cognitive load, it depends on the duration of the interval in which the data are collected, and is less susceptible to random variation in pupil size than the peak amplitude is (Ahern & Beatty, 1979; Beatty, 1982; Siegle et al., 2001). Therefore, the collapsing to singular value methods can be appropriate if researchers do not expect meaningful changes in effect within the chosen time-window. In other words, the use of the averaging approach is based on the premise that the effect (if it can be detected) is relevant to the entire time-window examined. However, this is not always the case (see detailed discussion in the *Temporal analysis approach* section below).

Another typically used analysis refers to peak or dip amplitude of the pupils during the measured time-window (Beatty, 1982; Steinhauer et al., 2000). It has been suggested that peak amplitude (i.e., the actual maximum value) reflects the maximum task-evoked processing load, whereas the mean dilation reliably indicates the total processing load in a certain interval (Ahern & Beatty, 1979; Siegle et al., 2001). In this approach, the maximum or minimum value of the whole signal of the trial is calculated. Once the values are calculated, it is possible to compare the average peaks or dips in the various conditions. This method is based on a single value and therefore may be more susceptible to random variations compared with an averaging technique (Beatty & Lucero-Wagoner, 2000). In order to avoid using extreme values that may bias the results, it is recommended to set a time-window around the peak for analysis. In addition, as already described above, peak (as well as dip) analysis cannot be applied if the pattern of pupil dilation changes does not include distinct peaks or dips.

Peak latency is strongly associated with processing load (e.g.,, van der Meer et al., 2003). Therefore, another common analysis is of peak or dip latencies (Hess & Polt, 1964; Steinhauer et al., 2000). These dependent measures are defined as the duration from the stimulus onset until the peak (or dip) amplitude of the pupil. Thus, they can only be applied following the peaks (or dips) calculation. Like the other singular value methods described above, once peak (or dip) latencies are calculated, it is possible to compare the average values in the various conditions.

Despite the simplicity of the collapsing to a singular value methods and their ease of use, they may suffer from several weaknesses that might skew the results and obscure true effects (see the following section). To overcome these weaknesses, temporal methods avoid collapsing data into singular values. Instead, they focus on the pupil dilation changes along the timeline (Practice Box 7).

Practice Box 7

Once the preprocessing has been successfully carried out and the group analysis has been completed, CHAP allows users to get information about the typically used a single value parameters (i.e., mean, max, and min pupil sizes, as well as peak and dip latencies[3]). In the *Statistical Analysis* window of CHAP, the user is able to choose (red field) what statistical approach will be used for the analysis (i.e., classical or Bayesian). In addition, the user is able to decide what the time-window for the analysis will be by selecting onset (yellow field) and offset (green field) for the required time-window. Once the time-window has been selected, the user is able to decide what parameter to present/compare (i.e., mean, max, and min pupil sizes, as well as peak and dip latencies) by using the parameter field (purple field). After that, the user is able to decide whether to present descriptive information or inference information (the orange field).

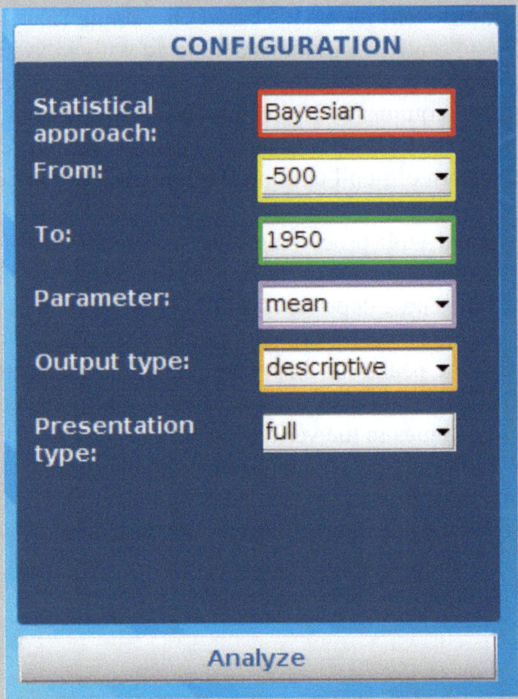

The descriptive information will include the number of participants in the measured condition, mean value of the required parameter, as well standard deviation, standard error and both the lower and upper confidence interval of 95%.

In the case of Bayesian analysis, the inference data will include information about paired samples t-tests including t-values, Bayes factors in favor of the alternative hypothesis (BF_{10}) as well in favor of the null hypothesis (BF_{01}), Cohen's d, and the number of participants in each condition.

In the case of classical analysis, the inference data will include information about repeated measures analysis of variance (ANOVA) including degree of freedom for the analysis (dfa), degree of freedom of the error, F value, mean square of error (MSE), p value, and partial eta square ($\text{pes} - \eta_p^2$).

The first row refers to the ANOVA itself and each subsequent row refers to an independent comparison.

All the data of each analysis can be save as a csv file that can be analyzed by any external statistical tool.

5.2 Temporal Analysis Approach

Unlike the methods discussed in the previous section, analyzing temporal information enables one to take into consideration all information of the pupil data along a timeline rather than using single data points only. Such temporal analysis provides more sensitive results compared with other methods. There are three vital benefits to keep in mind when considering whether it is worthwhile to perform a temporal analysis in addition to or instead of the traditional analyzes. First, if small effects are predicted, which may only be reflected in partial time-windows (i.e., relative to the total observed time-window), the singular value approaches might miss them. For example, inspection of Fig. 12 shows that the collapsing to a singular value comparison (e.g., of mean as well as peak amplitude) of congruent (green curve) and incongruent (red course) conditions between the beginning of the trial until 1100 ms post stimulus onset would suggest that there is no difference between the conditions. This analysis will take into account the time-window that is characterized mostly by no difference between the conditions and will provide minor weight to the time-window that includes meaningful differences between the conditions (i.e., between 750 and 1100 ms post stimulus onset). Importantly, a preliminary examination of the visualization of the results can shed light on the partial time-window in

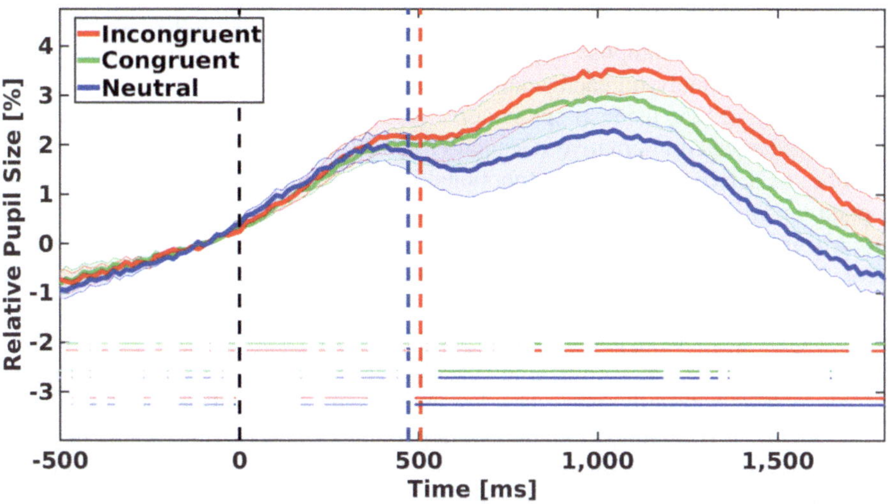

Fig. 12 Mean pupil dilation as a function of time in a two-color Stroop task run on 19 participants (Exp. 1 in Hershman & Henik, 2019). 0 represents the time of the Stroop stimulus onset, and the vertical lines close to 500 ms represent mean participant response. The shaded areas represent 1 standard error from the mean. The dark double horizontal lines represent meaningful differences (i.e., $BF_{10} \geq 3$) between two conditions (e.g., the dark double horizontal red and green lines represent meaningful differences between congruent and incongruent trials at the compared time points). The light double horizontal lines represent meaningful similarities (i.e., $BF_{01} \geq 3$) between two conditions (e.g., the light double horizontal red and green lines represent meaningful similarities between congruent and incongruent trials at the compared time points)

which the predicted effect is significant, but performing a traditional analysis on this time-window alone, without prior theoretical justification, may be defined as a misuse of data analysis. Differentiation between putative cognitive processes differing in their temporal characteristics is the second advantage suggested by temporal analysis. For example, inspection of Fig. 12 shows that the differences between the congruent trials (green curve) and the neutral trials (the blue curve) occur before the differences between the incongruent trials (red curve) and the congruent trials (green curve). Importantly, the temporal differences between different conditions may suggest different cognitive processes as well (in the presented case, one cognitive process that is associated with the differences between congruent and incongruent trials, and another cognitive process that is associated with the differences between congruent and neutral trials). In addition, the temporal separation between putative cognitive processes may clarify their temporal development over time (Hershman & Henik, 2019).

Temporal analysis can be applied not only for post-stimulus onset time-windows but for pre-stimulus time-windows, as well. That is, temporal analysis can be used for manipulation validity testing (as described above in the *Manipulation check* section).

Despite the theoretical advantage, the temporal perspective challenges researchers with significant statistical exertions. A variety of data analysis techniques have been proposed to meet this task. A hybrid strategy that attempts to combine the single value average method with a temporal approach suggests calculating the mean change in pupil diameter for small intervals and including time as a factor in the analysis. To decompose significant interactions, follow-up analysis could be conducted separately on each small interval (Henderson et al., 2018). Functional data analysis (FDA) using B-spline curve functions (Ramsay et al., 1997) offers a more sophisticated alternative (Geangu et al., 2011), namely, converting the total number of data points into one functional expression for each participant in each condition, which can be further used for analysis. Although the splines are required for a degree of smoothing, the curves can approximate the shape of the original data (for further details on FDA and its application to pupillometry, see Sirois & Brisson, 2014).

Analysis methods applied to other psychophysiological measures such as functional magnetic resonance imaging (fMRI) or event-related potential (ERP) have also been used for pupillometry data. For instance, the principal component analysis (PCA) method is intended to reduce a large number of data time points in pupillometry data to a small number of meaningful factors. In the context of the temporal approach discussed, this technique enables identifying the theoretical factors along the time axis that may account for unique variance in the data (Kuchinke et al., 2007; Nuthmann & Van Der Meer, 2005; Schluroff et al., 1986). In addition, the PCA method has been applied to differentiate between contributions of separate physiological components to the cognitive pupillary response (Steinhauer & Hakerem, 1992). That is, sympathetic and parasympathetic neural reactions differ in their intensity as well as in their temporal dynamics (Clewett et al., 2020; Wetzel et al., 2016).

The deconvolution method used mainly in fMRI analyses has been adopted for analyzing dilation time course, as well (Hoeks & Levelt, 1993; Wierda et al., 2012). In essence, the deconvolution technique can quantitate the behavior and magnitude of two (or more) component processes contributing to an observed outcome. Therefore, it may be suitable for constructing an approximate model of the dynamics of the various neural responses that drive the pupil. Several pupil-modeling methods have been suggested, such as the general linear models (GLMs) (de Gee et al., 2014; Denison et al., 2020) and the generalized additive models (GAMs) (Oliva & Anikin, 2018; van Rij et al., 2019). Hershman and Henik (2019, 2020) recently introduced an additional method in which each data point is analyzed separately by a t-test comparison so that a series of multiple t-tests are used for the time-window selected. This method allows us to look at the whole time-series for detecting differences (or similarities) between the investigated conditions in a more intuitive way without relying on the complex statistics mentioned above. In this method we start with identifying the specific period (start and end points) in which a specific difference between conditions may be of theoretical importance. Next, we use Bayesian statistics for the pairwise comparisons to prevent auto-correlated errors that may increase the probability of type I errors (i.e., false effects). This method is particularly appropriate and accessible for testing hypotheses for main effects (see Fig. 12) (Practice Box 8).

Practice Box 8

Once the preprocessing has been successfully carried out and the group analysis has been completed, CHAP allows users to get information about the temporal changes of the pupils during the experiment. In the *Statistical Analysis* window of CHAP, the user is able to choose (red field) what statistical approach will be used for the analysis (i.e., classical or Bayesian). If the user decides to use Bayesian statistic (yellow field), then three types of presentation will be available: full presentation, which will include both BF_{10} and BF_{01} as a function of time for each comparison separately; compact presentation, which will include both information about the observed pupil size as well information about the time-windows when there are differences /similarities between each of two compared conditions (see Fig. 12); or combine presentation, which will include all the comparisons in one figure (where the y-axis is logarithmic).

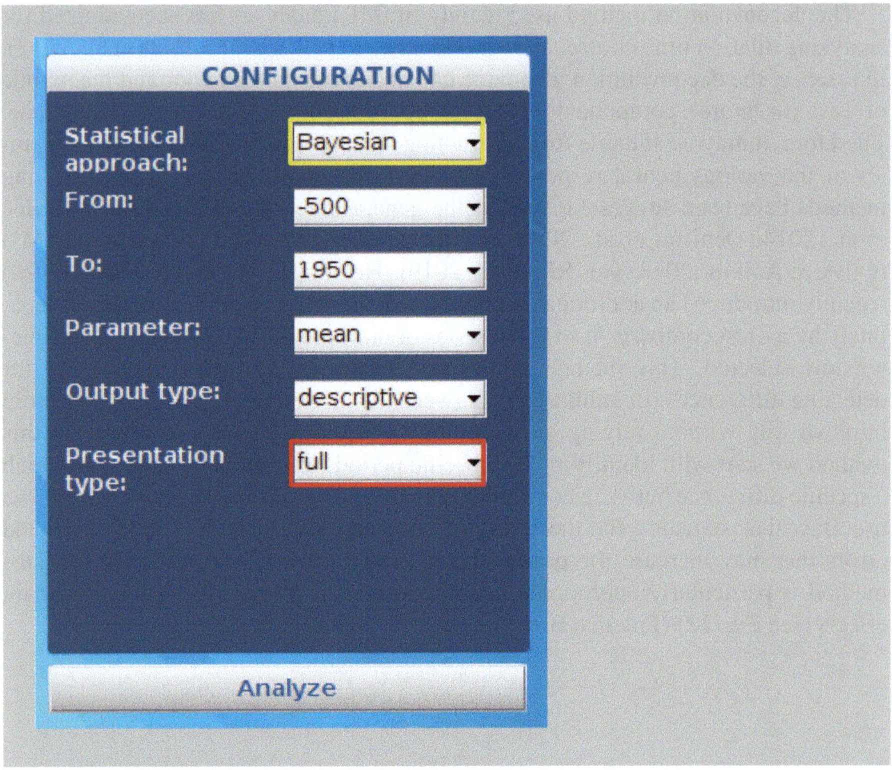

5.3 *Baseline Analysis*

The baseline concept (see discussions about baseline correction above) does not refer to a practical procedure for extracting the pupil pre-task size only. The baseline reflects a tonic mood of the participant. It has been proposed that the baseline should not be treated as simply noise to be discarded or normalized but rather it should be inspected for its own sake (Winn et al., 2018). For instance, increases in baseline pupil diameter may reflect disengagement from a task whereas reduced baseline diameter (but increases in task-evoked dilations) would indicate task engagement (Gilzenrat et al., 2010; Hopstaken et al., 2015). Moreover, tonic pupil size has been suggested as an index of general arousal (McGinley et al., 2015; Wass et al., 2015), as well as attention wandering (Konishi et al., 2017; Smallwood et al., 2011; Unsworth & Robison, 2018). In baseline analysis studies, the pupil size during the baseline interval is measured and is compared among different conditions. The analysis can be done by collapsing the whole data into one data point, or by temporal analysis (see discussion about temporal analysis above in the *temporal analysis approach* section).

While pupillometry data is often presented by changes over the time, the same data could be presented as a composition of waves in which each wave has a different power. Interestingly, the tonic baseline activation that is strongly associated with deep arousal state is typically associated with higher activation in large pupil fluctuations (or waves). In general, pupil response fluctuations are typically smaller than 1 Hz (i.e., wave lengths that are smaller than 1 s). While the phasic response to stimulus is relatively fast and is associated with fast frequencies, the tonic responses are relatively slow and associated with slow frequencies (Peysakhovich et al., 2015; Villalobos-Castaldi et al., 2016). Hence, frequency analysis of low frequencies (i.e., large wave lengths) may indicate differences in tonic state. The frequency analysis could be done by converting the data from the time space into the frequency space (e.g., by using Fourier transformation). Once, the data is converted into the frequency space, analysis of different frequencies as well as groups of frequencies could be conducted (Shechter et al., 2022) (Practice Box 9).

Practice Box 9

While in most cases it is highly recommended to look at relative changes compared to a baseline, frequency analysis should be done on uncorrected-to-baseline data. However, it is highly recommended to normalize data by using a Z-score or even millimeters (instead of arbitrary units). Once group analysis has been carried out, frequency analysis can be done by clicking on the *wave analysis* in the main window of CHAP. The user will be required to select a summary file that will be created in CHAP (*total.mat*) from the project folder (each group analysis creates a folder for the analysis). Then, the user will be required to select the relevant condition to compare, and the presentation type of the results (linear or logarithmic). The output of this analysis will be a figure that displays power as a function of frequency, as well as a csv file that will include all the information for each participant. This csv file could be used by external statistical tools.

6 Summary

The current chapter provides a tutorial for processing and analyzing the time course of pupillometric data and explains the variety of artifacts that need to be addressed in advance. Two main perspectives for analyzing changes in data pupil dilation are presented, namely, the collapsing to singular value methods and the temporal methods. The former are simple to implement and can be satisfactory if no differences in temporal dynamics between the observed conditions are expected. However, often this is not the state of affairs. Changes in pupil dilation may reflect various cognitive and perceptual processes. Thus, in a given time-window, the time course of pupil changes is dictated by different processes being chronologically differentiated.

In order to enable practice with real data, we have added the whole output data file from Hershman and Henik (2019; the data files are available here: https://osf.io/p74u8) and practical instructions for using the CHAP (Hershman et al., 2019) application that provides an accessible solution for performing some of the analyzes presented in this chapter. The software is open source, is written for MATLAB, and can be downloaded from the CHAP site (more details can be found at: https://in.bgu.ac.il/en/Labs/CNL/chap). The methodology guide presented in this chapter is written with the aim of supporting anyone interested in conducting a pupillometry study, regardless of its specific domain. Needless to say, research in different fields requires specific adjustments. However, formulation of uniform norms is a common goal for all researchers. There is also no doubt that a research field that is in accelerated development like pupillometry research offers innovations and developments frequently. Nevertheless we hope that at present, the guide proposed above is as up-to-date as possible and may assist in expanding the research circle of those conducting pupillometry studies.

Funding This research was supported by the Israel Science Foundation grant No. 359/22 to Avishai Henik.

References

Ahern, S., & Beatty, J. (1979). Pupillary responses during information processing vary with scholastic aptitude test scores. *Science, 205*(4412), 1289–1292. https://doi.org/10.1126/science.472746

Akdoğan, B., Balcı, F., & van Rijn, H. (2016). Temporal expectation indexed by pupillary response. *Timing & Time Perception, 4*(4), 354–370. https://doi.org/10.1163/22134468-00002075

Aston-Jones, G., & Cohen, J. D. (2005). An integrative theory of locus coeruleus-norepinephrine function: Adaptive gain and optimal performance. In *Annual review of neuroscience* (Vol. 28, pp. 403–450). Annual Reviews. https://doi.org/10.1146/annurev.neuro.28.061604.135709

Attard-Johnson, J., Ó Ciardha, C., & Bindemann, M. (2019). Comparing methods for the analysis of pupillary response. *Behavior Research Methods, 51*(1), 83–95. https://doi.org/10.3758/s13428-018-1108-6

Beatty, J. (1982). Task-evoked pupillary responses, processing load, and the structure of processing resources. *Psychological Bulletin, 91*(2), 276–292. https://doi.org/10.1037/0033-2909.91.2.276

Beatty, J., & Lucero-Wagoner, B. (2000). The pupillary system. *Handbook of Psychophysiology, 2*, 142–162.

Binda, P., & Murray, S. O. (2015a). Keeping a large-pupilled eye on high-level visual processing. In *Trends in cognitive sciences* (Vol. 19, pp. 1–3). Elsevier Ltd. https://doi.org/10.1016/j.tics.2014.11.002

Binda, P., & Murray, S. O. (2015b). Spatial attention increases the pupillary response to light changes. *Journal of Vision, 15*(2), 1–1. https://doi.org/10.1167/15.2.1

Binda, P., Pereverzeva, M., & Murray, S. O. (2013). Attention to bright surfaces enhances the pupillary light reflex. *Journal of Neuroscience, 33*(5), 2199–2204. https://doi.org/10.1523/JNEUROSCI.3440-12.2013

Binda, P., Pereverzeva, M., & Murray, S. O. (2014). Pupil size reflects the focus of feature-based attention. *Journal of Neurophysiology, 112*(12), 3046–3052. https://doi.org/10.1152/jn.00502.2014

Binda, P., Straßer, T., Stingl, K., Richter, P., Peters, T., Wilhelm, H., Wilhelm, B., & Kelbsch, C. (2017). Pupil response components: Attention-light interaction in patients with Parinaud's syndrome. *Scientific Reports, 7*(1), 10283. https://doi.org/10.1038/s41598-017-10816-x

Bitsios, P., Szabadi, E., & Bradshaw, C. (2004). The fear-inhibited light reflex: Importance of the anticipation of an aversive event. *International Journal of Psychophysiology, 52*(1), 87–95. https://doi.org/10.1016/J.IJPSYCHO.2003.12.006

Blaser, E., Eglington, L., Carter, A. S., & Kaldy, Z. (2014). Pupillometry reveals a mechanism for the autism spectrum disorder (ASD) advantage in visual tasks. *Scientific Reports, 4*, 4301. https://doi.org/10.1038/srep04301

Bradley, M. M., Miccoli, L., Escrig, M. A., & Lang, P. J. (2008). The pupil as a measure of emotional arousal and autonomic activation. *Psychophysiology, 45*(4), 602–607. https://doi.org/10.1111/j.1469-8986.2008.00654.x

Braem, S., Coenen, E., Bombeke, K., van Bochove, M. E., & Notebaert, W. (2015). Open your eyes for prediction errors. *Cognitive, Affective, & Behavioral Neuroscience, 15*(2), 374–380. https://doi.org/10.3758/s13415-014-0333-4

Brisson, J., Mainville, M., Mailloux, D., Beaulieu, C., Serres, J., & Sirois, S. (2013). Pupil diameter measurement errors as a function of gaze direction in corneal reflection eyetrackers. *Behavior Research Methods, 45*(4), 1322–1331. https://doi.org/10.3758/s13428-013-0327-0

Caffier, P. P., Erdmann, U., & Ullsperger, P. (2003). Experimental evaluation of eye-blink parameters as a drowsiness measure. *European Journal of Applied Physiology, 89*(3), 319–325. https://doi.org/10.1007/s00421-003-0807-5

Chiew, K. S., & Braver, T. S. (2013). Temporal dynamics of motivation-cognitive control interactions revealed by high-resolution pupillometry. *Frontiers in Psychology, 4*, 15. https://doi.org/10.3389/fpsyg.2013.00015

Clewett, D., Gasser, C., & Davachi, L. (2020). Pupil-linked arousal signals track the temporal organization of events in memory. *Nature Communications, 11*(1), 1–14. https://doi.org/10.1038/s41467-020-17851-9

Cohen, N., Moyal, N., & Henik, A. (2015). Executive control suppresses pupillary responses to aversive stimuli. *Biological Psychology, 112*, 1–11. https://doi.org/10.1016/j.biopsycho.2015.09.006

Costa, V. D., & Rudebeck, P. H. (2016). More than meets the eye: The relationship between pupil size and locus coeruleus activity. In Neuron (Vol. 89, 1, pp. 8–10). Cell Press. https://doi.org/10.1016/j.neuron.2015.12.031.

de Gee, J. W., Knapen, T., & Donner, T. H. (2014). Decision-related pupil dilation reflects upcoming choice and individual bias. *Proceedings of the National Academy of Sciences of the United States of America, 111*(5), E618–E625. https://doi.org/10.1073/pnas.1317557111

Denison, R. N., Parker, J. A., & Carrasco, M. (2020). Modeling pupil responses to rapid sequential events. *Behavior Research Methods, 52*(5), 1991–2007. https://doi.org/10.3758/s13428-020-01368-6

Einhauser, W., Stout, J., Koch, C., & Carter, O. (2008). Pupil dilation reflects perceptual selection and predicts subsequent stability in perceptual rivalry. *Proceedings of the National Academy of Sciences, 105*(5), 1704–1709. https://doi.org/10.1073/pnas.0707727105

Ellis, C. J. (1981). The pupillary light reflex in normal subjects. *The British Journal of Ophthalmology, 65*(11), 754–759. https://doi.org/10.1136/BJO.65.11.754

Feigl, B., Zele, A. J., Fader, S. M., Howes, A. N., Hughes, C. E., Jones, K. A., & Jones, R. (2012). The post-illumination pupil response of melanopsin-expressing intrinsically photosensitive retinal ganglion cells in diabetes. *Acta Ophthalmologica, 90*(3), e230–e234. https://doi.org/10.1111/j.1755-3768.2011.02226.x

Forbes, S. H. (2020). PupillometryR: An R package for preparing and analysing pupillometry data. *Journal of Open Source Software, 5*(50), 2285. https://doi.org/10.21105/joss.02285

Fukuda. (2001). Eye blinks: New indices for the detection of deception. *International Journal of Psychophysiology, 40*(3), 239–245. https://doi.org/10.1016/S0167-8760(00)00192-6

Gagl, B., Hawelka, S., & Hutzler, F. (2011). Systematic influence of gaze position on pupil size measurement: Analysis and correction. *Behavior Research Methods, 43*(4), 1171–1181. https://doi.org/10.3758/s13428-011-0109-5

Geangu, E., Hauf, P., Bhardwaj, R., & Bentz, W. (2011). Infant pupil diameter changes in response to others' positive and negative emotions. *PLoS One, 6*(11), e27132. https://doi.org/10.1371/journal.pone.0027132

Geller, J., Winn, M. B., Mahr, T., & Mirman, D. (2020). GazeR: A package for processing gaze position and pupil size data. *Behavior Research Methods, 52*(5), 2232–2255. https://doi.org/10.3758/s13428-020-01374-8

Geva, R., Zivan, M., Warsha, A., & Olchik, D. (2013). Alerting, orienting or executive attention networks: Differential patters of pupil dilations. *Frontiers in Behavioral Neuroscience, 7*, 145. https://doi.org/10.3389/fnbeh.2013.00145

Gilzenrat, M. S., Nieuwenhuis, S., Jepma, M., & Cohen, J. D. (2010). Pupil diameter tracks changes in control state predicted by the adaptive gain theory of locus coeruleus function. *Cognitive, Affective, & Behavioral Neuroscience, 10*(2), 252–269. https://doi.org/10.3758/CABN.10.2.252

Goldinger, S. D., & Papesh, M. H. (2012). Pupil dilation reflects the creation and retrieval of memories. *Current Directions in Psychological Science, 21*(2), 90–95. https://doi.org/10.1177/0963721412436811

Goldwater, B. C. (1972). Psychological significance of pupillary movements. *Psychological Bulletin, 77*(5), 340–355. https://doi.org/10.1037/h0032456

Hayes, T. R., & Petrov, A. A. (2016). Mapping and correcting the influence of gaze position on pupil size measurements. *Behavior Research Methods, 48*(2), 510–527. https://doi.org/10.3758/s13428-015-0588-x

Henderson, R. R., Bradley, M. M., & Lang, P. J. (2014). Modulation of the initial light reflex during affective picture viewing. *Psychophysiology, 51*(9), 815–818. https://doi.org/10.1111/psyp.12236

Henderson, R. R., Bradley, M. M., & Lang, P. J. (2018). Emotional imagery and pupil diameter. *Psychophysiology, 55*(6), e13050. https://doi.org/10.1111/psyp.13050

Hepach, R., & Westermann, G. (2016). Pupillometry in infancy research. *Journal of Cognition and Development, 17*(3), 359–377. https://doi.org/10.1080/15248372.2015.1135801

Hershman, R., & Henik, A. (2019). Dissociation between reaction time and pupil dilation in the Stroop task. *Journal of Experimental Psychology: Learning Memory and Cognition, 45*(10), 1899–1909. https://doi.org/10.1037/xlm0000690

Hershman, R., & Henik, A. (2020). Pupillometric contributions to deciphering Stroop conflicts. *Memory and Cognition, 77*(8), 2601–2610. https://doi.org/10.3758/s13421-019-00971-z

Hershman, R., Henik, A., & Cohen, N. (2018). A novel blink detection method based on pupillometry noise. *Behavior Research Methods, 50*(1), 107–114. https://doi.org/10.3758/s13428-017-1008-1

Hershman, R., Henik, A., & Cohen, N. (2019). CHAP: Open-source software for processing and analyzing pupillometry data. *Behavior Research Methods, 51*(3), 1059–1074. https://doi.org/10.3758/s13428-018-01190-1

Hershman, R., Levin, Y., Tzelgov, J., & Henik, A. (2020). Neutral stimuli and pupillometric task conflict. *Psychological Research, 85*(3), 1084–1092. https://doi.org/10.1007/s00426-020-01311-6

Hershman, R., Levin, Y., Tzelgov, J., & Henik, A. (2021). The contribution of meaning to the detection of task conflict. *Quarterly Journal of Experimental Psychology, 74*(9), 1553–1561. https://doi.org/10.1177/17470218211001331

Hershman, R., Beckmann, L., & Henik, A. (2022). Task and information conflicts in the numerical Stroop task. *Psychophysiology, 59*(9), e14057. https://doi.org/10.1111/psyp.14057

Hershman, R., Milshtein, D., & Henik, A. (2023). The contribution of temporal analysis of pupillometry measurements to cognitive research. *Psychological Research, 87*(1), 28–42. https://doi.org/10.1007/S00426-022-01656-0

Hess, E. H., & Polt, J. M. (1964). Pupil size in relation to mental activity during simple problem-solving. *Science, 143*(3611), 1190–1192. https://doi.org/10.1126/science.143.3611.1190

Hoeks, B., & Levelt, W. J. M. (1993). Pupillary dilation as a measure of attention: A quantitative system analysis. *Behavior Research Methods, Instruments, & Computers, 25*(1), 16–26. https://doi.org/10.3758/BF03204445

Höfle, M., Kenntner-Mabiala, R., Pauli, P., & Alpers, G. W. (2008). You can see pain in the eye: Pupillometry as an index of pain intensity under different luminance conditions. *International Journal of Psychophysiology, 70*(3), 171–175. https://doi.org/10.1016/j.ijpsycho.2008.06.008

Hopstaken, J. F., van der Linden, D., Bakker, A. B., & Kompier, M. A. J. (2015). The window of my eyes: Task disengagement and mental fatigue covary with pupil dynamics. *Biological Psychology, 110*, 100–106. https://doi.org/10.1016/j.biopsycho.2015.06.013

Joshi, S., & Gold, J. I. (2020). Pupil size as a window on neural substrates of cognition. In Trends in cognitive sciences (Vol. 24, 6, pp. 466–480). Elsevier Ltd. https://doi.org/10.1016/j.tics.2020.03.005

Joshi, S., Li, Y., Kalwani, R. M., & Gold, J. I. (2016). Relationships between pupil diameter and neuronal activity in the locus coeruleus, colliculi, and cingulate cortex. *Neuron, 89*(1), 221–234. https://doi.org/10.1016/j.neuron.2015.11.028

Kaakinen, J. K., & Simola, J. (2020). Fluctuation in pupil size and spontaneous blinks reflect story transportation. *Journal of Eye Movement Research, 13*(3), 10. https://doi.org/10.16910/jemr.13.3.6

Kahneman, D. (1973). *Attention and effort*. Prentice-Hall.

Kang, O., & Wheatley, T. (2015). Pupil dilation patterns reflect the contents of consciousness. *Consciousness and Cognition, 35*, 128–135. https://doi.org/10.1016/j.concog.2015.05.001

Karson, C. N. (1983). Spontaneous eye-blink rates and dopaminergic systems. *Brain, 106*(Pt 3), 643–653. https://doi.org/10.1093/brain/106.3.643

Klingner, J., Tversky, B., & Hanrahan, P. (2011). Effects of visual and verbal presentation on cognitive load in vigilance, memory, and arithmetic tasks. *Psychophysiology, 48*(3), 323–332. https://doi.org/10.1111/j.1469-8986.2010.01069.x

Kloosterman, N. A., Meindertsma, T., van Loon, A. M., Lamme, V. A. F., Bonneh, Y. S., & Donner, T. H. (2015). Pupil size tracks perceptual content and surprise. *European Journal of Neuroscience, 41*(8), 1068–1078. https://doi.org/10.1111/ejn.12859

Koelewijn, T., Zekveld, A. A., Festen, J. M., Rönnberg, J., & Kramer, S. E. (2012). Processing load induced by informational masking is related to linguistic abilities. *International Journal of Otolaryngology, 865731*. https://doi.org/10.1155/2012/865731

Koelewijn, T., de Kluiver, H., Shinn-Cunningham, B. G., Zekveld, A. A., & Kramer, S. E. (2015). The pupil response reveals increased listening effort when it is difficult to focus attention. *Hearing Research, 323*, 81–90. https://doi.org/10.1016/J.HEARES.2015.02.004

Konishi, M., Brown, K., Battaglini, L., & Smallwood, J. (2017). When attention wanders: Pupillometric signatures of fluctuations in external attention. *Cognition, 168*, 16–26. https://doi.org/10.1016/j.cognition.2017.06.006

Kostic, C., Crippa, S. V., Martin, C., Kardon, R. H., Biel, M., Arsenijevic, Y., & Kawasaki, A. (2016). Determination of rod and cone influence to the early and late dynamic of the pupillary light response. *Investigative Ophthalmology and Visual Science, 57*(6), 2501–2508. https://doi.org/10.1167/iovs.16-19150

Kret, M. E., & Sjak-Shie, E. E. (2019). Preprocessing pupil size data: Guidelines and code. *Behavior Research Methods, 51*(3), 1336–1342. https://doi.org/10.3758/s13428-018-1075-y

Kuchinke, L., Võ, M. L.-H., Hofmann, M., & Jacobs, A. M. (2007). Pupillary responses during lexical decisions vary with word frequency but not emotional valence. *International Journal of Psychophysiology, 65*(2), 132–140. https://doi.org/10.1016/J.IJPSYCHO.2007.04.004

Kuchinsky, S. E., Ahlstrom, J. B., Vaden, K. I., Cute, S. L., Humes, L. E., Dubno, J. R., & Eckert, M. A. (2013). Pupil size varies with word listening and response selection difficulty in older adults with hearing loss. *Psychophysiology, 50*(1), 23–34. https://doi.org/10.1111/j.1469-8986.2012.01477.x

Laeng, B., Ørbo, M., Holmlund, T., & Miozzo, M. (2011). Pupillary Stroop effects. *Cognitive Processing, 12*(1), 13–21. https://doi.org/10.1007/s10339-010-0370-z

Laeng, B., Sirois, S., & Gredebäck, G. (2012). Pupillometry: A window to the preconscious? *Perspectives on Psychological Science, 7*(1), 18–27. https://doi.org/10.1177/1745691611427305

Leigh, R., & Zee, D. (2015). *The neurology of eye movements*. Oxford University Press, Oxford.

Lemercier, A., Guillot, G., Courcoux, P., Garrel, C., Baccino, T., & Schlich, P. (2014). Pupillometry of taste: Methodological guide-from acquisition to data processing-and toolbox for MATLAB. In *Quantitative methods in psychology* (Vol. 10(2)). Sage. https://hal.archives-ouvertes.fr/hal-01249252

Loewenfeld, I. E. (1958). Mechanisms of reflex dilatation of the pupil – Historical review and experimental analysis. *Documenta Ophthalmologica, 12*(1), 185–448. https://doi.org/10.1007/BF00913471

Loewenfeld, I. E. (1993). *The pupil: Anatomy, physiology, and clinical applications* (Vol. 2). Iowa State University Press.

MacLachlan, C., & Howland, H. C. (2002). Normal values and standard deviations for pupil diameter and interpupillary distance in subjects aged 1 month to 19 years. *Ophthalmic and Physiological Optics, 22*(3), 175–182. https://doi.org/10.1046/j.1475-1313.2002.00023.x

Mathôt, S. (2013). *A simple way to reconstruct pupil size during eye blinks*. https://doi.org/10.6084/m9.figshare.688001.

Mathôt, S. (2018). Pupillometry: Psychology, physiology, and function. *Journal of Cognition, 1*(1), 16. https://doi.org/10.5334/joc.18

Mathôt, S., & Vilotijević, A. (2022). Methods in cognitive pupillometry: Design, preprocessing, and statistical analysis. *Behavior Research Methods, 55*, 1–23. https://doi.org/10.3758/S13428-022-01957-7

Mathôt, S., van der Linden, L., Grainger, J., & Vitu, F. (2013). The pupillary light response reveals the focus of covert visual attention. *PLoS One, 8*(10), e78168. https://doi.org/10.1371/journal.pone.0078168

Mathôt, S., Grainger, J., & Strijkers, K. (2017). Pupillary responses to words that convey a sense of brightness or darkness. *Psychological Science, 28*(8), 1116–1124. https://doi.org/10.1177/0956797617702699

Mathôt, S., Fabius, J., Van Heusden, E., & Van der Stigchel, S. (2018). Safe and sensible preprocessing and baseline correction of pupil-size data. *Behavior Research Methods, 50*(1), 94–106. https://doi.org/10.3758/s13428-017-1007-2

McGinley, M. J., David, S. V., & McCormick, D. A. (2015). Cortical membrane potential signature of optimal states for sensory signal detection. *Neuron, 87*(1), 179–192. https://doi.org/10.1016/j.neuron.2015.05.038

Murphy, P. R., Robertson, I. H., Balsters, J. H., & O'connell, R. G. (2011). Pupillometry and P3 index the locus coeruleus-noradrenergic arousal function in humans. *Psychophysiology, 48*(11), 1532–1543. https://doi.org/10.1111/j.1469-8986.2011.01226.x

Murphy, P. R., O'Connell, R. G., O'Sullivan, M., Robertson, I. H., & Balsters, J. H. (2014). Pupil diameter covaries with BOLD activity in human locus coeruleus. *Human Brain Mapping, 35*(8), 4140–4154. https://doi.org/10.1002/hbm.22466

Naber, M., & Nakayama, K. (2013). Pupil responses to high-level image content. *Journal of Vision, 13*(6), 7. https://doi.org/10.1167/13.6.7

Nuthmann, A., & Van Der Meer, E. (2005). Time's arrow and pupillary response. *Psychophysiology, 42*(3), 306–317. https://doi.org/10.1111/j.1469-8986.2005.00291.x

Nyström, M., Hansen, D. W., Andersson, R., & Hooge, I. (2016). Why have microsaccades become larger? Investigating eye deformations and detection algorithms. *Vision Research, 118*, 17–24. https://doi.org/10.1016/j.visres.2014.11.007

Oliva, M., & Anikin, A. (2018). Pupil dilation reflects the time course of emotion recognition in human vocalizations. *Scientific Reports, 8*(1), 4871. https://doi.org/10.1038/s41598-018-23265-x

Otero, S. C., Weekes, B. S., & Hutton, S. B. (2011). Pupil size changes during recognition memory. *Psychophysiology, 48*(10), 1346–1353. https://doi.org/10.1111/j.1469-8986.2011.01217.x

Partala, T., & Surakka, V. (2003). Pupil size variation as an indication of affective processing. *International Journal of Human-Computer Studies, 59*(1–2), 185–198. https://doi.org/10.1016/S1071-5819(03)00017-X

Peysakhovich, V., Causse, M., Scannella, S., & Dehais, F. (2015). Frequency analysis of a task-evoked pupillary response: Luminance-independent measure of mental effort. *International Journal of Psychophysiology, 97*(1), 30–37. https://doi.org/10.1016/j.ijpsycho.2015.04.019

Preuschoff, K., T Hart, B. M., & Einhauser, W. (2011). Pupil dilation signals surprise: Evidence for noradrenaline's role in decision making. *Frontiers in Neuroscience, 5*, 115. https://doi.org/10.3389/fnins.2011.00115

Ramsay, J. O., Heckman, N., & Silverman, B. W. (1997). Spline smoothing with model-based penalties. *Behavior Research Methods, Instruments, and Computers, 29*(1), 99–106. https://doi.org/10.3758/BF03200573

Sapir, A., Hershman, R., & Henik, A. (2021). Top-down effect on pupillary response: Evidence from shape from shading. *Cognition, 212*, 104664. https://doi.org/10.1016/J.COGNITION.2021.104664

Satterthwaite, T. D., Green, L., Myerson, J., Parker, J., Ramaratnam, M., & Buckner, R. L. (2007). Dissociable but inter-related systems of cognitive control and reward during decision making: Evidence from pupillometry and event-related fMRI. *NeuroImage, 37*(3), 1017–1031. https://doi.org/10.1016/j.neuroimage.2007.04.066

Schluroff, M., Zimmermann, T. E., Freeman, R. B., Hofmeister, K., Lorscheid, T., & Weber, A. (1986). Pupillary responses to syntactic ambiguity of sentences. *Brain and Language, 27*(2), 322–344. https://doi.org/10.1016/0093-934X(86)90023-4

Schmidtke, J. (2014). Second language experience modulates word retrieval effort in bilinguals: Evidence from pupillometry. *Frontiers in Psychology, 5*, 137. https://doi.org/10.3389/fpsyg.2014.00137

Shechter, A., Hershman, R., & Share, D. L. (2022). A pupillometric study of developmental and individual differences in cognitive effort in visual word recognition. *Scientific Reports, 12*(1), 1–7. https://doi.org/10.1038/S41598-022-14536-9

Siegle, G. J., Granholm, E., Ingram, R. E., & Matt, G. E. (2001). Pupillary and reaction time measures of sustained processing of negative information in depression. *Biological Psychiatry, 49*(7), 624–636. https://doi.org/10.1016/S0006-3223(00)01024-6

Siegle, G. J., Steinhauer, S. R., & Thase, M. E. (2004). Pupillary assessment and computational modeling of the Stroop task in depression. *International Journal of Psychophysiology, 52*(1), 63–76. https://doi.org/10.1016/J.IJPSYCHO.2003.12.010

Siegle, G. J., Ichikawa, N., & Steinhauer, S. (2008). Blink before and after you think: Blinks occur prior to and following cognitive load indexed by pupillary responses. *Psychophysiology, 45*(5), 679–687. https://doi.org/10.1111/j.1469-8986.2008.00681.x

Sirois, S., & Brisson, J. (2014). Pupillometry. In *Wiley interdisciplinary reviews: Cognitive science* (Vol. 5, 6, pp. 679–692). Wiley-Blackwell. https://doi.org/10.1002/wcs.1323.

Sirois, S., & Jackson, I. R. (2012). Pupil dilation and object permanence in infants. *Infancy, 17*(1), 61–78. https://doi.org/10.1111/j.1532-7078.2011.00096.x

Smallwood, J., Brown, K. S., Tipper, C., Giesbrecht, B., Franklin, M. S., Mrazek, M. D., Carlson, J. M., & Schooler, J. W. (2011). Pupillometric evidence for the decoupling of attention from perceptual input during offline thought. *PLoS One, 6*(3), e18298. https://doi.org/10.1371/journal.pone.0018298

Snell, J., Mathôt, S., Mirault, J., & Grainger, J. (2018). Parallel graded attention in reading: A pupillometric study. *Scientific Reports, 8*(1), 3743. https://doi.org/10.1038/s41598-018-22138-7

Steinhauer, S. R., & Hakerem, G. (1992). The pupillary response in cognitive psychophysiology and schizophrenia. *Annals of the New York Academy of Sciences, 658*(1), 182–204. https://doi.org/10.1111/j.1749-6632.1992.tb22845.x

Steinhauer, S. R., Condray, R., & Kasparek, A. (2000). Cognitive modulation of midbrain function: Task-induced reduction of the pupillary light reflex. *International Journal of Psychophysiology, 39*(1), 21–30. https://doi.org/10.1016/S0167-8760(00)00119-7

Steinhauer, S. R., Siegle, G. J., Condray, R., & Pless, M. (2004). Sympathetic and parasympathetic innervation of pupillary dilation during sustained processing. *International Journal of Psychophysiology, 52*(1), 77–86. https://doi.org/10.1016/J.IJPSYCHO.2003.12.005

Stern, J. A., Walrath, L. C., & Goldstein, R. (1984). The endogenous eyeblink. *Psychophysiology, 21*(1), 22–33. https://doi.org/10.1111/j.1469-8986.1984.tb02312.x

Tkacz-Domb, S., & Yeshurun, Y. (2018). The size of the attentional window when measured by the pupillary response to light. *Scientific Reports, 8*(1), 1–7. https://doi.org/10.1038/s41598-018-30343-7

Unsworth, N., & Robison, M. K. (2018). Tracking arousal state and mind wandering with pupillometry. *Cognitive, Affective, & Behavioral Neuroscience, 18*(4), 638–664. https://doi.org/10.3758/s13415-018-0594-4

van der Wel, P., & van Steenbergen, H. (2018). Pupil dilation as an index of effort in cognitive control tasks: A review. *Psychonomic Bulletin & Review, 25*(6), 2005–2015. https://doi.org/10.3758/s13423-018-1432-y

van der Meer, E., Friedrich, M., Nuthmann, A., Stelzel, C., & Kuchinke, L. (2003). Picture–word matching: Flexibility in conceptual memory and pupillary responses. *Psychophysiology, 40*(6), 904–913. https://doi.org/10.1111/1469-8986.00108

van Rij, J., Hendriks, P., van Rijn, H., Baayen, R. H., & Wood, S. N. (2019). Analyzing the time course of pupillometric data. *Trends in Hearing, 23*, 2331216519832483. https://doi.org/10.1177/2331216519832483

van Steenbergen, H., & Band, G. P. H. (2013). Pupil dilation in the Simon task as a marker of conflict processing. *Frontiers in Human Neuroscience, 7*, 215. https://doi.org/10.3389/fnhum.2013.00215

Villalobos-Castaldi, F. M., Ruiz-Pinales, J., Kemper-Valverde, N. C., Flores-Flores, M., Ramírez-Sánchez, L. G., & Ortiz-Hernández, M. G. (2016). Spontaneous pupillary oscillation signal analysis applying hilbert huang transform. In *Proceedings of the 9th international joint conference on biomedical engineering systems and technologies* (pp. 67–77). https://doi.org/10.5220/0005697400670077

Vogelzang, M., Hendriks, P., & van Rijn, H. (2016). Pupillary responses reflect ambiguity resolution in pronoun processing. *Language, Cognition and Neuroscience, 31*(7), 876–885. https://doi.org/10.1080/23273798.2016.1155718

Wainstein, G., Rojas-Líbano, D., Crossley, N. A., Carrasco, X., Aboitiz, F., & Ossandón, T. (2017). Pupil size tracks attentional performance in attention-deficit/hyperactivity disorder. *Scientific Reports, 7*(1), 8228. https://doi.org/10.1038/s41598-017-08246-w

Wang, C.-A., Brien, D. C., & Munoz, D. P. (2015). Pupil size reveals preparatory processes in the generation of pro-saccades and anti-saccades. *European Journal of Neuroscience, 41*(8), 1102–1110. https://doi.org/10.1111/ejn.12883

Wang, C.-A., McInnis, H., Brien, D. C., Pari, G., & Munoz, D. P. (2016). Disruption of pupil size modulation correlates with voluntary motor preparation deficits in Parkinson's disease. *Neuropsychologia, 80*, 176–184. https://doi.org/10.1016/j.neuropsychologia.2015.11.019

Wass, S. V., de Barbaro, K., & Clackson, K. (2015). Tonic and phasic co-variation of peripheral arousal indices in infants. *Biological Psychology, 111*, 26–39. https://doi.org/10.1016/j.biopsycho.2015.08.006

Wetzel, N., Buttelmann, D., Schieler, A., & Widmann, A. (2016). Infant and adult pupil dilation in response to unexpected sounds. *Developmental Psychobiology, 58*(3), 382–392. https://doi.org/10.1002/dev.21377

Wierda, S. M., Van Rijn, H., Taatgen, N. A., & Martens, S. (2012). Pupil dilation deconvolution reveals the dynamics of attention at high temporal resolution. *Proceedings of the National Academy of Sciences of the United States of America, 109*(22), 8456–8460. https://doi.org/10.1073/pnas.1201858109

Winn, M. B., Edwards, J. R., & Litovsky, R. Y. (2015). The impact of auditory spectral resolution on listening effort revealed by pupil dilation. *Ear and Hearing, 36*(4), e153–e165. https://doi.org/10.1097/AUD.0000000000000145

Winn, M. B., Wendt, D., Koelewijn, T., & Kuchinsky, S. E. (2018). Best practices and advice for using pupillometry to measure listening effort: An introduction for those who want to get started. *Trends in Hearing, 22*, 233121651880086. https://doi.org/10.1177/2331216518800869

Zekveld, A. A., & Kramer, S. E. (2014). Cognitive processing load across a wide range of listening conditions: Insights from pupillometry. *Psychophysiology, 51*(3), 277–284. https://doi.org/10.1111/psyp.12151

Zekveld, A. A., Heslenfeld, D. J., Johnsrude, I. S., Versfeld, N. J., & Kramer, S. E. (2014). The eye as a window to the listening brain: Neural correlates of pupil size as a measure of cognitive listening load. *NeuroImage, 101*, 76–86. https://doi.org/10.1016/j.neuroimage.2014.06.069

Correction to: A Primer on Design and Data Analysis for Cognitive Pupillometry

Jamie Reilly, Bonnie Zuckerman, and Alexandra E. Kelly

Correction to:
Chapter 14 in: M. H. Papesh, S. D. Goldinger (eds.), *Modern Pupillometry*, https://doi.org/10.1007/978-3-031-54896-3_14

The original version of the chapter "A Primer on Design and Data Analysis for Cognitive Pupillometry" was inadvertently published with an incorrect figure (Figures 1 and 2). The correct figures has now been updated in the chapter.

The updated version of this chapter can be found at
https://doi.org/10.1007/978-3-031-54896-3_14

M. H. Papesh, S. D. Goldinger (eds.), *Modern Pupillometry*,
https://doi.org/10.1007/978-3-031-54896-3_16

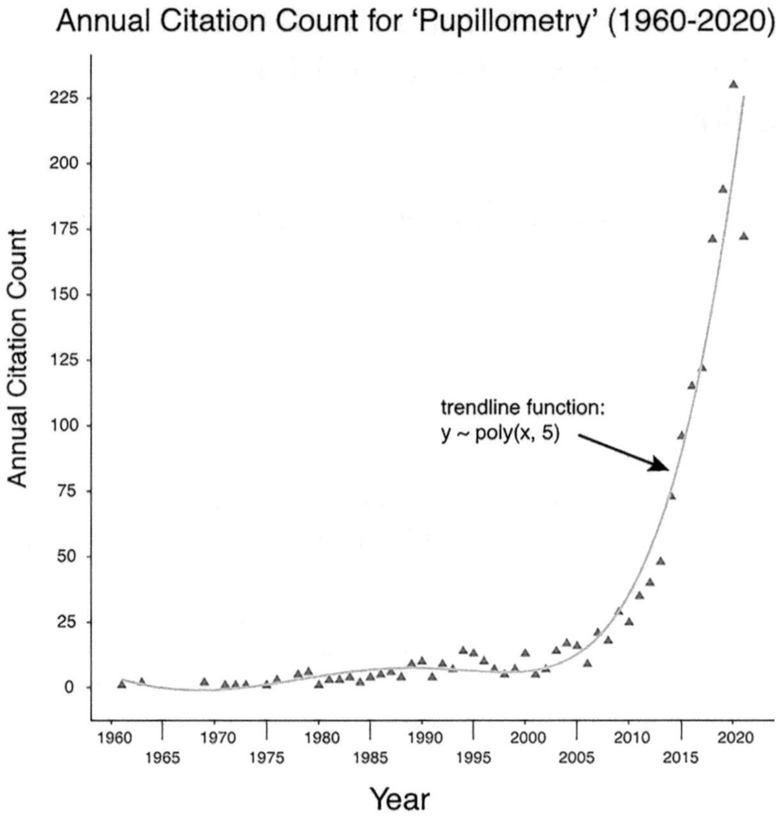

Fig. 1 PubMed citation counts by year for pupillometry-related search terms

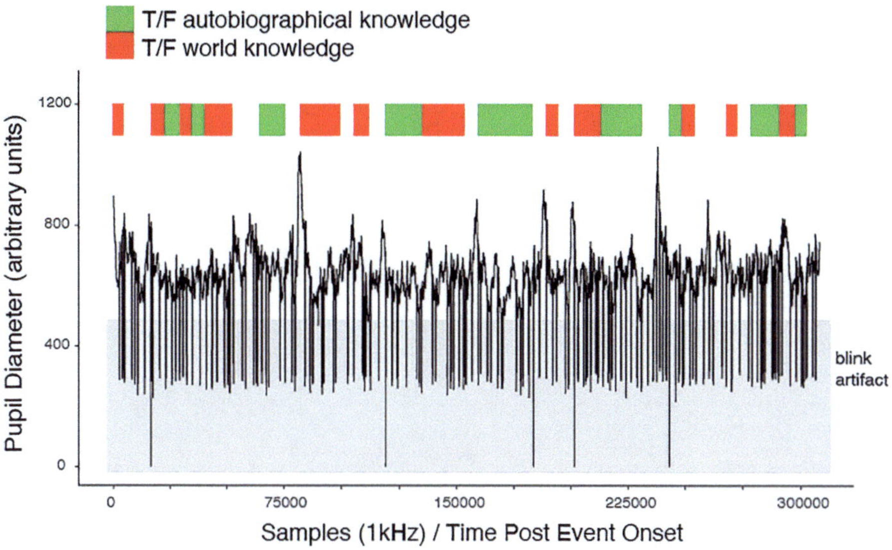

Fig. 2 Raw single-subject data in an event-related pupillometry study*Note*: These data reflect a session of a single-subject raw pupil time series sampled from the participant's left eye via an Eyelink 1000 Plus eyetracker (1000 Hz sampling rate). The color bar reflects points in the time series corresponding to event onsets within two experimental conditions. The experimental conditions involve true or false statements about the world versus the self

Index

M. H. Papesh, S. D. Goldinger (eds.), *Modern Pupillometry*,
https://doi.org/10.1007/978-3-031-54896-3

MIX
Papier aus verantwortungsvollen Quellen
Paper from responsible sources
FSC® C105338

If you have any concerns about our products,
you can contact us on
ProductSafety@springernature.com

In case Publisher is established outside the EU,
the EU authorized representative is:
Springer Nature Customer Service Center GmbH
Europaplatz 3, 69115 Heidelberg, Germany

Printed by Libri Plureos GmbH
in Hamburg, Germany